Submarine Design and Construction Fundamentals

Richard Skiba

AFTER MIDNIGHT PUBLISHING

Skiba, Richard (author)

Submarine Design and Construction Fundamentals

ISBN 978-1-7637173-3-6 (Paperback) 978-1-7637173-4-3 (eBook) 978-1-7637173-5-0 (Hardcover)

Non-fiction

Contents

Preface

Submarine construction is a complex process that requires specialized shipyards, highly skilled labour, advanced technology, and stringent quality controls. Due to the complexity of building submarines, only a limited number of countries have the industrial, financial, and technical capacity to produce them. These nations build submarines for their own navies and, in some cases, for export to other countries. The global demand for submarine construction has grown in recent years, driven by strategic needs, national security concerns, and maritime defence policies.

Due to the specialized nature of submarine construction, only a limited number of countries possess the necessary industrial, financial, and technical capabilities to produce these vessels. This exclusivity is often a reflection of national security priorities and strategic military needs. For instance, Indonesia's strategic plan emphasizes the necessity of developing indigenous submarine capabilities to enhance its maritime defence posture [1]. Similarly, the global landscape of submarine construction is shaped by geopolitical dynamics, where nations not only build submarines for their own naval forces but also engage in export activities to bolster their defence partnerships [1, 2]. The demand for submarines has surged in recent years, driven by heightened national security concerns and evolving maritime defence policies including the importance of sea power in contemporary defence strategies [3].

The growing complexity of maritime security challenges has led to an increased emphasis on submarine capabilities as a deterrent and a means of asserting national interests in contested waters. The strategic implications of submarine construction are further explored in the context of international relations, where nations seek to enhance their maritime capabilities in response to perceived threats [3]. This trend underscores the critical role that submarines play in modern naval warfare and the broader context of national defence strategies.

Submarine construction is a critical aspect of naval defence capabilities for many countries around the world. The United States, United Kingdom, France, Russia, China, Germany, South Korea, India, and Australia are among the key players in this field, each with specific facilities and strategic focuses.

In the United States, submarine construction is predominantly carried out at two major shipyards: General Dynamics Electric Boat in Connecticut and Newport News Shipbuilding in Virginia. These facilities are responsible for the production of advanced nuclear-powered submarines, including the Virginia-class and Columbia-class submarines, which are integral to the U.S. Navy's strategic capabilities. The collaboration between these shipyards and the Navy ensures that the U.S. maintains one of the most sophisticated submarine fleets globally, focusing on both attack and ballistic missile submarines.

The United Kingdom's submarine construction is centred at BAE Systems' shipyard in Barrow-in-Furness, where the focus is primarily on nuclear-powered submarines. The UK is currently developing the Astute-class SSNs and the Dreadnought-class SSBNs, which are set to replace the aging Vanguard-class submarines. These submarines are essential components of the UK's strategic deterrent force, underscoring the importance of maintaining a robust submarine fleet.

France's submarine construction capabilities are based at Naval Group's shipyards in Cherbourg. The French Navy produces both nuclear-powered and conventionally powered submarines, with notable projects including the Suffren-class (Barracuda) SSNs and the Scorpène-class diesel-electric submarines, which are also exported to various countries, including India and Brazil. This dual focus on domestic and export markets illustrates France's strategic approach to submarine technology and defence.

In Russia, submarine construction occurs at the Sevmash shipyard in Severodvinsk and Admiralty Shipyards in Saint Petersburg. The Russian submarine program features a diverse array of nuclear-powered and diesel-electric submarines, such as the Yasen-class SSNs and Borei-class SSBNs. Russia continues to advance its submarine technologies while also exporting submarines like the Kilo-class to nations including India and Vietnam, reflecting its ongoing commitment to maintaining a formidable naval presence.

China's submarine construction is primarily conducted at state-owned shipyards, notably the Bohai Shipyard and Wuchang Shipyard. The Chinese Navy focuses on both nuclear and conventional submarines, including the Type 093 SSN and Type 039A/B diesel-electric submarines. China's expanding submarine fleet is a crucial element of its maritime strategy, aimed at establishing dominance in the Indo-Pacific region.

Germany specializes in the production of advanced diesel-electric submarines, particularly the Type 212 and Type 214 classes, built by ThyssenKrupp Marine Systems (TKMS). These submarines are recognized for their air-independent propulsion (AIP) systems, which enhance their operational capabilities by allowing them to remain submerged for extended periods. Germany's submarine exports to countries like Norway and Israel further highlight its technological advancements in this field.

In South Korea, submarine construction is carried out at Daewoo Shipbuilding & Marine Engineering (DSME) and Hyundai Heavy Industries (HHI). The country has developed advanced diesel-electric submarines, such as the KSS-III class, for both domestic use and export, including contracts with Indonesia. This development reflects South Korea's growing presence in the global submarine market.

India's submarine construction efforts are spearheaded by state-run shipyards, particularly Mazagon Dock Shipbuilders (MDL) in Mumbai. India is focused on developing both conventional and nuclear-powered submarines, including the domestically-built Arihant-class SSBNs and the licensed production of Scorpène-class submarines in collaboration with France's Naval Group. This strategic focus aims to enhance India's naval capabilities and self-reliance in defence manufacturing.

Australia is currently modernizing its submarine fleet under the AUKUS agreement, which will enable it to acquire nuclear-powered submarines. Although Australia lacks the capability to build nuclear submarines domestically at present, it is investing in submarine infrastructure and workforce development to facilitate co-production with the United States and the United Kingdom.

The global demand for submarine construction is significantly influenced by various factors, including strategic deterrence, conventional defence needs, geopolitical tensions, and technological advancements. This demand reflects the evolving nature of maritime security and the strategic imperatives of nations worldwide.

- **Strategic Deterrence and Nuclear Capability:** The role of submarines, particularly ballistic missile submarines (SSBNs), is paramount in the nuclear deterrence strategies of countries with nuclear arsenals, such as the United States, Russia, China, France, and the United Kingdom. These submarines provide a credible second-strike capability, which is essential for maintaining strategic stability. The construction and modernization of SSBNs are critical to ensuring that these nations can effectively deter adversaries through assured retaliation [4]. The integration of advanced technologies into these submarines enhances their operational effectiveness, thereby sustaining the demand for their construction [5].

- **Conventional Defence and Maritime Patrol:** For countries that do not possess nuclear capabilities, the focus shifts towards acquiring conventional submarines, such as diesel-electric and air-independent propulsion (AIP) submarines. These vessels are increasingly sought after for their effectiveness in coastal defence, surveillance, and maritime patrol operations. Their quieter operation and smaller size make them particularly suited for littoral engagements, which are becoming more relevant in contemporary naval warfare [6]. Nations like Japan and South Korea are actively expanding their submarine fleets to enhance their maritime security and operational readiness [7].

- **Geopolitical Tensions and Regional Rivalries:** The rise of geopolitical tensions, especially in the Indo-Pacific region, has led to an escalation in submarine procurement and modernization efforts among regional powers. Countries such as India, Australia, and South Korea are investing heavily in their submarine capabilities to counter perceived threats from China and to assert their maritime sovereignty [8]. This regional arms race is characterized by a strategic emphasis on enhancing naval capabilities, which directly correlates with increased demand for submarine construction [9].

- **Export Market for Submarines:** The global submarine market is also shaped by the export activities of countries with advanced submarine manufacturing capabilities, such as France, Germany, and South Korea. These nations actively engage in selling submarines to countries looking to bolster their naval forces, including Brazil, Egypt, and Indonesia [10]. The export of submarines not only supports the economies of these manufacturing countries but also contributes to the proliferation of advanced naval technologies globally [11].

- **Technological Advancements and Modernization:** The demand for next-generation submarines is being fuelled by technological advancements, particularly in areas such as air-independent propulsion (AIP) and unmanned underwater vehicles (UUVs). These innovations are critical for enhancing the stealth and operational capabilities of submarines, making them more effective in modern warfare scenarios [6]. Countries are increasingly focused on modernizing their fleets to incorporate these advanced systems, which further drives the demand for submarine construction [5].

- **Increased Focus on Stealth and Underwater Warfare:** The operational effectiveness of submarines is heavily reliant on their stealth capabilities. As nations recognize the importance of covert operations in naval warfare, there is a growing investment in stealth technologies, including advanced sonar systems and acoustic coatings. This emphasis on stealth not only enhances the survivability of submarines but also increases their strategic value, thereby contributing to the rising demand for their construction [6].

The purpose of this book on submarine design is to cater to a broad spectrum of readers, each with specific needs and interests related to submarine engineering, operations, and strategic applications. By exploring the fundamental and advanced principles of submarine design, this book seeks to equip a variety of professionals and enthusiasts with the knowledge required to engage with this highly specialized field.

Foremost, naval architects and marine engineers will find this book invaluable. It provides in-depth insights into the principles of modern submarine design, addressing structural considerations, propulsion systems, and the integration of advanced technologies. For those tasked with designing, constructing, and maintaining submarines, this book will serve as an essential resource to stay abreast of innovations that drive the future of submarine capabilities.

Military personnel and defence analysts will also benefit greatly from this book. It sheds light on how submarine design impacts tactical operations, including stealth, endurance, and weapon deployment. For naval officers, strategists, and defence experts, this book provides information on the interplay between submarine capabilities and naval warfare, strategic deterrence, and power projection.

This book is equally vital for submarine technicians and maintenance crews. By offering detailed explanations of how submarines operate and how various systems function, the book enables technicians to better troubleshoot, repair, and upgrade submarines. Understanding the intricacies of submarine architecture can lead to more efficient maintenance and higher operational readiness.

For students and academics in naval engineering or defence studies, this book serves as both a foundational text and a deep dive into advanced concepts in submarine design. It supports academic development by offering a comprehensive overview of engineering practices, historical innovations, and the future of submarine technologies, making it an excellent reference for research or coursework.

Additionally, defence industry professionals and contractors will find this book particularly useful. It highlights the technical requirements, challenges, and innovations involved in submarine construction, helping these professionals contribute to and manage projects more effectively. Understanding the complexities of submarine design enables better decision-making and successful project execution.

Beyond professionals, this book caters to submarine enthusiasts and historians who are interested in the evolution of submarine technology and its role in military history. The book provides a thorough examination of how design innovations have influenced naval outcomes, offering a deeper appreciation of the strategic and technological milestones in submarine development.

For policy makers and government officials involved in defence procurement, maritime strategy, or military policies, the book offers a clear understanding of submarine design complexities. This knowledge supports more informed decision-making when approving defence budgets, choosing submarine technologies, or negotiating international treaties.

With the growing relevance of autonomous underwater vehicles (AUVs) and unmanned systems, researchers in this field will also benefit from the book's exploration of how traditional submarine design principles apply to unmanned technologies. The insights provided can inform the development of next-generation underwater drones and robotics.

Environmental scientists and marine conservationists will find value in understanding how submarine design interacts with marine ecosystems. This book offers a perspective on the environmental impact of submarine operations, helping advocate for more sustainable practices that minimize underwater disruption.

Finally, investors and stakeholders involved in naval defence projects will gain from this book's technical insights into submarine design. Equipped with this knowledge, they can better evaluate the potential risks, innovations, and successes of projects they are funding or supporting.

This book serves as a comprehensive resource for a wide-ranging audience, including engineers, military professionals, students, industry leaders, policymakers, and enthusiasts, all seeking a deeper understanding of the complexities and innovations shaping submarine technology today.

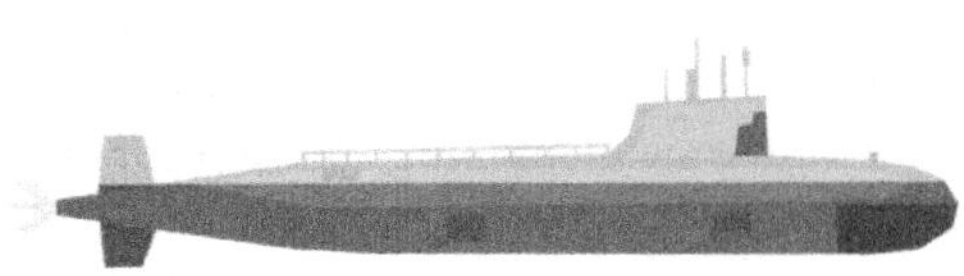

Chapter 1

Introduction to Submarine Engineering

Submarines are underwater self-propelled crafts that are designed and built to perform underwater operations for a stipulated amount of time [12].

A submarine is a specialized watercraft designed to operate underwater. Unlike surface ships, submarines can submerge and travel beneath the ocean for extended periods. They are typically used for military purposes, such as reconnaissance, stealth attacks, and deterrence, but also serve in scientific research, underwater exploration, and deep-sea salvage operations.

Figure 1: Remora 2000 research submarine of the Hellenic Centre for Marine Research. Montereypine, CC BY-SA 4.0, via Wikimedia Commons.

Submarines are equipped with advanced navigation, communication, and propulsion systems that allow them to manoeuvre underwater. They maintain buoyancy through ballast tanks, which can be filled with water to submerge or filled with air to surface. Many submarines are nuclear-powered, allowing them to stay submerged for months at a time without needing to refuel. Others are diesel-electric, which requires surfacing more frequently to recharge batteries.

Figure 2: The SS-083 ROKS Dosan Ahn Chang-ho Cruising above Water. ROK Ministry of National Defense's public work is used according to Korea Open Government License (KOGL) (KOGL Type 1, via Wikimedia Commons.

Their hulls are built to withstand the extreme pressure of deep-sea environments, and they are often armed with torpedoes or missiles, making them a strategic asset in naval warfare. The submarine design consists of a single or double-hull system that houses all the necessary systems and manpower required for completion of their mission.

When designing a submarine, the first and foremost objective is to ensure that the submarine meets the functional requirements of the customer. This includes considering what the submarine will be used for, such as military defence, scientific exploration, or commercial purposes like deep-sea mining or salvage operations. The functional design must support the specific operational needs, whether it's stealth, speed, endurance, or specialized equipment for research. Therefore, the design process involves a deep understanding of the intended mission of the submarine and aligning the design features—such as size, shape, propulsion, and onboard systems—to achieve that purpose effectively [12].

The next critical objective is ensuring that the submarine's design can be constructed with the resources that are available. This involves considering the capabilities of the shipbuilding facilities, the availability of materials, and the technical expertise required for the construction. Even if the design is perfect in theory, it must be feasible in practice, which means using materials and technologies that are accessible and familiar to the manufacturing team. This often includes balancing cutting-edge innovations with tried-and-tested techniques to ensure the submarine can be built efficiently and without unnecessary delays [12].

Lastly, the cost of the submarine project must be acceptable to the customer. Submarine construction is an expensive and complex undertaking, so the budget plays a crucial role in design decisions. Designers must carefully consider the cost of materials, labour, testing, and any technological systems that will be incorporated into the submarine. The goal is to deliver a design that not only meets the customer's operational needs but also stays within a reasonable budget. Balancing functionality, feasibility, and cost ensures that the final submarine design satisfies the customer both in terms of performance and affordability [12].

Submarine design is critically important because it directly impacts the vessel's functionality, safety, and effectiveness in fulfilling its mission. Several key reasons highlight the importance of submarine design:

- **Operational Effectiveness:** The design of a submarine determines how well it can perform its intended mission, whether it's military operations like surveillance, reconnaissance, and deterrence, or civilian applications such as deep-sea research and exploration. A well-designed submarine ensures that it meets specific functional requirements, such as stealth, speed, endurance, and the ability to deploy weapons or scientific instruments. For military submarines, effective design allows them to operate undetected, project power, and serve as strategic deterrents, while for civilian submarines, it enables safe and effective exploration of underwater environments.

- **Survivability and Safety:** Submarines operate in one of the most hostile environments on Earth: deep underwater, where high pressures, limited oxygen, and the risk of mechanical failure can be deadly. Proper design ensures that the submarine can withstand the extreme conditions of deep-sea operations, including pressure resistance, structural integrity, and reliable life support systems. Submarine design must also account for crew safety during extended missions, with provisions for emergency systems, escape routes, and redundant systems to prevent catastrophic failures.

- **Stealth and Detection Avoidance:** For military submarines, stealth is one of the most critical aspects of design. Submarines must be able to operate undetected, avoiding sonar detection, radar, and other surveillance systems. This involves designing the hull shape to minimize noise and hydrodynamic drag, using soundproofing materials, and incorporating advanced propulsion systems to reduce acoustic signatures. Effective submarine design gives navies a strategic advantage by enabling covert operations and reducing vulnerability to enemy attacks.

- **Technological Integration:** Submarine design must integrate a wide range of advanced technologies, including navigation systems, sonar, communication equipment, weapons systems, and propulsion technologies. These systems must be seamlessly integrated into the submarine's design to ensure they function efficiently and reliably under demanding conditions. Effective design ensures that the submarine can accommodate new technologies, allowing for upgrades over time without significant alterations to the vessel's structure.

- **Resource Efficiency and Cost Management:** Designing submarines involves balancing the need for cutting-edge technology with the constraints of available resources and budget. Proper design allows for efficient use of materials, energy, and manpower while keeping construction and operational costs within acceptable limits. This includes considerations for ease of construction, maintenance, and the long-term lifecycle of the submarine. Effective design helps prevent cost overruns and ensures that the vessel delivers value over its operational lifespan.

- **Environmental Impact:** Submarine design also plays a role in minimizing the environmental impact of underwater operations. This includes the use of clean, efficient propulsion systems (such as nuclear reactors or advanced battery technologies) that reduce emissions and environmental disturbances. Additionally, civilian submarine designs may focus on non-invasive exploration techniques to preserve fragile underwater ecosystems.

- **Strategic and Political Implications:** Submarine design influences a nation's naval power and strategic capabilities. Submarines are often a key component of national defence, especially in maintaining a nuclear deterrent. The design of a country's submarine fleet can impact international relations, military alliances, and strategic posturing. A well-designed submarine force strengthens a nation's ability to project power, protect its interests, and maintain security in contested regions.

Submarine Types

Submarines can be broadly categorized into military and civilian types, each designed with distinct purposes, capabilities, and features to meet their respective objectives. They can be classified into several types based on their design, propulsion systems, and operational roles. The primary categories include military submarines, research submarines, and tourist submarines, each with distinct functionalities and technological requirements.

Military Submarines

These are further divided into nuclear-powered and diesel-electric submarines. Nuclear submarines, such as the Russian "Alpha" class, utilize liquid metal coolants like Lead-Bismuth Eutectic (LBE) for their reactors, allowing for extended underwater endurance and speed [13]. They are designed for stealth and can carry out a range of missions, including anti-submarine warfare, reconnaissance, and strategic deterrence through nuclear capabilities. Diesel-electric submarines, while generally slower and with limited underwater endurance compared to their nuclear counterparts, are often quieter and less expensive to operate, making them suitable for coastal defence and operations in shallow waters [14].

Military submarines are engineered primarily for defence and offensive operations, serving as strategic assets in naval warfare. The main types include:

1. **Attack Submarines (SSNs and SSKs)**: These are designed for hunting and destroying enemy submarines and surface ships. They can be powered by nuclear reactors (SSNs) or diesel-electric engines (SSKs). Nuclear-powered attack submarines can operate for extended periods underwater due to their virtually unlimited range and endurance, while diesel-electric submarines, though quieter, require more frequent surfacing to recharge batteries. Attack submarines are often equipped with torpedoes, cruise missiles, and sometimes special operations forces for covert missions.

2. **Ballistic Missile Submarines (SSBNs)**: These submarines are built to carry and launch nuclear ballistic missiles. They are the core of a nation's nuclear deterrence strategy, capable of remaining hidden underwater for long durations to ensure a second-strike capability in the event of a nuclear attack. SSBNs are usually large, heavily armed, and designed to maintain stealth while on patrol in deep ocean waters.

3. **Cruise Missile Submarines (SSGNs)**: These submarines are designed to launch a large number of conventional or nuclear-tipped cruise missiles, typically targeting land-based facilities. SSGNs can also support special operations forces and intelligence-gathering missions. Like SSBNs, they are often nuclear-powered to provide extended

endurance and operational range.

4. **Midget Submarines**: Smaller in size, midget submarines are used for special operations, including reconnaissance, insertion of special forces, or laying mines. They have limited operational range and are often carried by larger submarines or surface ships to their mission areas.

5. **Unmanned Underwater Vehicles (UUVs)**: These are autonomous or remotely operated submarines used for reconnaissance, mine detection, or other missions without risking crewed vessels. UUVs are becoming more prevalent in modern navies due to advancements in underwater drone technology.

The acronyms for various types of submarines stand for the following:

- **SSN: Ship Submersible Nuclear** - A nuclear-powered attack submarine.
- **SSK: Ship Submersible Killer** - A diesel-electric attack submarine (also referred to as a hunter-killer submarine).
- **SSBN: Ship Submersible Ballistic Nuclear** - A nuclear-powered submarine that carries and launches ballistic missiles.
- **SSGN: Ship Submersible Guided Missile Nuclear** - A nuclear-powered submarine designed to carry and launch cruise missiles.

Each acronym reflects the type of propulsion system (nuclear or diesel-electric) and the primary function of the submarine (attack, ballistic missile, or cruise missile).

The classification of submarines into categories such as Attack Submarines (SSNs and SSKs), Ballistic Missile Submarines (SSBNs), Cruise Missile Submarines (SSGNs), Midget Submarines, and Unmanned Underwater Vehicles (UUVs) is widely acknowledged by many navies and defence organizations around the world. However, these classifications are not entirely universal. Different nations may adopt variations in how they categorize their submarine fleets or use alternative terminology, depending on the specific roles or capabilities that their submarines are designed to fulfill. This classification system is crucial for understanding the strategic roles and capabilities of different submarine types. For instance, Attack Submarines (SSNs and SSKs) are primarily designed for anti-submarine warfare and land attack missions, while Ballistic Missile Submarines (SSBNs) serve as a strategic deterrent by carrying nuclear missiles [15]. Similarly, Cruise Missile Submarines (SSGNs) are equipped to launch precision strikes against land targets, showcasing the versatility of modern submarine warfare [15].

Attack Submarines (SSNs and SSKs) are classified based on their power source and intended mission. Nuclear-powered attack submarines (SSNs) are common in the fleets of major powers like the United States, Russia, the United Kingdom, and France. These submarines offer extended operational endurance due to their nuclear propulsion. In contrast, diesel-electric attack submarines (SSKs) are more prevalent in smaller or regional navies, as they are quieter and less expensive to operate. Despite this general classification, some countries may have unique designations for their attack submarines, particularly when differences in size, range, or operational focus come into play.

Figure 3: Mooring lines secure the nuclear-powered attack submarine USS ALBANY (SSN 753) to the pier for the vessel's commissioning ceremony. The U.S. National Archives, Public Domain, via Picryl.

Ballistic Missile Submarines (SSBNs), sometimes referred to as "boomers," form a key component of the nuclear triad in major powers like the United States, Russia, China, France, and the United Kingdom. This classification is widely accepted across the globe because SSBNs play a crucial role in maintaining nuclear deterrence. Their strategic purpose—carrying and launching nuclear ballistic missiles—is well-defined, making the SSBN classification nearly universal among countries that operate them.

Figure 4: USS Kentucky (SSBN 737) returns to homeport at Naval Base Kitsap-Bangor, Wash. U.S. Navy, CC BY 2.0, via Flickr.

Cruise Missile Submarines (SSGNs) are also a significant category, primarily utilized by countries with advanced naval capabilities such as the United States and Russia. These submarines are either converted from SSBNs or specifically designed to launch conventional or nuclear-tipped cruise missiles. Although the SSGN classification is widely recognized, not all countries operate submarines exclusively designated as SSGNs. Instead, some nations may rely on multi-role submarines that fulfill various tasks, including launching cruise missiles, without assigning a specific SSGN classification.

Midget Submarines are smaller in size and are employed by several nations for special operations, coastal defence, and intelligence-gathering missions. Countries such as North Korea, Iran, and some European states make use of these submarines for specific tactical purposes. The classification and terminology of midget submarines may vary by country, with some navies referring to them as "mini-submarines" or using other localized terms based on their operational role, size, or technology.

Unmanned Underwater Vehicles (UUVs) represent an evolving class of submarines in modern naval operations. These autonomous or remotely operated underwater vehicles are increasingly used for reconnaissance, mine detection, and surveillance missions without putting human crews at risk. While the UUV classification is becoming more established, terms like "Autonomous Underwater Vehicles" (AUVs) or "Remotely Operated Vehicles" (ROVs) are sometimes used interchangeably, depending on the level of autonomy and the specific mission profile.

While these submarine classifications—SSNs, SSKs, SSBNs, SSGNs, Midget Submarines, and UUVs—are widely recognized and utilized by the world's leading naval forces, there can be slight variations in terminology and classification depending on a nation's naval doctrine, fleet capabilities, and operational focus. Nevertheless, these general categories remain consistent across major submarine-operating nations, reflecting the distinct roles that different types of submarines play in naval operations.

Civilian Submarines

Civilian submarines are designed for peaceful purposes, including research, exploration, and commercial activities. They are typically smaller and less heavily armed but feature advanced technology for their specific uses. The main types include:

1. **Research Submarines**: These submarines are used by scientific institutions to explore and study underwater environments, including deep-sea ecosystems, geology, and underwater archaeology. They are equipped with scientific instruments, cameras, and robotic arms to collect data, samples, and footage. Examples include Alvin and DSV Limiting Factor, which are designed for deep ocean exploration. These submarines are equipped with advanced scientific instruments and are used for oceanographic research, underwater exploration, and environmental monitoring. They can operate at various depths and are designed to withstand extreme underwater conditions. The wakes produced by these vehicles in stratified fluids are of particular interest, as they influence hydrodynamic performance and can affect the accuracy of data collected during missions [16]. Research submarines often collaborate with underwater sensor networks to gather data on marine ecosystems and geological formations [17].

2. **Tourism Submarines**: Designed for short, shallow dives, these submarines provide a unique experience for tourists, allowing them to explore coral reefs, shipwrecks, or marine life in areas where diving would be challenging. These submarines are pressurized to ensure passenger safety and offer panoramic views through large windows. These are designed for civilian use, allowing tourists to explore underwater environments safely. They are typically smaller and equipped with large viewing windows, providing an immersive experience of marine life. Unlike military and research submarines, tourist submarines operate in shallower waters and are not equipped for deep-sea exploration [14].

3. **Commercial Submarines**: These submarines are used in industrial applications such as underwater construction, maintenance, and repair, typically in the oil and gas industry. They are used to inspect pipelines, deep-sea drilling operations, and subsea infrastructure. Some are also used for underwater salvage operations, recovering sunken ships or treasure from the ocean floor.

4. **Personal Submarines**: Personal or recreational submarines are small, privately owned submarines used by individuals for underwater exploration. These submarines typically have limited range and depth capabilities but offer private owners the ability to explore underwater locations for leisure or personal projects.

Hybrid and Emerging Types

With advancing technology, new types of submarines are emerging that blur the lines between military and civilian uses. These include multipurpose submarines designed for dual-use operations, capable of performing both scientific research and military reconnaissance. Additionally, innovations in autonomous underwater vehicles are pushing the boundaries of what submarines can achieve in both civilian and military sectors, leading to increasingly sophisticated and capable submarines across the spectrum of uses.

The development of submarines has been significantly influenced by advancements in materials science and communication technologies. For instance, the use of fibre-reinforced composites has improved the structural integrity and performance of submarine components [18]. Moreover, underwater communication remains a challenge due to the absorption of electromagnetic waves in seawater, leading to the exploration of acoustic and optical communication methods to enhance operational capabilities [19, 20].

History of Submarine Development

Submarines first became a significant force in naval warfare during World War I (1914–18), primarily due to Germany's effective use of them to target and destroy surface merchant vessels. Their main weapon was the torpedo, a self-propelled underwater missile that allowed submarines to strike ships without being detected. This strategic use of submarines continued into World War II (1939–45), where they played an even larger role in both the Atlantic, where Germany's U-boats targeted Allied shipping, and the Pacific, where American submarines attacked Japanese vessels. The 1960s marked a major advancement in submarine design with the advent of nuclear-powered submarines. These vessels could stay submerged for extended periods, sometimes months, and were capable of launching long-range nuclear missiles without surfacing, solidifying their place as a cornerstone of strategic military power. Armed with torpedoes as well as anti-ship and anti-submarine missiles, the nuclear attack submarine has since become a critical component of naval warfare [21].

The concept of a "submarine," a craft designed to navigate underwater, first emerged in 1578 when British mathematician William Bourne proposed a fully enclosed boat that could submerge and be rowed underwater. His design featured a wooden frame covered in waterproof leather and was intended to submerge by contracting the sides to reduce volume using hand vises. Although Bourne never constructed his design, the idea was later brought to life by Dutch inventor Cornelis Drebbel, who built the first functioning submarine between 1620 and 1624. Drebbel's craft, made of a greased leather-covered wooden frame, was propelled by oars and could reach depths of 12 to 15 feet during trials in the Thames River. King James I of England is said to have even taken a short ride in the craft. Drebbel's work laid the foundation for future submarine designs, and his success was followed by two larger versions of his original craft [21].

The early 18th century saw a proliferation of submarine concepts, with numerous designs and patents emerging. By 1727, 14 types of submarine boats had been patented in England alone. In 1747, an inventor proposed an innovative submersion method using goatskin bags that filled with water to sink the craft and were then emptied using a twisting rod to bring it back to the surface. This design was an early precursor to the modern submarine ballast tank. By the mid-18th century, over a dozen patents for submarine designs had been granted in England. Nathaniel Symons is credited with building the first working ballast tank system in 1747, using leather bags filled with water to submerge the submarine and a mechanism to force the water out, bringing it back to the surface. Although these early designs showed promise, further advancements stagnated for more than a century until technological progress in propulsion and stability reignited interest [21].

The submarine was first used as an offensive weapon during the American Revolution (1775–83) with David Bushnell's one-man craft, the Turtle. Built in the shape of a walnut and powered by propellers cranked by hand, the Turtle was designed to approach enemy ships underwater, attach a gunpowder charge, and retreat before the explosion. Although the Turtle's attack on a British warship was unsuccessful, it marked the first use of a submarine in naval warfare [21].

Figure 5: Full-size model of the Turtle submarine on display at the Royal Navy submarine museum. Geni, CC BY-SA 4.0, via Wikimedia Commons.

Years later, American inventor Robert Fulton experimented with submarine design. In 1800, while in France, Fulton built the Nautilus, a copper submarine with iron ribs, a collapsible mast and sail for surface propulsion, and a hand-turned propeller for underwater movement. The Nautilus could submerge using ballast tanks and maintain depth with a horizontal rudder, the precursor to modern diving planes. Despite successfully sinking a schooner during trials, Fulton's attempts to use the Nautilus in warfare were thwarted when both France and Britain lost interest in the project. Although he continued his efforts in the United States, Fulton died before his steam-powered submarine could be completed [21].

During the War of 1812, a copy of the Turtle was built and used in an attack on the British ship HMS Ramillies off the coast of Connecticut. The operator managed to bore a hole in the copper sheathing of the warship but failed to attach the explosive due to the screw breaking loose. Despite these early setbacks, the pioneering efforts of inventors like Bushnell and Fulton laid the groundwork for future submarine development, proving the potential of underwater vessels in naval warfare [21].

During the American Civil War (1861–65), the Confederacy sought unconventional methods to counter the Union Navy's superior strength, particularly the blockade of Southern ports. One such attempt involved the use of submarines.

In 1862, Horace L. Hunley from Mobile, Alabama, financed the construction of a Confederate submarine named *Pioneer*. This 34-foot craft was powered by a hand-cranked propeller operated by three men. Although some records suggest the *Pioneer* was lost during a dive, it was likely scuttled to avoid capture when Union forces took control of New Orleans [21].

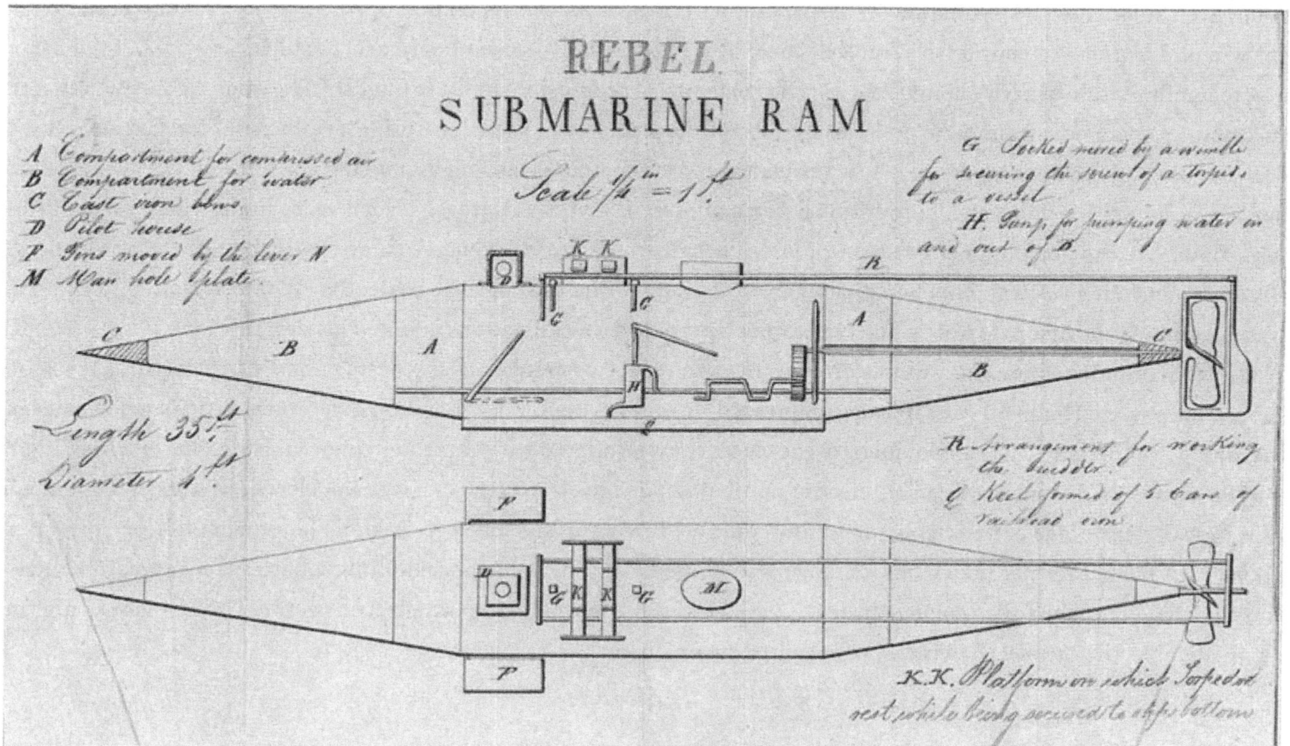

Figure 6: The Confederate submarine Pioneer's measured drawings given to William Shock, Fleet Engineer of the Western Gulf Blockading Squadron in 1862. William Shock, Public domain, via Wikimedia Commons.

Following the *Pioneer*, the same builders developed a second, more advanced submarine. This 25-foot iron boat was designed to be powered by a battery and electric motors, but suitable motors could not be found. Instead, a hand-cranked propeller operated by four men was again used. Unfortunately, this submarine sank in heavy seas off Mobile Bay while attempting to engage the enemy, though no lives were lost in the incident [21].

The third Confederate submarine was the *H.L. Hunley*, a modified iron boiler that was lengthened to about 36 to 40 feet. It featured ballast tanks and a system of weights to submerge the vessel. Powered by eight men cranking the propeller, the *Hunley* could reach speeds of four miles per hour. Its weapon was a "torpedo" containing 90 pounds of gunpowder, which was towed behind the submarine on a 200-foot line. The plan was for the submarine to dive beneath an enemy warship and drag the torpedo into its hull. After a successful test against a barge, the *Hunley* was transported to Charleston, South Carolina. Despite suffering several accidents, including three sinkings that resulted in numerous crew deaths, including Hunley himself, the *Hunley* continued its operations.

On February 17, 1864, the *Hunley* attacked the Union warship *Housatonic* in Charleston Harbor. It used a torpedo mounted on the end of a long spar to strike the warship. The explosion ignited the *Housatonic*'s magazines, sinking the ship in shallow water and killing five men. However, the *Hunley* was also destroyed in the explosion, resulting in the loss of its entire crew [21].

Around the same time, Wilhelm Bauer, a Bavarian artillery non-commissioned officer, was making his own advances in submarine design. Bauer built two submarines: *Le Plongeur-Marin* (1851) and *Le Diable-Marin* (1855). The first boat sank in Kiel Harbor on February 1, 1851, but Bauer and his two assistants managed to escape from a depth of 60 feet after being trapped for five hours. His second submarine, constructed for the Russian government, proved more successful, reportedly completing 134 dives before being lost at sea. In September 1856, during the coronation of Tsar Alexander II, Bauer even submerged his submarine in Kronshtadt harbor with musicians on board, who played the Russian national anthem underwater—an impressive feat that could be clearly heard by those aboard nearby ships.

A major limitation of early submarines was the challenge of propulsion. In 1880, English clergyman George W. Garrett successfully operated a submarine powered by steam from a coal-fired boiler, featuring a retractable smokestack. However, the fire had to be extinguished before the craft could submerge, as leaving it lit would consume all the available air inside the submarine. Even so, the remaining steam allowed the vessel to travel several miles underwater. Around the same time, Swedish gun designer Torsten Nordenfelt constructed a steam-powered submarine driven by twin propellers. His design included vertical propellers that enabled the submarine to submerge to depths of 50 feet. Nordenfelt's craft was also among the first to feature a practical torpedo tube, and several nations adopted his design.

Efforts to improve submarine propulsion continued. In 1864, two French naval officers built the *Le Plongeur*, a 146-foot submarine powered by an 80-horsepower compressed-air engine. However, the air tanks were quickly depleted whenever the craft got underway. The development of the electric motor offered a breakthrough in propulsion. The *Nautilus*, built in 1886 by two Englishmen, was an all-electric submarine powered by two 50-horsepower electric motors and operated using a 100-cell storage battery. This submarine could achieve a surface speed of six knots (approximately 6.9 miles per hour or 11.1 kilometres per hour), but its range was limited to 80 miles before requiring a battery recharge. In France, Gustave Zédé launched the *Gymnote* in 1888, which was also propelled by an electric motor. Though it was highly manoeuvrable, the *Gymnote* had issues with control when diving.

Figure 7: Photo of the French submarine Gymnote in Page's Magazine No. 2 from 1902. That issue had an article about the development of submarines. Note that this photo shows the small conning tower that was added in 1898. archive.org, Public domain, via Wikimedia Commons.

The late 19th century was a period of rapid development in submarine technology, particularly in France. Zédé collaborated on various designs sponsored by the French navy, leading to the successful launch of the *Narval* in 1899. Designed by naval engineer Maxime Laubeuf, the *Narval* was a double-hulled submarine, 111.5 feet long, that used a steam engine for surface propulsion and electric motors when submerged. The ballast tanks were located between the double hulls, a feature still in use today. The *Narval* completed numerous successful dives and marked a significant step forward in submarine design. French advancements continued with the Sirène-class steam-driven submarines completed between 1900 and 1901, and the *Aigrette*, launched in 1905 as the first diesel-powered submarine in any navy [21].

In the United States, rival inventors John P. Holland and Simon Lake also made significant contributions to submarine design. Holland, an Irish immigrant, launched his first submarine in 1875. His designs were notable for combining water ballast with horizontal rudders for diving. In 1895, the U.S. Navy commissioned Holland to build a submarine, the *Plunger*, which was to be powered by steam on the surface and electricity when submerged. However, the project encountered numerous challenges and was eventually abandoned. Holland later built another submarine, the *Holland*, at his own expense. Launched in 1897 and accepted by the U.S. Navy in 1900, this 53.25-foot submarine was powered by an electric motor underwater and a gasoline engine on the surface. The *Holland* was armed with a bow torpedo tube and two dynamite guns, and it proved to be a successful vessel, undergoing several modifications to test different configurations of propellers, diving planes, rudders, and other equipment [21].

Holland's chief competitor, Simon Lake, constructed his first submarine, the *Argonaut I*, in 1894. Powered by a gasoline engine and electric motor, the *Argonaut I* was designed primarily for undersea research. In 1898, it became the first submarine to operate extensively in open sea conditions, traveling under its own power from Norfolk, Virginia, to New York City. This achievement preceded the voyages of the French *Narval* and demonstrated the potential for submarines in long-range missions. Lake's second submarine, the *Protector*, was launched in 1901.

Figure 8: The submarine Narval (S 64) docked at the port of Melilla; it is the fourth and final unit of the Delfín Class. Bruno Cleries, upload by Basilio, CC BY-SA 3.0, via Wikimedia Commons.

At the turn of the century, most major naval powers were investing in submarine development, with Britain being a notable exception until 1901 when the Royal Navy ordered five submarines based on Holland's designs. Germany, too, entered the submarine race, completing its first submarine, the *U-1*, in 1905. The *U-1* was 139 feet long and powered by a heavy oil engine on the surface and an electric motor when submerged. It was armed with one torpedo tube. These early submarines set the stage for the 20th-century submarine, which would be propelled by diesel engines on the surface and electric motors underwater, submerging by taking on water ballast and using diving planes, and armed with torpedoes to sink enemy ships. Despite these advancements, life aboard early submarines was far from comfortable; the quarters were cramped, often wet, and smelled of diesel oil [21].

Figure 9: German submarine U 1 on trials. navyphotos.co.uk, Public domain, via Wikimedia Commons.

By the outbreak of World War I, all major naval powers had incorporated submarines into their fleets. However, these early submarines were relatively small and primarily intended for coastal operations, with their military value still viewed with some scepticism. One notable exception to this coastal focus was the German *Deutschland* class of merchant U-boats. These submarines were significantly larger, measuring 315 feet in length and featuring two spacious cargo compartments. Capable of carrying 700 tons of cargo, they could travel at speeds of 12 to 13 knots on the surface and seven knots while submerged. Eventually, the *Deutschland* itself was converted into the U-155 by fitting it with torpedo tubes and deck guns, and along with seven similar submarines, it played a combat role during the later stages of the war. In contrast, the standard submarine of World War I was smaller, typically just over 200 feet long and displacing less than 1,000 tons on the surface [21].

Before the war, submarines had been primarily armed with self-propelled torpedoes designed to attack enemy vessels. However, as the war progressed, submarines were also equipped with deck guns, which allowed them to surface and signal enemy merchant ships to halt for inspection (a policy used earlier in the war) or to sink smaller or unarmed ships without expending torpedoes. Most war-built submarines had one or two deck guns of three- or four-inch calibre, though some later German submarines, including the *Deutschland* class in its military configuration, carried larger 150-millimeter guns.

An important variation in submarine armament during the war was the development of mine-laying submarines. The Germans constructed several specialized submarines equipped with vertical mine tubes that allowed them to covertly lay mines in enemy harbors. Some U-boats could carry as many as 48 mines in addition to their torpedoes, greatly enhancing their strategic impact.

Another significant innovation during World War I was the introduction of submarines designed to combat enemy submarines. British submarines sank 17 German U-boats during the conflict, leading to the development of the R-class submarine, specifically designed for anti-submarine warfare. These submarines were relatively small, measuring 163 feet in length and displacing 410 tons on the surface, and featured a single propeller (while most submarines of the time had two). The R-class submarines could reach speeds of nine knots on the surface using diesel engines, but once submerged, their large batteries powered electric motors that allowed them to achieve an impressive underwater speed of 15 knots for up to two hours. This made them both highly manoeuvrable and fast, particularly in comparison to other submarines of the era, which typically achieved submerged speeds of around 10 knots until after World War II. The R-class submarines were also equipped with advanced underwater listening equipment, known as asdic or sonar, and were armed with six forward torpedo tubes, making them formidable weapons. Although they arrived too late to have a significant impact on the war, they pioneered a new concept in submarine development [21].

With the exception of the British *Swordfish* and K-class submarines, all World War I-era submarines used diesel engines for surface propulsion and electric motors for submerged operations. The *Swordfish* and K-class submarines were intended to operate as scouts for surface warships and required higher speeds, which were achievable only with steam turbines at the time. These submarines could reach surface speeds of 23.5 knots, while their electric motors allowed for submerged speeds of up to 10 knots [21].

Figure 10: German Cargo Submarine Deutschland 1916-22. tormentor4555, Public Domain, via Flickr.

During the period between World War I and World War II, interest in submarines remained strong among the world's navies. Countries such as Britain, France, and Japan continued to build improved submarine models, while the U.S. Navy introduced its first large, long-range submarine, the *Argonaut,* in 1928. At 381 feet long and with a displacement of 2,710 tons on the surface, the *Argonaut* was armed with two six-inch guns, four forward torpedo tubes, and could carry 60 mines. This submarine became the largest non-nuclear submarine ever built by the U.S. Navy and paved the way for the highly successful *Gato*- and *Balao*-class submarines used during World War II.

In the 1930s, the Soviet Union also focused on building submarines, primarily small coastal craft, as part of an effort to strengthen its naval power without investing heavily in surface warships. However, despite producing a large number of submarines, the Soviet vessels were unsuitable for operations against the German Navy. The crews were often poorly trained, and their bases were frequently blocked by ice, limiting their effectiveness [21].

World War II saw widespread submarine campaigns across all the world's oceans. In the Atlantic, the primary German U-boat was the Type VII, a relatively small yet highly effective craft when employed properly. The Type VIIC variant was 220.25 feet long, with a surface displacement of 769 tons. Powered by diesel-electric machinery, it could reach speeds of 17 knots on the surface and 7.5 knots submerged. Its armament included a 90-millimeter deck gun, various anti-aircraft guns, and five torpedo tubes. These submarines, manned by crews of 44, had a surface endurance of 6,500 miles at 12 knots but could remain submerged for less than a day when traveling at four knots [21].

The German Type XXI represented the ultimate development of diesel-electric submarines during the war. Measuring 250 feet long and with a displacement of 1,600 tons, this submarine could achieve submerged speeds of 17.5 knots for over an hour, travel underwater at six knots for two days, or "creep" at slower speeds for four days. Fitted with snorkel devices, the Type XXI did not need to fully surface to recharge its batteries. Its operating depth was an impressive 850 feet, more than double the norm at the time, and it was armed with four 33-millimeter guns and six forward torpedo tubes, carrying 23 torpedoes. Had the war continued past the spring of 1945, Allied forces would have faced significant challenges in combating these advanced submarines [21].

Another notable German innovation during the war was the Walter turbine propulsion plant. Developed by scientist Hellmuth Walter, this system allowed submarines to use steam turbines while submerged by employing oxygen generated from hydrogen peroxide. The *V-80*, an experimental submarine built in 1940, could reach submerged speeds of over 26 knots for short periods. Later, Walter-propelled Type XVII combat submarines achieved 25 knots underwater, with trials showing speeds of 20 knots for five and a half hours. However, these advanced submarines, like the Type XXI, were not ready for widespread use by the time the war ended.

One of the most significant German innovations of the war was the schnorchel (or snorkel) device, originally conceived by Dutch officer Lieutenant Jan J. Wichers in 1933. The schnorchel allowed submarines to operate their diesel engines while submerged, drawing in fresh air through a tube. The Dutch Navy began using this technology in 1936, and after falling into German hands in 1940, it was widely adopted by German U-boats. The schnorchel enabled U-boats to recharge their batteries at periscope depth while minimizing the risk of detection by Allied radar.

In the Pacific, the Japanese Navy deployed a diverse range of submarines, including aircraft-carrying submarines, midget submarines, and "human torpedoes" launched from larger submarines. One notable Japanese class was the *I-201*, a high-speed submarine that was 259 feet long with a displacement of 1,291 tons. This submarine could reach 15 knots on the surface using diesel engines and 19 knots submerged for nearly an hour, thanks to its large batteries and electric motors. The *I-201* was armed with two 25-millimeter guns and four forward torpedo tubes, carrying ten torpedoes.

Figure 11: Japanese Submarine HA-201(SENTAKA-SHO-class) near Sasebo. Unknown author, Public Domain, via Wikimedia Commons.

The U.S. submarine campaign in the Pacific during World War II was highly successful, primarily utilizing the *Gato*- and *Balao*-class submarines. These submarines were around 311.5 feet long, displaced 1,525 tons, and were powered by diesel-electric machinery, allowing for surface speeds of 20 knots and underwater speeds of nine knots. The main difference between the two classes was their operating depth, with the *Gato* class reaching depths of 300 feet and the *Balao* class capable of descending to 400 feet. Manned by crews of 65 to 70, these submarines were armed with one or two five-inch deck guns, smaller anti-aircraft weapons, and ten torpedo tubes (six forward and four aft), carrying a total of 24 torpedoes.

After World War II, the Allies quickly adopted advanced German submarine technology. The British experimented with two peroxide turbine-propelled submarines, but the concept fell out of favor due to the unstable nature of hydrogen peroxide and the success of nuclear propulsion in the United States. Meanwhile, the Soviet Union began producing modifications of the German Type XXI submarine, leading to the creation of the Whiskey and Zulu class submarines. Between 1950 and 1958, the Soviets completed 265 of these submarines—more than all other navies combined produced between 1945 and 1970. In total, Soviet shipyards produced 560 new submarines during this period [21].

The U.S. Navy also studied German technology, converting 52 war-built submarines into the Guppy configuration (standing for "greater underwater propulsive power"). These submarines had their deck guns removed, received streamlined conning towers, and were equipped with larger batteries and snorkels. In some cases, one of the diesel engines and some torpedoes were removed, resulting in increased underwater speeds of 15 knots and improved endurance [21].

While the major powers eventually transitioned to nuclear-powered submarines after the war, many of the world's navies continued to use diesel-electric submarines derived from the fast U-boats of World War II. Several of these submarines were designed and built in West Germany. Though these postwar submarines still use snorkels, advances in radar have made it possible to detect even the small head of a snorkel, just as primitive radars detected surfaced U-boats during the war.

The most significant postwar advancements have been in weaponry and sensors. Deck guns have largely been replaced by antiship missiles, while torpedoes have become much faster, capable of exceeding 50 knots. These torpedoes are guided by self-contained sonar or by electronic commands sent through a wire paid out behind the torpedo. Many submarines are now equipped with cruise missiles or antiship missiles, capable of targeting both land and sea. Submarine sonar technology has also seen significant improvements, enhancing the ability to detect both surface ships and other submarines. Additionally, on the most advanced submarines, traditional periscopes are being replaced by photonic or optronic masts, which project sensors to the surface and relay information electronically to the control room. This eliminates the need for any hardware to pierce the submarine's hull, and these masts are operated using a simple joystick with data displayed on screens throughout the submarine.

While the maximum submerged speed of modern submarines has increased slightly to over 20 knots, endurance at top speed remains similar to what it was at the end of World War II. Improvements in lead-acid battery design have increased endurance at lower speeds, allowing modern submarines to remain submerged at around three knots for up to a week to 10 days. This capability is significant, as it allows submarines to take advantage of changing sea conditions or force surface hunters to disperse. The development of air-independent propulsion (AIP), particularly through fuel cells that generate electricity using stored hydrogen and oxygen, has further extended underwater endurance, with some AIP-capable submarines able to operate at low speeds for as long as a month [21].

As a result, diesel-electric submarines remain quiet and effective platforms, operating stealthily and conserving energy for post-attack escapes. Their electric motors are quieter than nuclear units and can even be shut off to allow the submarine to wait silently for enemy submarines to pass. AIP technology offers additional possibilities, such as long-term operations under ice in polar seas, monitoring coastal shipping in antiterrorist missions, or deploying special operations forces to foreign shores. Modern diesel-electric submarines, whether equipped with AIP or not, continue to be affordable and effective tools for navies worldwide, allowing them to defend their coastal waters against all potential threats, even from nuclear powers [21].

In 1954, with the commissioning of the *USS Nautilus*, nuclear power became a reality for submarines. The introduction of nuclear propulsion allowed submarines to operate without the need for oxygen, meaning a single power plant could be used both on the surface and while submerged. Moreover, nuclear fuel (enriched uranium) provided power for extended periods, enabling submarines to remain submerged at high speeds indefinitely. This revolutionized submarine warfare. Previously, submarines would approach their targets on the surface to conserve battery power and would only submerge just before reaching the target. Underwater, they had to move slowly—often at two to three knots—to avoid depleting their batteries. Following an attack, submarines would use full underwater power (seven to 10 knots) to evade counter-attacks, but this would drain the batteries within one or two hours. As a result, diesel-electric submarines were limited in their ability to engage fast warships, such as aircraft carriers and battleships.

Figure 12: USS Nautilus' (SSN-571). Official U.S. Navy Photograph, Public domain, via Wikimedia Commons.

Nuclear-powered submarines, however, were a game-changer. These vessels could operate freely before and after attacks without worrying about battery depletion and could keep up with fast surface ships. This advantage was demonstrated during the Falklands War in 1982 when the British nuclear submarine *HMS Conqueror* followed the Argentine cruiser *General Belgrano* for over 48 hours before closing in and sinking it. Such performance was beyond the capability of pre-nuclear submarines. Nuclear propulsion gave submarine commanders unprecedented freedom to manoeuvre underwater, making fast surface ships vulnerable to attack.

Initially, many navies continued building diesel-electric submarines alongside their nuclear counterparts, but the cost of maintaining both types eventually led some nations to transition entirely to nuclear-powered fleets. The U.S. Navy ceased non-nuclear submarine construction in 1959. Similarly, the Royal Navy, which completed its first nuclear submarine *HMS Dreadnought* in 1963, eventually focused exclusively on nuclear-powered vessels after briefly building the diesel-electric Upholder class in the 1980s and 1990s. France launched its first nuclear submarine, *Le Redoutable*, in 1971 and stopped building diesel-electric submarines for its own navy in 1976, though it continued producing conventional submarines for export. The Soviet Union followed a similar path, building both nuclear and diesel submarines, though their primary focus shifted to nuclear power after their first nuclear-powered submarines, the *November* class, entered service in 1958. Russia, the successor state to the Soviet Union, continued to maintain a mixed nuclear-conventional submarine force post-1991. Meanwhile, China began building nuclear submarines in 1968 while still producing large numbers of diesel-electric vessels, a pattern India followed, launching its first nuclear submarine in 1998.

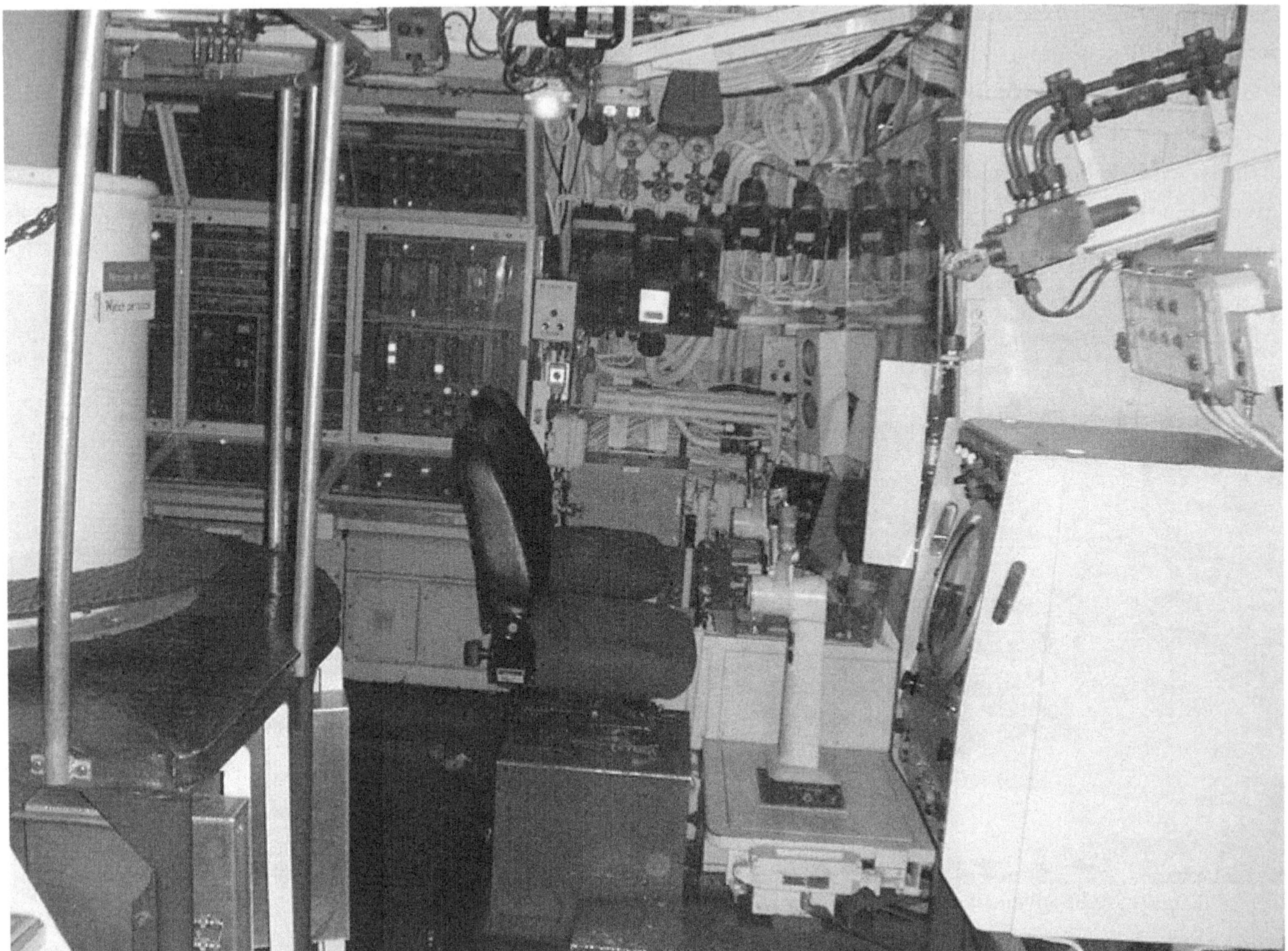

Figure 13: Le Redoutable at the Cherbourg Maritime Museum, France, 28 May 2008. Hugh Llewelyn, CC BY-SA 2.0, via Wikimedia Commons.

Nuclear reactors power submarines by producing heat through nuclear fission. In a reactor, uranium fuel is surrounded by a moderator (usually water) that slows neutrons and helps sustain the reaction. This water, called primary loop water, is pressurized to prevent boiling and carries the heat to a heat exchanger. The heat is then transferred to a secondary water circuit, which produces the steam that powers the turbine. The primary loop remains sealed, preventing contamination of the rest of the plant.

Water in the primary loop is usually pumped, but some reactors use natural circulation based on temperature differences. In these reactors, cooler water from the heat exchanger is fed into the bottom of the reactor, where it heats up as it passes through the fuel elements. Another type of reactor, the liquid-metal-cooled reactor, uses molten metal to carry more heat than water, enabling more compact turbines. However, this introduces challenges, such as the potential for radioactive leaks and the risk of the molten metal solidifying in pipes, leading to catastrophic failures.

Under the direction of Admiral Hyman Rickover, the U.S. Navy developed both pressurized-water and liquid-metal reactor prototypes. The first two nuclear submarines, the *Nautilus* and *Seawolf*, tested these systems, but issues with the *Seawolf*'s liquid-metal reactor led to the abandonment of that approach. The U.S. Navy subsequently adopted pressurized-water reactors for its attack submarines, while its Ohio-class strategic submarines use quieter natural-circulation reactors. Other nuclear navies also rely on pressurized-water or natural-circulation reactors, with one exception: the Soviet Union's *Alfa*-class submarines, which used liquid-metal reactors for high-speed operation.

Figure 14: A starboard bow view of a Soviet Alfa class attack submarine underway. NARA & DVIDS, Public Domain, Public Domain Archive.

The advent of nuclear submarines had two major consequences. First, it led to the development of strategic submarines, which carry submarine-launched ballistic missiles (SLBMs). These submarines are highly valuable due to their stealth, making them difficult to detect and destroy. Second, it revolutionized antisubmarine warfare, as nuclear attack submarines became the primary tools for hunting enemy submarines. Armed with torpedoes and antiship missiles, these submarines also support special operations and electronic intelligence missions.

Strategic submarines play a crucial role in national defence by providing a secure platform for SLBMs, which are difficult for enemies to locate and destroy. Early strategic submarines appeared in the 1950s, with the U.S. and Soviet navies developing missile-carrying diesel-electric submarines. However, these submarines had to surface to launch their missiles, limiting their effectiveness. The introduction of nuclear-powered strategic submarines with dive-launched missiles allowed these vessels to remain submerged for extended periods, significantly enhancing their survivability [21].

The first modern strategic submarines were the U.S. *George Washington*-class submarines, which became operational in 1959 and carried Polaris missiles. The Soviet Union followed with the *Yankee* class in 1967 and later the *Delta* and *Typhoon* classes. The *Typhoon* class submarines, launched in 1982, were the largest ever built and remain in service with the Russian navy. As these vessels age, Russia has introduced the *Borey* class, launched in 2007, equipped with the new Bulava SLBMs [21].

Other nuclear powers also developed strategic submarines. The U.S. Navy commissioned 18 *Ohio*-class submarines between 1981 and 1997, each capable of carrying 24 Trident missiles. The British *Resolution*-class submarines, launched in 1967, were replaced by the *Vanguard* class in the 1990s. France replaced its *Redoutable*-class submarines with the *Triomphant* class, while China and India both developed their own strategic submarine programs, launching the *Jin* class and *Arihant*, respectively [21].

Nuclear-powered attack submarines emerged as the primary antisubmarine weapons in the postwar era. These submarines are designed to destroy enemy surface ships and submarines, while also supporting additional roles such as land attacks, minelaying, and intelligence gathering. The U.S. Navy's four generations of nuclear attack submarines—the *Sturgeon*, *Los Angeles*, *Seawolf*, and *Virginia* classes—have evolved to meet the changing demands of naval warfare. The *Seawolf*-class submarines, built during the Cold War, were optimized for submarine hunting, while the more versatile *Virginia*-class submarines are capable of various missions near shore and in open ocean [21].

Figure 15: The Virginia-class attack submarine Pre-commissioning. NARA & DVIDS, Public Domain, Public Domain Archive.

The Soviet navy also developed a range of attack submarines, including the *Victor* and *Oscar* classes. These vessels were equipped with a mix of torpedoes and cruise missiles, including the long-range SS-N-19 *Shipwreck* missile. More recent Russian attack submarines, such as the *Akula* and *Yasen* classes, carry advanced antisubmarine and land-attack missiles. Other nations, including Britain, France, and China, have similarly developed nuclear attack submarines with capabilities for both antisubmarine and land-attack missions [21].

Nuclear submarines, both strategic and attack, have fundamentally reshaped naval warfare, offering unparalleled stealth, endurance, and firepower. They remain central to the defence strategies of the world's major naval powers [21].

During the Cold War, three major trends emerged in the design of nuclear attack submarines among NATO and Soviet forces: increased speed, greater diving depth, and enhanced silencing. These developments were driven by the need for greater tactical advantage in underwater warfare and were exemplified by the submarine fleets of the United States, Britain, and the Soviet Union.

One of the primary objectives in submarine design was to increase speed, which required greater power. However, increasing power presented challenges. The resistance a submarine encounters is related to its surface area, so achieving higher speeds without expanding the size or weight of the power plant was ideal. A more powerful engine could increase noise, but silencing measures would require enlarging the submarine, potentially offsetting any speed gains. This balance of trade-offs was illustrated in the evolution from the *Sturgeon* class to the *Los Angeles* class submarines, where reactor power doubled, but surface displacement increased significantly from 3,600 to 6,000 tons. The Soviet *Alfa* class, which achieved speeds of about 40 knots, faster than the 30 knots typical of Western submarines, likely sacrificed quiet operation for this speed [21].

Speed was important for several reasons. Initially, fast submarines were developed as antiship weapons. For example, the *USS Nautilus* could reach over 20 knots submerged, sufficient for evasion but inadequate for counterattacks. To address this, nuclear power was adapted to the streamlined "tear-drop" hull of the *Albacore*, resulting in the *Skipjack* class, which had a top speed of more than 30 knots. In a Cold War incident, a Soviet submarine demonstrated this capability by tracking the nuclear aircraft carrier *USS Enterprise* in 1968, maintaining enough speed to keep the carrier within range of its weapons.

As the Soviet Union advanced its nuclear submarine program, the U.S. Navy shifted focus to dual-purpose vessels capable of targeting both submarines and surface ships. The high speed of the *Los Angeles* class submarines, introduced in the 1970s and '80s, allowed them to keep up with the fast surface ships the Soviets were expected to target. Speed also helped submarines deploy more efficiently to distant patrol areas, as their supplies limited them to missions of 60 to 90 days. The faster they reached their patrol stations, the more effective their missions became [21].

However, speed alone wasn't a guarantee of evasion. Submarines that sacrificed quiet operation for speed became easier to detect, and by the mid-1950s, the main antisubmarine weapons—homing torpedoes and nuclear depth bombs—became faster than the submarines they targeted. Speed remained a critical factor in underwater combat, but it came with trade-offs, particularly in terms of noise and detection.

Greater diving depth was another important design trend, offering several advantages. Deep diving allowed for better evasion and improved sonar performance, as submarines could operate in different layers of the sea. This was reflected in changes to U.S. submarine sonar systems around 1960, when new spherical sonars replaced the earlier cylindrical units. The new sonars produced narrower beams that could detect targets at different depths and take advantage of sonar reflections from the sea bottom and surface.

Increased diving depth also became crucial at high speeds, where there was a risk that a submarine might inadvertently descend below safe operating depths. The Soviet *Alfa* class combined the greatest reported diving depth (about 2,800 feet) with the highest speed (43 knots). However, achieving greater depth required stronger, heavier hulls and more powerful power plants. This created a challenge: larger hulls added resistance, reducing the speed advantage gained from more powerful engines. For example, the *Los Angeles* class submarines traded some diving depth for higher speed, while the *Alfa* class used an expensive titanium-alloy hull to save weight and maintain both speed and depth [21].

Silencing became a crucial focus of submarine design in the late 1950s. Until then, submarines were typically detected by active sonar, which bounced sound waves off their hulls. However, active sonar had limitations in range, and the sound waves it generated could be detected by the hunted submarine, alerting it to potential danger. By the early 1950s, passive sonar emerged as a new detection method, relying on the noise emitted by the submarine's own machinery. Early nuclear submarines were particularly susceptible to this form of detection due to their noisy machinery, especially the pumps used to circulate coolant [21].

As a result, silencing became a priority in submarine design. Pressurized-water reactor pumps were redesigned to be quieter, and many submarines mounted their machinery on sound-absorbing mounts, separated from the hull. These measures added to the size, weight, and cost of the submarines but significantly reduced noise levels. Some submarines also adopted natural-circulation reactors, which operated more quietly because they required no pumps at low to moderate power.

Further efforts to reduce noise included coating hulls with sound-absorbing material, which also helped counter homing torpedoes. These coatings, combined with other silencing techniques, greatly improved the stealth capabilities of modern submarines.

The Cold War era saw the development of submarines with increased speed, greater diving depths, and advanced silencing technologies. These improvements allowed submarines to operate more effectively in underwater combat, whether hunting enemy ships or evading detection themselves [21].

In summary, the history of submarine development is marked by significant technological advancements and strategic military applications, evolving from primitive designs to sophisticated vessels capable of stealth operations and advanced warfare capabilities. The inception of submarines can be traced back to the late 17th century, with early designs such as the Dutch "duikboot" and later innovations like the American "Turtle" during the Revolutionary War. However, it was during the late 19th and early 20th centuries that submarines began to gain prominence, particularly with the introduction of the first practical submarines powered by internal combustion engines, which allowed for longer underwater endurance and greater operational range [22].

The role of submarines expanded dramatically during World War I, where they were employed primarily for blockades and disrupting enemy shipping. The German U-boats exemplified this shift, utilizing torpedoes to target Allied merchant vessels, which led to significant advancements in submarine technology, including improved hull designs and stealth capabilities [22]. The interwar period saw further developments, as nations began to recognize the strategic importance of submarines in naval warfare, leading to increased investment in research and development [22].

World War II marked a pivotal point in submarine history, with technological innovations such as sonar and radar significantly enhancing detection and combat capabilities. The United States and Germany, in particular, engaged in intense submarine warfare, leading to the development of more advanced submarines, including the American Gato-class and the German Type VII U-boats. The war underscored the importance of submarines in naval strategy, as they were used effectively for both offensive and defensive operations [23].

Post-World War II, the Cold War era brought about a new phase in submarine development, characterized by the introduction of nuclear-powered submarines. These vessels, such as the U.S. Navy's Nautilus, revolutionized naval warfare by allowing submarines to remain submerged for extended periods without the need for surfacing, thus enhancing their stealth and operational capabilities [23]. The development of ballistic missile submarines (SSBNs) further transformed naval strategy, providing a second-strike capability that became a cornerstone of nuclear deterrence [23].

In recent decades, advancements in materials science, hydrodynamics, and digital technologies have continued to shape submarine design and functionality. Modern submarines are equipped with sophisticated sensors and communication systems, enabling them to operate effectively in complex environments while minimizing their acoustic signatures [24, 25]. The integration of advanced user interfaces and ergonomic designs has also improved operational efficiency, allowing smaller crews to manage increasingly complex systems [25]. Furthermore, the focus on non-acoustic detection methods, such as wake vortex analysis, reflects the ongoing evolution of submarine technology in response to emerging threats and operational requirements [26]. In conclusion, the history of submarine development is a testament to the interplay between technological innovation and military strategy. From their humble beginnings to the cutting-edge vessels of today, submarines have continually adapted to meet the challenges of modern warfare, underscoring their critical role in naval operations.

Parts of a submarine

A submarine is a highly complex vessel composed of numerous systems and components that enable it to operate underwater for extended periods. Below is an outline of the general parts of a submarine:

1. Hull

- **Pressure Hull**: The inner hull that withstands the external pressure of deep water and protects the crew. It is typically cylindrical to maximize strength and minimize stress.
- **Outer Hull**: Also known as the light hull, this provides hydrodynamic shape and houses ballast tanks but does not need to withstand high pressure.

2. Ballast and Trim Systems

- **Ballast Tanks**: Located between the inner and outer hulls, these tanks control the submarine's buoyancy by filling with water to submerge or with air to surface.
- **Trim Tanks**: Smaller tanks used to adjust the submarine's balance (trim) to maintain level positioning underwater.

3. Propulsion System

- **Nuclear Reactor (or Diesel Engines)**: Provides power for the submarine's propulsion and electrical systems. Nuclear submarines use a nuclear reactor, while conventional submarines may use diesel engines.
- **Electric Motors**: Propels the submarine when submerged (in nuclear submarines, the motor is powered by the reactor; in diesel-electric submarines, by batteries).
- **Propeller (Screw)**: Provides thrust to move the submarine through the water.

4. Energy Storage

- **Batteries**: Store electrical power, particularly in diesel-electric submarines, to operate the electric motors when the submarine is submerged and not using the engines.
- **Fuel Cells (AIP systems)**: In some modern diesel-electric submarines, these provide additional power underwater for longer periods without surfacing.

5. Sonar and Sensor Systems

- **Sonar (Sound Navigation and Ranging)**: Used for detecting other vessels, navigating underwater, and mapping the ocean floor. Includes active sonar (which emits sound) and passive sonar (which listens for sounds).
- **Periscope or Photonic Mast**: Used to observe the surface while the submarine remains submerged at periscope depth. Modern submarines may use optronic masts, which are sensor arrays without the traditional tube that penetrates the hull.

6. Control Surfaces

- **Rudders**: Control the direction of the submarine (left or right).
- **Diving Planes**: Horizontal fins located at the bow and/or stern that control the submarine's depth (up or down) by adjusting the pitch of the vessel.

7. Crew and Command Spaces

- **Control Room (Bridge)**: The nerve centre of the submarine where navigation, sonar, and weapons systems are monitored and controlled.

- **Crew Quarters**: Living spaces for the crew, including sleeping areas, dining facilities, and recreational areas.
- **Mess Hall/Galley**: The dining area where food is prepared and served for the crew.

8. **Weapons Systems**
   - **Torpedo Tubes**: Located at the bow (and sometimes stern) for launching torpedoes.
   - **Missile Launch Systems**: In ballistic missile submarines (SSBNs) and cruise missile submarines (SSGNs), vertical launch systems or missile tubes are used to carry and launch long-range missiles.
   - **Deck Guns**: Some submarines, particularly older ones, have external guns for surface combat.
9. **Life Support Systems**
   - **Oxygen Generation and CO2 Scrubbing**: Systems that generate oxygen and remove carbon dioxide from the air, allowing the crew to breathe while submerged for extended periods.
   - **Water Distillation**: Converts seawater into drinking water for the crew.
   - **Air Conditioning and Ventilation**: Maintains comfortable and safe atmospheric conditions inside the submarine.
10. **Navigation and Communication Systems**
    - **Navigation Systems**: Includes inertial navigation systems, GPS, and sonar-based navigation for submerged operations.
    - **Communication Systems**: Various radio and satellite communication systems for contact with other vessels and command structures. Submarines may also use buoyant antennas or trailing wire antennas for communication while submerged.
11. **Steering and Diving Controls**
    - **Helm**: The controls for steering the submarine, adjusting speed, and controlling depth.
    - **Ballast Control Panel**: Used to manage the ballast tanks for surfacing, submerging, and maintaining proper trim.
12. **Emergency Systems**
    - **Escape Pods/Hatches**: Safety systems that allow the crew to exit the submarine in emergencies.
    - **Emergency Buoy**: Deployed to signal the submarine's position in distress situations.

These components work together to enable the submarine to carry out its missions, whether for military operations, scientific exploration, or other purposes.

Figure 16: The Ohio-class ballistic-missile submarine USS Tennessee (SSBN 734). Defense Visual Information Distribution Service, Public Domain, via Picryl.

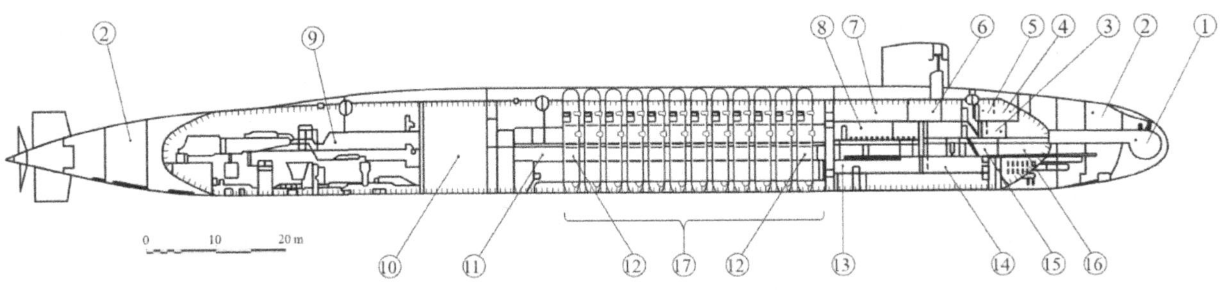

Figure 17: Trident/Ohio class submarine drawing. Voytek S, CC BY 3.0, via Wikimedia Commons.

Components:

1. Sonar dome
2. Main ballast tanks
3. Computer room
4. Integrated radio room

5. Sonar room
6. Command and control centre
7. Navigation Center
8. Missile control centre
9. Engine room
10. Reactor compartment
11. Auxiliary machinery room no.2
12. Crew's berthing
13. Auxiliary machinery room no.1
14. Torpedo room
15. Wardroom
16. Chief Petty Officer quarters
17. Missile compartment

Outer Hull and Pressure Hull

Submarines are typically designed with two main hulls: the outer hull and the pressure hull. The pressure hull is the internal structure that houses all the critical components of the submarine, including accommodation spaces, weapons systems, communication and control systems, battery banks, and both the main and auxiliary machinery. The primary function of the pressure hull is to provide a safe, enclosed environment for the crew and sensitive equipment, protecting them from the immense water pressure encountered at significant depths [12].

The pressure hull earns its name because it is specifically engineered to withstand the extreme hydrostatic pressure that builds as the submarine descends deeper into the ocean. As the submarine operates at greater depths, the surrounding water exerts more force on the hull. The pressure hull is constructed from robust materials, often steel or titanium, to endure these conditions, ensuring that the submarine can maintain its structural integrity at maximum operating depth.

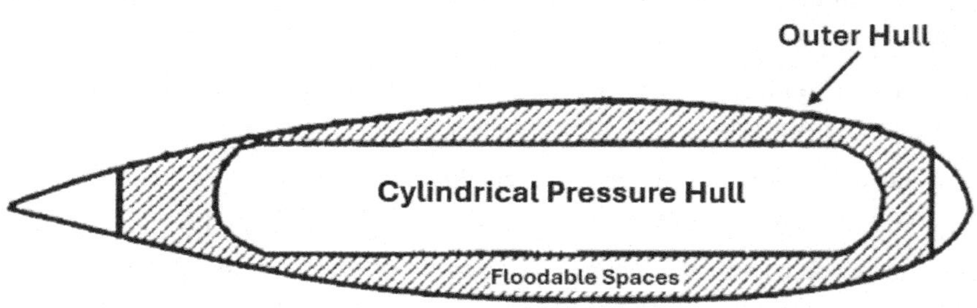

Figure 18: Cylindrical Pressure hull and Outer hull of a submarine [12].

Surrounding the pressure hull is the outer hull, sometimes referred to as the light hull. This hull provides the submarine with its hydrodynamic shape, allowing it to move efficiently through the water. The space between the outer hull and the pressure hull often houses the ballast tanks, which are critical for controlling the submarine's buoyancy, allowing it to submerge or surface as needed. Unlike the pressure hull, the outer hull is not designed to withstand deep-sea pressures but serves to streamline the vessel and enhance its performance in underwater environments. Together, the pressure and outer hulls play essential roles in enabling the submarine to function safely and effectively at depth.

The pressure hull of a submarine is housed within the outer hull, which is not designed to be pressure-tight. The reason for this lies in the difference between the functions of the two hulls and the unique environment in which submarines operate. When a submarine is submerged, the space between the outer hull and the pressure hull is intentionally flooded with seawater. This space, often referred to as the ballast tank area, is filled with water as part of the process that controls the submarine's buoyancy, allowing it to dive or surface as needed [12].

Since this space is constantly flooded when the submarine is underwater, the outer hull experiences the same external water pressure on both its inner and outer surfaces. As a result, the hydrostatic pressure on the outer hull is negligible because there is no pressure differential across it. Essentially, the outer hull does not need to resist the intense pressures of deep-sea environments because it is already equilibrated with the surrounding seawater. Its primary function is to provide the submarine with a streamlined, hydrodynamic shape that helps it move efficiently through the water, rather than to protect the submarine from water pressure [12].

The pressure hull, on the other hand, is the critical component designed to protect the submarine's interior spaces and its crew. This inner hull is built to withstand the immense hydrostatic pressure encountered at great depths, keeping the submarine's interior safe and habitable. By separating the structural responsibilities of the pressure hull and the outer hull, submarine design ensures that the submarine can perform its underwater missions safely while maintaining efficiency in movement and control [12].

Main Ballast Tanks

In submarines, the spaces between the outer hull and the pressure hull are divided into compartments known as Main Ballast Tanks (MBTs). These tanks are a crucial component of the submarine's ability to control buoyancy and submerge or surface as needed. The configuration and distribution of these ballast tanks depend on the submarine's overall design, particularly the relationship between the outer hull and the pressure hull. The placement of the MBTs is carefully considered based on the submarine's shape and the interaction between the two hulls [12].

The operation of the MBTs is tied closely to the process of submerging a submarine and ensuring its stability underwater. By filling the ballast tanks with water, the submarine becomes denser, losing buoyancy and sinking beneath the surface. Conversely, when the submarine needs to surface, the water in the MBTs is expelled, often using compressed air, allowing the submarine to become more buoyant and rise to the surface. The design of the MBTs plays a crucial role in how smoothly and efficiently this process can be achieved.

In some submarine designs, the MBTs are located only at the forward and aft sections of the vessel. This means that the areas at the front and back of the submarine have dedicated ballast tanks, while the space between the pressure hull and the outer hull along the midsection is minimized or even eliminated. In these designs, the pressure hull is essentially "flushed" with the outer hull in the middle section of the submarine, streamlining the structure [12].

Other submarine designs feature a more distinct separation between the outer and pressure hulls along the entire length of the submarine. In these cases, the space between the two hulls is utilized for ballast, and the MBTs extend along the length of the submarine. This design allows for more even distribution of ballast and may improve the submarine's ability to control its depth and stability during operations. The choice of MBT configuration depends on the specific needs of the submarine's mission and the overall engineering objectives of the design [12].

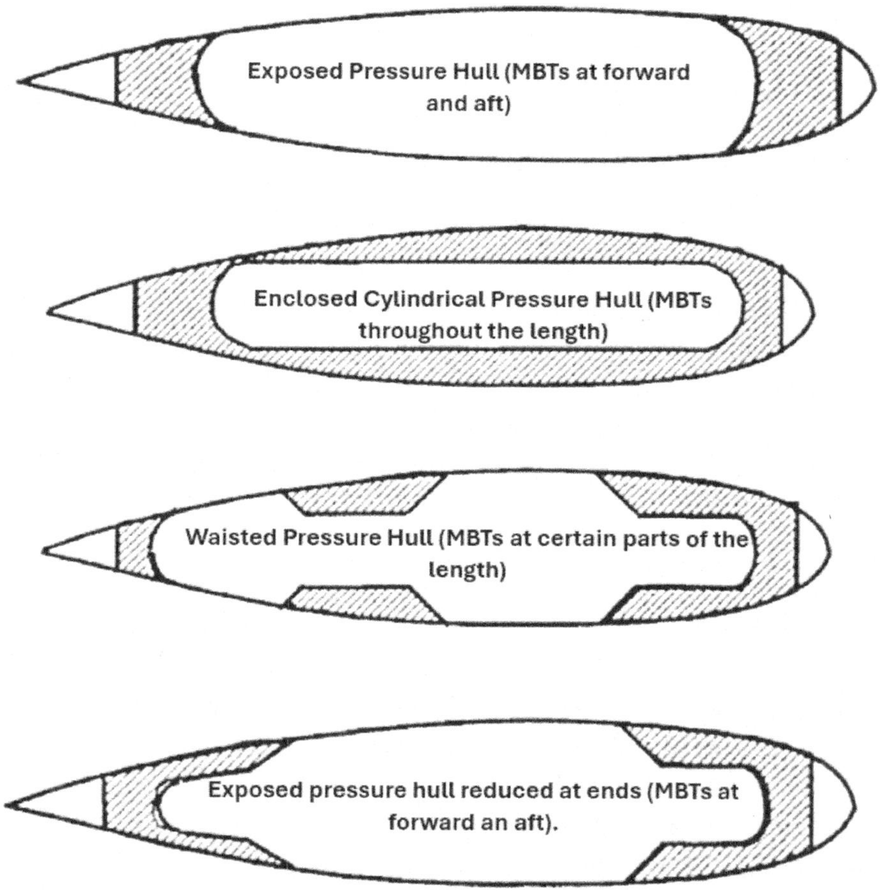

Figure 19: Some arrangements of MBTs [12].

Sail or Bridge Fin

The sail of a submarine, also known as the conning tower or bridge fin, is the streamlined structure that rises above the outer hull. Unlike the pressure hull, the sail is not designed to withstand deep underwater pressure; instead, it is shaped to minimize resistance while the submarine is moving just below the surface or with its sail above the waterline. The sail houses various masts and sensors that can be deployed when the submarine is operating near the surface, allowing it to monitor the environment while maintaining a degree of stealth [12].

Different types of masts are stored within the sail and raised when needed. These include the periscope mast, communications mast, radar mast, and weapon sensor mast, among others. These masts are crucial for a submarine's operations, especially when it needs to gather intelligence, communicate with other vessels or command centres, or detect threats while remaining hidden just beneath the surface. By deploying these masts from the sail while the submarine itself stays submerged, the crew can maintain situational awareness without fully exposing the vessel.

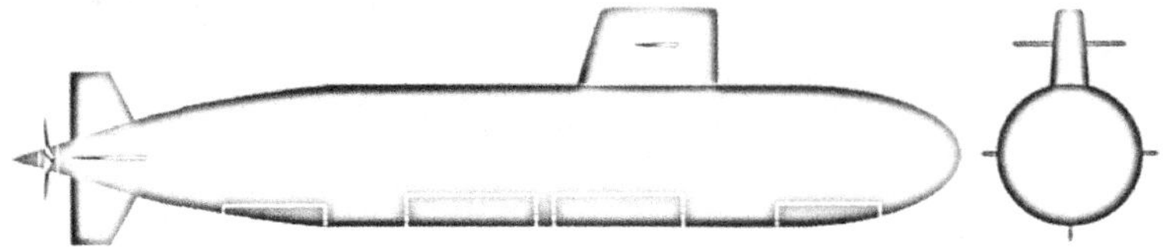

Figure 20: Bridge Fin or Sail in a submarine.

The sail itself is designed with an aerofoil shape, which acts as a hydrofoil when the submarine is traveling near the surface. This aerodynamic design reduces drag on the submarine, improving its speed and efficiency while minimizing disturbances in the water. Keeping drag to a minimum is crucial, as it reduces the formation of water eddies around the sail. This, in turn, helps to lower the submarine's acoustic signature, making it harder to detect by sonar. The stealthy design of the sail ensures that the submarine can operate more quietly and efficiently, enhancing its ability to perform covert missions [12].

Control Surfaces

When a submarine is submerged, it relies on hydroplanes to control its direction and depth. These hydroplanes function as control surfaces, similar to the control surfaces of an aircraft. To fully understand their role, it's important to first recognize the nature of the motions experienced by a submarine underwater. Unlike surface ships, submarines are less affected by surface wave effects, so they experience less heave (vertical motion) and pitch (tilting forward and backward). The more stable underwater environment allows for more precise control of the submarine's movements using hydroplanes [12].

A typical submarine is equipped with pairs of hydroplanes, or fins, located at the forward and aft sections of the vessel. These hydroplanes work to control the submarine's heave and pitch independently. The forward hydroplanes help manage the submarine's depth, while the aft hydroplanes maintain the pitch or tilt of the submarine. Additionally, there are vertical hydroplanes at the aft section that function similarly to rudders on a ship. These hydroplanes allow the submarine to change its lateral direction while in motion. Unlike surface ships, however, the rudders on a submarine are positioned forward of the propeller.

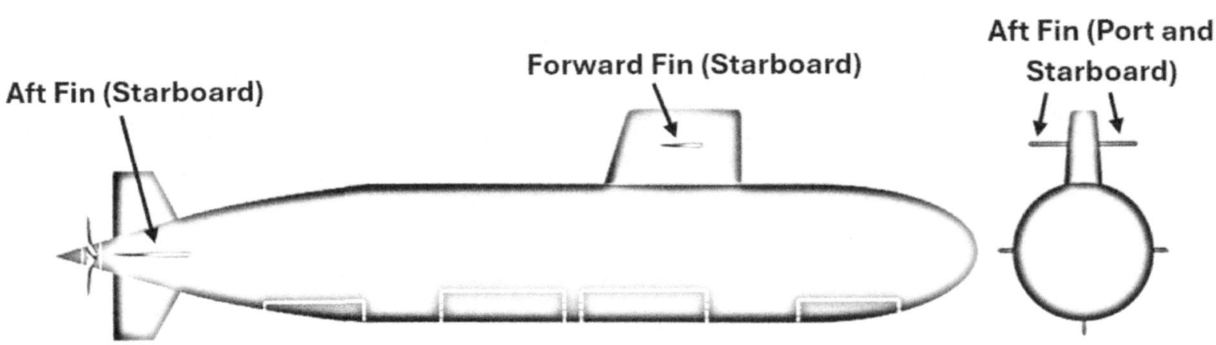

Figure 21: Fins on a submarine.

The reason for this difference in design is that, on surface ships, the rudder is placed behind the propeller to take advantage of the propeller's outflow, which generates maximum lift and steering efficiency. In a submarine, however, the entire hull is submerged, and the flow of water around the vessel is smoother and more streamlined. By placing the rudder forward of the propeller, the submarine benefits from this undisturbed flow, which allows the rudder to operate more efficiently. If the rudder were placed behind the propeller, the flow of water would be more turbulent, increasing the likelihood of cavitation, which can reduce control efficiency and create noise that might give away the submarine's position [12].

It's important to note that hydroplanes are most effective at higher speeds. At slower speeds, the efficiency of the hydroplanes decreases, which can make it more challenging to control the submarine's movements. At higher speeds, however, the hydroplanes are able to generate more lift, making them highly effective for precise adjustments in depth and direction. This is why submarines rely on maintaining a certain speed when they need to manoeuvre underwater.

General Arrangement

The general arrangement of a submarine involves the strategic spatial distribution of its main compartments and systems along the length and breadth of the hull. The submarine's design features two distinguishable hulls: the pressure hull, which houses the vital systems and is built to withstand the pressure at operational depths, and the outer hull, which provides hydrodynamic shape and encloses the flooded spaces where some systems like sensors are located.

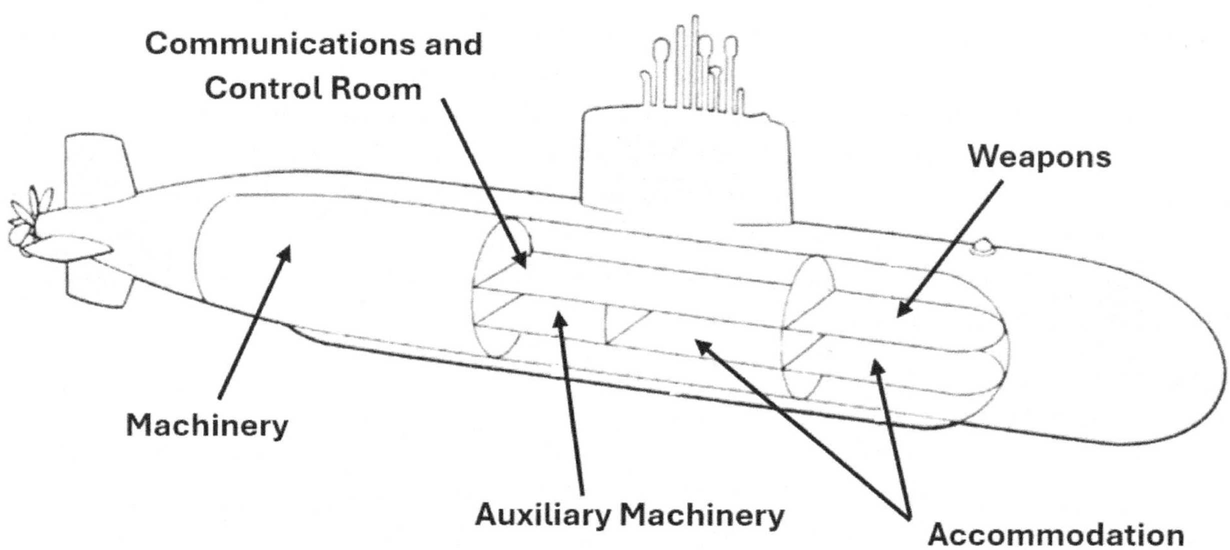

Figure 22: Schematic General Arrangement of a Diesel Electric Submarine.

The forward section of the pressure hull is dedicated to the weapon systems and sensors. Sensors are housed in the flooded space between the outer hull and the forward section of the pressure hull. This placement is critical because locating sensors at the forward minimizes interference from noise generated by machinery at the aft, and avoids turbulence caused by water flow near the stern. The forward area also houses the submarine's torpedo tubes, torpedoes, and associated launching and operating systems. The torpedo tubes extend from within the pressure hull to the outer hull's periphery, making it easier to load and launch weapons [12].

The midship section of the submarine serves several key functions. One of the primary compartments in this section is the ship and weapon control centre. This compartment houses all the essential control systems for navigation, weapon firing, machinery operation, and communication. Modern submarines are highly automated, allowing nearly all operational tasks during patrols or wartime missions to be controlled from this central command area without requiring crew presence elsewhere on the submarine. This control compartment also manages external communications, linking the submarine with its base or other external data sources [12].

Additionally, the midship section houses accommodation and life support systems, including crew quarters, toilets, the galley, and food storage areas such as cool and cold rooms. This placement provides easy access to both the forward and aft sections of the submarine, improving functionality and emergency evacuation. Since this section is also located beneath the sail, it offers a strategic escape route for the crew in case of emergencies.

Another critical component located in the midship section is the battery bank, which is crucial for powering the submarine, particularly in diesel-electric submarines. The batteries, charged by diesel alternators, consist of hydrogen cell units stacked in arrays. For redundancy and safety, these batteries are often distributed across multiple watertight compartments. Hydrogen elimination and proper ventilation are key concerns in these compartments because hydrogen buildup could lead to dangerous explosions [12].

Toward the aft section of the submarine is the machinery and auxiliary machinery compartment. The main machinery comprises systems such as the diesel alternators that charge the batteries, the air conditioning plant, and high-pressure air systems. These systems are essential for maintaining the submarine's operations and the crew's living conditions. A watertight bulkhead separates the main machinery compartment from the auxiliary machinery compartment, which contains auxiliary propulsion motors, an auxiliary air conditioning system, and other critical backup systems. These auxiliary systems support the main operations of the submarine and ensure continuous functionality.

Finally, the propulsion compartment, located at the very aft of the submarine, houses the main electric propulsion motor, the propulsion shaft, and associated systems. This section also contains the tail shaft and the forward and aft glands, which ensure watertight integrity at the hull openings. In diesel-electric submarines, this compartment may also house the reduction gearbox, which is a vital component for translating engine power into propulsion. The arrangement of these systems ensures that the submarine can move efficiently and maintain propulsion even in complex underwater environments [12].

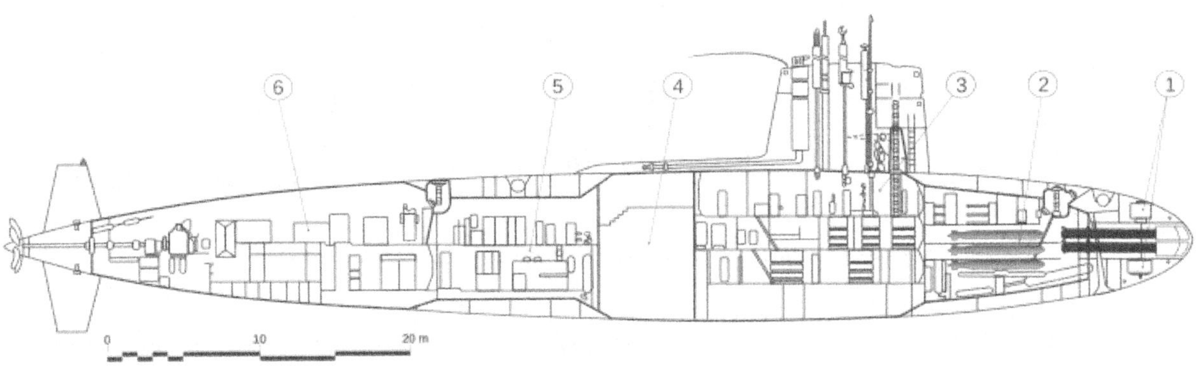

Figure 23: Skipjack class submarine arrangement. Voytek S, CC BY 3.0, via Wikimedia Commons.

Skipjack class arrangements:

1. Sonar arrays

2. Torpedo room

3. Operations compartment

4. Reactor compartment

5. Auxiliary machinery space

6. Engine room

This general arrangement of the submarine's compartments and systems allows for efficient operation, safety, and survivability during both peacetime patrols and wartime missions.

Hullform of a Submarine Design

The early submarines had a hull form that differed significantly from modern designs. Originally, the hull shape was based on the ideal streamlined form, with a parabolic bow and elliptical stern, which minimized drag and required less power due to the reduction in flow separation around the hull. This design, first used in the 1940s, aimed to achieve maximum hydrodynamic efficiency. However, it became clear that the streamlined shape came with significant drawbacks. The narrow hull shape, which tapered significantly aft and forward of the midship region, resulted in limited usable volume. This restricted interior space made it difficult to incorporate multiple decks and increased production costs.

By the late 1970s, a shift occurred in submarine hull design. The streamlined form was replaced by a long cylindrical mid-body with elliptical bow and stern sections. Although this design increased drag and power requirements, the advantages in terms of production and internal space outweighed these disadvantages. Cylindrical sections are easier and cheaper to construct, leading to lower overall production costs. Moreover, the cylindrical shape allows for the incorporation of multiple decks, greatly increasing the submarine's usable internal volume and enhancing operational efficiency.

The shape and geometry of a submarine's hull are crucial in the design process, as they affect various factors beyond internal volume. For instance, a cylindrical hull improves the submarine's manoeuvrability due to the larger hydrodynamic forces generated by the hydroplanes. Research has shown that an optimal Length-to-Breadth (L/B) ratio of 6 to 8 provides the best balance of low drag and high manoeuvrability. This ratio becomes a guiding principle in determining the overall dimensions of the submarine.

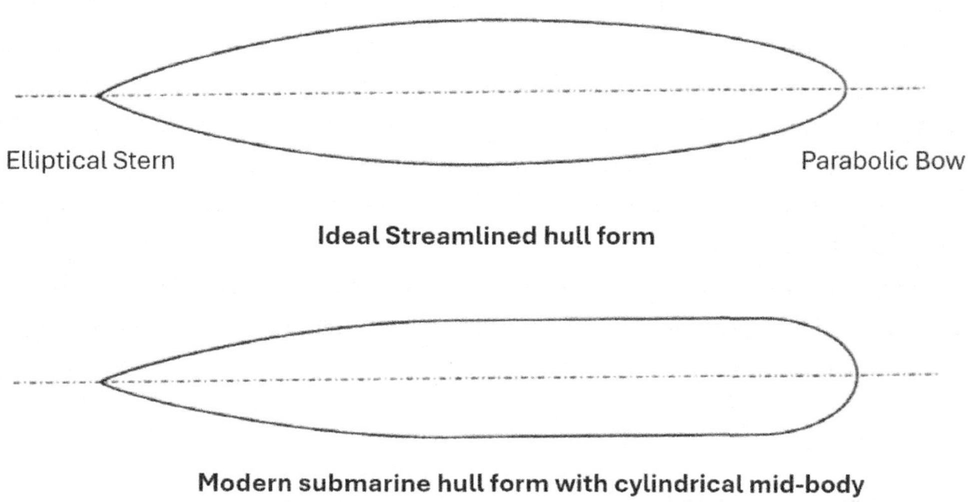

Figure 24: Submarine hull forms.

The diameter of the submarine is another key design factor, often dictated by its required length and pressure hull volume. The diameter influences how many decks can be incorporated within the pressure hull, which, in turn, affects the number of functional levels within the submarine. Smaller submarines with hull diameters between 4 and 7 meters typically have one deck, resulting in two accessible levels (above and below deck). Submarines with hull diameters of 7 to 8 meters can accommodate twin decks and three accessible levels, a common design for larger diesel-electric submarines.

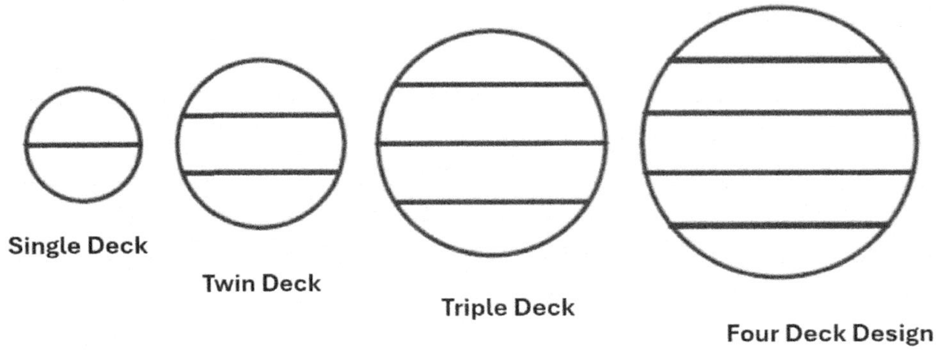

Figure 25: Possible deck levels for different hull diameters.

For even larger submarines, such as those powered by nuclear energy, diameters between 9 and 13 meters allow for three or four decks, providing substantial vertical space. This is especially important for housing large components like nuclear reactors, which require significant room for safe operation. The ability to incorporate multiple decks into the design enhances volumetric efficiency, a key consideration in maximizing the submarine's capabilities.

In the design process, a good submarine designer must balance the volumetric requirements of various compartments and systems. For example, some spaces, like main ballast tanks, have very specific volume requirements, while others, like battery banks, are defined by specific dimensions. Certain systems, such as torpedo operating tanks and ballast tanks, may have volume demands but are more flexible in shape. A skilled designer will prioritize these needs at different stages of the design process, ensuring that the submarine meets all operational requirements while optimizing space and functionality.

Stability is another critical aspect of submarine design. Although it might seem less complicated than surface ships, submarine stability is much more nuanced because the submarine must operate in both surfaced and submerged conditions. The stability characteristics of a submarine change dramatically when it dives or resurfaces, creating situations where the vessel is at risk of floundering or tipping over. Maintaining stability during these transitions is vital for the safe and effective operation of the submarine.

Ultimately, designing a submarine involves finding a delicate balance between hydrodynamic performance, internal volume, structural integrity, and stability. Achieving maximum volumetric efficiency, while meeting the specific demands of various systems and maintaining operational stability, is the hallmark of a well-designed submarine.

Key Design Considerations for Submarines

Designing a submarine is a complex process that requires balancing numerous factors to ensure that the vessel operates effectively underwater, remains stealthy, and performs its mission-critical functions. One of the primary considerations is hull shape and geometry, which plays a crucial role in determining the submarine's drag, manoeuvrability, and internal space. A streamlined hull minimizes drag, improving efficiency, while a cylindrical hull offers more usable internal volume, which is essential for housing equipment, weapons, and crew accommodations. Designers must carefully balance hydrodynamic performance with spatial efficiency, taking into account the optimal length-to-breadth ratio, which typically ranges between 6 and 8, to ensure stability and manoeuvrability without excessive drag.

Another important aspect is pressure hull design, which must be strong enough to withstand the immense hydrostatic pressure at deep depths. The hull is typically constructed from materials such as steel, titanium, or advanced composites, with each material choice influencing cost, weight, and maximum operational depth. Submarines are also divided into multiple watertight compartments within the pressure hull. This compartmentalization enhances safety by ensuring that, in the event of a breach, damaged sections can be sealed off, allowing the submarine to maintain buoyancy and avoid catastrophic failure.

Buoyancy and ballast systems are critical for controlling a submarine's ability to submerge and surface. Main Ballast Tanks (MBTs), which are located between the outer and pressure hulls, are responsible for controlling the submarine's buoyancy by filling with water or air. Their configuration and placement are crucial for ensuring balance and stability during dive and surface operations. Precise buoyancy control is essential for maintaining neutral buoyancy, which allows the submarine to remain at a consistent depth without expending excess energy.

The propulsion system is another key consideration. Nuclear-powered submarines offer extended underwater endurance and range without the need to surface, while diesel-electric submarines are less costly but require periodic surfacing or snorkel use to recharge batteries. The choice of propulsion system affects the submarine's operational endurance, speed, stealth, and logistical requirements. Silent operation is also vital for stealth, so propulsion systems must minimize noise by reducing vibrations and sound emissions from components like the propeller, engine, and cooling systems.

Stability and manoeuvrability are critical factors for submarines, as they must maintain control both on the surface and when submerged. Submarines rely on hydroplanes and rudders to control their depth and direction, with the size and placement of these control surfaces influencing the submarine's ability to manoeuvre efficiently. Designers must ensure that the submarine remains stable during transitions between surfaced and submerged conditions, accounting for changes in buoyancy and trim to prevent loss of control or capsizing.

Stealth and acoustic signature are central to a submarine's effectiveness in avoiding detection. Acoustic quieting involves reducing the noise generated by onboard machinery, minimizing propeller cavitation, and using sound-dampening materials on the hull to reduce detection by enemy sonar systems. The hull's shape and special coatings, such as anechoic tiles, are also designed to absorb sonar signals and reduce noise, further enhancing stealth capabilities.

Another important consideration is the weapons systems and payload capacity. Submarines are equipped with torpedo tubes and missile launchers, with the placement and number of these systems optimized for combat effectiveness while minimizing impacts on stealth and manoeuvrability. Modern submarines are designed with payload flexibility in mind, allowing them to carry and deploy a variety of munitions, including torpedoes, cruise missiles, mines, and unmanned underwater vehicles (UUVs).

Submarines also require advanced sensors and communication systems. Sonar systems are critical for navigation and situational awareness, while communication systems enable the submarine to maintain contact with surface forces or command bases even when submerged. These systems must be carefully integrated to avoid compromising the submarine's stealth while ensuring reliable performance in various underwater environments.

Crew accommodation and life support are essential for long missions. Submarines must be designed to optimize the use of space for crew quarters, food storage, and sanitation facilities, ensuring the crew's comfort and well-being during extended deployments. Life support systems, including oxygen generation, CO_2 scrubbing, and waste management systems, are crucial for maintaining a habitable environment even in the harshest conditions.

The submarine's endurance and power management are also significant design considerations. Diesel-electric submarines rely on battery systems for energy storage between recharges, while nuclear-powered submarines need effective power management to operate onboard systems efficiently. A submarine's endurance is determined by its size, power systems, and ability to remain at sea without resupply, which influences its operational capability.

Finally, safety and emergency systems are critical to ensuring the survival of the crew in the event of an emergency. Watertight integrity, compartmentalization, and damage control systems are essential for maintaining the submarine's operational capability after an incident. Redundant systems ensure that the submarine can continue to function even if critical components are damaged.

Balancing these key design considerations is a significant challenge in submarine design. Each decision impacts the vessel's performance, operational capability, and overall cost, making the design process a specialized and demanding field in naval engineering.

Chapter 2

Basic Principles of Submarine Architecture

Hull Design and Structural Integrity

Submarine design is a delicate art that requires striking a balance between functionality and survivability. At the core of a submarine's structure is the pressure hull, a watertight boundary that not only keeps the ocean from flooding the interior but also ensures the vessel's structural integrity. This hull provides a safe and habitable environment for the crew, even in the crushing depths of the ocean. Surrounding the pressure hull is an external structure designed for hydrodynamic efficiency, ensuring the submarine can navigate through underwater terrain with minimal resistance. This external structure is floodable, allowing the submarine to dive by filling its ballast tanks with water, and then be emptied with pressurized air when it's time to surface [27].

The pressure hull is the submarine's primary barrier against the intense underwater pressures it encounters at depths of hundreds of meters. The hull is typically made of high-strength steel, chosen for its ability to resist both the immense external pressures and the repetitive fatigue that results from the constant cycles of diving and surfacing. However, the thick plates required to withstand these forces add considerable weight to the submarine, creating a new challenge: ensuring that the vessel remains buoyant and capable of surfacing despite its massive weight [27].

This is where T-frames come into play. T-frames are the hidden, yet essential, reinforcements within a submarine's structure, akin to the skeleton of the vessel. They provide critical support against the compressive forces encountered during deep-sea travel, helping the submarine maintain its structural integrity while keeping the overall weight in check. These T-frames are visible in cross-sectional views of submarine pressure hulls and are essential for distributing the stresses encountered during operation [27].

T-frames, also referred to as ring stiffeners, can be placed either internally or externally along the submarine's pressure hull. These stiffeners consist of a web and a flange and are spaced intermittently along the cylindrical structure of the hull, adding strength and support. External stiffening generally offers slightly more strength—about 5% more—than internal stiffening, and it also frees up additional space within the pressure hull. However, external stiffeners are more difficult to inspect and maintain, making them a less attractive option in some designs. The spacing of these stiffeners directly influences the submarine's maximum operating depth, as they help determine how much pressure the hull can withstand.

Due to the sensitivity of this information, details about stiffener placement and submarine depth capabilities are often closely guarded secrets [27].

In summary, the design of a submarine is a careful balance between performance and survivability. The pressure hull serves as the critical barrier against deep-sea pressures, while the external structure contributes to hydrodynamic efficiency and buoyancy. T-frames reinforce the hull, enabling the vessel to withstand the stresses of underwater travel while minimizing weight. These design elements work together to ensure that the submarine can operate effectively and safely in some of the most extreme environments on Earth [27].

Hull design is one of the most critical elements in the construction of a submersible, involving several key decisions that must be carefully considered. These include the shape, dimensions, and the construction materials of the hull. Each of these aspects plays a significant role in determining the submersible's functionality, strength, cost, and overall performance.

First, let's talk about the shape. The shape of the hull must be chosen to maximize the internal volume in which scientists can work. However, it's not simply about making the hull as large as possible. While it might be tempting to maximize the volume to create more space, practical constraints—such as size limitations, budget, and material costs—prevent us from building an excessively large submersible. Instead, the goal is to find a balance that provides ample internal space while adhering to cost and performance restrictions.

Another key consideration is cost minimization, which is closely tied to the surface area of the hull. The hull must be constructed from expensive materials, such as titanium alloy, which is necessary to withstand the pressures of deep-sea environments. The greater the surface area, the more material is required, leading to increased costs. Therefore, an ideal hull shape would have a minimal surface area to keep costs down, but without compromising the internal volume.

Strength is also a paramount concern in hull design. The hull must be strong enough to resist the immense pressures of deep-sea conditions. While it might seem logical to simply build thick walls for maximum strength, doing so would be inefficient and expensive. Instead, the hull shape can be designed to inherently provide strength, reducing the need for excessively thick walls. A well-chosen shape distributes pressure evenly across the structure, enabling the hull to resist collapse without needing to reinforce it excessively.

Additionally, the hull's shape influences drag, which affects the submersible's speed and efficiency. Reducing drag allows the submersible to move through water more easily, conserving fuel and oxygen, and ultimately increasing the submersible's operational efficiency. Therefore, an ideal hull design minimizes drag while still maintaining internal volume and strength.

While the spherical shape is the most efficient in terms of minimizing surface area and maximizing strength, it has a significant drawback: high drag. Spheres experience considerable resistance as they move through water, which reduces speed and increases energy consumption. Many submersibles have used spherical shapes for their strength and simplicity, but the high drag coefficient makes them less ideal for dynamic underwater movement.

To address this, designers often consider more elongated shapes, such as cylinders with hemispherical ends. As the hull elongates, the drag coefficient decreases, improving hydrodynamics. However, elongation also increases frictional drag and reduces the inherent strength provided by a pure spherical shape. To solve this, a hybrid structure consisting of intersecting spheres has been proposed. This design maintains the strength advantages of spheres while benefiting from the hydrodynamic efficiency of a more cylindrical form.

The intersecting spheres are reinforced at their connections by rings, which can either be made from the same material as the hull or from a lighter material to increase buoyancy. This configuration provides structural strength and improves buoyancy while maintaining the necessary internal space for equipment and operations. The buoyancy of this structure depends on various factors, including the radius of the spheres, the angle of intersection, the thickness of the shell, and the number of reinforcing rings.

An additional advantage of this hybrid structure is that it provides a more practical internal layout. In a large sphere, it can be difficult to efficiently utilize the vertical space, as the upper half of the sphere may go unused if scientists are primarily working on the lower half. The intersecting sphere design allows for more natural horizontal distribution of equipment, making it easier to organize the workspace. Horizontal floors can be added over the intersection rings, providing flat surfaces for scientists to move around and work effectively. These joints between spheres can be secured through shrink-fitting or press-fitting, ensuring a strong and reliable connection between the sections.

Hull design for submersibles requires careful consideration of factors such as internal volume, cost, strength, and drag. While spherical shapes offer benefits in terms of minimizing surface area and maximizing strength, their high drag makes them less ideal for practical applications. Hybrid structures, such as intersecting spheres, offer a promising solution by combining strength with improved hydrodynamics and internal layout efficiency.

The thickness of a submarine's hull varies depending on the type of vessel and its specific operational requirements. Nuclear-powered submarines generally have thicker hulls than their diesel-powered counterparts. This is due to the higher internal pressure that nuclear reactors generate, which requires additional structural strength. The thickness of these hulls can range from 10 to 25 centimetres (4 to 10 inches), though the exact specifications are often classified for security reasons [28].

The materials used to construct a submarine's hull must be incredibly strong to withstand the intense pressure exerted by the surrounding water at great depths. Early submarines were originally built with wooden hulls, as it was initially believed that wood could better endure the pressures of deep-sea environments than steel. However, this proved incorrect, and steel quickly became the preferred material for submarine hulls due to its superior strength and durability. Today, steel remains the standard material, although it is often reinforced with other materials, such as titanium or fiberglass, to enhance its resistance to extreme pressures [28].

Submarines are known for their ability to dive to extraordinary depths, sometimes reaching over two miles below the surface of the ocean. The key to withstanding the crushing pressure at such depths lies in the hull's construction. The hull serves as the watertight shell of the submarine, safeguarding the crew and equipment inside from the immense forces exerted by the water. These hulls are made from thick, solid steel, specifically engineered to handle extreme pressure without collapsing [28].

To achieve even greater depths, some submarine hulls are reinforced with additional layers of materials, such as rubber. This layer provides increased strength and flexibility, allowing the hull to better endure the intense pressure experienced at deeper levels. Moreover, submarines are equipped with specialized air filters that maintain a clean and breathable atmosphere inside the vessel. These air filters are particularly important at great depths, where the air must be continuously purified to ensure the safety and well-being of the crew.

Structural Integrity

Submarine hull structural integrity is fundamental to the vessel's ability to survive and function in the extreme underwater environment. Ensuring that the hull can withstand the immense pressures of deep-sea operation is a complex engineering challenge, and the design of the hull involves numerous considerations that balance strength, weight, material properties, and safety. Let's delve into the details of what ensures a submarine's hull integrity.

Pressure Hull vs. Outer Hull: The structural integrity of a submarine primarily centres on the pressure hull, which is the innermost, watertight compartment of the vessel designed to endure extreme hydrostatic pressures at depth. While the outer hull provides hydrodynamic efficiency and houses systems like ballast tanks, it is not designed to resist the pressures of the deep sea. The pressure hull's primary function is to protect the crew and systems by maintaining a safe internal environment despite the external pressure, which can reach several hundred atmospheres at operational depths.

Material Selection: The choice of material for the pressure hull is critical. It must be capable of withstanding the compressive forces at depth while resisting fatigue over numerous dive cycles. Modern submarine pressure hulls are typically made from high-strength steels, such as HY-80, HY-100, or HY-130, where the number indicates the yield strength of the material in kilopounds per square inch (ksi). For instance, HY-80 has a yield strength of 80 ksi (approximately 550 MPa). Some advanced submarines also use titanium, which offers higher strength-to-weight ratios and superior corrosion resistance, although at significantly greater cost.

The selection of materials for submarine pressure hulls is a critical aspect of submarine design, as these materials must endure significant hydrostatic pressures and fatigue over numerous dive cycles. Modern submarines predominantly utilize high-strength steels, such as HY-80, HY-100, and HY-130, which are specifically engineered to withstand the compressive forces encountered at great depths. The designation of these materials reflects their yield strength, with HY-80 having a yield strength of approximately 80 ksi (around 550 MPa) [29, 30]. These steels exhibit a combination of high tensile strength and toughness, which is essential for maintaining structural integrity under extreme conditions [30].

In addition to high-strength steels, advanced submarines increasingly incorporate titanium into their designs. Titanium offers superior strength-to-weight ratios and excellent corrosion resistance, making it an attractive alternative despite its higher cost [31, 32]. The use of titanium can significantly enhance the performance and longevity of submarine hulls, particularly in environments where corrosion from seawater is a concern [31, 32]. However, the economic implications of using titanium must be carefully considered, as the material's cost can substantially increase the overall expenditure of submarine construction [31, 32].

The design of pressure hulls also involves careful consideration of geometric configurations and material properties to optimize performance. Factors such as hull shape, wall thickness, and stiffener layout are crucial in ensuring that the hull can withstand the external pressures experienced during operation [33-35]. Research has shown that the structural performance of pressure hulls can be maximized by optimizing these parameters, which directly influence the hull's ability to resist hydrostatic pressure and fatigue [34, 35]. Furthermore, numerical analyses and finite element methods (FEM) are commonly employed to simulate the behaviour of different materials and designs under various loading conditions, providing insights into the most effective configurations for deep-sea applications [36, 37].

Fiberglass and other composite materials are sometimes used in small-scale submersibles or in non-pressure components, but for deep-sea military submarines, steel and titanium remain the primary materials.

Shape and Geometry: The geometry of the pressure hull is another crucial factor in maintaining structural integrity. The most efficient shape for resisting external pressure is a sphere, as it distributes stress evenly across its surface. However, submarines are generally designed with a cylindrical shape with hemispherical ends. This cylindrical design is a compromise between maximizing internal volume, minimizing drag, and ensuring strength.

A cylindrical shape provides good resistance to pressure, although not as effective as a sphere, but it offers better internal space utilization, which is critical for housing the crew, systems, and weaponry. The hemispherical ends ensure that stress concentrations at the ends of the cylinder are minimized, reducing the risk of structural failure.

Hull Thickness and Pressure: The thickness of the pressure hull is designed to withstand the specific operational depth of the submarine. At greater depths, the external pressure increases significantly, and the hull must be thick enough to resist this pressure without deforming. Hull thickness can range from 10 cm to 25 cm (4 to 10 inches) or more, depending on the material used, the submarine's intended depth, and other design considerations.

The design and structural integrity of submarine pressure hulls are critical for ensuring safety and operational effectiveness at significant depths. Hydrostatic pressure increases by approximately 1 atmosphere (14.7 psi or 101 kPa) for every 10 meters of depth, leading to pressures that can exceed 5,000 psi (34.5 MPa) at depths of several hundred meters [31, 38, 39]. This necessitates that pressure hulls are engineered to withstand not only the normal operational pressures but also extreme conditions, such as "crush depth," where the hull experiences maximum pressure beyond typical operating conditions [32].

To accommodate these pressures, designers incorporate a substantial safety margin into the hull's design, often up to 50% [32, 38]. This safety margin is essential to prevent catastrophic failure, which could result from unexpected increases in external pressure or structural weaknesses. The materials and geometries used in pressure hull construction are chosen to optimize strength while minimizing weight, which is crucial for enhancing the submarine's performance and payload capacity [31, 35, 36]. Advanced materials, such as composite structures, are increasingly utilized due to their favourable strength-to-weight ratios and resistance to buckling under high external pressures [40, 41].

Furthermore, the design process involves sophisticated numerical modelling techniques to predict the behaviour of the hull under various loading conditions. Finite Element Analysis (FEA) is commonly employed to assess the structural integrity and failure modes of pressure hulls, ensuring that they can withstand the dynamic and static loads encountered during operation [36, 42]. The integration of these methodologies allows for the optimization of hull shapes and material configurations, enhancing the overall safety and reliability of submarine operations [34, 43].

Ring Stiffeners and Frames: The pressure hull is further reinforced with ring stiffeners, also known as T-frames, which are spaced along the length of the hull. These stiffeners act like ribs in a skeleton, providing additional support against compressive forces. The ring stiffeners are welded to the hull's surface and help distribute stress evenly around the hull, preventing buckling or deformation.

There are two main types of stiffeners: internal and external. Internal stiffeners are placed inside the pressure hull and contribute to the structural integrity by supporting the hull from within. External stiffeners, on the other hand, are placed outside the pressure hull. While external stiffeners provide slightly more strength, they are harder to maintain and inspect, which can be a disadvantage. The spacing of these stiffeners directly impacts the hull's ability to withstand pressure, and their arrangement is carefully calculated to optimize structural integrity without adding unnecessary weight.

Fatigue and Cyclic Loading: Submarines are subjected to repetitive cycles of pressure loading and unloading as they dive and surface. This cyclic loading can lead to fatigue, where micro-cracks develop in the material over time. If not properly managed, fatigue can compromise the structural integrity of the hull and lead to failure.

To mitigate fatigue, engineers design the hull to be as resilient as possible, selecting materials with good fatigue resistance and incorporating safety factors into the design. Regular inspections and maintenance are also crucial for detecting any signs of fatigue or stress damage before they become critical. This is particularly important for older submarines or those that have undergone numerous dive cycles.

Welding and Fabrication: The integrity of the hull also depends heavily on the quality of welding and fabrication. The welding of submarine pressure hulls must be flawless, as any defects can act as stress concentrators, leading to cracks or failures under pressure. The welds are subjected to rigorous testing, including non-destructive testing techniques like ultrasound and X-rays, to ensure there are no hidden defects.

In advanced submarines, electron beam welding or laser welding is often used to create strong, uniform bonds between sections of the pressure hull, minimizing the risk of failure at the seams.

Inspection and Maintenance: Given the harsh operating environment, regular inspections of the hull are essential to maintaining structural integrity. Non-destructive testing (NDT) methods, such as ultrasonic testing and magnetic particle inspection, are commonly used to detect any cracks or weaknesses in the hull. Additionally, any components subject to high stress, such as stiffeners and weld seams, are closely monitored for signs of fatigue or damage.

Scheduled maintenance and inspection intervals are strictly adhered to, particularly for submarines that have experienced heavy use or operate at greater depths. Detecting and addressing issues early is key to preventing catastrophic failures.

Crush Depth: Every submarine has a crush depth, which is the depth at which the pressure hull can no longer withstand the external pressure and begins to collapse. Crush depth is typically well beyond the submarine's operational depth, with a substantial safety margin built in. However, the exact crush depth of military submarines is usually classified information. The design of the pressure hull must ensure that the submarine operates safely well above this critical limit.

Redundancy and Safety Features: Submarines are designed with redundant systems to ensure that even if part of the hull is compromised, the vessel can maintain its integrity. Compartmentalization, where the submarine is divided into multiple watertight sections, ensures that a breach in one part of the submarine doesn't necessarily lead to the loss of the entire vessel. These compartments can be sealed off in the event of flooding, providing the crew with additional time to either repair the damage or surface safely.

The structural integrity of a submarine's hull is the result of careful consideration of materials, geometry, reinforcements, and safety factors. The pressure hull must withstand immense pressures at depth, resist fatigue from repetitive dives, and remain robust in the face of operational stresses. Through the use of advanced materials, sophisticated design techniques, and meticulous construction standards, modern submarines are engineered to operate safely in some of the harshest environments on Earth. Ensuring the hull's strength and integrity is essential for the survival of the crew and the successful completion of missions, making it a cornerstone of submarine design.

Loads

Approximately 40% of the focus in the submarine design process is dedicated to its structural design, reflecting the critical importance of ensuring the vessel's integrity under extreme conditions. The structural design process is not only concerned with the strength of the submarine but is also intricately linked with various functional aspects, making it one of the most time-consuming phases of the design process. Structural design begins by identifying the different loads that the submarine will be subjected to during its missions, with each type of load informing key design decisions [44].

The first major load type is loading due to diving pressure. Depth is one of the most significant structural design criteria for submarines. The pressure hull, as the main structural element, must withstand the immense external hydrostatic pressure encountered at depth. The hull is specifically designed to withstand collapse at a particular depth, known as the collapse depth, which is calculated using a safety factor. This safety factor is applied to the maximum operable depth (MOD) or service depth to ensure the submarine can survive beyond its normal operational limits. For example, a safety factor of 1.5 means the submarine should not exceed its service depth, while designs with a higher safety factor, such as 2.5, allow the submarine to dive deeper in emergency situations. The hydrostatic pressure at collapse depth becomes the design pressure, and all calculations for the pressure hull are based on this pressure [44].

Another significant load that the submarine must withstand is shock loads. Submarines must be able to endure the forces generated by underwater detonations, such as those from mines or large underwater gas bubbles. The physics of underwater explosions differ significantly from explosions in air. Upon detonation, a shockwave radiates outward, expanding until the internal pressure equals the external hydrostatic pressure of the water. The resulting pressure imbalance causes the explosion to contract inward, creating a cloud of gas bubbles that expands and contracts repeatedly until the explosion's energy is dissipated. This process results in a series of shockwaves that the submarine must withstand, not just a single impact [44].

Moreover, these recurring shockwaves can create a unique hazard for submarines. The explosion cloud can create a suction effect, pulling the submarine toward the explosion centre. This effect is particularly dangerous if the explosion occurs below the submarine, as it may draw the vessel deeper into the ocean, increasing the hydrostatic pressure and putting additional stress on the hull. The vibrations generated by the shockwaves also propagate through the submarine's structure, reducing the fatigue life of the materials and potentially causing resonance, which can lead to catastrophic structural failure [44].

In addition to diving pressure and shock loads, submarines must also contend with other loads. Like surface ships, submarines experience longitudinal bending loads when surfaced, as well as transverse shear forces acting on the hull due to wave action. Torsional loads may also arise from the twisting motion caused by waves. The submarine's engines generate local longitudinal and torsional vibrations, which must be kept within acceptable limits to prevent damage or excessive wear to the structure. These forces contribute to the overall design challenges, requiring the submarine to maintain its structural integrity under various environmental and operational conditions [44].

The structural design of a submarine is an intricate process that must account for numerous load types, from the hydrostatic pressure at depth to the shockwaves generated by underwater explosions and the forces experienced on the surface. The pressure hull must be designed to withstand these extreme forces while balancing weight, material strength, and safety factors.

Strength of the Pressure Hull

The strength of a submarine's pressure hull is critical to its ability to withstand the immense forces encountered underwater, particularly the hydrostatic pressure at depth. The pressure hull is typically cylindrical, and when the submarine is submerged, it is subjected to various stresses—primarily longitudinal and circumferential (hoop) stresses. These stresses must be carefully managed to prevent structural failure, such as buckling, which could be catastrophic for the vessel.

Longitudinal Stress on the Pressure Hull

In the dived condition, the cylindrical pressure hull experiences longitudinal compressive stress. This longitudinal stress is about half the magnitude of the hoop stress, which is the circumferential stress acting around the perimeter of the cylinder. The expression used to calculate the longitudinal stress helps determine the required thickness of the hull and the dimensions (scantlings) of the stiffeners needed to prevent buckling and failure of the pressure hull.

The longitudinal stress is a function of the external pressure (which increases with depth), the radius of the pressure hull, and the thickness of the hull plate. Mathematically, it can be expressed as:

$$\sigma_{\text{long}} = \frac{p \cdot r}{2t}$$

Where:

- σ_{long} is the longitudinal stress,
- p is the external pressure (hydrostatic pressure),
- r is the radius of the pressure hull,
- t is the thickness of the hull plate.

Role of the Submarine Designer

The submarine designer plays a key role in determining the thickness of the pressure hull. Several factors influence this decision, but two of the primary variables in the equation are the external pressure and the radius of the pressure hull.

1. **External Pressure (Hydrostatic Pressure):** This is determined by the depth at which the submarine operates, particularly the collapse depth, which is specified in the contract. The collapse depth is typically a fixed value based on the submarine's intended operational depth and a safety margin. The hydrostatic pressure at this depth becomes the external pressure used in the calculation.

2. **Radius of the Pressure Hull:** The radius of the pressure hull is another fixed entity. It is specified by the client based on the submarine's size requirements. The designer, therefore, cannot change this parameter but must work within its constraints.

The designer's main task is to calculate the minimum thickness of the pressure hull required to keep the stresses within safe limits, given a specific material and operating conditions. This thickness is determined by the yield strength of the material, which is the stress level at which the material begins to deform permanently. Using the yield strength as a reference, the designer calculates the thickness necessary to prevent the longitudinal compressive stress from exceeding the material's limits.

Observations from the Equation

Several important observations can be made from the relationship between stress, hull thickness, and the other factors:

1. **Effect of Diameter on Hull Thickness:** For a fixed maximum operable depth (MOD), a submarine with a larger diameter requires a thicker pressure hull than a smaller diameter submarine. This is because a larger radius increases the magnitude of the stresses for the same external pressure, necessitating a thicker hull to ensure structural integrity.

2. **Material Selection:** The minimum thickness of the pressure hull can be reduced by using materials with higher yield strength. This is advantageous because reducing thickness decreases the overall weight of the submarine, improving buoyancy and potentially reducing fuel consumption. However, materials with higher yield strength, such as titanium, are significantly more expensive than standard materials like steel. Therefore, the choice of material is a balance between cost, performance, and weight considerations.

Ring Stiffeners and Buckling Resistance

While the pressure hull absorbs all the forces in the longitudinal direction, it is vulnerable to buckling under circumferential stresses, particularly at depth. To prevent this, ring stiffeners, often in the form of T-profiles, are welded to the pressure hull. These stiffeners provide additional structural support, particularly against the buckling loads that act circumferentially.

The ring stiffeners work in conjunction with the pressure hull shell to absorb these stresses and distribute them evenly across the hull, reducing the likelihood of buckling. Even though the shell itself absorbs the longitudinal stresses, the ring stiffeners are crucial for maintaining the hull's shape and integrity under circumferential compression. Together, the hull and stiffeners form a unified structure capable of withstanding the extreme pressures of the deep sea.

Failure modes

Submarines are meticulously designed to operate within specific depth limits, known as their design envelope, which includes a range of operational depths and a designated maximum operating depth (MOD). This MOD is based on the material strength, pressure hull design, and numerous safety factors built into the submarine's construction. However, should a submarine exceed its design envelope and approach its crush depth, catastrophic consequences could ensue [27].

The Concept of Crush Depth and Failure Modes

The crush depth is the point at which the external hydrostatic pressure becomes greater than the pressure hull can withstand, leading to structural failure. For modern submarines, the collapse of the pressure hull typically occurs through buckling, which can manifest in several distinct modes depending on the distribution of forces and the spacing of the ring stiffeners [27].

Mode 1: Symmetrical Buckling

In Mode 1, the pressure hull between the T-frames (ring stiffeners) experiences symmetrical buckling, where the hull compresses inward evenly around its circumference. This type of failure is usually localized between two closely spaced T-frames, meaning the pressure hull bends inward in a small, confined area but doesn't immediately lead to total collapse. The deformation is often contained, and while it compromises the integrity of the hull in that localized region, it may not result in the complete destruction of the submarine. Symmetrical buckling in this mode is more likely when the T-frames are placed relatively close to each other, providing added support to the hull.

Mode 2: Waveform Buckling

Mode 2 buckling occurs when the T-frames are spaced farther apart. In this scenario, the pressure hull buckles in a wave-like pattern over two or more consecutive frames. One segment of the hull buckles inward while the next section buckles outward, creating a repeating pattern of inward and outward deformation. While this mode of failure is still localized and might not lead to immediate catastrophic failure, it poses a greater risk than Mode 1 because it involves larger sections of the hull. The repeated wave-like buckling between frames could eventually lead to further deformations, especially under continued pressure, which might compromise the entire structure over time.

Mode 3: Complete Failure

In Mode 3, both the T-frames and the pressure hull itself deform simultaneously. This mode is not localized; instead, it involves large sections of the hull collapsing at once, usually over multiple ring stiffeners. This failure mode results in complete structural failure of the pressure hull, leading to a sudden and catastrophic collapse. Once Mode 3 buckling occurs, the pressure hull is unable to maintain its shape, and the submarine will implode. This type of failure generally leads to the total loss of the vessel and all onboard.

Contributing Factors to Failure Beyond Design

While the design of the pressure hull is critical to ensuring a submarine's integrity under pressure, other factors can contribute to structural failure, even before the submarine reaches its design limits. These factors include [27]:

- **Manufacturing Defects:** Non-circular frames, improper welds, and poor construction techniques can introduce weak points in the pressure hull. These flaws may reduce the hull's ability to withstand pressure, making it more prone to buckling.

- **In-Service Issues:** Over time, submarines experience wear and tear from repeated dive cycles, which can lead to the development of cracks and other stress-related damages. Fatigue and cycling stress, combined with inadequate repairs or maintenance, can exacerbate these issues. Excessive corrosion, particularly in regions with poor access, can weaken the hull's structural integrity.

- **Stress Concentrations:** Sudden transitions in hull diameter, such as at connection points between different compartments, can create stress concentrations, which are points where the forces acting on the hull are amplified. These areas are particularly vulnerable to buckling if not properly designed or maintained.

Real-World Example: The USS Thresher Incident

A tragic example of a submarine collapse is the loss of USS Thresher (SSN-593) during deep dive trials on April 9, 1963. The Thresher was undergoing tests at deep depths (around 400 meters) when it experienced catastrophic failure, killing all 129 crew members. While the exact cause of the failure remains unknown, the submarine was unable to surface, and acoustic analysis from 2013 revealed that the pressure hull likely imploded at 730 meters—nearly twice its test depth [27].

The incident highlighted the dangers of exceeding a submarine's design envelope and emphasized the importance of structural integrity in submarine design. Despite the disaster, the Thresher's failure also demonstrated the level of redundancy built into its design, as the submarine withstood pressures far beyond its operational limits before collapsing. This incident led to the creation of the SUBSAFE program, a comprehensive safety program implemented by the U.S. Navy to ensure that submarines undergo rigorous testing, maintenance, and quality control to prevent such failures in the future [27].

Exceeding a submarine's design depth is a life-threatening scenario, as the submarine's pressure hull is subjected to extreme forces that it was not built to withstand. The potential failure modes—ranging from localized symmetrical buckling to complete collapse—underscore the critical importance of precise engineering, robust materials, and stringent quality control during construction. While design factors play a significant role in preventing failure, issues such as manufacturing defects, fatigue, and stress concentrations must be carefully managed throughout the vessel's lifespan. The lessons learned from incidents like the USS Thresher continue to inform the design and operation of submarines, ensuring the highest possible standards of safety and performance in deep-sea environments.

In addition to the three primary modes of failure previously discussed, a pressure hull can experience several other types of structural failure that are often equally concerning. These additional failure modes, though less frequently highlighted, represent critical risks in the design and operational life of a submarine [44].

General Instability of Pressure Hull Components

The overall stability of a submarine's pressure hull is highly dependent on the precision of its design, including proper welding techniques and accurate scantling calculations (the dimensions of structural elements). General instability refers to potential failures within the pressure hull shell, circular frames, bulkheads, or decks due to inaccuracies in design or construction. Improper welding, for example, can create weak spots or defects in the structure, leading to local failures that may initially be confined to a small area but can spread across multiple frames or components over time. Similarly, errors in the calculation of scantlings may result in insufficient reinforcement of critical sections, causing the structure to buckle or deform under pressure. These local instabilities can propagate along the hull, compromising the entire submarine.

Snap-Through Buckling of Bulkheads

Another failure mode is snap-through buckling, which can occur specifically at the forward elliptical bulkhead (dome) or the aft conical bulkhead. These components of the submarine, which differ in shape from the main cylindrical hull, are subjected to varying compressive loads during deep dives. The difference in geometry and the way forces are distributed across these curved surfaces make them more vulnerable to sudden buckling, especially under high-pressure conditions. Snap-through buckling happens when these bulkheads abruptly collapse or buckle under compressive loads, leading to a rapid failure. Due to the high stress concentrations in these areas, failures here can have serious consequences, potentially leading to the deformation of adjoining structures and compromising the pressure hull's ability to maintain its integrity.

Low-Cycle Fatigue and Crack Propagation

Submarines undergo repeated cycles of stress as they dive and surface, exposing the hull to low-cycle fatigue. This type of fatigue occurs over a relatively small number of cycles but with high stress variations, which is common in pressure hulls that repeatedly experience changes in pressure during submersion and resurfacing. Over time, this fatigue can lead to the development of cracks in the structure. These cracks typically start small but can propagate if not detected and addressed. As cracks spread, they reduce the structural integrity of the hull, making it more susceptible to catastrophic failure under the extreme pressures of deep-sea operation. Monitoring and repairing fatigue-induced cracks is a crucial part of submarine maintenance to prevent this failure mode from becoming a major issue.

Stress Concentration at Discontinuities

Stress concentrations occur in areas where there are abrupt changes in the shape or geometry of the submarine's structure. One common example is the joint between the cylindrical pressure hull and the conical aft or elliptical forward ends. At these junctions, the change in shape creates discontinuities that focus stresses into smaller areas. This localized increase in stress can lead to structural failure at these joints, especially if the hull is subjected to high external pressures or if there are defects in the materials or welds at these critical points. The design of these transitions must be carefully optimized to minimize stress concentrations, often through the use of reinforcing materials or more gradual changes in geometry to distribute the stresses more evenly.

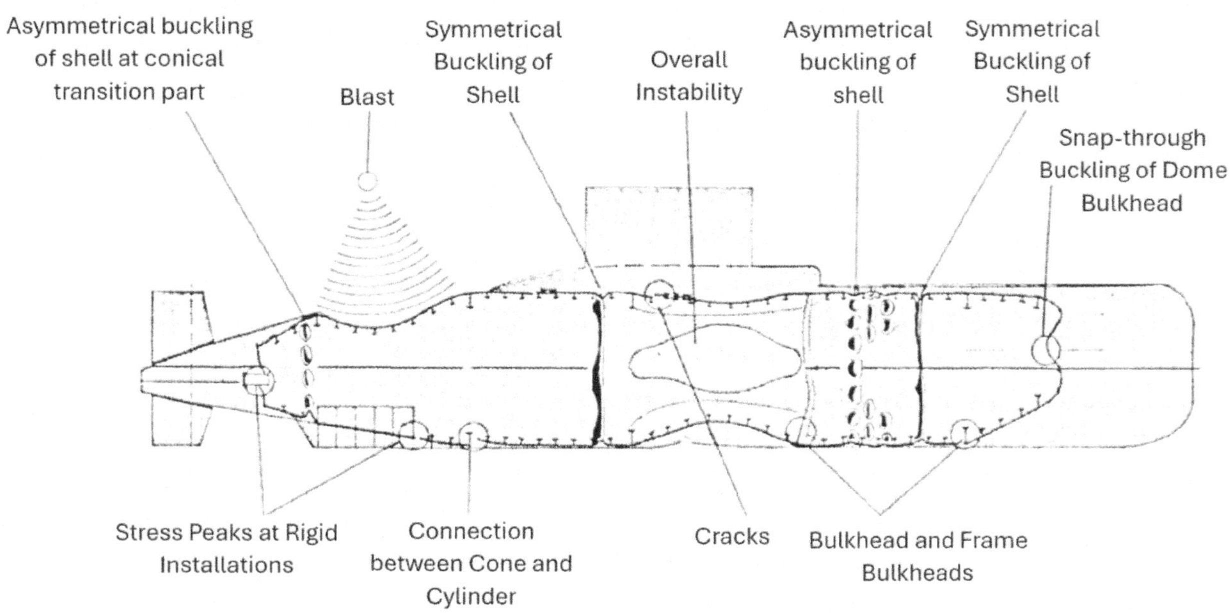

Figure 26: Forms of failure of a submarine pressure hull.

These additional failure modes underscore the complexity and precision required in submarine design and construction. Each potential failure point must be accounted for during both the design phase and throughout the submarine's operational life. From ensuring the stability of the pressure hull and its components to managing stress concentrations and the risk of fatigue-induced cracks, engineers must balance a range of factors to ensure the submarine can operate safely under the intense pressures of the deep sea. Proper design, quality control in manufacturing, and rigorous maintenance are essential to preventing these failure modes from compromising the safety and effectiveness of the vessel.

Pressure Hull and External Hull Dynamics

The pressure hull and external hull are two of the most critical components of a submarine's structure, each playing a distinct but interconnected role in the vessel's overall functionality, performance, and survivability. Their design and interaction are vital for ensuring that the submarine can safely operate at depth, maintain buoyancy, navigate efficiently, and withstand immense underwater pressure.

The dynamics of pressure hulls and external hulls are critical in the design and operation of underwater vehicles, particularly submarines and autonomous underwater vehicles (AUVs). Pressure hulls are specifically engineered to withstand the immense hydrostatic pressures encountered at significant depths, while external hulls must contend with various environmental forces, including hydrodynamic loads and potential impacts from underwater obstacles.

The pressure hull is the core structural element of a submarine. It is designed to withstand the extreme hydrostatic pressures exerted on the submarine as it dives deeper into the ocean. The pressure hull is a watertight boundary that houses the crew, machinery, and systems in a protected environment, allowing the submarine to operate at great depths without being crushed by the external water pressure.

Pressure hulls are primarily designed to resist compressive forces due to hydrostatic pressure, which can reach levels as high as 65 Bar in deep-sea environments [36]. The structural integrity of these hulls is influenced by several factors, including geometric configurations such as hull shape, stiffener layout, and wall thickness, as well as the choice of construction materials [33]. The most effective geometries for resisting external pressures typically involve annular cross-sections, often realized through a combination of cones and ring-stiffened cylinders, capped with torispherical or spherical domes [36]. The design optimization of these hulls is essential not only for safety but also for enhancing operational efficiency, as minimizing weight while maximizing strength can significantly improve a vehicle's buoyancy factor and overall performance [31, 38].

The dynamics of the pressure hull primarily involve its ability to resist compressive forces. As a submarine dives deeper, the water pressure increases significantly, and the pressure hull experiences longitudinal (along the length of the hull) and circumferential (hoop) stresses. The cylindrical shape of the hull is critical in distributing these forces evenly around its surface, providing structural integrity against the compressive loads.

However, the pressure hull must also account for several dynamic factors:

1. **Material Strength and Thickness:** The material used for the pressure hull, typically high-strength steel or titanium, is chosen for its ability to withstand the external pressure without deforming. The thickness of the hull is a function of the submarine's operational depth. As the submarine dives, the external pressure increases exponentially, and the pressure hull must have sufficient thickness to resist this pressure without collapsing.

2. **Fatigue and Stress Cycles:** Submarines operate under cyclic loading, meaning they undergo repeated cycles of pressure as they dive and surface. This repetitive loading can lead to fatigue in the material, where microscopic cracks may form and propagate over time, potentially leading to failure. Fatigue management is a critical aspect of pressure hull dynamics, with materials and design techniques chosen to extend the service life of the hull under these conditions.

3. **Buckling Risks:** The pressure hull is also vulnerable to buckling, particularly as it approaches its crush depth—the point at which the external pressure exceeds the hull's structural limits. Proper stiffening, often in the form of T-frames or ring stiffeners, is integrated into the design to prevent buckling and ensure that the hull maintains its shape under pressure.

4. **Temperature and Environmental Factors:** The pressure hull is subjected to varying environmental conditions, including temperature changes as the submarine moves through different layers of the ocean. Cold temperatures at extreme depths can affect the material properties of the hull, making it more brittle and susceptible to cracking or buckling. Thus, the pressure hull must be designed to operate effectively in a range of thermal conditions.

The susceptibility of pressure hulls to buckling under extreme conditions is a significant concern. Research has shown that the roundness of the hull can impact its buckling strength, with deviations from ideal geometries leading to increased vulnerability under high compressive stresses [45]. Furthermore, the design process must account for various failure criteria, including yielding and instability, to ensure that the hull can withstand the operational pressures without compromising structural integrity [31, 38]. Advanced computational tools, such as finite element analysis (FEA), are often employed to simulate and predict the behaviour of pressure hulls under various loading conditions, allowing for more informed design decisions [36, 38].

In contrast, the external hull dynamics involve the interaction of the hull with surrounding water, which can lead to complex hydrodynamic behaviours. The external hull must be designed to manage these interactions effectively, as the structural capacity must exceed the external forces to prevent catastrophic failure, particularly in extreme sea conditions [46]. The dynamics of ship propulsion and the hull-water interaction are also crucial, as they influence the overall stability and manoeuvrability of the vessel [47]. Additionally, the effects of environmental factors, such as fouling from microorganisms and corrosion, can compromise the integrity of the external hull, necessitating regular maintenance and monitoring [48, 49].

The external hull, also known as the light hull, serves a different purpose from the pressure hull. It is typically not pressure-resistant and is designed to provide hydrodynamic efficiency and house non-pressurized systems, such as ballast tanks and sonar equipment. The external hull encloses the pressure hull and other subsystems, forming the outer shell of the submarine.

Several dynamics influence the design and function of the external hull:

1. **Hydrodynamic Performance:** The external hull is crucial for reducing drag as the submarine moves through the water. A streamlined shape minimizes resistance, allowing the submarine to travel efficiently, conserve energy, and remain quiet. The hydrodynamics of the external hull also affect the submarine's manoeuvrability and speed, especially during submerged operations. Designers often use an aerofoil profile for the sail and fin structures to optimize performance and reduce water turbulence.

2. **Floodable Spaces and Buoyancy Control:** Unlike the pressure hull, the spaces between the external hull and the pressure hull, known as floodable spaces, are often filled with water when the submarine is submerged. These areas are compartmentalized into main ballast tanks (MBTs), which play a critical role in controlling the submarine's buoyancy. When the submarine dives, the MBTs are flooded with water, increasing the vessel's weight and allowing it to sink. When surfacing, compressed air is used to expel the water from the ballast tanks, restoring buoyancy and bringing the submarine to the surface.

3. **Interaction with Pressure Hull:** While the external hull is not designed to withstand pressure, its interaction with the pressure hull is critical for the overall stability of the submarine. The external hull provides structural support for various systems and protects the pressure hull from direct contact with the environment, such as rough seabed surfaces or floating debris. The external hull's shape also influences the flow of water around the pressure hull, affecting acoustic stealth by reducing noise generated by water turbulence.

4. **Material and Structural Considerations:** The external hull is usually made from lighter materials compared to the pressure hull, often incorporating fiberglass, composites, or aluminium alloys. These materials are chosen for their strength-to-weight ratio, ensuring that the external hull provides the necessary structural support without adding excessive weight. Additionally, the design of the external hull must allow for easy maintenance of the submarine's systems, particularly those located between the external and pressure hulls, such as sensors,

antennas, and ballast tanks.

The interaction between the pressure hull and the external hull is a critical factor in submarine design. These two components must work together to ensure the submarine's operational effectiveness, survivability, and stealth capabilities. The pressure hull provides the strength necessary to withstand the ocean's crushing forces, while the external hull ensures the submarine can move through the water efficiently and quietly.

One important aspect of their combined dynamics is the distribution of buoyancy and stability. The centre of buoyancy (the point where the buoyant force acts) and the centre of gravity (where the submarine's weight is concentrated) must be carefully balanced to maintain stability both on the surface and underwater. The distribution of MBTs, their interaction with the pressure hull, and the overall shape of the external hull all contribute to this balance, ensuring the submarine can dive, surface, and maintain its depth with precision.

The pressure hull and external hull serve distinct yet interconnected roles in submarine design. The pressure hull's structural integrity is paramount for withstanding deep-sea pressures, while the external hull focuses on hydrodynamic performance and buoyancy control. Together, they form the foundation of a submarine's ability to operate safely, efficiently, and stealthily in the challenging underwater environment.

Despite being a sealed, watertight environment, the pressure hull must still allow for controlled access between the inside and outside of the submarine, both in surfaced and submerged conditions. This is achieved through the use of circular hatches and penetrations for pipelines and cables, but each access point introduces significant design challenges, particularly in terms of structural integrity and safety [44].

Circular Hatches for Personnel Access

To allow personnel to enter and exit the pressure hull, circular hatches are installed at strategic locations. Typically, there are three primary hatches: one at the conning tower (centre of the submarine) and one each at the forward and aft sections of the submarine. These hatches must be pressure-tight to prevent water ingress when submerged, and they must also be robust enough to withstand both the internal atmospheric pressure and the external hydrostatic pressure at operating depths.

The circular design of the hatches is crucial for maintaining structural integrity. Circular shapes distribute stress evenly around the edges, minimizing the risk of deformation or failure under pressure. These hatches are essential for personnel access during routine operations, maintenance, or emergency evacuations. However, incorporating such hatches into the pressure hull introduces discontinuities—areas where the otherwise uniform cylindrical structure is interrupted, creating points of potential weakness. These areas are prone to stress concentrations, where forces acting on the hull are focused, potentially leading to structural failure if not properly managed [44].

Penetrations for Pipelines and Cables

In addition to personnel access, penetrations are necessary for pipelines and cables that connect equipment located outside the pressure hull to systems inside the submarine. These penetrations are critical for the functioning of external equipment such as ballast tanks, sensors, and weapons systems, all of which require control and power from within the pressure hull. The design of these penetrations must ensure that they remain watertight while accommodating the passage of pipes and cables [44].

Each penetration introduces another discontinuity in the pressure hull, which, like the hatches, creates stress concentration points. These penetrations are often circular or elliptical in shape, and the areas around the edges of these openings are particularly vulnerable to high stress levels, especially at deep diving depths where external pressure is extreme. To mitigate these risks, careful attention is given to the material selection, thickness, and reinforcement around these openings.

Forward Elliptical Dome and Torpedo Tube Penetrations

The forward elliptical dome bulkhead of the submarine presents additional design challenges because it accommodates multiple penetrations, particularly for torpedo tubes and weapon compensating tanks. The torpedo tubes, which extend through the forward dome, require large penetrations that must be reinforced to withstand the forces exerted during both submerged operations and torpedo launches. These tubes are the primary weapons system for many submarines, and their structural integrity is paramount to the submarine's combat effectiveness [44].

In addition to the torpedo tubes, secondary penetrations are required for pipelines that serve the weapon compensating tanks, which help maintain the submarine's balance and stability after torpedoes are fired. These penetrations are essential for the function of the weapons system but also increase the complexity of the structural design at the forward bulkhead.

Stress Concentrations and Weld Integrity

The areas around these penetrations—whether for hatches, pipelines, or torpedo tubes—are subject to high stress concentrations due to the interruptions in the otherwise smooth surface of the pressure hull. These stress concentrations make the penetrations vulnerable to cracking, deformation, or buckling under pressure. As a result, the design and construction of these penetrations require the highest levels of precision and care.

To mitigate the risks associated with these stress concentrations, welding processes used for attaching and reinforcing penetrations are subjected to intense scrutiny. The quality of these welds is critical because a failure in the weld could compromise the watertight integrity of the pressure hull, leading to catastrophic consequences during submerged operations.

To ensure the integrity of the welds, multiple non-destructive testing (NDT) methods are employed, including ultrasonic testing, radiographic testing, and dye penetrant testing. These tests allow engineers to detect any flaws, such as cracks or voids in the welds, without damaging the structure. Given the critical nature of these penetrations, more than one type of NDT is usually performed to verify that the welds meet the stringent safety and performance standards required for deep-sea operations [44].

The pressure hull's ability to accommodate hatches and penetrations while maintaining its structural integrity under extreme conditions is a testament to the advanced engineering involved in submarine design. Every penetration introduces a potential weak point that must be carefully reinforced and monitored. By using advanced materials, precision welding techniques, and rigorous testing protocols, submarine designers ensure that these critical areas can withstand the immense pressures encountered in deep-sea environments, while allowing for the necessary functionality and accessibility that modern submarines require.

Buoyancy, Ballast, and Stability

Buoyancy, ballast, and stability are critical concepts in submarine design and operation, each playing a vital role in the underwater vehicle's ability to navigate and maintain its position in the water column. Buoyancy, ballast, and stability are fundamental concepts that play crucial roles in the vessel's ability to dive, surface, and maintain control while submerged. Each of these terms relates to how a submarine interacts with its surrounding environment, particularly in terms of controlling its depth and maintaining its orientation.

Buoyancy

Buoyancy refers to the upward force exerted on a submarine when it is submerged in water, which is governed by Archimedes' Principle. This principle states that the buoyant force acting on an object submerged in a fluid is equal to the weight of the fluid that the object displaces [50]. For submarines, buoyancy is essential for achieving neutral buoyancy, where the weight of the submarine is equal to the weight of the water it displaces, allowing it to neither sink nor float. The design of submarines must carefully consider buoyancy to ensure that they can ascend and descend effectively, which is influenced by their shape and the distribution of mass within the vessel [51].

In submarines, buoyancy is actively controlled by adjusting the amount of water in the ballast tanks:

- When a submarine takes in water into its ballast tanks, its overall weight increases, reducing buoyancy and causing the vessel to sink.
- When air is pumped into the ballast tanks to expel water, the submarine becomes lighter, increasing buoyancy and causing the vessel to rise toward the surface.

Maintaining neutral buoyancy is critical for submerged operations, allowing the submarine to remain at a fixed depth without sinking or rising.

Ballast

Ballast is a mechanism used to control a submarine's buoyancy and trim. By adding or removing water from ballast tanks, submarines can adjust their weight and, consequently, their buoyancy. This adjustment allows submarines to dive, surface, or maintain a specific depth. The effective management of ballast is crucial for operational flexibility, enabling submarines to respond to varying operational conditions and maintain stability during manoeuvres [52]. The use of variable ballast systems, which can quickly adjust the amount of water in the tanks, enhances the submarine's ability to control its ascent and descent rates, thereby improving overall manoeuvrability [52].

Submarines are equipped with main ballast tanks (MBTs) that are located between the outer hull and the pressure hull. These tanks can be filled with seawater or air, depending on whether the submarine needs to submerge or surface.

- When the submarine is at the surface, the ballast tanks are filled with air, making the submarine buoyant enough to float.
- When the submarine needs to dive, the ballast tanks are flooded with seawater, increasing the submarine's weight and allowing it to sink.

The process of controlling ballast is crucial for achieving the right balance between weight and buoyancy, ensuring that the submarine can dive, surface, and operate effectively at various depths.

Stability

Stability in submarines refers to the ability of the vessel to maintain its orientation and resist unwanted rolling or pitching motions. Stability is influenced by the submarine's design, including the placement of its centre of mass and centre of buoyancy. A submarine is considered stable if it returns to its original position after being disturbed [53]. The dynamics of stability are complex, as they involve the interplay of buoyancy forces, ballast adjustments, and hydrodynamic forces acting on the submarine during its operation [54]. Research has shown that the design of the submarine's hull and the configuration of its ballast systems are critical for enhancing stability during various manoeuvres, such as ascending from great depths or navigating through turbulent waters [55, 56].

Stability is influenced by the centre of gravity (the point where the submarine's weight is concentrated) and the centre of buoyancy (the point where the upward buoyant force acts).

There are two types of stability:

1. **Static Stability** – This refers to the submarine's ability to return to its original position after being tilted by external forces (e.g., waves, currents). If the centre of buoyancy is positioned above the centre of gravity, the submarine will have positive stability, meaning it will naturally right itself after being tilted.

2. **Dynamic Stability** – This relates to the submarine's ability to maintain a steady course and avoid uncontrolled rolling or pitching during movement. Submarines use hydroplanes (control surfaces on the bow and stern) to maintain dynamic stability while changing depth and direction.

Submarine stability is particularly complex because the vessel operates in both surfaced and submerged conditions, which affect the relationship between buoyancy, weight, and hydrodynamic forces. Maintaining proper stability is essential for preventing the submarine from rolling, pitching excessively, or losing control, particularly during high-speed manoeuvres or when transitioning between different depths.

Submarine stability is uniquely different from that of surface ships because submarines operate in two distinct states: surfaced stability and submerged stability. These two conditions require separate analyses, as the behaviour of the submarine differs greatly depending on whether part of the vessel is above or below the waterline [57].

Surfaced Stability

In the surfaced condition, part of the submarine is above the waterline, similar to how a ship floats on the surface. In this state, the submarine's buoyancy (the upward force keeping it afloat) is determined by the volume of the submerged part of the hull. The relationship between the submarine's weight and buoyancy is governed by Archimedes' Principle, which states that the upward buoyant force is equal to the weight of the displaced water.

In this condition [57]:

- The submarine is partially submerged, so its centre of buoyancy is lower than when fully submerged.

- The centre of gravity plays a crucial role in maintaining stability, particularly when waves or other forces cause the submarine to tilt. If the centre of buoyancy is above the centre of gravity, the submarine will naturally right itself and be stable. This is referred to as positive static stability.

- The metacentric height (distance between the centre of gravity and the metacentre, which is a theoretical point where buoyant forces act) determines the submarine's stability in the surfaced state. A higher metacentric height generally means greater stability on the surface.

For a submarine to remain stable in the surfaced state, the difference between the centre of buoyancy and the centre of gravity needs to be managed, ensuring that the vessel does not tip excessively due to external forces like waves.

Submerged Stability

In the submerged condition, the entire submarine is beneath the waterline, with no part of the hull or appendages exposed to the surface. In this state, the stability dynamics are different because the centre of buoyancy shifts upward as the entire volume of the submarine is now submerged, and the vessel's interaction with external forces, such as waves, is minimized.

In the submerged state [57]:

- The entire hull is subjected to hydrostatic pressure, and the submarine relies on neutral buoyancy—a condition where the buoyant force and the weight of the submarine are exactly equal. This allows the submarine to remain at a constant depth without rising or sinking.

- The distribution of weight within the submarine is critical for maintaining submerged stability. Factors like fuel consumption, movement of crew, or the use of weapons can change the submarine's balance, so careful management of internal systems and ballast is necessary.

- Hydroplanes (control surfaces on the bow and stern) are used to control pitch and depth, ensuring that the submarine can manoeuvre while maintaining stability.

One of the key challenges in submerged stability is ensuring that the submarine does not experience excessive pitching or rolling during manoeuvres or when external forces (such as underwater currents) act on it [57].

Transition Between Surfaced and Submerged States

The transition between surfaced and submerged states involves managing the buoyancy by controlling the submarine's main ballast tanks (MBTs).

- When the submarine is on the surface, it is lighter than the total volume of water it could displace. To submerge, the submarine needs to increase its weight to balance out the buoyant force acting on the entire submerged volume.
- This is achieved by flooding the MBTs with seawater. As water fills the tanks, the overall weight of the submarine increases, reducing its net buoyancy and causing the submarine to sink.
- The process of neutral buoyancy is achieved by careful control of how much water is taken into the ballast tanks. Once submerged, the tanks are adjusted to ensure that the buoyant force exactly matches the submarine's weight, allowing it to remain at a desired depth without continuously sinking or rising.

In reverse, when the submarine needs to surface, compressed air is blown into the MBTs, forcing out the water and reducing the submarine's weight. This increases its buoyancy and causes it to rise back to the surface.

Weight and Buoyancy Balance

In both surfaced and submerged conditions, the principle of buoyancy is based on the idea that the weight of the submarine must equal the buoyant force for stable operation. When surfaced, the volume of the submarine above the waterline does not contribute to the buoyant force, so the vessel needs to add weight (via flooding the MBTs) to dive. In contrast, when submerged, the entire hull volume contributes to buoyancy, and adjustments are made to maintain depth and balance.

Diving and Surfacing of Submarines

Diving and surfacing of submarines rely on the careful management of buoyancy, which is controlled through the use of main ballast tanks (MBTs). These tanks are a fundamental component of the submarine's ability to dive below the water and resurface. The MBTs use a combination of flood ports and air vents to manipulate the water and air within the tanks, enabling the submarine to control its buoyancy as needed.

The two critical parts of the main ballast tanks are [57]:

1. **Flood Ports**: These are located at the bottom of the outer hull and are openings that allow seawater to flow into or out of the tanks. The flood ports play an essential role in the diving process by letting water in to increase the submarine's weight, and during surfacing, they let the water be expelled from the tank when air is forced in.
2. **Air Vents**: These are located at the top of the ballast tanks and are connected to the pressure hull. Each side of the submarine (port and starboard) has air vent risers that extend from the ballast tank to the main air vent. These air vents allow air to escape from the tanks when water is taken in and play an essential role in the surfacing process by allowing high-pressure air to push water out of the tanks.

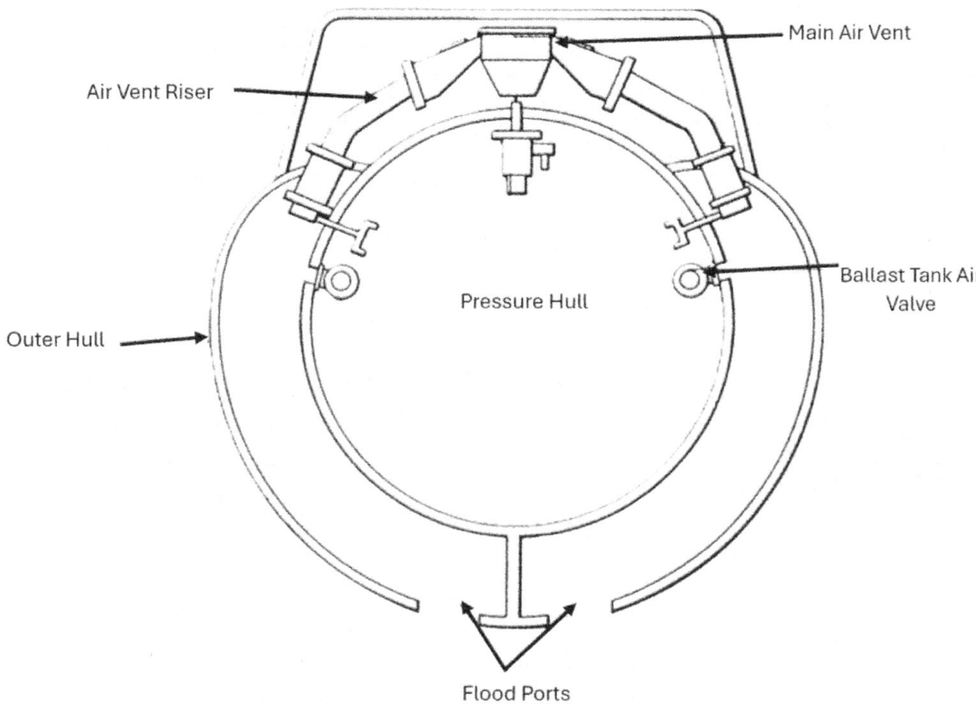

Figure 27: Transverse Section of Main Ballast Tank [57].

When a submarine needs to dive, the sequence of operations involves opening the air vents to release the air from the tanks and allowing water to flood in through the flood ports. The steps are as follows [57]:

1. **Opening the Air Vents**: To begin the diving process, the air vents at the top of the MBTs are opened. These vents allow the air trapped inside the tanks to escape.

2. **Flooding the Ballast Tanks**: Once the air has an exit path through the vents, seawater floods into the ballast tanks through the flood ports located at the bottom of the submarine. As the tanks fill with water, the overall weight of the submarine increases.

3. **Negative Buoyancy Achieved**: The added weight from the water in the ballast tanks causes the submarine to become negatively buoyant, meaning the force of gravity (weight) is greater than the buoyant force from the displaced water, causing the submarine to sink.

In this way, the submarine controls its descent by adding ballast water into the tanks until it reaches the desired depth.

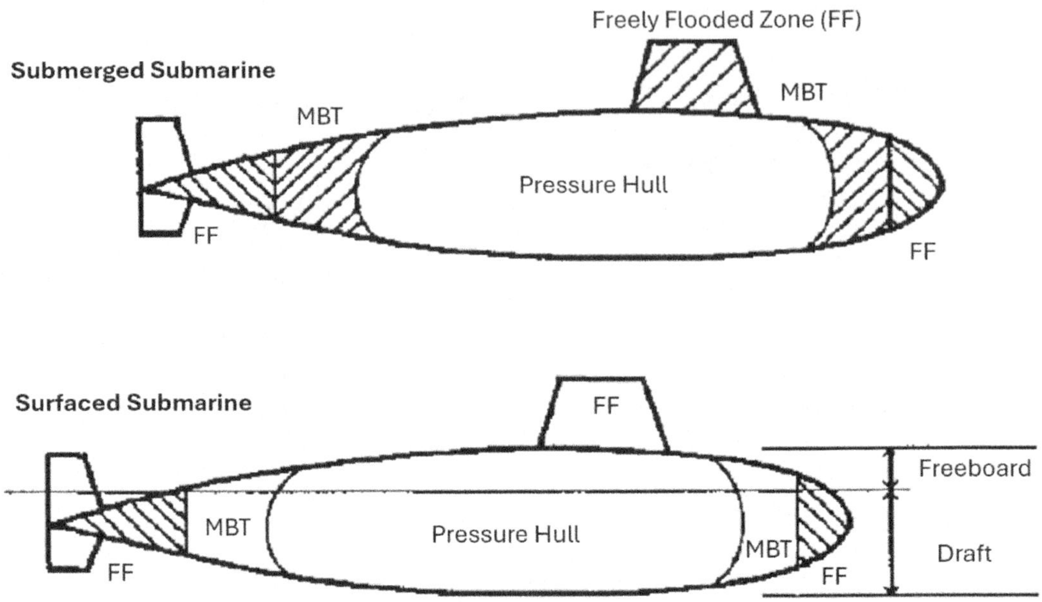

Figure 28: Submarine in Submerged and Surfaced Conditions [57].

The process of surfacing is the reverse of diving and requires expelling water from the MBTs to make the submarine lighter and regain positive buoyancy. This is achieved using high-pressure air to displace the water. The steps involved are [57]:

1. **Reducing Depth Using Hydroplanes**: Before surfacing fully, the submarine first uses its hydroplanes (control surfaces) to reduce its depth to around 3 to 4 meters below the waterline. This ensures that the submarine is close enough to the surface to use compressed air to expel water from the ballast tanks efficiently.

2. **Introducing High-Pressure Air**: At this reduced depth, high-pressure air (approximately 15 bar) is introduced into the ballast tanks through the air valve. The pressurized air forces the seawater out through the flood ports at the bottom of the submarine.

3. **Water Expulsion and Buoyancy Control**: As the water is expelled from the ballast tanks, the submarine's weight decreases, making it positively buoyant. Positive buoyancy occurs when the buoyant force is greater than the weight of the submarine, allowing it to rise to the surface.

4. **Rising to Surface Condition**: With the weight of the water removed from the tanks, the submarine naturally rises to the surface, achieving surfaced stability once again.

Modern submarines are designed to operate at depths of around 300 to 450 meters. At these depths, the water pressure is immense, and surfacing requires precise control of the ballast system. The introduction of high-pressure air is crucial for overcoming the external pressure at depth, as it provides enough force to displace the water within the ballast tanks and allow the submarine to become buoyant enough to rise [57].

Once the submarine has reduced its depth using the hydroplanes, the high-pressure air provides the force necessary to push the water out and initiate the surfacing process.

Alternative Submarine Surfacing Methods

Submarines typically surface by utilizing electrically operated valves that control the Main Ballast Tanks (MBT) to expel water and increase buoyancy. However, alternative methods are available in situations where the standard system fails or power is lost. These alternative systems are crucial for the submarine's safety and ability to return to the surface [58].

Emergency Main Ballast Tank (EMBT) Blow System: In the event of a power failure, submarines are equipped with an Emergency Main Ballast Tank (EMBT) blow system. This system is a completely mechanical (pneumatic) alternative that operates independently of electrical power. It uses compressed air stored in dedicated safety air banks to blow water out of the ballast tanks. Unlike the standard system, which relies on electrically operated valves, the EMBT blow system rapidly forces air directly into the ballast tanks from the air banks, allowing the submarine to surface quickly. This method is a vital safety feature, ensuring that submarines can surface even if power systems or electrical components are compromised [58].

Airless Surfacing Using Propulsion: Another alternative to surfacing involves using the submarine's propulsion system to power the vessel to the surface without using air to blow out the water from the ballast tanks. Known as an "airless surface," this method allows the submarine to rise to the surface by generating enough forward thrust. However, this technique has a significant limitation. If the water in the ballast tanks is not expelled, the submarine will remain negatively buoyant, making it unable to stay on the surface without continuous propulsion. Essentially, the submarine would surface but would not be able to stop moving, as it would start sinking again once the propulsion stops [58].

Using Diesel Generator Exhaust to Blow Ballast Tanks: In some older submarines equipped with diesel generators, an additional alternative method was available. The exhaust from the diesel generator could be redirected to blow the ballast tanks. This method takes advantage of the pressure created by the exhaust gases, allowing the tanks to be emptied of water and the submarine to surface. While this method is less common in modern submarines, it provided a useful backup system in the past, especially in situations where the main air systems were unavailable or compromised [58].

It is essential to note that these alternative methods have limitations. For example, if a submarine experiences more flooding than its reserve buoyancy is designed to handle, or if there is a total loss of propulsion, these methods may not be sufficient to bring the submarine to the surface. Reserve buoyancy plays a crucial role in ensuring the submarine can remain afloat once it surfaces. If reserve buoyancy is exceeded due to excessive flooding, surfacing may become impossible, and the submarine could remain trapped underwater [58].

These alternative surfacing methods highlight the need for redundancy and backup systems in submarine design, ensuring that even in critical situations, the crew has options for safely returning to the surface.

Surface Stability of Submarines

In the surfaced condition, a submarine must adhere to the same fundamental stability principles as a surface ship, ensuring it remains buoyant and stable even in the event of damage. One key concept in this regard is the Reserve of Buoyancy (ROB). Understanding ROB is crucial for submarine designers because it helps determine the necessary volume of Main Ballast Tanks (MBTs) in relation to the pressure hull volume [57].

ROB refers to the percentage of the hull's volume above the waterline when the submarine is in a surfaced condition. This reserve provides the necessary buoyancy to keep the submarine afloat after any potential damage, ensuring the vessel doesn't fully submerge. In practical terms, ROB is the ratio between the effective volume of all the MBTs (i.e., the volume of water that can be displaced by air) and the volumetric displacement of the submarine in a surfaced condition.

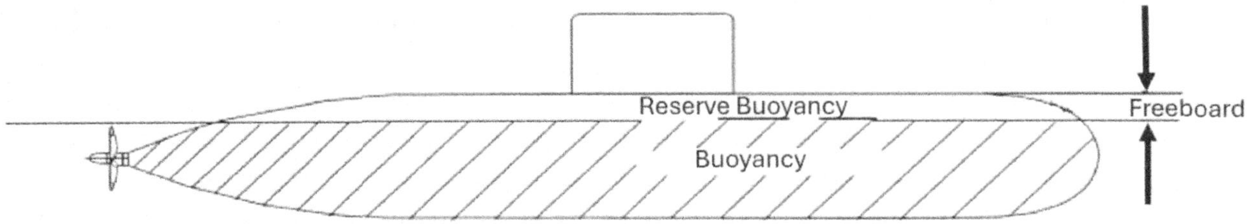

Figure 29: Volume of Hull contributing to ROB of a submarine [57].

Key Points About ROB [57]:

- Effective volume refers to the total "blowable" volume of the MBTs, meaning the volume that must be filled with air to make the submarine positively buoyant and allow it to surface.
- Volumetric displacement in surfaced condition is defined as the weight of the submarine minus any free flood water (seawater that floods into areas outside the pressure hull, such as spaces between the pressure hull and the outer hull).

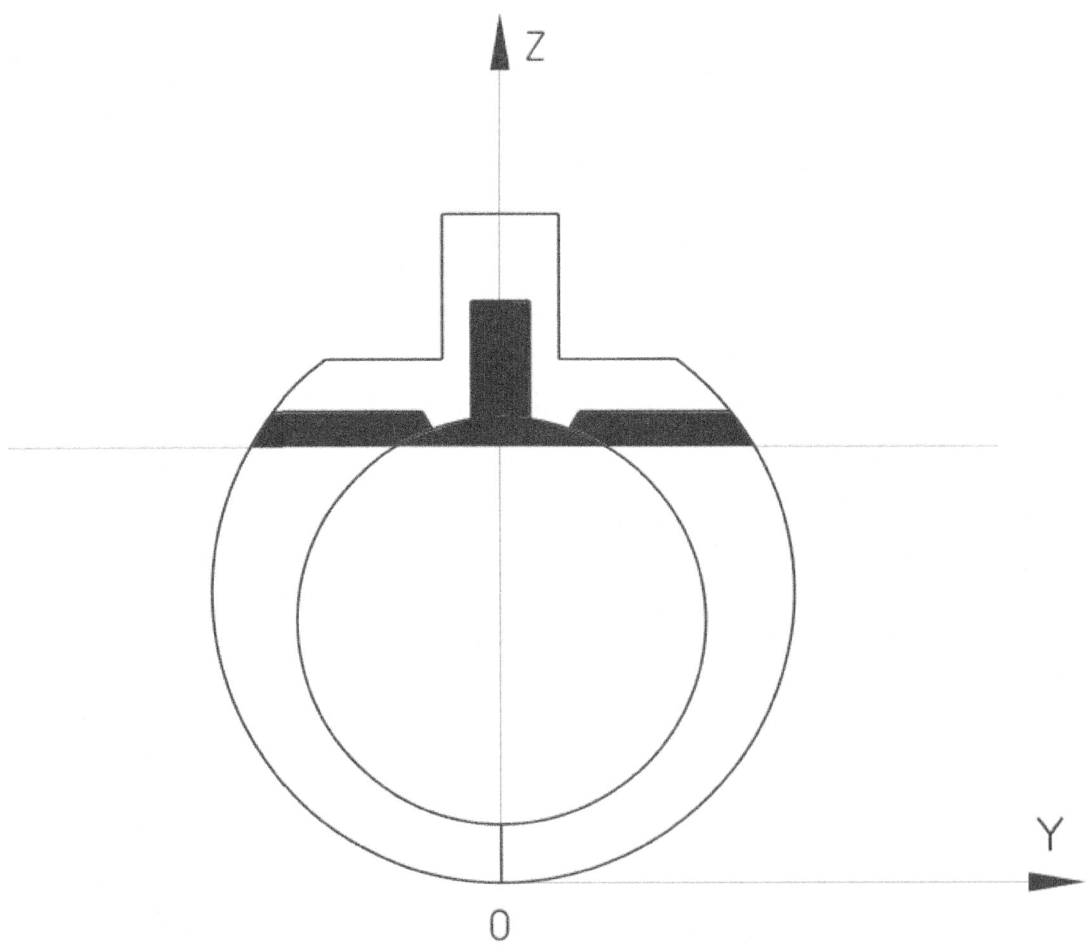

Figure 30: Reserve of buoyancy. Alex Rave, CC BY-SA 3.0, via Wikimedia Commons.

To understand the relation between ROB and the pressure hull, it's essential to recognize that in a submerged condition, only the pressure hull contributes to the buoyancy, as the space between the pressure hull and the outer hull is flooded. Hence, the buoyancy is provided solely by the pressure hull. To return to the surface, the submarine must expel water from the MBTs by forcing air into them, reducing the overall weight and creating enough positive buoyancy for surfacing.

The designer's task is to ensure the correct amount of ROB by calculating the required volume of the MBTs. The ROB depends on factors like submarine size, and smaller submarines generally require a larger ROB due to their limited freeboard (the distance from the waterline to the deck), typically ranging between 10% and 20%. Submarines with double hulls, which are more complex, may have even higher ROB percentages due to their additional internal and external hull spaces [57].

The transverse stability (side-to-side stability) of a surfaced submarine is similar to that of a surface ship. Both follow hydrostatic principles, where the centre of buoyancy (B) shifts sideways when the submarine heels (tilts to one side). As the centre of buoyancy moves to a new position (B1), the metacentre (M)—a point above the centre of buoyancy—creates a righting moment if it is higher than the centre of gravity (G). This righting moment helps return the submarine to an upright position.

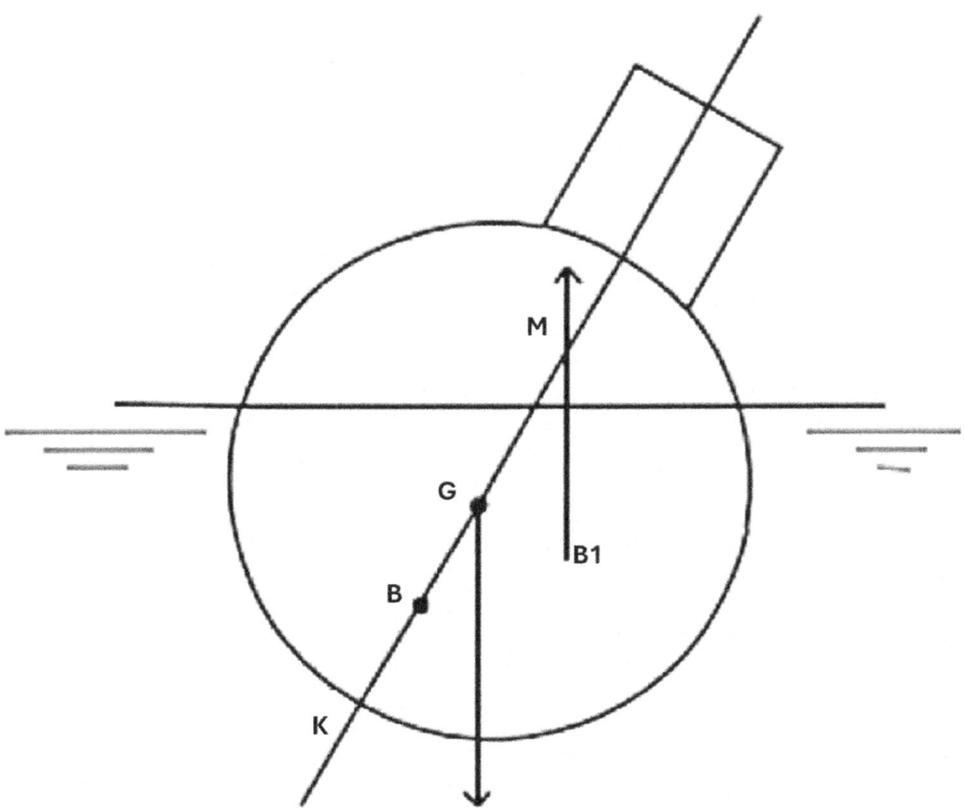

Figure 31: Surfaced Stability of Submarine [57].

The submarine's stability is critically dependent on maintaining an appropriate relationship between the centre of buoyancy and the centre of gravity. If the centre of gravity is too high, the metacentric height (distance between the centre of gravity and the metacentre) decreases, reducing the submarine's stability and making it prone to capsizing.

Submarines are extremely weight-sensitive, meaning any shift in weight can significantly impact their balance and stability. Even slight changes in the longitudinal centre of gravity (LCG) can cause a trimming moment (a force that affects the submarine's pitch or bow-to-stern angle). This trimming moment alters the water plane area (the surface area of the waterline), which in turn affects the longitudinal metacentric height.

A reduction in the water plane area causes a rapid decrease in the metacentric height, thereby reducing the submarine's longitudinal stability. To maintain proper stability, submarine operations must be carefully managed to minimize any shifts in the longitudinal centre of gravity. Ensuring that the LCG remains balanced is essential for maintaining trim stability, which prevents the submarine from tilting forward (bow down) or aft (stern down) during operations.

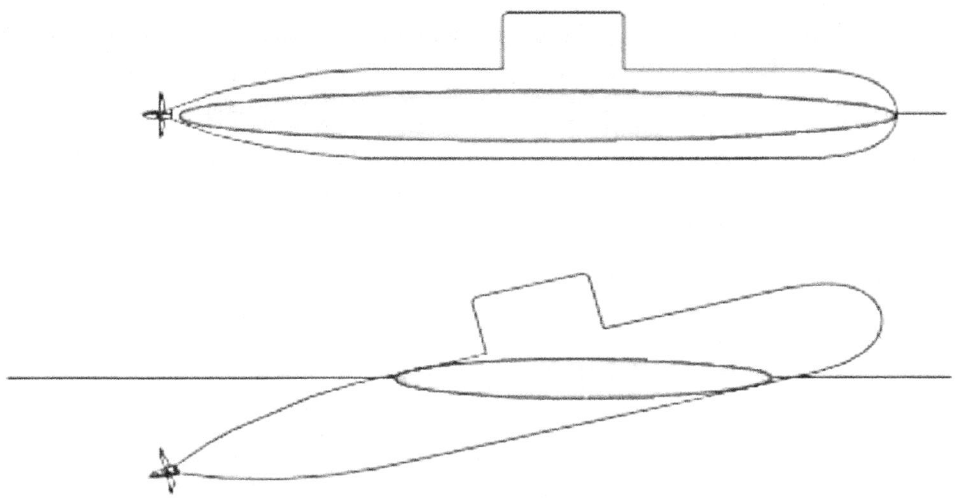

Figure 32: Change in waterplane area due to trimming moment [57].

The concept of Reserve of Buoyancy (ROB) is vital in designing a submarine's buoyancy and stability systems. ROB ensures that a submarine remains afloat and stable even after sustaining damage. Designers must carefully calculate the appropriate volume of the Main Ballast Tanks (MBTs) to achieve the correct ROB, especially considering the submarine's size and operational needs.

In both transverse and longitudinal stability, the relationship between buoyancy, the centre of gravity, and the metacentric height is critical. Submarine stability is not only a factor of proper design but also of careful operational management, ensuring that the submarine maintains balance while surfaced or submerged [57].

Submerged Stability of Submarines

When a submarine's Main Ballast Tanks (MBTs) are filled with water (blowed and pressed full), the submarine submerges and enters an environment vastly different from that of surface ships. In this submerged state, the submarine exhibits a unique characteristic: it can manoeuvre in six degrees of freedom, meaning it can move forward, backward, sideways, rotate, and change depth. Additionally, unlike surface ships or airplanes, a submarine doesn't rely on forward motion to support its weight. This ability to remain stationary in a submerged state, without the use of its fins or any forward motion, distinguishes it from airplanes [57].

The key to this unique capability lies in the balance between weight and buoyancy. Submarines operate based on three different buoyancy conditions [57]:

1. Negatively Buoyant: When the submarine's weight exceeds its buoyancy, it sinks. The submarine will continue

to descend until the weight is reduced (e.g., by blowing water out of the MBTs) or buoyancy is increased.

2. Positively Buoyant: In this state, the buoyancy is greater than the weight of the submarine, causing it to float or rise toward the surface. Corrective measures would need to be taken to increase the weight, such as taking in more water into the MBTs, to stop the ascent.

3. Neutrally Buoyant: Submarines in submerged condition typically operate in this state. Here, the weight and buoyancy are perfectly balanced, allowing the submarine to remain at a constant depth without sinking or rising. This neutral buoyancy enables the submarine to stay static in water without forward motion or reliance on its control surfaces like fins.

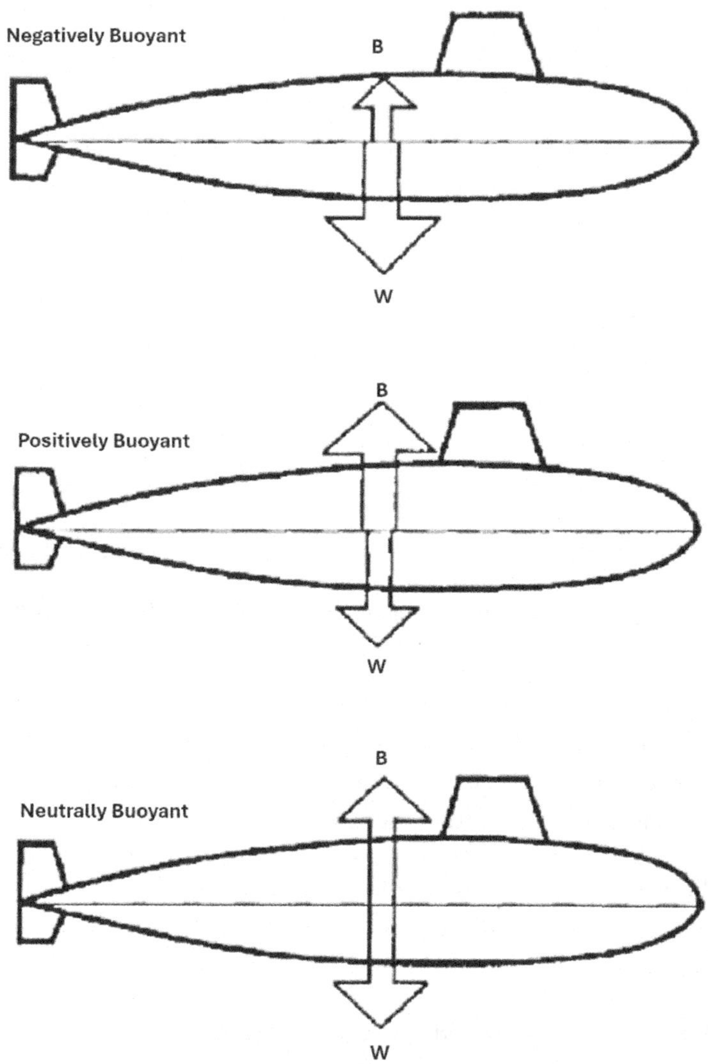

Figure 33: Submarine in Positively, Negatively, and Neutrally Buoyancy Conditions [57].

The transverse stability of a submarine—its ability to resist rolling or tilting—differs significantly in submerged conditions compared to when it is surfaced. On the surface, the metacentric height (GM) determines the stability, but in a submerged state, the metacentric height transforms into the distance between the centre of buoyancy (B) and the centre of gravity (G). This distance is now referred to as BG, and the dynamics change as follows [57]:

- When the submarine is submerged, an angle of heel (tilt) does not change the submerged volume. This means that the centre of buoyancy (B) remains fixed, unlike on the surface where it shifts with the waterline.

- In this case, the metacentre coincides with the centre of buoyancy (B), and the righting moment—the force that brings the submarine back to its upright position—depends on the relationship between the centre of gravity (G) and centre of buoyancy (B).

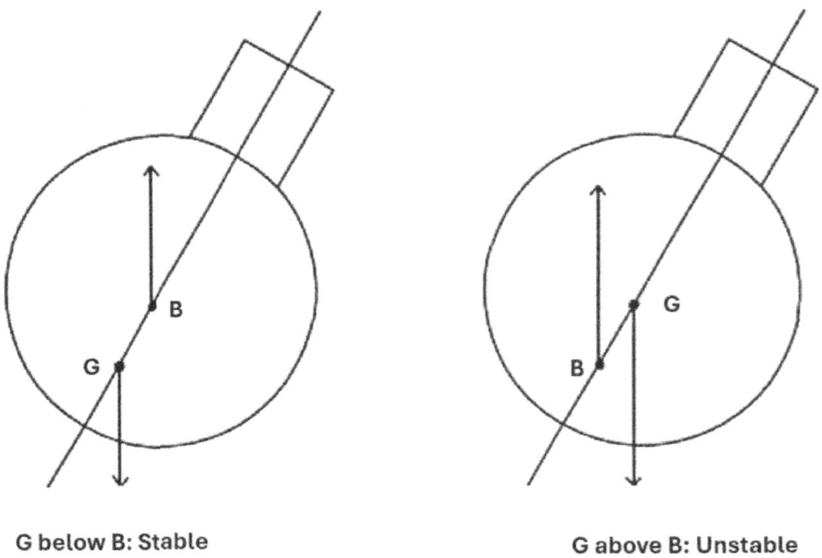

Figure 34: Stable and Unstable Conditions in a Submerged Submarine [57].

Two cases determine the stability of a submarine underwater:

1. **G below B (Stable Condition)**: If the centre of gravity (G) is located below the centre of buoyancy (B), a righting moment is created when the submarine tilts. This moment works to bring the submarine back to its upright position, ensuring stability. This is the desired condition in submarine design.

2. **G above B (Unstable Condition)**: If the centre of gravity (G) is above the centre of buoyancy (B), the submarine becomes unstable and can capsize. In this scenario, there is no righting moment, and the submarine is likely to tip over in the water.

For a submarine to remain stable in the submerged condition, it is essential that the centre of gravity (G) remains below the centre of buoyancy (B). To achieve this, the position of the centre of buoyancy is usually fixed based on the design of the submarine, so the centre of gravity must be carefully managed. Special ballast tanks and trim systems are utilized to ensure that the G remains below B, which helps maintain the submarine's stability in all conditions.

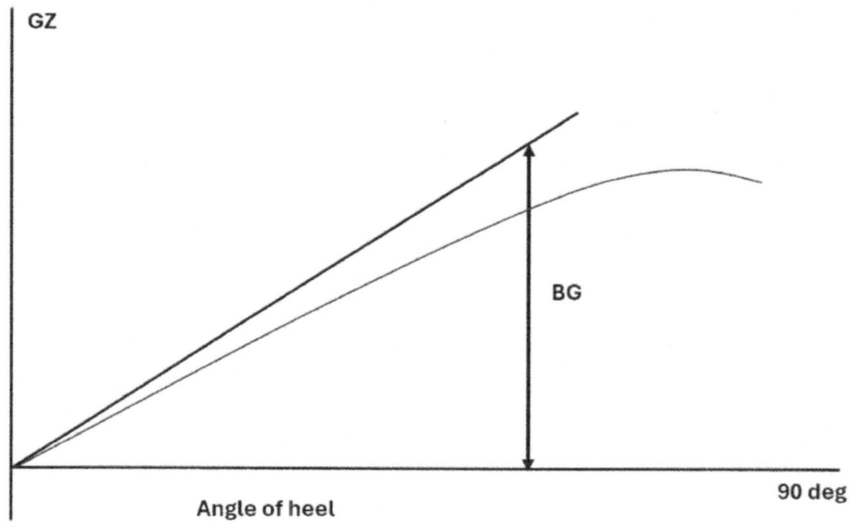

Figure 35: Stability curve of a submerged submarine [57].

The submerged submarine's stability is governed by the BG criterion, and the variation of the righting lever (GZ) with the angle of heel. In stable conditions, the righting lever creates a moment that helps the submarine regain its upright position after tilting. Designers pay close attention to these stability parameters to ensure that the submarine can withstand various underwater conditions and remain safe for its crew and mission.

Buoyancy and Weight

To understand how submersibles and submarines manipulate their buoyancy, we first need to explore the fundamental concept of buoyancy. Buoyancy, or upthrust, is a gravitational force exerted by a fluid (like water) that pushes upward on an object submerged in it. This force opposes the weight of the object. The scientific basis for buoyancy is explained by Archimedes' principle, which states that any object submerged in a fluid experiences an upward force equal to the weight of the fluid displaced by the object [58].

To visualize this, think of placing a cork in a glass of water. The cork floats because the water exerts an upward buoyant force greater than the cork's weight. This upward force, or buoyancy, is equivalent to the weight of the water displaced by the cork. On the other hand, if you drop a steel ball into the same water, it sinks because the weight of the ball is greater than the buoyant force acting on it [58].

Buoyancy is determined by the density of the object in relation to the fluid it is placed in. Density is defined as mass per unit volume, meaning that an object's ability to float or sink depends on whether its density is less or more than the fluid's density. For instance, even if an object weighs a lot, if it has a lower density than the fluid, it will float. This explains why a large, thin steel bowl can float on water while a smaller, denser steel ball of the same weight will sink. The difference in their distribution of mass relative to volume plays a crucial role in their buoyancy [58].

Understanding buoyancy is essential for designing submersibles and submarines, as their ability to float, dive, and surface relies on manipulating the forces of buoyancy to balance against their weight and density in the surrounding water. Further, in submarine design and operation, understanding the components contributing to buoyancy and weight is critical for maintaining control, stability, and safety underwater. These components are divided into buoyancy elements and weight elements, each playing a vital role in the submarine's ability to dive, surface, and manoeuvre efficiently.

Buoyancy Elements

Buoyancy elements are parts of the submarine that displace water, generating an upward force in accordance with Archimedes' Principle. Correctly identifying and calculating the impact of these elements is crucial for determining the submarine's overall buoyancy and its centre of buoyancy.

- **Pressure Hull Volume**: The primary buoyancy element is the pressure hull itself. This sealed, watertight compartment houses the crew and equipment, displacing a large volume of water and providing substantial buoyant force. However, it is essential to note that components inside the pressure hull, such as machinery, equipment, and crew, do not directly contribute to buoyancy as they do not come into contact with seawater and do not displace additional water.

- **Displaced Volume of the Outer Hull**: Surrounding the pressure hull, the outer hull shapes the submarine hydrodynamically. The material of the outer hull, usually steel plating, displaces water and thus contributes to the buoyancy. However, the space between the outer and pressure hull is typically flooded and does not contribute to buoyancy. Only the volume displaced by the outer hull material itself is considered a buoyancy element.

- **Non-Flooded External Tanks**: In some designs, certain external tanks located in the forward and aft sections remain unflooded even when the submarine is submerged. These tanks, filled with air or other substances, displace water without adding much weight, thus increasing buoyancy. Their enclosed volumes are included in buoyancy calculations.

- **Sail and Internal Structures**: The sail, also known as the conning tower or fin, is a vertical structure on the submarine's top that houses equipment like periscopes and masts. While the external part of the sail is freely flooded when submerged, internal structures—such as the conning tower and masts—remain sealed and displace water. The volumes occupied by these structures contribute to buoyancy.

- **Torpedo Tubes in Freely Flooded Regions**: In the forward section, portions of the torpedo tubes extend beyond the pressure hull into the freely flooded area. These sections of the torpedo tubes, being airtight, displace water and act as buoyancy elements, adding to the overall buoyant force without increasing weight.

- **Propeller Shaft and Appendages**: The propeller shaft, which passes through a sealed tube into the freely flooded aft section, also displaces water and contributes to buoyancy. Similarly, other appendages like the propeller, rudders, forward and aft fins, and sonar arrays contribute to buoyancy by displacing water with their physical volumes.

- **Air Pockets in Ballast Tanks**: Occasionally, unintended air pockets may form in the ballast tanks, which are meant to be flooded when the submarine is submerged. These air pockets displace water without adding weight, unexpectedly increasing buoyancy. Although they contribute to the overall buoyant force, they are undesirable as they can disrupt the submarine's neutral buoyancy and affect depth control.

By calculating the volume of these buoyancy elements and their placement, submarine engineers ensure that the vessel remains stable, controllable, and capable of executing its dives and ascents effectively.

Weight Elements

Weight elements are the components that contribute to a submarine's total weight, exerting a downward force due to gravity. Proper accounting of these elements, including their individual weights and centres of gravity, is essential for maintaining the submarine's balance, stability, and overall performance.

Fixed Weight Elements:

- **Pressure Hull Structure**: This includes the pressure hull plating, circular stiffening frames, tanks, brackets, and other structural components. These elements are critical for providing the necessary strength to withstand the immense external pressures at depth, but they also add significant weight to the submarine. The weight distribution of the hull structure impacts the vessel's stability and performance.

- **Main Propulsion Plant**: The propulsion system is one of the heaviest parts of the submarine. It consists of diesel alternators, electric propulsion motors (for diesel-electric submarines), shafting systems, thrust blocks, thrust bearings, and the propeller. These components are essential for powering the submarine and driving it through the water, but their weight requires careful management to ensure proper balance and trim.

- **Battery Banks**: Submarines rely heavily on large battery banks, especially for silent underwater operations. These batteries, typically made of dense materials such as lead-acid or lithium-ion cells, add considerable weight to the vessel. The placement and distribution of the battery banks must be optimized to avoid compromising the submarine's stability.

- **Weapons**: The weight of torpedoes, missiles, and other onboard munitions is significant. The storage of these weapons within the submarine directly influences its centre of gravity, which must be carefully calculated to maintain balance and proper trim.

- **Other Machinery and Permanent Fixtures**: In addition to the propulsion system, submarines contain a range of other equipment, including life support systems, navigation and communication tools, hydraulic systems, and interior fittings. While these individual components may not be as heavy as propulsion machinery, their collective weight contributes significantly to the submarine's overall mass.

Variable Weight Elements:

- **Crew**: The weight of the crew is variable depending on the number of personnel on board and their movements throughout the submarine. Shifts in crew positions can affect the submarine's trim and balance, especially during critical manoeuvres.

- **Stores**: Freshwater, food supplies, and other consumables are stored on board and are gradually used throughout the mission. As these supplies are consumed, the overall weight of the submarine decreases, impacting its buoyancy and stability, which must be managed during operations.

- **Consumables**: Fuel (such as diesel oil), lubricating oil, and other operational fluids are consumed as the submarine operates. The depletion of these resources reduces the vessel's weight, requiring adjustments to maintain neutral buoyancy.

- **Bilge and Solid Waste**: Bilge water and solid waste can accumulate during operations, adding weight to the submarine. Submarines are equipped with systems to manage and expel waste, but variations in these factors can affect the overall weight distribution and stability of the vessel.

- **Ballast**: Ballast water is used to control buoyancy and trim the submarine. Taking in or expelling ballast water directly alters the submarine's weight, requiring precise monitoring to ensure proper balance and operational performance.

Balancing Weight and Buoyancy

Achieving the right balance between weight and buoyancy is crucial for a submarine's ability to dive, surface, and maintain depth control. This balance also directly impacts the vessel's stability and manoeuvrability underwater, ensuring safe and efficient operations.

Total Buoyancy and Total Weight:

- **Total Buoyancy** is calculated by summing the contributions of all buoyancy elements. Each element's enclosed volume and centre of buoyancy are considered to determine the overall buoyant force acting on the submarine.

- **Total Weight** is determined by adding up all fixed and variable weight elements. This includes not only their individual weights but also their centres of gravity, which influence the overall balance of the vessel.

Centres of Buoyancy and Gravity:

The centre of buoyancy is the point where the upward buoyant force acts, while the centre of gravity is where the downward weight force acts. For a submarine to remain stable, especially in a submerged condition, the centre of gravity should ideally be positioned below the centre of buoyancy. This configuration creates a righting moment that helps keep the submarine upright and stable.

Calculating Buoyancy and Weight:

- **Listing Buoyancy Elements**:
    - Identify all components that contribute to buoyancy.
    - Determine the enclosed volume for each element, which represents the volume of water displaced by the element.
    - Calculate the centre of volume (buoyancy), the 3D coordinates representing where the buoyant force of each element acts.
- **Listing Weight Elements**:
    - Catalogue all components contributing to weight.
    - For each element, record the weight and the centre of gravity, the point where the weight force of the element acts.
- **Summation and Analysis**:
    - **Total Buoyancy**: Sum the volumes of all buoyancy elements and calculate the overall centre of buoyancy by finding the weighted average based on each element's volume and position.
    - **Total Weight**: Sum all weight elements and determine the overall centre of gravity in a similar manner.
- **Correlating Weight and Buoyancy**:

- Compare the total weight and buoyancy to ensure they are balanced, achieving neutral buoyancy for optimal underwater performance.
- Evaluate the positions of the centres of gravity and buoyancy to confirm stability. The vertical distance between them is crucial as it affects the righting moment, which maintains the submarine's stability.

If imbalances between weight and buoyancy are identified, adjustments are made. Ballast tanks may be used to compensate for any weight discrepancies, or the distribution of weight elements may be altered. Since variable weight elements (such as fuel and crew) change throughout a mission, continuous monitoring and adjustments are necessary to maintain balance and neutral buoyancy.

Practical Considerations:

- **Trim and Stability**: Proper distribution of weight elements ensures the submarine maintains its trim, or longitudinal balance. This prevents unintended pitching or rolling, which could compromise the vessel's performance.
- **Dynamic Operations**: As the submarine operates, fuel consumption, depth changes, and other factors dynamically affect weight and buoyancy. Systems onboard adjust ballast accordingly to maintain neutral buoyancy and stability throughout these changes.
- **Design Optimization**: During the design phase, the placement of heavy machinery and buoyancy elements is carefully planned to optimize the submarine's balance and overall performance. A well-designed vessel reduces the need for excessive adjustments during operations.
- **Safety Margins**: Engineers incorporate safety factors into the design and operational calculations to ensure the submarine remains controllable even under extreme or unexpected conditions. This adds a layer of security, ensuring that the vessel can handle varying pressures and conditions safely.

Balancing weight and buoyancy in submarine design and operation is a continuous process involving precise calculations and dynamic adjustments. Stability and safety depend on accurate assessments of both buoyant forces and weight distribution, ensuring the submarine remains effective in all operational conditions.

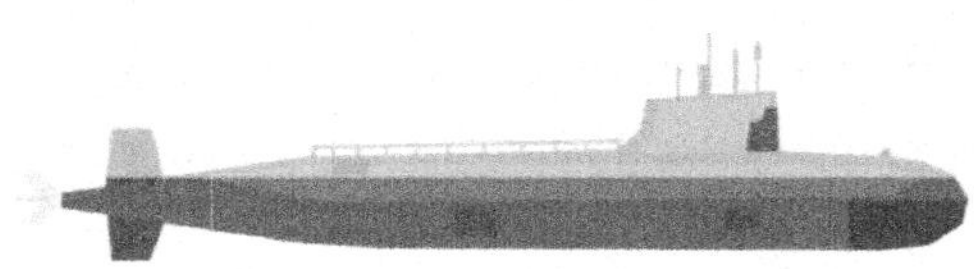

Chapter 3

Hydrodynamics and Submarine Performance

Hydrodynamic Principles in Submarine Design

Hydrodynamic principles are fundamental to the design of submarines, as they directly affect a submarine's ability to move efficiently, remain stable, and minimize detection while operating underwater. Submarine designers must carefully balance factors like drag, lift, flow resistance, and buoyancy to ensure that the vessel performs optimally in diverse underwater conditions.

Hydrodynamic principles govern how fluids, particularly water, move around objects like submarines and the forces that arise from these interactions. These principles are crucial in submarine design, as they determine how efficiently and effectively a submarine can navigate underwater. The main hydrodynamic principles influencing submarine design are as follows:

1. Drag

Drag refers to the resistance a submarine encounters as it moves through water. This resistance arises from the friction between the water and the submarine's surface, as well as turbulence behind the vessel. There are several types of drag:

- **Frictional Drag**: Caused by water molecules moving along the hull's surface.
- **Pressure Drag**: Results from differences in water pressure between the bow (front) and stern (rear) of the submarine.
- **Wave Drag**: Occurs near the surface where the submarine creates waves, adding resistance. To minimize drag, submarines are designed with streamlined hulls, allowing water to flow smoothly over their surface, reducing both turbulence and wake.

2. Buoyancy

Buoyancy is the upward force exerted by a fluid, such as water, that opposes the weight of an object immersed in it. This principle is governed by Archimedes' law, which states that any object submerged in a fluid experiences an upward force equal to the weight of the fluid displaced by the object. Submarines manage buoyancy with ballast tanks that can take in or expel water, enabling the vessel to dive, surface, or maintain neutral buoyancy. Achieving neutral buoyancy is critical for energy-efficient and stable movement underwater.

3. Streamlining

Streamlining involves shaping the submarine to minimize drag. A streamlined object allows water to flow over its surface smoothly, which reduces the amount of energy needed to propel it forward. Modern submarines often have a teardrop or cigar-shaped hull, which balances internal volume and hydrodynamic efficiency, reducing water resistance and improving speed while conserving energy.

4. Lift and Control Surfaces

Lift is a force that allows submarines to change their depth and direction. It is generated by control surfaces like hydroplanes (horizontal fins) or rudders (vertical fins). Submarines use hydroplanes to control their depth and rudders to steer. These surfaces create lift as water flows around them, helping the submarine change position within the water column without relying on buoyancy adjustments alone.

5. Reynolds Number

The Reynolds number is a dimensionless value representing the ratio of inertial forces to viscous forces in a fluid. It helps predict whether the water flow around the submarine will be smooth (laminar) or chaotic (turbulent). Submarine designers aim to minimize turbulent flow, which increases drag, by selecting hull shapes that encourage laminar flow, optimizing the submarine's efficiency and stealth.

6. Pressure Distribution

Pressure distribution refers to how water pressure is distributed along the submarine's surface. Higher pressure at the bow and lower pressure at the stern can create resistance, known as pressure drag. By designing the hull to reduce these pressure differences, submarines can move more efficiently through the water, improving their overall hydrodynamic performance.

7. Cavitation

Cavitation occurs when water pressure drops to a point where vapor bubbles form, usually near moving parts like propellers. When these bubbles collapse, they create noise and vibration, which can affect stealth and damage components. Submarine designers work to prevent cavitation, especially in propeller design, to minimize noise and maintain the vessel's stealth capabilities.

8. Wake

Wake refers to the turbulent water left behind a moving object. Excessive wake increases drag and makes the submarine more detectable. To reduce wake formation, submarines are designed to move quietly and efficiently through the water, which helps maintain stealth and reduces the energy required for propulsion.

9. Propeller Efficiency

Propeller efficiency is the ability of the propulsion system to convert energy into forward movement with minimal energy loss. The design and placement of the propeller are critical to generating sufficient thrust while minimizing drag and cavitation, ensuring the submarine moves efficiently through the water.

Streamlining and Drag Reduction

Drag is a critical factor in the performance of submarines, as it directly opposes their motion through water. The design of a submarine's hull is meticulously crafted to minimize drag, thereby enhancing speed, fuel efficiency, and stealth capabilities. The hydrodynamic characteristics of a submarine's hull significantly influence its operational effectiveness, particularly in terms of resistance encountered during movement through water.

To achieve optimal hydrodynamic performance, submarines are typically designed with a streamlined hull shape, often resembling a teardrop. This design allows water to flow smoothly over the hull, minimizing turbulence and the resultant drag. Research indicates that a well-designed hull can significantly reduce the turbulent wake generated as the submarine moves, which is crucial for maintaining higher speeds with lower energy consumption [24, 59]. The reduction of drag not only enhances fuel efficiency but also minimizes noise, which is essential for stealth operations [60].

The teardrop shape, characterized by a rounded bow and a tapering stern, is particularly effective in minimizing drag. The rounded front reduces pressure drag, while the tapered rear decreases wake formation, leading to less turbulent water trailing behind the submarine [24, 61]. This hydrodynamic efficiency is vital for submarines, especially during covert missions where stealth is paramount. Studies have shown that various hull shapes can lead to different resistance levels, with conical and cylindrical shapes being evaluated for their hydrodynamic properties [24, 61, 62].

While the teardrop shape is ideal for drag reduction, designers must also consider the internal volume of the submarine. Modern submarines often utilize a cylindrical hull with elliptical or rounded ends to strike a balance between hydrodynamic efficiency and the need for internal space for equipment and crew [59, 63]. This design approach allows submarines to maintain a streamlined profile while accommodating the necessary operational components. Computational Fluid Dynamics (CFD) has become a crucial tool in this design process, enabling engineers to simulate and analyse the hydrodynamic characteristics of various hull shapes and configurations [24, 59, 64].

When a submarine moves underwater, it encounters resistance against its forward motion due to the forces of lift and drag. These forces are influenced by the shape and velocity of the submarine as it interacts with the surrounding water.

Lift and Drag Forces

When a body, like a submarine, moves with velocity "V" in a fluid (water), making an angle with the streamlines of the fluid flow, it experiences a force "F" acting on it. This force can be decomposed into two components [65]:

- **Drag**: The force acting along the direction of the fluid flow (parallel to the streamlines). Drag is a resistance that opposes the motion of the submarine.

- **Lift**: The force acting perpendicular to the fluid flow (orthogonal to the streamlines). Lift, in general, can cause the submarine to rise or fall in the water.

The drag and lift forces are represented by their respective coefficients—coefficient of drag (Cd) and coefficient of lift (Cl). These coefficients are defined as the ratio of the corresponding forces (drag or lift) to the dynamic force on the projected or planform area of the submarine. Mathematically, these relationships are expressed as [65]:

Coefficient of Drag (Cd):

$$Cd = \frac{D}{\frac{1}{2}\rho V^2 A}$$

Where:

- D is the drag force,
- ρ is the fluid density,
- V is the velocity of the submarine,
- A is the reference area of the submarine.

Therefore, the drag force is given by:

$$D = Cd \cdot \frac{1}{2}\rho V^2 A$$

Coefficient of Lift (Cl):

$$Cl = \frac{L}{\frac{1}{2}\rho V^2 A}$$

Where:

- L is the lift force.

The lift force is similarly given by:

$$L = Cl \cdot \frac{1}{2}\rho V^2 A$$

Interaction with Fluid Flow

When a submarine moves at an angle to the streamlines of water flow (often referred to as an angle of attack), the fluid separates and moves around the submarine's hull. Because of the shape of the hull and the variation in distance travelled by water around different parts of the body, the streamlines on the two sides of the submarine will have different speeds. The streamline covering more distance will move faster to conserve mass, and according to Bernoulli's principle, the faster-moving streamline will exert lower pressure compared to the slower-moving one. This pressure difference across the submarine results in a net force, creating both lift and drag [65].

Additionally, the angle of attack alters the momentum of the fluid. By applying the momentum equation in a control volume around the submarine, a net force (F) acting on the body is found. The combined effects of the conservation of mass (Bernoulli) and the conservation of momentum result in the forces of lift and drag acting on the submarine [65].

Managing Lift in Submarines

While lift is crucial for aircraft, it is not a desirable force for submarines when moving underwater. If lift is produced while the submarine is submerged, it would cause the vessel to rise to the surface, which is undesirable during underwater operations. Submarines are designed to move and operate efficiently at various depths, and generating lift can interfere with this function [65].

To counteract unnecessary lift generated by the movement of the submarine's pressure hull, submarines are equipped with hydroplanes. These hydroplanes are typically mounted on the conning tower and act like control surfaces or flaps on an aircraft wing. They have an aerofoil shape, and by tilting them (changing the angle of attack), the submarine can generate negative lift—a downward force that neutralizes the unwanted upward lift produced by the hull.

Hydroplanes play a crucial role in controlling the submarine's depth and movement underwater. By adjusting the angle of attack of the hydroplanes, the crew can control the submarine's depth and pitch without relying on forward motion to support its weight, unlike aircraft. This ensures that the submarine can remain submerged and manoeuvre efficiently [65].

Coordination with Navigation

Hydroplanes also work in conjunction with the submarine's navigation system to maintain precise control over its motion. By adjusting the hydroplanes, the submarine can move forward while maintaining its desired depth and stability, allowing it to perform various operations underwater without surfacing unnecessarily [65].

In summary, as a submarine moves underwater, it experiences lift and drag forces due to its interaction with water flow. While drag acts as a resistance to forward motion, lift can potentially cause the submarine to ascend. To manage these forces and maintain control, submarines use hydroplanes to negate unnecessary lift, ensuring stable and efficient underwater operations.

Lift and Control Surfaces

Submarines, akin to aircraft, utilize control surfaces to manage their movement; however, the mechanisms employed differ significantly due to the unique challenges posed by the underwater environment. Unlike airplanes, which rely on constant lift generated by wings to remain airborne, submarines primarily depend on hydrodynamic control surfaces, such as hydroplanes and rudders, to adjust their depth and maintain stability underwater. The hydrodynamic forces acting on these surfaces are influenced by the submarine's speed and the surrounding water conditions, necessitating a distinct approach to manoeuvrability compared to aerial vehicles [66, 67].

Hydroplanes, also known as diving planes, are critical components of submarine design, positioned at both the bow and stern. These movable fins allow for precise control over the submarine's pitch and depth. By altering the angle of the hydroplanes, the crew can generate lift in either an upward or downward direction, facilitating ascent, descent, or maintenance of a specific depth. For instance, when the bow planes are angled upward, they produce downward lift, causing the submarine's front to rise, while the opposite configuration results in a descent [66]. The effectiveness of these hydroplanes is contingent on the submarine's speed; as the vessel moves faster, the hydrodynamic forces acting on the hydroplanes increase, enhancing their control capabilities [67].

In addition to hydroplanes, submarines are equipped with rudders, typically located at the stern, which are essential for controlling yaw, or lateral movement. The rudder's deflection alters the water flow around the submarine, generating lateral forces that enable steering. This capability is crucial for navigating complex underwater environments, whether for evasion or pursuit [68]. The interaction between the rudder and the submarine's propulsion system is vital for effective manoeuvring, particularly in scenarios requiring rapid directional changes [14].

Both hydroplanes and rudders rely on water flow over the submarine's hull to generate the necessary forces for depth and directional control. This reliance on hydrodynamic principles means that these control surfaces are most effective when the submarine is in motion. However, submarines possess the unique ability to achieve neutral buoyancy, allowing them to maintain a stationary position at a specific depth without the need for forward motion. This capability is achieved through careful ballast adjustments, which balance the submarine's weight and buoyancy [69]. In contrast, aircraft must maintain continuous forward motion to generate lift, highlighting a fundamental difference in operational dynamics between the two types of vehicles [55].

Buoyancy and Stability

As previously outlined, buoyancy and stability are fundamental principles in submarine design, playing critical roles in controlling the submarine's movement and maintaining its position underwater. These principles are essential for ensuring that submarines can float, dive, hover, and surface safely while remaining stable in both submerged and surfaced conditions.

Buoyancy, as articulated by Archimedes' Principle, is the upward force that allows an object to float or sink based on the volume of water it displaces. In submarines, the buoyant force is equal to the weight of the water displaced, which is manipulated through ballast tanks that can be filled with air or water to adjust the submarine's buoyancy state. Submarines can achieve three states of buoyancy: positive, negative, and neutral.

1. Positive Buoyancy: This state occurs when the submarine is lighter than the water it displaces, allowing it to float. This is achieved by expelling water from the ballast tanks and replacing it with air, which is crucial for operations at the surface or preparing to ascend after a dive [50].

2. Negative Buoyancy: To dive, submarines take on water in their ballast tanks, increasing their weight and allowing them to sink. The careful management of ballast water is vital for controlling the rate of descent and ensuring safe operations underwater [50].

3. Neutral Buoyancy: Achieving neutral buoyancy is critical for submarines to maintain a constant depth without rising or sinking. This state is energy-efficient as it allows submarines to hover underwater without expending energy for vertical movement [50].

Stability is equally important in submarine design, ensuring that the vessel remains balanced and upright in various conditions.

1. Surfaced Stability: When on the surface, submarines rely on reserve buoyancy to keep a significant portion of the hull above water. This reserve buoyancy acts as a safety margin, allowing submarines to remain afloat even if parts of the hull are compromised. The ballast tanks play a crucial role in maintaining this stability, particularly in rough sea conditions [50].

2. Submerged Stability: Underwater, stability is determined by the relationship between the centre of buoyancy (the point where the buoyant force acts) and the centre of gravity (the point where the submarine's weight is concentrated). For a submarine to remain stable, the centre of gravity must be below the centre of buoyancy, creating a righting moment that helps return the submarine to an upright position if it tilts or rolls [70]. If the centre of gravity rises too high, the submarine risks becoming unstable, which could lead to capsizing. Therefore, careful attention is given to weight distribution and the operation of ballast systems to maintain stability throughout various operational scenarios [71].

The principles of buoyancy and stability are integral to submarine design, enabling controlled movement and ensuring safety during operations. By manipulating buoyancy through ballast systems and maintaining a favourable relationship between the centre of buoyancy and centre of gravity, submarines can effectively navigate underwater environments while remaining stable.

Propulsion and Cavitation

Submarines utilize either nuclear or diesel-electric propulsion systems to generate thrust, enabling them to navigate underwater effectively. The propeller, often referred to as the screw, is the primary component responsible for propelling the submarine forward. The design and function of the propeller are governed by various hydrodynamic principles, which aim to optimize movement through water while simultaneously minimizing noise to maintain stealth. The significance of propeller design in submarines is critical, as it directly impacts performance, efficiency, and the vessel's ability to evade detection by enemy sonar systems [72, 73].

The design of a submarine's propeller is crucial for achieving a balance between thrust, speed, and noise reduction. Unlike conventional ship propellers, submarine propellers are engineered to minimize cavitation—a phenomenon where vapor bubbles form around the blades due to rapid movement through water. This cavitation can produce noise detectable by sonar, compromising the submarine's stealth capabilities [74]. To mitigate cavitation, submarine propellers are often designed with specific shapes, sizes, and blade configurations. For instance, longer blades with lower rotation speeds can distribute thrust over a larger area, reducing the likelihood of cavitation [75, 76]. Additionally, modern submarines may employ pump-jet propulsors, which encase the propeller within a nozzle, further decreasing noise emissions [72].

In the context of silent running, submarines must operate in a stealth mode to avoid detection. This state requires careful management of the propulsion system, particularly the propeller, to limit cavitation and mechanical noise [77]. Operating at lower speeds is essential, as higher speeds tend to increase cavitation and mechanical vibrations, which can compromise stealth [78]. Furthermore, contemporary submarines are equipped with advanced noise-reduction technologies, including insulation and vibration-dampening systems, to minimize onboard noise from machinery and equipment [79]. The precision engineering of propulsion components, including the propeller, is vital for reducing vibrations and noise, allowing submarines to traverse underwater quietly [80].

By optimizing propeller design and propulsion systems, submarines can achieve efficient movement while maintaining a low acoustic profile, which is particularly critical during covert operations. The integration of effective propeller designs and silent running strategies enables submarines to perform at high levels while remaining undetected by adversaries [81]. In summary, the interplay of hydrodynamic principles, innovative design, and advanced technologies in submarine propeller systems is essential for enhancing operational effectiveness and stealth capabilities.

Hydrodynamic Forces and Manoeuvrability

Submarines operate under the influence of various hydrodynamic forces that significantly affect their manoeuvrability, particularly at different depths and speeds. These forces, including heave, pitch, and yaw, dictate how submarines navigate through water, impacting their ability to dive, turn, and maintain depth. The hydrodynamic characteristics of submarines are critical for their operational effectiveness, especially in tactical scenarios where manoeuvrability is paramount.

Heave refers to the vertical movement of a submarine, while pitch describes the rotation about its lateral axis, which is essential for controlling the angle of ascent or descent. Hydroplanes, located at the bow and stern, are the primary control surfaces that manage these motions. When a submarine needs to change depth, the hydroplanes are adjusted to create lift. For instance, to dive, the hydroplanes are angled downward, facilitating a forward pitch that allows the submarine to descend. Conversely, to ascend, they are tilted upward, lifting the bow. This control is vital for maintaining the correct orientation during underwater travel, ensuring smooth vertical movement [66].

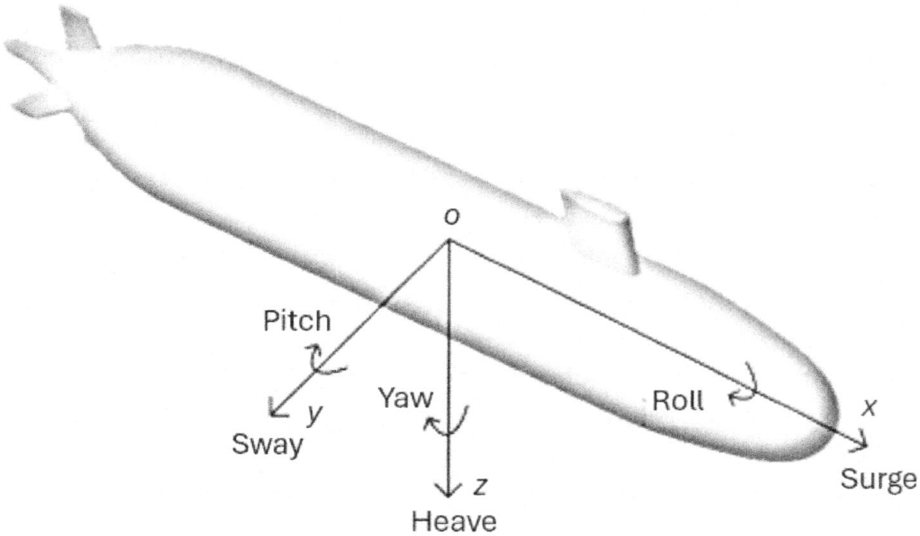

Figure 36: Submarine six degrees of freedom motion.

Yaw, on the other hand, pertains to the horizontal rotation or turning motion of the submarine, which is controlled by the rudder at the stern. The rudder's design and efficiency are crucial for enabling precise turns and direction changes, particularly during evasive manoeuvres where rapid course adjustments are necessary. Effective yaw control is essential for maintaining stability and control, especially in high-speed scenarios where quick responses are required to avoid obstacles or enemy detection [66]. At lower speeds, however, rudder adjustments must be minimal yet precise to prevent destabilization, highlighting the importance of control surface efficiency across varying operational conditions [66].

The shape of a submarine's hull, its weight distribution, and the efficiency of its control surfaces are integral to its manoeuvrability. Submarines are typically designed with a streamlined, teardrop-shaped hull to minimize hydrodynamic resistance and enhance handling capabilities. A well-balanced weight distribution contributes to stability during movement, diving, or turning. The hydroplanes and rudder serve as the primary control surfaces, and their design significantly influences the submarine's agility and responsiveness to hydrodynamic forces [66].

Manoeuvrability is also highly dependent on the speed of the submarine. At higher speeds, the hydrodynamic forces acting on the submarine increase, facilitating sharper turns and rapid dives due to enhanced lift generated by the control surfaces. This characteristic is particularly advantageous during tactical manoeuvres in combat situations where quick, responsive movements are critical [82]. Conversely, at lower speeds, such as during stealth operations, manoeuvrability can be compromised due to reduced water flow over the control surfaces. In these scenarios, submarines may rely more on buoyancy adjustments and careful control surface manipulation to maintain precise positioning and movement [66].

Wave-Making Resistance

Wave-making resistance is a significant form of drag that submarines encounter when operating on the surface. This resistance is primarily caused by the waves generated as the submarine's hull moves through the water, which can adversely affect performance metrics such as speed, fuel efficiency, and overall energy consumption. When a submarine is at the surface, part of its hull is submerged while the rest is above water, leading to the displacement of water and the creation of waves at both the bow and stern. This wave generation contributes to wave-making resistance, which adds to the overall drag opposing the submarine's motion through the water [83-85].

The magnitude of wave-making resistance is influenced by several factors, including the submarine's speed, hull shape, and the position of the sail. Research indicates that as the submarine accelerates, the waves it generates become larger, necessitating more energy to maintain forward motion [63, 83]. For instance, simulations using the Havelock source panel method have demonstrated the relationship between speed variations and wave-making resistance, emphasizing the importance of hull design in managing drag forces [86]. Additionally, variations in speed and draft can significantly affect wave-making resistance, with optimized hull designs shown to mitigate these effects [83].

To minimize wave-making resistance, submarine designers focus on adopting specific hull shapes that reduce wave formation. A streamlined hull, typically in a teardrop or cigar shape, allows water to flow more smoothly around the submarine, thereby decreasing the size and intensity of the waves generated [63]. Furthermore, the placement and design of the sail are crucial; a well-positioned sail minimizes disruption to water flow, reducing the formation of additional waves and preventing turbulent wakes that can further increase drag [86]. Experimental investigations have confirmed that methods such as the Havelock source panel method can effectively evaluate wave resistance for various submarine forms, supporting the notion that hull and sail design are pivotal in enhancing hydrodynamic efficiency [86].

While wave-making resistance is a significant factor during surfaced operations, it becomes negligible when the submarine is fully submerged. In submerged conditions, the submarine does not interact with the water surface in a manner that generates waves, leading to a drastic reduction in drag forces and enhancing energy efficiency [84, 85]. This reduction in resistance is a primary reason submarines are designed to operate predominantly underwater, where they can travel more stealthily and efficiently. The increased wave drag on the surface not only slows the vessel but also elevates energy requirements, adversely affecting operational range and endurance [84, 85].

Effectively managing wave-making resistance is essential for submarines, especially when they must cover long distances at the surface or operate at low speeds. By optimizing hull design and sail placement, designers can improve the submarine's performance and fuel efficiency, ensuring that it can maintain reasonable speeds with minimal energy expenditure [63, 86]. Wave-making resistance is a crucial consideration for submarines in surfaced conditions, and through careful design and engineering, it can be mitigated to enhance overall operational effectiveness.

Manoeuvrability and Speed Optimization

Manoeuvrability and speed optimization are critical aspects of submarine performance, both influenced significantly by hydrodynamic principles. The shape of the hull, the design of control surfaces like rudders and hydroplanes, and the efficiency of the propulsion system all play essential roles in determining how well a submarine can manoeuvre and achieve optimal speeds under various operational conditions. These factors must be carefully balanced to ensure that a submarine can carry out its missions while remaining stealthy, fuel-efficient, and agile.

Submarines need to manoeuvre effectively in both three dimensions—horizontally (yaw), vertically (pitch), and with depth control (heave). Hydrodynamic forces acting on the submarine influence all these movements, and control surfaces such as rudders, hydroplanes, and the hull shape are crucial for maintaining agility.

1. **Yaw and Horizontal Manoeuvring**: The submarine's rudder, typically located at the stern, controls its yaw, or horizontal turning motion. Hydrodynamic forces act on the rudder as water flows past it, enabling the submarine to change direction smoothly. The effectiveness of yaw control is greatly influenced by the streamlining of the hull

and the rudder's placement. A well-streamlined hull allows water to flow smoothly over the rudder, maximizing its efficiency. Submarines must be able to make precise turns, particularly in confined or shallow waters, where rapid, accurate yaw control can be crucial for evasion or stealth.

2. **Pitch and Vertical Manoeuvring**: Submarines also need to control their pitch, or vertical angle of motion. Hydroplanes, mounted either at the bow or stern, play a critical role in adjusting the submarine's pitch and depth. By changing the angle of these fins, submarines can ascend or descend smoothly while maintaining a stable course. When moving at higher speeds, hydrodynamic forces acting on the hydroplanes generate significant lift, allowing for rapid depth changes. However, at slower speeds, the submarine must rely on buoyancy control systems to manage vertical movement, as the lift generated by the hydroplanes is reduced.

3. **Heave and Depth Control**: Heave refers to the submarine's ability to move up or down in the water column. While hydroplanes assist with vertical movement, achieving neutral buoyancy is the key factor that allows submarines to maintain a steady depth. By balancing the forces of weight and buoyancy, submarines can hover at a desired depth with minimal energy expenditure. Efficient heave control is essential for staying undetected during operations, as submarines often need to remain at a specific depth for extended periods while conserving energy.

The shape of the submarine's hull is significant for reducing drag and optimizing speed. Submarines are typically designed with a teardrop or cigar-shaped hull, which minimizes resistance from water flow. This streamlined design reduces frictional drag, allowing the submarine to move more efficiently through the water. The smoother the hull and the less disturbance it creates in the surrounding water, the less energy the propulsion system needs to maintain speed.

1. **Reducing Drag for Speed**: Frictional drag occurs when water flows along the surface of the submarine's hull. The goal in submarine design is to reduce surface roughness and streamline the hull so that water flows smoothly, minimizing resistance. The pressure drag, caused by differences in water pressure at the bow and stern, also needs to be minimized. Submarines achieve this through tapered hull designs that prevent high-pressure buildup at the front and avoid turbulent wakes at the rear. A well-optimized hull can travel faster with less energy, enhancing both speed and fuel efficiency.

2. **Cavitation and Speed Constraints**: One of the challenges submarines face when optimizing for speed is cavitation—the formation of vapor bubbles around the propeller at high speeds. These bubbles collapse violently, creating noise and potentially damaging the propeller. This noise can also compromise the submarine's stealth by making it detectable by enemy sonar. To address cavitation, submarine designers focus on propeller design, shaping the blades to operate efficiently at lower speeds while reducing the likelihood of cavitation. Advanced propulsion systems also help submarines achieve high speeds while maintaining quiet operations.

There is often a trade-off between manoeuvrability and speed in submarine design. A faster submarine experiences stronger hydrodynamic forces, which can make it more difficult to manoeuvre quickly or turn sharply. This is particularly relevant during high-speed pursuits or evasive manoeuvres, where a submarine needs to balance agility with the need to move at high speeds.

1. **High-Speed Manoeuvring**: At high speeds, submarines generate more lift on their control surfaces, making depth and pitch changes easier. However, turning becomes more challenging because of the increased resistance. High-speed turns require a greater force from the rudder, which can introduce stability challenges and potentially

increase the drag. As a result, submarines must carefully balance speed with the need to retain agile manoeuvring, especially during combat or stealth operations where avoiding detection is critical.

2. **Low-Speed Manoeuvring**: At lower speeds, submarines become more manoeuvrable, but controlling depth and direction becomes more reliant on buoyancy adjustments rather than hydrodynamic forces. The effectiveness of hydroplanes and rudders diminishes at low speeds because less water flows over these surfaces to generate lift. In these scenarios, the submarine's ballast systems play a larger role in maintaining depth and ensuring precise manoeuvring.

A submarine's propulsion system, whether nuclear or diesel-electric, directly impacts its speed and manoeuvrability. Nuclear-powered submarines can maintain higher speeds for longer periods because they do not need to surface as frequently as diesel-electric submarines, which must recharge their batteries. This gives nuclear submarines a performance edge in long-distance operations and high-speed engagements.

Submarines are also equipped with speed control systems that help manage propulsion in relation to hydrodynamic forces. For instance, they may operate in "silent running" mode, where they move slowly to minimize noise from the propeller and other systems, thus enhancing their stealth. Balancing speed with stealth is a key consideration in submarine operations, especially in combat scenarios where being undetected is paramount.

In summary, submarine manoeuvrability and speed optimization are deeply interconnected with hydrodynamic forces. A well-designed submarine must balance streamlining, control surface efficiency, and propulsion to achieve both high speeds and agile manoeuvring. These factors are critical to the submarine's overall performance, influencing its ability to perform stealth operations, engage in combat, and navigate challenging underwater environments efficiently.

Submarine Resistance and Propulsion Systems

Submarine resistance refers to the various forces that oppose a submarine's forward motion through water, affecting its speed, fuel efficiency, and overall performance. The hydrodynamic design of a submarine aims to minimize these resistive forces to optimize its operational capabilities. Submarine resistance is primarily caused by drag, which is the friction between the water and the submarine's hull as well as the turbulence created in its wake. Understanding the types of resistance that submarines face is crucial for improving speed, stealth, and energy efficiency.

Types of Submarine Resistance

- **Frictional Resistance (Skin Friction)**: This type of resistance arises due to the friction between the water molecules and the surface of the submarine's hull. As the submarine moves forward, water flows over its surface, creating a layer of water molecules that stick to the hull. This results in friction that resists the submarine's motion. The magnitude of this resistance depends on the hull's surface roughness, the velocity of the submarine, and the viscosity of the water.
    - **Smooth Hulls for Reduced Friction**: Submarines are designed with smooth, streamlined hulls to minimize frictional resistance. Any roughness or irregularity on the surface increases the friction, which in turn increases the resistance, leading to higher energy consumption and reduced efficiency.
    - **Laminar vs. Turbulent Flow**: In ideal conditions, the flow of water over the hull should remain laminar, meaning smooth and orderly, which minimizes frictional drag. However, at higher speeds, the flow becomes

turbulent, creating more resistance. Submarine designers aim to delay this transition from laminar to turbulent flow for as long as possible to reduce frictional resistance.

- **Pressure (Form) Resistance**: Pressure resistance, also known as form drag, is caused by the pressure differences between the front (bow) and the rear (stern) of the submarine. As water moves around the submarine, it creates a high-pressure area at the bow and a low-pressure area at the stern. This pressure imbalance causes drag, which opposes the forward motion.
    - **Streamlined Design**: Submarines are designed with a teardrop-shaped hull to minimize pressure drag. The streamlined bow reduces the buildup of high pressure at the front, and the tapered stern helps reduce low pressure at the rear. This design allows water to flow smoothly around the submarine, minimizing turbulence and pressure resistance.
- **Wave-Making Resistance**: Wave-making resistance occurs when a submarine is moving near the surface. As the submarine moves forward, it creates waves at the interface between the water and the air. These waves absorb energy from the submarine, increasing drag and reducing speed.
    - **Surface vs. Submerged Operation**: Wave-making resistance is a significant concern for submarines when they are in surfaced condition. When submerged, the absence of surface waves eliminates this form of resistance, making the submarine more energy-efficient.
- **Eddy Resistance**: Eddy resistance is caused by the formation of vortices or eddies in the wake of the submarine. As water flows around the hull and its appendages (such as the rudders and hydroplanes), it can become turbulent and form swirling water patterns, which generate additional drag.
    - **Reducing Eddy Formation**: Submarines are designed with rounded edges and smooth transitions between the hull and the control surfaces to minimize the creation of eddies. The goal is to keep the water flow as streamlined as possible to reduce the drag caused by turbulence.
- **Induced Drag**: Induced drag is a result of the lift generated by the submarine's control surfaces, such as hydroplanes and rudders. These surfaces create lift to control the depth and direction of the submarine, but the lift also produces drag as a byproduct. The amount of induced drag depends on the angle of attack of the control surfaces and the speed of the submarine.
    - **Minimizing Induced Drag**: Submarines are designed to operate with minimal induced drag by optimizing the size and shape of the control surfaces. The goal is to generate the required lift for manoeuvring while keeping the drag as low as possible.
- **Cavitation Resistance**: Cavitation occurs when water pressure drops to such an extent that vapor bubbles form around the propeller blades. These bubbles collapse violently, creating noise and vibration, which not only affects propulsion efficiency but also increases drag and compromises the submarine's stealth.
    - **Propeller Design**: To minimize cavitation, submarine propellers are designed with special blade shapes and low rotation speeds to prevent the formation of vapor bubbles. Cavitation resistance is essential for maintaining high-speed efficiency while reducing noise levels.

Factors Affecting Submarine Resistance

Submarine resistance is influenced by several critical factors, including hull shape, surface roughness, speed, depth of operation, and propulsion system design. Each of these elements plays a significant role in determining the overall hydrodynamic performance of a submarine.

Hull Shape: The shape of a submarine's hull is paramount in minimizing resistance. Submarines are typically designed with streamlined forms, such as teardrop or cigar shapes, which significantly reduce both frictional and pressure resistance. Research indicates that an optimized hull shape can lead to a substantial reduction in total resistance, encompassing both viscous and wave-making components [63, 87]. The design of the hull must balance internal volume with hydrodynamic efficiency, as a well-optimized hull can enhance performance and operational capabilities [24, 61]. Additionally, studies have shown that specific hull configurations can lead to improved resistance characteristics, particularly when considering the interaction of the hull with water at various speeds and depths [85, 86].

Surface Roughness: The smoothness of the submarine's hull surface is another critical factor affecting resistance. Surface imperfections, such as roughness or damage, can increase frictional resistance, thereby reducing the submarine's efficiency [88]. To combat this issue, submarines are often treated with specialized coatings that maintain a smooth surface, which is essential for minimizing drag and enhancing hydrodynamic performance [88, 89]. The impact of biofouling on surface roughness has also been studied, revealing that fouling can significantly increase the power required for propulsion, further emphasizing the importance of maintaining a smooth hull surface [88].

Speed: The relationship between speed and resistance is complex; as a submarine increases its speed, both frictional and pressure drag rise. This necessitates a careful balance between the desire for speed and the associated increase in resistance [90]. Submarines often operate in "silent running" modes, where they travel at lower speeds to minimize drag and noise, thereby enhancing stealth capabilities [90]. The design of the propulsion system must therefore accommodate these operational modes, ensuring efficiency at varying speeds [90].

Depth of Operation: The operational depth of a submarine significantly affects the resistance it encounters. Near the surface, wave-making resistance becomes a major concern due to the generation of surface waves, which adds to the overall drag [85, 86]. Conversely, when submerged at greater depths, wave-making resistance is eliminated, allowing submarines to operate more efficiently, as frictional and pressure drag become the primary forces acting upon them [32]. Research has shown that submarines can achieve better energy efficiency at deeper depths, where the absence of surface waves reduces overall resistance [32].

Propulsion System: The design and efficiency of a submarine's propulsion system, particularly the propeller, are crucial in determining resistance. A well-designed propeller can generate thrust efficiently while minimizing cavitation and noise, which are critical for stealth operations [91]. The interaction between the propeller and the hull also affects the overall resistance; thus, optimizing the propeller design is essential for enhancing both performance and energy efficiency during underwater operations [91]. Studies have indicated that the structural and acoustic responses of the hull due to propeller forces must be carefully considered to minimize resistance and noise [91, 92].

The interplay of hull shape, surface roughness, speed, depth of operation, and propulsion system design collectively influences the resistance faced by submarines. Each factor must be meticulously optimized to enhance the submarine's efficiency, stealth, and overall performance in various operational conditions.

Wave-Making Resistance

When a submarine descends to a depth that is four or more times its maximum diameter or hull depth, the impact of surface disturbances caused by its forward motion becomes minimal. This significantly reduces wave-making resistance, one of the major factors affecting the efficiency and speed of surface vessels. By diving deeper, submarines essentially eliminate this form of resistance, which is especially beneficial when operating at high speeds. However, there is a trade-off: fully submerging the submarine increases the wetted surface area of the hull, which in turn leads to an increase in frictional resistance due to the continuous contact with the water. Despite this increase in friction, the overall benefit of eliminating wave-making resistance often outweighs the downside, making submerged travel more efficient, especially at higher velocities [93].

Older submersibles were not as streamlined as modern submarines. These older designs included various equipment and fittings above the waterline, such as flat decks, rails, anchors, capstans, and chocks. These structures, necessary for surface operations, presented significant challenges for underwater performance because they increased drag. The irregular shapes and protrusions made it difficult to achieve a low-resistance profile when submerged. Streamlining these elements was not always feasible, which added to the overall drag and reduced the hydrodynamic efficiency of the craft [93].

Modern submarines, designed primarily for extended underwater operations, particularly under nuclear power, have eliminated most of these surface irregularities. These true submarines are built with minimal external fittings and are designed to operate almost entirely below the surface, where their streamlined shapes allow for greater speed and efficiency. In addition, advancements in submarine design have led to changes in the length-to-diameter ratio. By optimizing this ratio, designers have been able to reduce frictional resistance, especially at higher speeds. A lower length-to-diameter ratio helps balance internal volume needs while minimizing drag, allowing submarines to travel faster and more efficiently underwater [93].

The elimination of wave-making resistance at deeper depths and the streamlined designs of modern submarines allow them to operate with greater hydrodynamic efficiency. While fully submerging the craft increases frictional resistance, the overall gains in performance and speed, especially at high velocities, far outweigh this downside. Modern submarines have evolved to eliminate surface irregularities, allowing them to be highly optimized for underwater travel.

Reducing Resistance in Submarine Design

Reducing resistance in submarine design is a multifaceted challenge that encompasses various engineering principles and innovative technologies. The primary strategies for achieving this goal include streamlining the hull, optimizing control surfaces, advancing propeller design, and employing specialized hull coatings.

Streamlining the Hull: One of the most effective methods for reducing hydrodynamic resistance in submarines is the design of a streamlined hull. A streamlined shape minimizes both frictional and pressure drag by ensuring that water flows smoothly over the submarine's surface, thereby reducing turbulence and the formation of eddies. This design approach is supported by computational fluid dynamics (CFD) studies, which demonstrate that a well-optimized hull can significantly decrease resistance and improve overall efficiency [64]. Furthermore, streamlined hulls help mitigate cavitation, a phenomenon that occurs when vapor bubbles form due to low pressure around the hull, leading to increased noise and resistance [24, 60]. The reduction of pressure differentials between the bow and stern is crucial for maintaining smooth water flow, which is essential for stealth operations and energy efficiency [61].

Optimizing Control Surfaces: The control surfaces of a submarine, including hydroplanes and rudders, play a vital role in manoeuvrability but can also contribute to hydrodynamic resistance if not properly designed. Research indicates that optimizing the size, shape, and placement of these surfaces can significantly reduce induced drag while maintaining necessary lift and directional control [80, 94]. Efficiently designed control surfaces allow submarines to adjust depth and direction with minimal resistance, which is critical for maintaining speed and stealth during operations [95]. Studies have shown that the hydrodynamic performance of submarines can be enhanced by carefully considering the interaction between the hull and control surfaces, leading to improved manoeuvrability and reduced energy consumption [24, 96].

Advanced Propeller Design: The design of the submarine's propeller is another critical factor influencing resistance and operational efficiency. Propellers must be engineered to minimize cavitation, which can occur when water pressure around the blades drops, leading to noise and reduced efficiency [94]. The optimization of blade shape and pitch angle is essential for achieving efficient thrust across various speeds, thereby enhancing the submarine's performance while maintaining a low acoustic signature [80, 94]. Experimental investigations have shown that innovative propeller designs can lead to significant improvements in propulsion efficiency and noise reduction, which are vital for stealth operations [94].

Hull Coatings: The application of specialized hull coatings is a modern approach to reducing frictional resistance in submarines. These coatings create a smoother surface that minimizes drag as water flows over the hull, thereby enhancing operational efficiency and stealth [97, 98]. Advanced antifouling coatings are particularly important, as they prevent the accumulation of marine organisms that can increase drag and fuel consumption [98, 99]. Research has demonstrated that the use of such coatings can lead to substantial reductions in energy requirements and improvements in stealth capabilities, making submarines more effective in various operational contexts [97, 99].

Propulsion Systems

Submarine propulsion is a complex process that enables submarines to navigate underwater while maintaining efficiency, speed, and stealth. The propulsion system converts energy into forward motion and is essential to a submarine's performance. Submarines typically use either nuclear propulsion or diesel-electric propulsion, depending on their design and operational requirements. Let's break down how these systems work in detail:

1. Diesel-Electric Propulsion

This system is commonly used in conventional submarines and operates with two main power sources: diesel engines and electric batteries.

- **Diesel Engines**: When the submarine is on the surface or at periscope depth, the diesel engines are used to power the submarine's propellers directly or to recharge its batteries. The diesel engines draw in air through a snorkel or when surfaced, combust fuel, and generate power. However, diesel engines cannot be used when the submarine is fully submerged because they require oxygen to operate.

Figure 37: U-995 diesel engine room. Arjun Sarup, CC BY-SA 4.0, via Wikimedia Commons.

- **Electric Batteries**: Once submerged, diesel engines are turned off, and the submarine relies on large batteries to power its electric propulsion motors. These batteries are charged by the diesel engines when the submarine is surfaced or at periscope depth. The electric motors, which drive the propellers, enable silent underwater operation, providing stealth advantages. The batteries provide limited operational time underwater, so submarines must eventually resurface to recharge.

- **Propeller**: The propeller (or screw) is driven by the electric motor when submerged. Diesel-electric submarines are quieter than nuclear submarines due to their simpler mechanical systems when running on battery power. However, their submerged endurance is limited by the battery capacity, so they periodically need to resurface to recharge the batteries.

Figure 38: Diesel engine room of submarine Wilhelm Bauer (ex U 2540). BigBen212, CC BY-SA 3.0, via Wikimedia Commons.

2. Nuclear Propulsion

Nuclear-powered submarines use a nuclear reactor to generate energy, which provides several advantages over diesel-electric propulsion. The key difference is that a nuclear reactor can operate continuously without the need for surfacing or oxygen.

- **Nuclear Reactor**: The heart of a nuclear submarine's propulsion system is its nuclear reactor. This reactor uses nuclear fission (splitting of uranium or plutonium atoms) to generate heat. The heat is used to create steam, which drives turbines connected to a generator. The turbines can either directly drive the propeller shaft or generate electricity to power electric motors connected to the propeller.

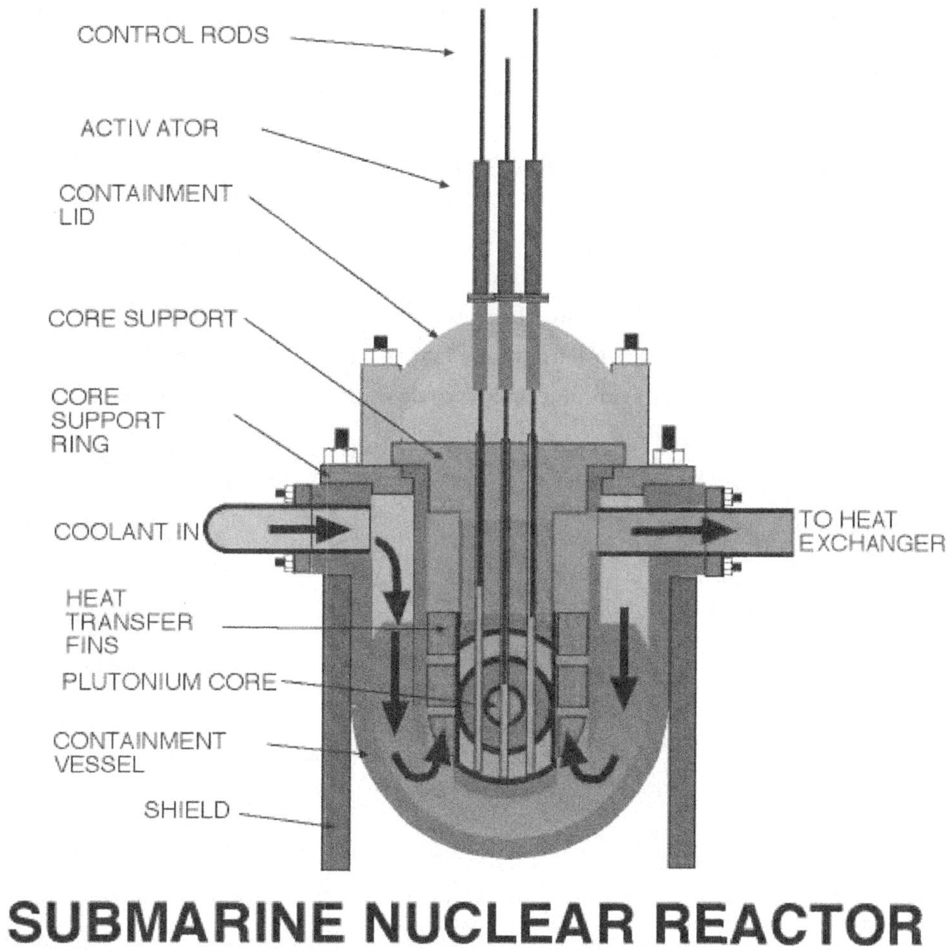

Figure 39: Nuclear Submarine Reactor diagram Author: Webber. No machine-readable author provided. Webber assumed (based on copyright claims)., Public domain, via Wikimedia Commons.

- **Steam Turbines**: In most nuclear-powered submarines, the steam produced by the reactor turns the turbine generators. These turbines are connected either directly to the propulsion shaft or indirectly through electric motors that drive the propellers. This system allows the submarine to operate for extended periods without surfacing, as the nuclear reactor provides nearly limitless power.
- **Electric Motor**: Some nuclear submarines also use electric motors driven by power generated from the turbines. These motors are highly efficient and allow for more control over the speed and stealth of the submarine.
- **Submerged Endurance**: Unlike diesel-electric submarines, nuclear-powered submarines can stay submerged for months at a time, as they don't rely on external oxygen for power generation. The submarine only needs to surface for crew supplies like food, not for fuel or air.

3. Propeller (Screw) and Silent Running

Regardless of the power source, the propeller (or screw) plays a critical role in submarine propulsion. The propeller converts the rotational power generated by the engine or electric motor into forward thrust, allowing the submarine to move through the water. However, the design of the propeller must also consider stealth because noise from the propeller can be detected by enemy sonar systems.

- **Propeller Design**: Submarine propellers are specially designed to reduce cavitation, a phenomenon where vapor bubbles form around the propeller blades at high speeds. When these bubbles collapse, they generate noise that can reveal the submarine's position. Propellers with fewer, larger blades and lower rotational speeds help reduce cavitation and noise.

- **Silent Running**: To minimize noise during operations, especially in stealth missions, submarines can enter a "silent running" mode. This involves reducing the speed of the propeller and other onboard machinery to a minimum, which reduces noise and makes it harder for enemy sonar to detect the submarine. Nuclear submarines, in particular, are optimized for silent running since they do not need to surface frequently.

Figure 40: HDMS Sælen (S323) propeller. AndrejS.K, CC BY-SA 4.0, via Wikimedia Commons.

4. Auxiliary Systems and Manoeuvring

Submarine propulsion systems also include auxiliary systems and control surfaces that assist in manoeuvring.

- **Hydroplanes (Diving Planes)**: Submarines use hydroplanes, which are control surfaces mounted at the bow and stern, to control their depth and angle while moving. By adjusting the angle of these planes, the crew can change the submarine's pitch and make it dive or surface.

- **Rudder**: A rudder, typically located at the stern, helps control the yaw (side-to-side movement) of the submarine, allowing it to steer and make precise turns while submerged.

Figure 41: Rudder of the french submarine Flore (S645). Photograph by Rama, Wikimedia Commons, Cc-by-sa-2.0-fr, CC BY-SA 2.0 FR, via Wikimedia Commons.

5. Energy Efficiency and Stealth

A critical consideration in submarine propulsion is balancing energy efficiency with stealth. In diesel-electric submarines, this involves conserving battery power by minimizing energy use when submerged. In nuclear-powered submarines, it means operating the propulsion system at low speeds and using quieting technologies to avoid detection.

Designing an Efficient, Reliable, and Stealthy Submarine Propulsion System

Designing a submarine propulsion system involves carefully balancing several factors to ensure the vessel performs efficiently, reliably, and with minimal detectability. Here are the key considerations for creating such a propulsion system [100]:

1. Type of Propulsion The first decision in designing a submarine propulsion system is selecting the most suitable type of propulsion based on the submarine's mission and technical specifications. There are several options, each with its own strengths and limitations:

- **Diesel-Electric Propulsion**: Diesel-electric submarines are cost-effective and simple to operate. However, they require surfacing or snorkelling to recharge their batteries, which can compromise stealth by exposing the submarine to detection.

- **Nuclear Propulsion**: This type of propulsion offers superior endurance and speed, allowing submarines to operate submerged for long periods without refuelling. However, nuclear submarines are expensive, complex, and require extensive maintenance.

- **Air-Independent Propulsion (AIP)**: AIP systems, which use alternative oxygen sources like hydrogen or liquid oxygen, extend the submerged time of non-nuclear submarines, but they have limitations in fuel capacity and power output compared to nuclear submarines.

- **Hybrid Propulsion**: Combining different propulsion methods (e.g., diesel-electric with AIP) provides flexibility and optimizes performance, but these systems are more complex and costly to design and maintain.

Each type of propulsion system must be carefully evaluated for efficiency, reliability, cost, and the stealth requirements of the submarine's mission.

2. Design of the Propeller The propeller is central to the propulsion system, providing the thrust that moves the submarine while also being a primary source of noise. Designing an efficient, low-noise propeller requires attention to several key factors:

- **Shape, Size, and Pitch**: The geometry of the propeller blades, including their size and angle (pitch), must be optimized to minimize cavitation. Cavitation occurs when bubbles form around the propeller blades due to pressure differences, causing noise and potential damage. Minimizing cavitation is crucial for maintaining stealth.

- **Number of Blades and Tip Clearance**: The number of blades and the distance between the blade tips and the hull (tip clearance) must be chosen to ensure the propeller operates efficiently and quietly.

- **Advanced Propeller Technologies**: In some modern designs, technologies such as pump-jet, contra-rotating, or rim-driven propellers are used to further reduce noise and improve efficiency. These systems generate thrust with minimal cavitation, making them ideal for stealth operations.

- **Bladeless Propeller Systems**: Emerging technologies, inspired by bladeless fans like the Dyson fan, could potentially revolutionize submarine propulsion by eliminating cavitation and enhancing stealth. This concept is still under development but holds promise for future submarine design.

3. Integration of Components The next step in propulsion system design is integrating all the components, such as the engine, generator, motor, shaft, gearbox, and propeller. This process involves balancing power and torque requirements while minimizing losses from friction, heat, and vibration. Key considerations include:

- **Noise and Vibration Isolation**: To maintain stealth, the components must be mounted using shock-absorbing materials, rubber bearings, and acoustic insulation to reduce noise and vibrations.

- **Space and Weight Constraints**: Submarines have limited space and strict weight requirements. The propulsion system must fit within these constraints without compromising performance or safety.

- **Maintainability**: Components must be arranged in a way that allows easy access for routine maintenance and repairs, ensuring reliability over long missions.

4. Testing and Validation Once the propulsion system is designed and assembled, it undergoes rigorous testing and validation. This includes both laboratory testing and real-world trials to ensure that the system meets all performance and stealth requirements:

- **Performance Testing**: The propulsion system is tested for speed, power output, fuel efficiency, and endurance to ensure that it operates within the desired parameters.

- **Stealth Testing**: Noise, vibration, and other signatures (magnetic, thermal, etc.) are measured to confirm that the system operates quietly and discreetly.

- **Reliability Testing**: Stress tests and endurance trials help identify potential weaknesses or failures, such as leaks, wear, or overheating, allowing designers to resolve these issues before the submarine enters service.

5. Adaptation and Innovation Based on the results of the testing and validation phase, designers may need to make adjustments or introduce new innovations to optimize the propulsion system further:

- **Modifications**: Changes may be made to individual components, such as improving the propeller design or upgrading materials to increase durability and performance.

- **New Technologies**: Advances in marine engineering, such as improved noise-reduction technologies or more efficient energy storage systems, can be incorporated to enhance the propulsion system.

The propulsion system must be continually adapted to meet changing mission requirements, technological advancements, and environmental standards.

6. Future Trends and Challenges Looking ahead, the submarine propulsion system must be designed to meet future challenges and opportunities:

- **Increased Endurance and Range**: Future submarines will require propulsion systems that allow them to operate for longer periods without resupply or refuelling.

- **Greater Stealth and Agility**: As detection technologies improve, submarines must become even quieter and more agile, which requires ongoing improvements in propulsion system design.

- **Environmental Impact**: Submarine propulsion systems will need to comply with stricter environmental regulations, reducing emissions and minimizing their impact on marine ecosystems.

- **Technological Threats**: Submarines must also be equipped to handle new threats, including cyberattacks and advanced anti-submarine warfare systems. The propulsion system must be resilient and adaptable to evolving operational environments.

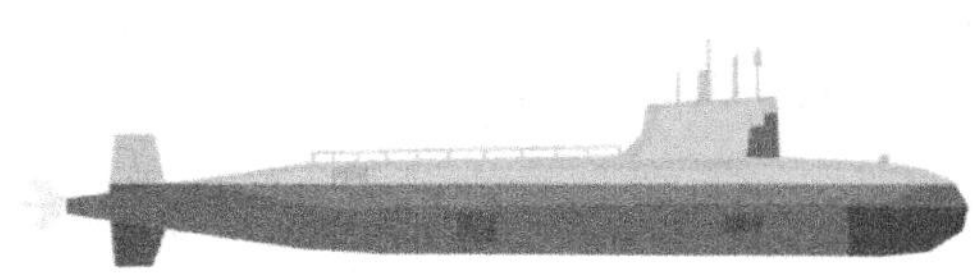

Chapter 4

Propulsion Systems for Submarines

Diesel-Electric Propulsion

A diesel submarine is a prime example of a hybrid vehicle, utilizing both diesel engines and electric power for efficient operation. The design of a typical diesel submarine includes two or more diesel engines, which provide the necessary power for different functions, making the submarine versatile in its capabilities. These engines can either directly drive the propellers or power generators that recharge a large battery bank. This flexibility allows for various combinations of operations, where one engine might drive a propeller while the other is dedicated to charging the batteries, ensuring the submarine is prepared for different mission requirements.

When operating on diesel power, the submarine must either surface or stay just below the water's surface using a snorkel. This is because the diesel engines require air for combustion, which can only be drawn from the atmosphere. While on the surface or using the snorkel, the diesel engines run efficiently, charging the batteries and powering the propellers as needed. Once the battery bank is fully charged, the submarine can dive and operate underwater for extended periods without needing to surface.

Underwater, the submarine relies solely on its batteries, which power electric motors that drive the propellers. This mode of operation is essential for submerging since the diesel engines cannot function without an external air supply. Running on batteries allows the submarine to navigate and remain submerged in a stealthy, silent manner, which is critical for avoiding detection by enemy vessels. The submarine's operational range and duration underwater depend largely on the battery capacity and how much charge has been stored during diesel engine operation.

This hybrid diesel-electric system provides the submarine with a balance between power, endurance, and stealth. The diesel engines enable long-range surface cruising and battery recharging, while the electric motors allow the submarine to operate quietly underwater, making it a highly adaptable vessel for various naval missions. However, the need to surface or use a snorkel for air limits the diesel submarine's ability to remain submerged for prolonged periods compared to nuclear-powered submarines.

Diesel-electric submarines operate using a combination of diesel engines and electricity, as their name suggests. These submarines are equipped with large networks of batteries that store electricity generated by diesel engines, which act as the primary power source when the submarine is on the surface. To recharge the batteries, diesel-electric submarines must rise to the surface or travel just below the surface with a snorkel. The snorkel allows the diesel engine exhaust and intake to stay above water, enabling them to run without exposing the vessel. Once the batteries are fully charged, the submarine submerges and operates silently on battery power, with the diesel engines shut down to minimize noise. These submarines can stay submerged for several days until the batteries drain, after which they must surface again to recharge [101].

One of the key advantages of diesel-electric submarines is their ability to produce significantly less noise compared to their nuclear counterparts. Since diesel engines can be completely shut down when submerged, these submarines can run on batteries, making them almost undetectable. This makes them especially effective in shallow waters where stealth is paramount [101].

Diesel-electric submarines are also smaller and more agile than nuclear-powered submarines, giving them an edge in coastal or littoral regions. Their smaller size allows them to manoeuvre in confined spaces and stay hidden in shallow waters. Another advantage is the ability to switch off and restart the engines as needed, unlike nuclear submarines, which must keep their reactors running continuously. This flexibility in operation contributes to their stealth and operational agility [101].

Additionally, diesel-electric submarines are more affordable to procure, train crews for, and operate than nuclear-powered submarines. Smaller navies can afford to purchase and maintain diesel-electric submarines, making them an attractive option for countries that cannot invest in more expensive nuclear submarines. Modern diesel-electric submarines, often referred to as Sub Surface Killers (SSKs), have advanced significantly with sound suppression systems, land-attack capabilities, and the ability to remain nearly undetectable when their generators are turned off. For instance, Air Independent Propulsion Systems (AIP) have significantly extended the time diesel-electric submarines can stay submerged, from one or two days to nearly a week [101].

Modern examples like the Type 212, Improved Kilo, Scorpene, and Soryu class submarines showcase advanced diesel-electric submarine technology. The Type 212, known for being small and quiet, operates with a crew of around 30 and is highly effective in stealth missions [101]. The Improved Kilo class has additional firepower and can launch anti-ship missiles with ranges up to 300 kilometres. The Soryu class, though larger, offers better endurance and operational flexibility.

Figure 42: Malaysia's first Scorpene-class diesel-electric submarine docked at its Naval base in Port Klang on the outskirts of Kuala Lumpur on September 3, 2009. Mak Hon Keong, CC BY-SA 3.0, via Wikimedia Commons.

Despite their advantages, diesel-electric submarines have limitations. One major drawback is their top speed when submerged, which is typically around 15-20 knots. This relatively slow speed makes it difficult for them to outrun torpedoes or anti-submarine warfare (ASW) ships. Additionally, diesel-electric submarines have a limited diving depth, usually between 150-300 metres, making them more vulnerable in deeper waters where they cannot use depth to evade detection or attacks [101].

The underwater endurance of diesel-electric submarines is also a limitation. Without an AIP system, most diesel-electric submarines can only remain submerged for a few hours. Even with AIP, endurance extends to about a week, which is much shorter than the nearly unlimited underwater endurance of nuclear submarines.

Moreover, diesel-electric submarines carry a smaller weapons payload compared to nuclear submarines. They typically carry around 12-20 torpedoes or missiles, limiting their firepower. There are exceptions, like the Soryu class, which can carry more, but generally, these submarines are more constrained in terms of their armament [101].

Figure 43: JMSDF Sōryū class submarine. □□□, CC BY 4.0, via Wikimedia Commons.

Lastly, the internal space of diesel-electric submarines is often cramped, making long deployments uncomfortable for the crew. This can be a significant disadvantage during extended missions, especially compared to the more spacious nuclear submarines designed for long-term deployments.

In conclusion, diesel-electric submarines offer significant advantages in terms of stealth, agility, and cost-effectiveness, making them ideal for certain naval missions, especially in shallow waters. However, they face limitations in speed, diving depth, endurance, and firepower, which must be weighed against their advantages.

Development

The development of diesel engines has a rich history that dates back to the mid-19th century when the concept of internal combustion engines began to take shape. In 1862, French engineer Beau de Roches introduced the theoretical idea of a reciprocating internal combustion engine, which laid the groundwork for future innovations. A few years later, the German engineer Otto took these ideas and successfully implemented them in a practical engine that used a 4-stroke cycle with gas as the fuel. The 4-stroke cycle became known as the Otto cycle and set the stage for further advancements in engine technology [102].

In 1872, American engineer George Brayton made significant progress by introducing a new method of fuel injection, which involved controlled injection to prolong the combustion phase and achieve more power per unit of fuel. However, the lack of effective compression of the fuel mixture limited the efficiency gains. Subsequent improvements in engine design, such as the Hornsby-Ackroyd engine in England, utilized liquid fuels derived from crude oil. This engine incorporated the Brayton principle of controlled fuel injection and pre-compression of air, resulting in improved performance. The introduction of mechanical or solid injection during this time made it one of the earliest examples of modern diesel technology [102].

The major breakthrough came in 1893 when Bavarian scientist Dr. Rudolf Diesel patented his design for a new type of internal combustion engine that would operate on a completely different thermodynamic principle. His concept involved using only air in the cylinder during the intake stroke, compressing it to a high temperature, and then injecting fuel at a controlled rate during the combustion phase. This method was intended to keep the combustion temperature constant. The diesel engine also introduced air injection to facilitate fuel delivery against high compression pressures. Although Diesel's initial experiments with coal dust fuel failed due to explosions, his theoretical framework laid the foundation for what is now known as the diesel cycle. Subsequent modifications by various experimenters, particularly at MAN (Maschinenfabrik Augsburg-Nürnberg) in Germany, led to practical diesel engines using liquid fuels with constant pressure combustion phases, making them suitable for commercial production [102].

Submarine Engine Development

The application of internal combustion engines in submarines began with gasoline engines. The first U.S. submarines used 45-horsepower, 2-cylinder gasoline engines from the Otto Company, while the British employed 12- and 16-cylinder gasoline engines. The hazards associated with gasoline, including storage difficulties and high explosion risks, soon became apparent. Additionally, carbon monoxide fumes posed significant health risks to crews. This led to a shift towards diesel engines, which were safer and offered greater fuel efficiency [102].

MAN was one of the early pioneers in adapting diesel technology for submarines, initially experimenting with 2-stroke diesel engines. However, the technological limitations of the time, particularly in metallurgy, posed challenges for these engines, as the materials could not withstand the high heat and stress. MAN then shifted to developing 4-stroke cycle diesel engines, achieving partial success with a 1000-horsepower engine. By 1914, MAN had improved its design to create the SV45/42, a 1200-horsepower engine used extensively in German submarines during World War I. The U.S. Navy adopted these engines for use in its early S-class submarines, with the New York Navy Yard producing a similar design [102].

The Electric Boat Company, which had acquired licensing rights from MAN in the U.S., developed the NELSECO engine before and during World War I. NELSECO engines became the mainstay in U.S. submarines up until the 1930s. However, most naval powers, except Great Britain, were using 4-stroke diesel engines for submarines until around 1930. The U.S. Navy started exploring 2-stroke engines, equipping several submarines with Busch-Sulzer engines, and eventually adopted 2-stroke engines for their efficiency and power [102].

Transition to Mechanical Injection

Prior to 1929, all U.S. submarine engines utilized air injection for fuel delivery. The introduction of mechanical or solid injection on MAN engines brought several advantages, such as simplifying engine design, reducing weight by eliminating the air compressor, and improving fuel efficiency. The benefits also included a shorter engine length, better load distribution, and increased reliability, which collectively led to reduced maintenance requirements [102].

Fleet-Type Submarines and New Engine Designs

The development of fleet-type submarines necessitated more powerful engines. Several engine types emerged as suitable candidates, including the Winton V-type (later known as the General Motors engine), the Fairbanks-Morse opposed-piston engine, and the Hooven-Owen-Rentschler (HOR) double-acting engine. Ultimately, the General Motors and Fairbanks-Morse engines became the standard in U.S. submarines, providing the necessary power while maintaining efficiency. By standardizing on these two designs, the U.S. Navy was able to mass-produce engines with minimal delays [102].

Modern Submarine Engines

Today, the General Motors Corporation manufactures 16-cylinder, single-acting diesel engines rated at 1600 brake horsepower (bhp) for main engine installations, along with 8-cylinder engines for auxiliary uses. Fairbanks Morse continues to produce 9- and 10-cylinder opposed-piston engines, also rated at 1600 bhp for main installations, and 7-cylinder engines for auxiliary roles. These modern engines, with advanced materials and engineering techniques, offer an impressive combination of power, efficiency, and reliability, weighing as little as 15 to 20 pounds per bhp, including auxiliary equipment [102].

The history of diesel engine development, particularly in the context of submarine propulsion, demonstrates a continuous evolution driven by the need for safer, more efficient, and powerful engines. Technological advancements in metallurgy, fuel injection methods, and engine design have significantly improved the capabilities of diesel engines, enabling submarines to operate more effectively in both wartime and peacetime applications [102].

Operational Requirements

The unique operational requirements of submarines significantly influence the design and characteristics of their engines. As submarines operate both on the surface and underwater, there are specific constraints related to their size, hull design, shape, and total weight that must be taken into account when designing the propulsion system. These limitations affect the size, location, and weight of the engines, which must be proportionally balanced with the vessel's displacement and power requirements.

Influence of Hull Characteristics on Engine Design

Submarine hulls have a streamlined design to minimize resistance underwater, resulting in restrictions on engine size and the placement of engine compartments. The available space within the hull is limited, so the engine must be compact enough to fit within these constraints without compromising power output. Additionally, the weight of the engine needs to be balanced against the overall weight and displacement of the submarine to maintain proper buoyancy and stability. If the engine is too heavy, it could adversely affect the submarine's ability to dive, surface, and manoeuvre underwater.

In early submarine designs, the engines were directly connected to the propeller shafting through a configuration known as direct drive. However, this setup quickly revealed operational challenges because the hull design dictated a fixed angle for the propeller shafting, thereby limiting the positioning of the engines. Furthermore, the most efficient operating speeds of the engines did not always match the most efficient speeds for the propellers. This mismatch caused issues with vibrations and resonance (known as critical speeds or synchronous torsional vibrations), which could transfer through the direct drive system and cause mechanical stress or damage.

Transition to Diesel-Electric Drive Systems

To address these issues, submarine designers began exploring ways to decouple the engines from the propeller shafting to eliminate mechanical unity and allow for different rotational speeds for the engines and propellers. After experimenting with various types of drive systems, the diesel-electric drive emerged as the most practical solution. In a diesel-electric configuration, the engines drive generators that supply electrical power to electric motors, which, in turn, drive the propeller shafting. This setup allows for flexibility in engine speed and propeller operation, as the engines are no longer mechanically connected to the propellers.

With diesel-electric propulsion, engine vibrations are not transmitted to the propeller shafting, nor are the stresses encountered by the propellers transferred back to the engines. This separation improves the operational stability and reduces maintenance requirements. Additionally, diesel-electric systems provide greater freedom in engine placement within the submarine, as the engines do not need to be aligned with the propeller shafting. This flexibility is crucial for optimizing space utilization and weight distribution.

Key Requirements for Submarine Diesel Engines

The design of submarine diesel engines is driven by several critical requirements to ensure efficiency, reliability, and performance under demanding conditions. These requirements include [102]:

1. **Maximum Power with Minimum Weight and Space:** Submarine engines must deliver high power output

while occupying minimal space and adding as little weight as possible. This is essential for maintaining the vessel's operational efficiency and ensuring that the engine fits within the restricted confines of the hull.

2. **Capability to Develop Occasional Overload:** The engine should be able to produce more than its rated full load power when needed, such as in emergency situations or during rapid manoeuvring. This flexibility ensures that the submarine can respond quickly to changing operational demands.

3. **Continuous Operation Near Full Load:** Submarine engines often run at or near full load for extended periods. They must be capable of maintaining consistent performance under such conditions without overheating or experiencing excessive wear.

4. **Low Fuel Consumption:** Efficiency is critical for submarines, especially during extended missions where fuel supplies are limited. The engine must consume as little fuel as possible while generating the required power, ensuring the submarine's operational range is maximized.

5. **Minimal Lubricating Oil Consumption:** Submarines carry a limited supply of lubricating oil, so the engine must use it sparingly. This helps extend the time between necessary maintenance and reduces the need for onboard storage of additional oil.

6. **Ease of Maintenance:** All engine components should be easily accessible for maintenance and replacement. Since space inside a submarine is limited, engine design must allow for quick servicing to minimize downtime and ensure the submarine remains operational.

7. **Perfect Balance of Forces:** To reduce vibrations and mechanical stresses, the engine must be balanced with respect to primary and secondary forces and couples. This balance enhances engine longevity and helps minimize noise, which is crucial for submarine stealth.

8. **Elimination of Critical Speeds:** Critical speeds within the engine's operating range must be avoided to prevent harmful vibrations and resonance. By designing the engine to operate smoothly across its speed range, the risk of mechanical failure is reduced.

The adoption of diesel-electric propulsion systems solved many of the challenges posed by direct drive systems. With no direct mechanical connection between the engine and the propeller shafting, designers could optimize both engine and propeller speeds independently. This setup allows engines to operate at their most efficient speeds for generating electricity, while electric motors can drive the propellers at speeds that suit the desired vessel speed and manoeuvrability.

Diesel-electric systems also allow for greater latitude in engine placement, enabling better utilization of available space within the submarine. This flexibility contributes to improved weight distribution and stability, helping designers meet the stringent requirements for submarine operations.

The unique requirements of submarines—such as the need for compact size, efficient power generation, and reduced vibrations—greatly influence engine design. Diesel-electric propulsion has emerged as a practical solution, offering numerous advantages over traditional direct drive systems. By addressing the challenges associated with engine placement, power transmission, and vibration control, diesel-electric systems enable submarines to achieve higher levels of performance, efficiency, and stealth. The continued evolution of submarine diesel engines is driven by the need to meet these demanding requirements while adapting to new technologies and changing operational environments [102].

Operation

In diesel-electric submarines, the engine gets its air from the outside atmosphere, which is crucial for the combustion process. This can only happen when the submarine is at the surface or in snorkelling mode. During snorkelling, the submarine stays just a few meters below the surface, using a snorkel mast that extends above the waterline to draw in air. The air flows from the snorkel down into the engine room, providing the necessary oxygen for the diesel engine to operate. However, to minimize the risk of detection, submarines try to keep snorkelling sessions as brief as possible. Exhaust gases are expelled underwater through a pressurized exhaust system. For each meter of water above the exhaust outlet, the engine must generate an additional 100 mbar of pressure to prevent water from entering the engine. This is accomplished using specialized air systems designed for the submarine environment [103].

Submarine Propulsion

Diesel-electric propulsion in submarines relies on a diesel engine to charge the battery bank. The diesel generator set acts as a charger, supplying electric current to the batteries, which in turn power the electric propulsion motor that drives the propeller. The diesel engine operates only while the submarine is on the surface or snorkelling, as it requires air for combustion. When submerged, the electric motor uses the stored battery power for propulsion, allowing the submarine to operate silently. Besides diesel-electric configurations, there are also submarines powered by nuclear reactors, which provide a more continuous power source [103].

Fuel Cell Systems for Air-Independent Propulsion (AIP)

For extended underwater operations, some submarines are equipped with Air-Independent Propulsion (AIP) systems that use fuel cells. ThyssenKrupp Marine Systems is a leading supplier of such systems, which have been successfully installed on numerous submarines. The fuel cell system generates electricity through a chemical reaction between hydrogen and liquid oxygen carried on board. This process is essentially the reverse of electrolysis, resulting in electricity as well as pure water as a byproduct. The electricity generated can power the submarine's systems and propulsion motor, allowing the vessel to remain submerged for longer durations without needing to surface for air [103].

Submerged Endurance and Fuel Cells

Diesel-electric submarines generally need to surface every few days to recharge their batteries. However, with fuel cell technology, submarines can stay underwater for much longer. The current record for a fuel cell-equipped submarine is 14 consecutive days submerged, set by the HDW Class 212A. Regulations also mandate that if a submarine is unable to surface for any reason, the crew should be able to survive for at least six days using emergency supplies and systems [103].

Snorkelling in Heavy Seas

In rough seas, submarines face the challenge of water entering through the snorkel. To prevent this, a cap temporarily closes off the snorkel when waves pass over it. During this short closure, the air already inside the submarine serves as a reserve supply for the engine. This system allows submarines to continue operating without interruption, even in challenging sea conditions [103].

Requirements for Submarine Diesel Engines

Submarine diesel engines need to meet several specific requirements to ensure operational effectiveness and stealth. First, the engines must be exceptionally quiet to minimize the submarine's acoustic signature. Second, they should be compact to save space for other essential systems onboard. High power output is critical for charging batteries quickly, while fuel efficiency is important to extend the submarine's operational range. Additionally, compliance with current emissions standards has become an increasingly significant factor in the design and development of submarine diesel engines [103].

Engine Design Principles and Operation

Reciprocating internal combustion engines are a fundamental type of engine that convert the energy stored in fuel into mechanical work. This process occurs through a series of controlled explosions or combustions within the engine's cylinders, which feature a back-and-forth motion known as reciprocation. The diesel and gasoline engines are the most widely recognized examples of this engine type, using different methods to initiate combustion. Here's a detailed breakdown of how these engines operate, the cycles of operation, and the underlying thermodynamics principles involved.

The core principle behind an internal combustion engine is the conversion of chemical energy in the fuel into mechanical energy, which is used to perform work. The engine's combustion chamber is the space where fuel and air are mixed and ignited. When the fuel-air mixture burns, it releases heat, causing the gases in the chamber to expand. This expansion increases the pressure inside the chamber, forcing the piston downward. The movement of the piston is transferred to the crankshaft through a connecting rod, converting the reciprocating (back-and-forth) motion into rotational motion, which is then used to drive machinery or a vehicle. Once the combustion process is complete, the exhaust gases are expelled from the chamber, and a new cycle of intake, compression, combustion, and exhaust begins.

Each internal combustion engine operates based on a cycle that repeats for each power stroke delivered to the crankshaft. A complete cycle consists of four distinct phases, which occur in a specific order, regardless of the number of piston strokes involved [102]:

1. **Intake:** The engine draws in either air or a mixture of air and fuel, which is then compressed within the cylinder.

2. **Compression:** The air-fuel mixture is compressed to increase its pressure and temperature, preparing it for ignition. In diesel engines, only air is compressed, and fuel is injected at high pressure for ignition.

3. **Combustion (Power Stroke):** The compressed mixture is ignited (by a spark in gasoline engines or by the heat of compression in diesel engines), causing the gases to expand rapidly and push the piston down, creating mechanical work.

4. **Exhaust:** The spent gases are expelled from the cylinder, making way for a fresh intake of air or air-fuel mixture.

The specific sequence and characteristics of these phases define the type of cycle used by the engine. The three main cycles for internal combustion engines are [102]:

- **Otto Cycle:** Typically used in gasoline engines, where the air-fuel mixture is ignited by a spark plug.

- **Diesel Cycle:** Characterized by compression ignition, where the fuel is injected into highly compressed, hot air.

- **Modified Diesel Cycle:** Incorporates features of both the Otto and Diesel cycles to achieve specific performance characteristics.

Thermodynamics plays a crucial role in the operation of internal combustion engines, as it governs how heat energy is transformed into mechanical work. Some key thermodynamic principles include [102]:

- **Conservation of Energy:** According to the laws of thermodynamics, energy cannot be created or destroyed; it can only be converted from one form to another. In the case of engines, chemical energy from the fuel is converted into thermal energy through combustion and then into mechanical energy via the movement of the piston.

- **Force and Pressure:** The force generated by combustion acts on the piston, and the pressure is the force distributed over the area of the piston head, pushing it downward.

- **Work and Power:** The work done by the engine is the product of the force exerted on the piston and the distance it travels. Power is the rate at which this work is done, typically measured in horsepower (hp).

The mechanical equivalent of heat is an essential concept in understanding how internal combustion engines function. For instance, experiments have determined that one British thermal unit (Btu) of heat energy is equivalent to 778 foot-pounds of mechanical work. This relationship shows that while mechanical energy can be converted entirely into heat, the reverse process is not entirely efficient, as some heat energy is always lost to the environment [102].

Understanding the operation of internal combustion engines also involves familiarity with basic terms [102]:

- **Energy:** The ability to perform work, categorized into kinetic (energy of motion) and potential (stored energy).
- **Heat:** A form of energy derived from molecular activity; increasing molecular motion in a substance raises its temperature.
- **Temperature:** A measure of heat intensity, recorded in degrees.
- **Pressure:** The force per unit area exerted by gases in the combustion chamber.
- **Matter and Molecules:** Substances that occupy space and have weight, with molecules being the smallest units retaining the properties of a substance.

Reciprocating internal combustion engines operate on the fundamental principle of converting heat energy into mechanical work, using a cycle of intake, compression, combustion, and exhaust. The thermodynamics involved in these engines dictate how energy is transferred and utilized, while developments in engine technology have continuously refined the efficiency and performance of these power sources. Understanding these principles is key to grasping how diesel and gasoline engines function and their applications in various industries, including marine engineering and submarine propulsion.

Thermodynamic cycles are fundamental to understanding how internal combustion engines operate, including the Otto cycle, the diesel cycle, and the modified diesel cycle. In an internal combustion engine, heat energy from burning fuel is converted into mechanical energy. The fuel acts as the working substance that undergoes changes in the engine's cylinder. These changes are described by different thermodynamic processes, which include isothermal and adiabatic changes, where temperature and heat exchange play crucial roles [102].

The Otto cycle is the thermodynamic basis for gasoline engines and involves four main processes. First, there is an adiabatic compression phase, where the fuel-air mixture is compressed within the cylinder without any heat loss. This is followed by constant-volume combustion, where the mixture is ignited, increasing the pressure in the chamber. Next comes the adiabatic expansion, where the high-pressure gases push the piston down, performing work. Finally, there is a constant-volume rejection of gases, where the exhaust is expelled. This cycle repeats, producing one power stroke for every four strokes of the piston.

The original diesel cycle, developed by Rudolf Diesel, was designed to achieve combustion at constant temperature, but this concept was never fully realized in practical engines. Instead, early diesel engines used a constant-pressure combustion cycle, which became a standard for many years. In the diesel cycle, adiabatic compression raises the air temperature to ignite the injected fuel, which is added at a rate that maintains constant pressure during combustion. The gases then expand adiabatically, pushing the piston downward, followed by constant-volume exhaust where the gases are expelled. Although this cycle is less commonly used in modern engines, it laid the groundwork for subsequent diesel engine designs [102].

Most modern diesel engines, including those used in submarines, operate on a modified diesel cycle. In this cycle, the compression process is adiabatic, similar to the standard diesel cycle. However, combustion occurs in two stages: first at constant volume, followed by constant pressure, allowing for a more controlled burn and greater efficiency. The expansion of gases after combustion is adiabatic, and the cycle is completed with a constant-volume exhaust phase. This hybrid approach combines elements of both the Otto and diesel cycles, resulting in improved performance and efficiency.

The efficiency of any internal combustion engine is measured by its ability to convert the potential energy in fuel into mechanical energy. This efficiency is always less than 100% due to energy losses through heat dissipation, friction, and the exhaust system. The concept of thermal efficiency compares the actual work output of the engine to the theoretical maximum based on the energy content of the fuel. Although no engine can achieve perfect efficiency, optimizing the thermodynamic cycle can reduce energy losses and improve the overall performance.

The four-stroke diesel cycle involves four distinct piston movements: intake, compression, power, and exhaust. The cycle begins with the intake stroke, where the intake valve opens and air is drawn into the cylinder. During the compression stroke, the air is compressed to high pressures and temperatures. Fuel is then injected, igniting spontaneously due to the heat of compression, initiating the power stroke, where the expansion of combustion gases pushes the piston downward. Finally, the exhaust stroke occurs, in which the spent gases are expelled from the cylinder, completing the cycle. This type of engine is known for its reliability and efficiency, making it suitable for heavy-duty applications like submarines [102].

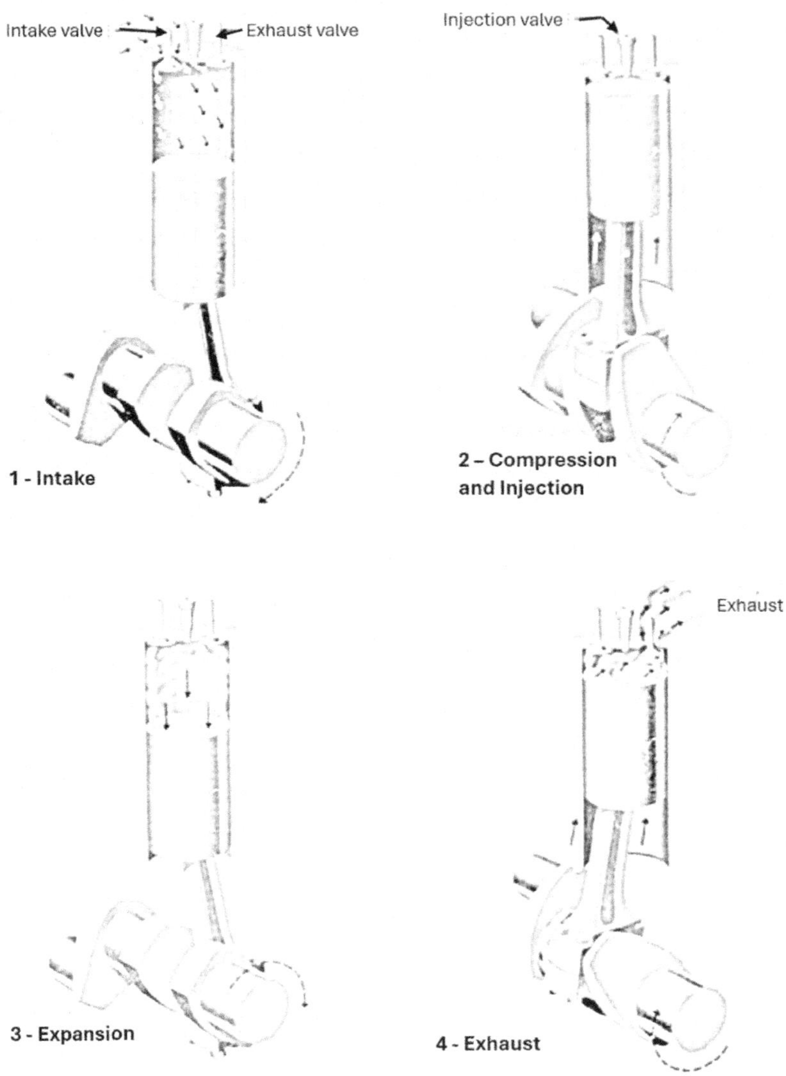

Figure 44: The 4-stroke diesel cycle.

The two-stroke diesel cycle compresses the four phases of the diesel cycle into two piston strokes, resulting in one power stroke per crankshaft revolution. After the compression phase, fuel is injected, and combustion occurs just before the piston reaches the top dead centre. The subsequent expansion phase forces the piston downward, driving the crankshaft. As the piston nears the bottom, the exhaust phase begins with the opening of the exhaust valve, allowing the exhaust gases to escape while the intake ports open to let in fresh air. A scavenging blower helps to clear the cylinder of exhaust gases, preparing it for the next cycle. Two-stroke engines produce more power for a given size, but they are typically less fuel-efficient and require additional systems for scavenging.

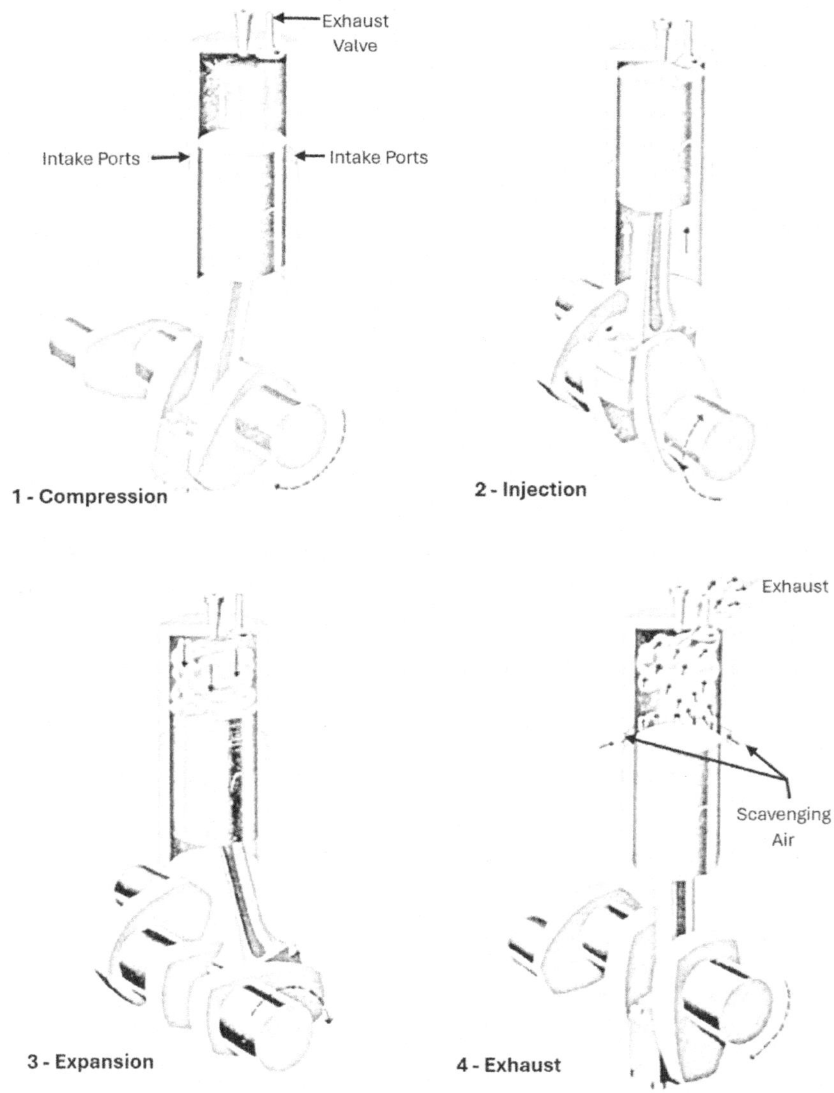

Figure 45: The 2-stroke diesel cycle.

The underlying principles of thermodynamics govern the conversion of heat energy to mechanical work in all these cycles. These principles include [102]:

- **Isothermal Process:** Occurs when the temperature remains constant during a thermodynamic change.
- **Adiabatic Process:** A change where the temperature may vary, but no heat is exchanged with the surroundings.
- **Heat Transformation:** Involves the conversion of chemical energy in the fuel into heat and then mechanical energy.

Thermodynamic laws also apply to concepts such as work (force exerted over a distance), power (rate of work), and efficiency, all of which are crucial in determining how effectively an engine converts fuel energy into useful output.

Practical Applications in Diesel Engines

While the theoretical cycles provide a framework, actual engine performance is influenced by real-world conditions such as friction, heat loss, and variations in fuel quality. In diesel engines, the use of high compression ratios results in better thermal efficiency compared to gasoline engines. However, advancements such as turbocharging, intercooling, and electronic fuel injection have further improved diesel engine performance, making them suitable for various applications including heavy-duty vehicles and submarines [102].

The Otto cycle, diesel cycle, and modified diesel cycle each represent different methods of converting heat energy into mechanical work. While they share some common thermodynamic principles, the differences in how compression, combustion, and expansion are handled result in variations in efficiency, power output, and fuel consumption. Understanding these cycles is essential for optimizing engine design and improving thermal efficiency across different engine types.

Diesel Engine Types

Single-Acting Diesel Engines

Single-acting diesel engines are common in both four-stroke and two-stroke cycle designs. In these engines, the combustion gases exert pressure on only one side of the piston, called the crown, which is the closed top end of the piston. The pistons used in single-acting engines are typically of the trunk type, meaning that their length is greater than their diameter. The connecting rod is attached to the piston pin, which extends through the open end of the piston, known as the skirt. In a four-stroke single-acting engine, the power stroke happens once every two revolutions of the crankshaft, while in a two-stroke single-acting engine, the power stroke occurs every revolution. This configuration is the standard for diesel engines in modern submarine fleets because it is relatively simpler and more compact compared to other engine types [102].

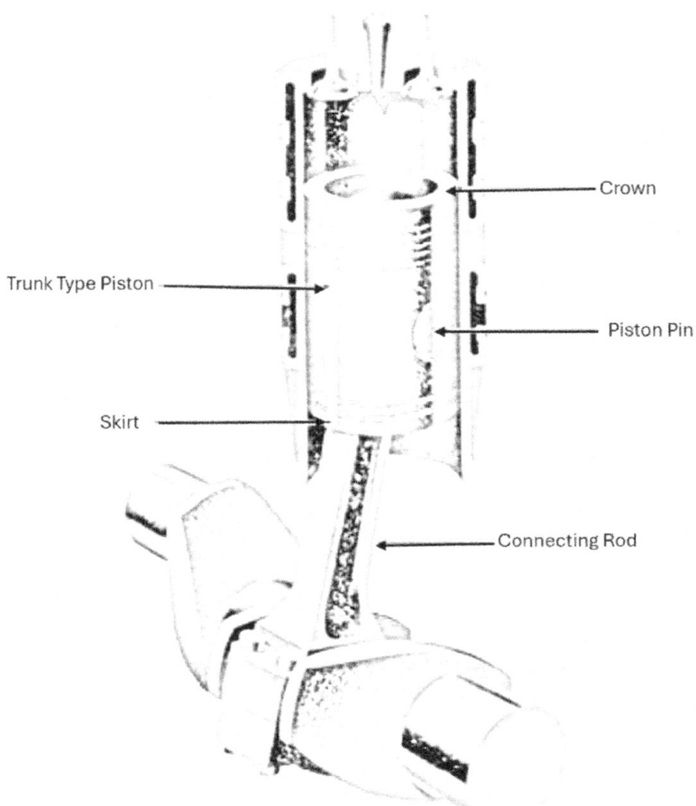

Figure 46: Single-acting diesel principle.

Double-Acting Diesel Engines

Double-acting diesel engines were used in some older submarine designs but have largely been phased out in favour of single-acting engines. In double-acting engines, the piston is subjected to combustion pressure on both its top and bottom surfaces, allowing for two power strokes per piston cycle. The piston in a double-acting engine is shorter and features a crosshead design, which includes a rigid piston rod extending from the lower end of the piston through the cylinder head. Combustion takes place alternately in the upper and lower combustion chambers, providing twice the number of power strokes compared to a single-acting engine. This design results in a smoother operation since the compression stroke in one chamber balances the expansion stroke in the other. However, double-acting engines are bulkier and more complex due to the need for additional length to accommodate the crosshead mechanism and for effective sealing where the piston rod passes through the cylinder head. These limitations make double-acting engines unsuitable for the confined spaces of modern submarines [102].

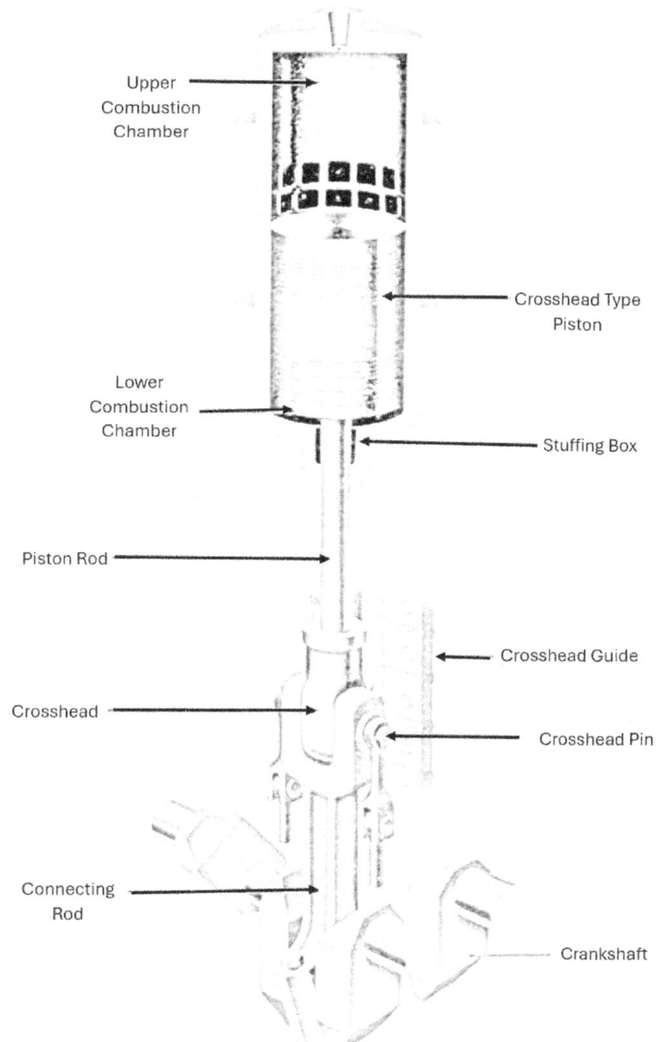

Figure 47: Double-acting diesel principle.

Opposed Piston Engines

The opposed piston engine is a unique design featuring two pistons within each cylinder. These pistons move toward each other, compressing the air between them to form the combustion chamber. When combustion occurs, the pistons are driven apart, converting the expansion of gases into mechanical work. The upper piston opens the scavenging air ports located near the top of the cylinder, while the lower piston controls the exhaust ports near the bottom. The upper and lower pistons are connected to separate crankshafts, which are linked via a vertical gear drive to synchronize their motion. This configuration allows the engine to achieve supercharging by maintaining positive air pressure within the cylinder during the scavenging process. The lower crankshaft typically leads the upper by 12 degrees, resulting in a larger share of power being delivered by the lower crankshaft. Opposed piston engines have several advantages, including higher thermal efficiency, the absence of cylinder heads and valve mechanisms, and fewer moving parts, making them well-suited for submarine applications [102].

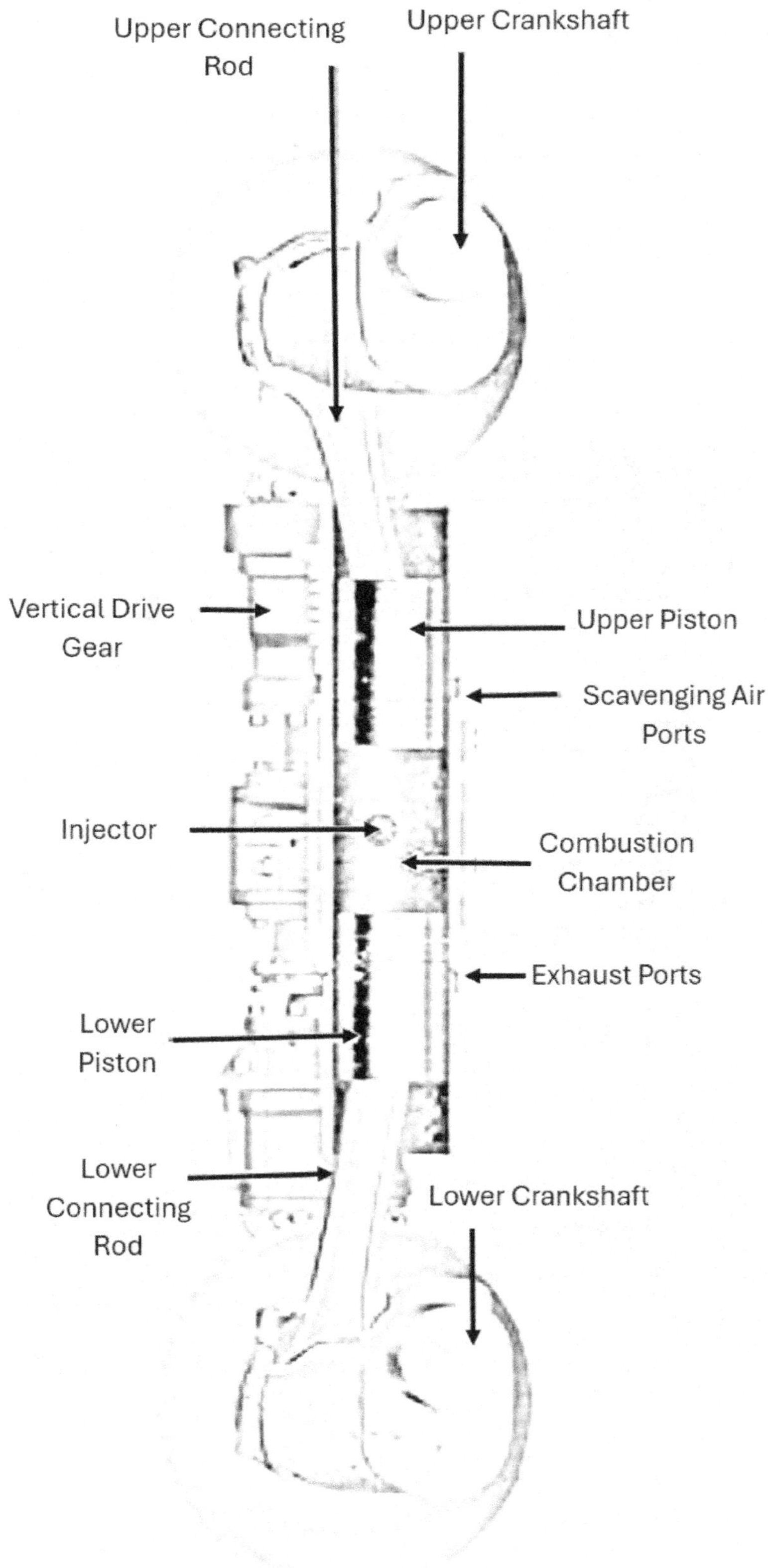

Figure 48: Opposed piston principle.

Modern Fleet-Type Submarine Diesel Engines

Modern submarines primarily use two-stroke cycle diesel engines due to their efficiency and ability to produce continuous power. The main types of engines found in fleet-type submarines include [102]:

1. **General Motors V-16 Engines:** These engines feature two banks of eight cylinders arranged in a V-configuration, with a power output of 1,600 brake horsepower (bhp) at 750 rpm. They utilize a uniflow valve and port scavenging system and are equipped with solid type fuel injection.

2. **Fairbanks-Morse Opposed Piston Engines (Model 38D 8 1/8):** Available in both 9-cylinder and 10-cylinder configurations, these engines are also rated at 1,600 bhp but operate at 720 rpm. Like the General Motors engines, they use a uniflow port system for scavenging.

Figure 49: General Motors Model 16-248 V16 diesel engine installed in the USS Bowfin. Brandon W. Smith, CC BY-SA 3.0 , via Wikimedia Commons.

Auxiliary engines on submarines typically include smaller models designed for tasks such as generating electrical power [102]:

1. **General Motors Model 8-268:** This 8-cylinder in-line engine generates 300 kilowatts when running at 1,200

rpm, using a uniflow scavenging system.

2. **Fairbanks-Morse Model 38E 5 1/4:** A 7-cylinder opposed piston engine capable of producing 300 kilowatts, also at 1,200 rpm.

Opposed piston engines provide three distinct benefits for submarine use. Firstly, they offer higher thermal efficiency compared to single-acting engines. This improved efficiency translates to better fuel economy and extended range for the submarine. Secondly, the design eliminates the need for traditional cylinder heads and associated valve mechanisms, which can present cooling and lubrication challenges. Thirdly, opposed piston engines have fewer moving parts, resulting in reduced maintenance requirements and increased reliability [102].

Despite their power output advantages, double-acting engines have drawbacks that limit their use in submarines. The need for a crosshead mechanism increases the engine's height, making it difficult to install in the limited space available on a submarine. Additionally, achieving an effective seal at the piston rod's stuffing box is challenging, increasing the risk of pressure leaks and reducing engine reliability.

Submarine diesel engines have evolved significantly from early designs that used direct mechanical drive systems. Modern submarine propulsion typically relies on two-stroke, single-acting diesel engines or opposed piston designs, which offer better power-to-weight ratios, improved efficiency, and greater operational flexibility. By using advanced configurations such as opposed piston engines, modern submarines achieve high performance with relatively simple mechanical systems, ensuring reliable and efficient underwater operation.

Installation

Submarine diesel engine installations on modern diesel-electric drive submarines typically consist of a main and auxiliary engine arrangement, designed to provide power for both propulsion and onboard systems. Each engine is coupled with an electrical generator to form a generator set, which plays a central role in the submarine's power generation and management. These generator sets can supply electricity to various systems on the submarine, offering flexibility in how power is utilized for propulsion and other operations [102].

The main engine-generator sets are responsible for generating electricity to power the submarine's main motors or charge the battery banks. Electricity generated by these main sets is routed through the main control cubicle, which serves as the central hub for directing electrical power to different parts of the submarine. The main control cubicle allows operators to choose whether the electrical power from the generators is used directly to drive the submarine's propulsion motors or diverted to recharge the batteries. This flexibility ensures that the submarine can switch between battery-powered silent running while submerged and diesel-powered battery recharging when near the surface.

In addition to the main engine-generator sets, a submarine also has an auxiliary generator set. This smaller generator set serves several important functions: it can charge the batteries, provide power to auxiliary systems, or even support the main motors indirectly. For example, if the main generators are offline or need supplemental power, the auxiliary generator can be used to ensure continuous operation of the submarine's systems. This setup provides redundancy and ensures that the submarine remains operational in various scenarios, including emergencies.

The main motors are the primary propulsion mechanism for the submarine, driving the propellers to propel the vessel through the water. These motors can be powered directly by the batteries, allowing the submarine to operate quietly and without the noise generated by running the diesel engines. Alternatively, the main motors can be powered by the electricity generated from the main engine-generator sets, enabling the submarine to conserve battery power while running near the surface.

The dual power source capability of the main motors gives diesel-electric submarines their distinctive operational flexibility. When submerged, the submarine can switch to battery power, enabling silent running and reducing the likelihood of detection by enemy sonar. When operating near the surface, the diesel engines can recharge the batteries and supply power to the main motors, extending the submarine's operational endurance.

Overall, the integration of main and auxiliary generator sets, batteries, and main motors within the submarine's diesel-electric propulsion system provides a versatile and efficient means of managing power and propulsion. The system's design ensures that the submarine can adapt to changing operational conditions, maintain stealth, and manage its energy resources effectively, making diesel-electric submarines highly capable in various tactical situations [102].

Nuclear Propulsion Technologies

Submarine nuclear propulsion involves using a nuclear reactor as the primary source of energy to power the submarine. The system allows submarines to operate underwater for extended periods without needing to surface for refuelling or recharging batteries, which gives them a significant strategic and operational advantage.

A nuclear submarine is powered by a nuclear reactor but is not necessarily equipped with nuclear weapons. The use of nuclear propulsion provides significant advantages over conventional submarines, which typically use diesel-electric propulsion. Unlike diesel-electric submarines, nuclear submarines do not require air for propulsion, allowing them to remain submerged for extended periods without surfacing to recharge batteries. This capability gives nuclear submarines superior endurance, as their operational limits are more related to supplies such as food for the crew than to fuel.

Nuclear submarines use nuclear reactors to produce power, either by generating electricity to run electric motors connected to the propeller or by using the reactor's heat to produce steam that drives steam turbines. Reactors used in submarines typically use highly enriched uranium, which provides a high energy density and allows for longer intervals between refuelling's. The reactor's position within the submarine's pressure hull makes refuelling a complex and infrequent task.

The heart of a nuclear-powered submarine is its nuclear reactor, which uses nuclear fission to generate heat. In nuclear fission, the nucleus of a heavy atom, such as uranium-235 or plutonium-239, is split into smaller nuclei when bombarded by neutrons, releasing a tremendous amount of heat energy in the process. This controlled chain reaction inside the reactor core continuously produces heat, which is then used to generate power for the submarine.

The heat produced by the reactor is transferred to a coolant, usually water, which circulates through the reactor core. This coolant absorbs the heat and is then pumped through a heat exchanger known as a steam generator. In the steam generator, the heated coolant transfers its energy to a separate loop of water, converting it into steam. This steam is produced under high pressure and is used to drive the submarine's propulsion system and electrical generators.

The high-pressure steam generated from the steam generator is channelled to steam turbines, which convert thermal energy into mechanical energy. The steam expands as it flows through the turbine blades, causing the blades to spin. This rotational motion drives a series of turbines connected to a reduction gearbox that turns the propeller shaft, providing the necessary thrust to propel the submarine through the water.

In addition to propulsion, the steam turbines are also used to generate electricity for onboard systems. An auxiliary turbine connected to a generator produces electrical power for essential systems, including life support, navigation, sensors, and combat systems. This electricity also powers other components of the submarine, such as lighting and communication equipment. The nuclear propulsion system ensures that the submarine has a continuous power supply, enabling it to stay submerged and operational for long periods.

A nuclear reactor produces an enormous amount of heat, so efficient cooling is essential to maintain safe and stable operation. The primary coolant loop circulates water through the reactor core to absorb the heat generated by nuclear fission. The heated coolant is then passed through the steam generator, where it releases its heat and cools down. Afterward, it returns to the reactor to absorb more heat, completing the loop. This closed-loop system ensures that the radioactive coolant is contained and does not come into contact with the submarine's external environment.

The nuclear reactor also provides power for the submarine's other essential subsystems. These include air quality maintenance, fresh water production by distilling seawater, and temperature regulation. All modern naval nuclear reactors have backup diesel generators to provide emergency power for reactor decay heat removal and emergency propulsion.

The main advantage of nuclear propulsion for submarines is the ability to remain submerged for extended durations. Unlike diesel-electric submarines, which need to surface periodically to recharge batteries, nuclear submarines can operate underwater almost indefinitely because the reactor does not need oxygen. The only limiting factor for a nuclear submarine's endurance is the need for food and supplies for the crew. Nuclear-powered submarines can travel at higher speeds for longer distances without surfacing, giving them superior strategic mobility, stealth, and endurance.

Nuclear propulsion allows submarines to operate independently of the atmosphere, as no air or oxygen is required for power generation. This feature, coupled with the submarine's ability to produce oxygen and fresh water from seawater, allows for extended underwater operations.

While nuclear submarines offer significant operational advantages, they also have some stealth limitations. The need to cool the reactor even when the submarine is stationary results in a continuous release of heat into the surrounding water, creating a "thermal wake" that can be detected using thermal imaging systems. Additionally, the reactor's continuous operation generates noise from steam production and reactor coolant pumps, potentially making the submarine more detectable. In contrast, diesel-electric submarines can operate on battery power alone, using silent electric motors.

Nuclear submarines are equipped with multiple safety systems to prevent radiation leaks and ensure safe reactor operation. These safety measures include reactor shielding to protect the crew from radiation, emergency shutdown systems that can quickly stop the fission process if necessary, and multiple cooling systems to ensure the reactor remains cool under all operating conditions. Additionally, submarines undergo regular maintenance and inspections to ensure the integrity of the reactor and propulsion system.

After passing through the steam turbine, the steam must be condensed back into water to be reused. This is achieved by cooling the steam using seawater in condensers. The waste heat from this process is expelled into the ocean. The closed-loop nature of the nuclear propulsion system ensures that there is no direct release of radioactive material into the environment, making it a relatively clean form of propulsion.

Nuclear submarines use highly enriched uranium, which allows the reactor to operate for many years without refuelling. The fuel has a high energy density, meaning a relatively small amount of uranium can power a submarine for decades. Refuelling a nuclear submarine is a complex process that involves replacing the reactor core and is usually done every 20-25 years.

Nuclear propulsion allows submarines to perform a variety of missions, from strategic deterrence and long-range patrols to covert intelligence gathering and anti-ship warfare. Because of their extended endurance and high-speed capabilities, nuclear submarines can be deployed worldwide, providing a significant advantage in terms of operational reach and versatility.

The continuous cooling requirement for the reactor creates a thermal wake, which may be detected by advanced thermal sensors. The constant reactor operation generates noise that could compromise stealth. Additionally, the high costs associated with building, operating, and decommissioning nuclear submarines restrict their use to a few countries with substantial defence budgets.

The useful life of a nuclear submarine is approximately 25 to 30 years, after which the vessel faces challenges such as component fatigue, corrosion, obsolescence, and increased maintenance costs. The decommissioning process is complex and expensive, involving the safe disposal of radioactive materials and components. Two main methods are used for disposal: the removal and burial of the reactor section in a designated disposal site, or deep-sea disposal, where the sealed submarine is flooded and sunk to settle on the sea floor. However, the latter option is less favoured due to regulatory concerns.

A nuclear reactor generates heat to power a steam turbine, which then drives the propeller. The three primary types of marine nuclear reactors are pressurized-water, natural-circulation, and liquid-metal reactors [104].

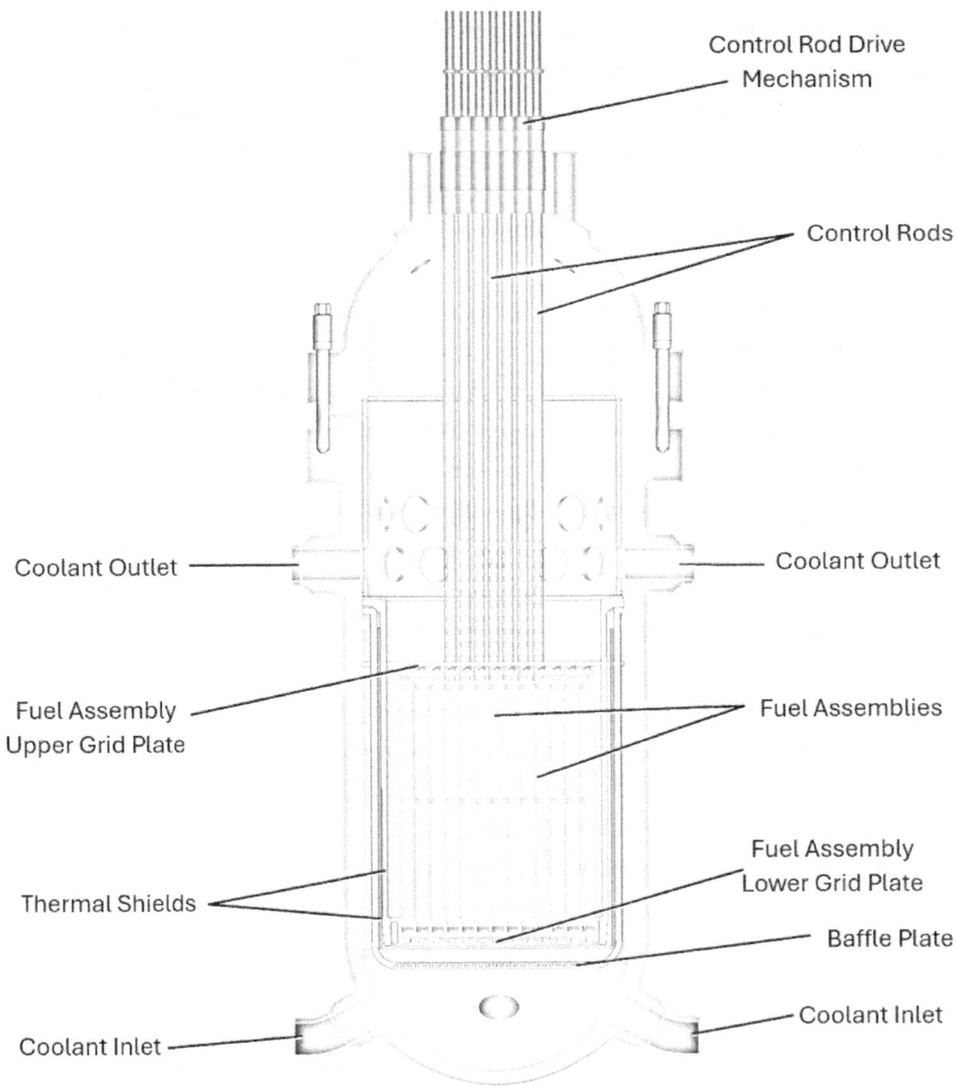

Figure 50: Section of a pressurized-water reactor, showing inlets and outlets for water coolant passing through the core.

In a nuclear reactor, uranium generates heat through the process of nuclear fission. When uranium atoms split, they release neutrons and energy in the form of heat. To make the fission process more efficient, the uranium fuel is surrounded by a material called a moderator, which slows down the released neutrons so they can more effectively interact with other uranium atoms. In most reactors, water serves as the moderator because it slows the neutrons and doubles as a heat-transfer medium, carrying away the heat produced during fission. This heated water is part of what is known as the "primary loop" [104].

The primary loop water is kept under high pressure to prevent it from boiling, even as it absorbs significant heat from the reactor core. It then travels through a heat exchanger, transferring its heat to a separate water circuit called the "secondary loop." The heat exchanger functions similarly to a boiler, converting the secondary loop water into steam. This steam is then used to drive the turbine, which generates electricity. The design of the heat exchanger ensures that the radioactive primary loop water does not mix with the non-radioactive secondary loop water, thus preventing contamination in other parts of the power plant [104].

In many reactors, pumps are used to circulate the primary loop water through the system. However, some reactor designs use natural circulation to move the coolant. In these reactors, temperature differences within the reactor create a natural flow: cooler water from the heat exchanger is introduced at the bottom of the reactor, and as it rises through the heated fuel elements, it absorbs heat and circulates naturally without the need for mechanical pumps [104].

Liquid-metal-cooled reactors operate on a different principle, using molten metal instead of water as the coolant. Molten metal can transfer much more heat than water, allowing for the use of a more compact turbine. However, this design comes with significant challenges. The molten metal can become highly radioactive, which makes any leaks in the system especially hazardous. Additionally, the pumps needed to circulate the dense liquid metal must be more powerful than those used in water-cooled reactors. Unlike water, which can serve as both a moderator and heat sink, the liquid-metal-cooled reactors lose this simplicity, requiring separate systems for moderation and heat transfer. There is also a critical risk that if too much heat is lost, the molten metal could solidify in the pipes, causing the reactor to seize up, which could result in catastrophic consequences [104].

Development

The development and implementation of nuclear reactor technology in the United Kingdom's submarine fleet have significantly enhanced its operational capabilities, particularly in terms of extended underwater operations, increased speed, and long-term deployment without the need for refuelling. This evolution can be traced through the introduction of various reactor designs, notably the PWR1, PWR2, and PWR3 reactors, each contributing to the UK's strategic naval capabilities.

The UK's first nuclear-powered submarine, HMS Dreadnought, was commissioned in 1963 and utilized an American Westinghouse S5W reactor, marking the inception of nuclear collaboration between the UK and the US under the 1958 Mutual Defence Agreement. This agreement facilitated the transfer of nuclear technology and laid the groundwork for the UK's own reactor development efforts [105]. Following this, the first British-designed reactor, PWR1 (Pressurized Water Reactor 1), was introduced. PWR1 was based on earlier US technology but included significant modifications to meet specific British naval requirements. The first submarine to be powered by PWR1 was HMS Valiant, which went critical in 1965. The design and production of these reactors were primarily managed by Rolls-Royce Marine Power Operations in Derby, which worked closely with the Ministry of Defence [105, 106].

The PWR1 reactors utilized highly enriched uranium (HEU) with enrichment levels between 93% and 97%, allowing for extended operational periods between refuelling cycles—approximately ten years per core, with refuelling typically required only twice during a submarine's lifetime [105, 106]. This capability significantly enhanced the operational flexibility and endurance of the Royal Navy's submarine fleet. The PWR1 reactor was employed across several classes of submarines, including the Valiant-class SSNs, Resolution-class SSBNs, and Churchill-class SSNs, among others, showcasing its versatility and effectiveness in various operational contexts [105, 106].

As the need for improved performance and safety emerged, the PWR2 reactor was developed as an advancement over PWR1, specifically designed for the larger Vanguard-class submarines, which were equipped to carry Trident ballistic missiles. The first PWR2 reactor was completed in 1985 and began testing in 1987. PWR2 featured enhanced safety and performance capabilities, including longer-lasting core designs (Core H) that extended reactor life to approximately 30 years, thereby eliminating the need for refuelling throughout the submarine's service life [105, 106]. However, safety assessments raised concerns regarding structural vulnerabilities and emergency procedures, particularly following a 2012 incident where radiation was detected in a test reactor's coolant water, prompting precautionary measures for the Vanguard-class and Astute-class submarines [105, 106].

The latest advancement in the UK's submarine reactor technology is the PWR3 reactor, developed to power the Dreadnought-class SSBNs, which are set to replace the Vanguard-class. The PWR3 design incorporates technology based on the US Navy's S9G reactor, adapted for British requirements, and aims to improve safety, efficiency, and maintenance while being cost-effective. Notably, PWR3 features 30% fewer parts than its predecessor, thereby simplifying the design and reducing potential failure points [105, 106]. The Ministry of Defence awarded contracts to Rolls-Royce for the production of PWR3 reactors in 2012, with significant refurbishments made to the Derby manufacturing plant to support this new generation of reactors [105, 106].

Safety considerations remain paramount in the design and operation of these nuclear reactors. The UK has implemented strict safety standards, including the separation of radioactive and non-radioactive systems, backup diesel generators for emergency power, and careful refuelling protocols. Despite these measures, challenges persist, as evidenced by the 2012 incident, underscoring the need for continuous monitoring and improvement in safety practices [105, 106]. The Vulcan Naval Reactor Test Establishment (NRTE) has played a critical role in testing and validating reactor designs, although advancements in computational modelling have reduced the need for extensive prototype testing in recent years [105, 106].

Design Principles of Nuclear Submarines

During the Cold War, three main trends emerged in nuclear attack submarine design, shaped by the rivalry between NATO and the Soviet Union. These trends were increased speed, greater diving depth, and enhanced silencing techniques. The U.S., U.K., and Soviet submarine forces each pursued these features to improve the performance of their submarines in underwater combat and strategic operations [104].

The drive for higher speed required more power, which presented challenges in balancing submarine size and engine noise. The resistance a submarine encounters underwater increases with its surface area, so adding more powerful engines usually meant increasing the submarine's size, thus affecting its speed and stealth capabilities. For example, between the U.S. Sturgeon-class and Los Angeles-class submarines, reactor power nearly doubled, but the size of the submarines also increased significantly from around 3,600 to 6,000 tons. The Soviet Union's Alfa-class submarines achieved extremely high speeds of approximately 40 knots, compared to just over 30 knots for Western submarines, likely at the cost of generating more noise at high speeds [104].

Figure 51: 571 USS Nautilus. Vonsky87, CC BY-SA 3.0, via Wikimedia Commons.

Submarine speed was highly valued for various reasons. Initially, faster submarines were developed primarily to target enemy ships. The USS Nautilus, with a speed of over 20 knots, could evade surface ships but was not fast enough to launch counterattacks. To address this, U.S. submarines adopted the streamlined "tear-drop" hull design of the experimental submarine USS Albacore, leading to the Skipjack class, which entered service in 1959 with speeds exceeding 30 knots [104].

Figure 52: Inside the USS Albacore submarine. Kristina D.C. Hoeppner, CC BY-SA 2.0, via Flickr.

A striking example of the importance of speed occurred in February 1968, when a Soviet submarine intercepted the nuclear aircraft carrier USS Enterprise. Although the submarine was not as fast as the carrier, it was able to keep up enough to stay within weapon range even as the Enterprise accelerated. As Soviet submarines became faster, the U.S. Navy focused on creating dual-purpose vessels capable of engaging both surface ships and other submarines. High sustained speeds also enabled submarines to deploy more efficiently to distant patrol areas, maximizing their mission time despite limited storage capacity for provisions. Faster speeds were also advantageous for evading enemy torpedoes and sonar, although louder operations at high speeds increased the risk of detection [104].

Greater diving depth provided several advantages, such as enhancing evasion by allowing submarines to operate in multiple underwater layers, improving sonar performance, and offering better protection. Deeper-diving submarines could use various sonar techniques to detect targets, including reflections off the seafloor and water surface, which could extend their sonar range. Around 1960, U.S. submarine sonars evolved from cylindrical units with broad beams to spherical units that produced narrow beams, enabling more precise detection of targets at different depths [104].

The need for deeper diving also became important for high-speed operations. At high speeds, a submarine could inadvertently tip downward and dive below its safe operating depth before corrective measures could be taken. The Soviet Alfa-class submarine, known for its speed and deep diving capabilities, had a reported maximum depth of about 2,800 feet (853 meters), significantly deeper than most modern attack submarines, which typically dive between 1,000 and 1,500 feet (304 to 457 meters). Achieving these depths required stronger, heavier hulls and more powerful engines, which led to design compromises. For instance, the U.S. Los Angeles class sacrificed some diving depth to achieve higher speeds, while the Alfa class used an expensive titanium-alloy hull and a compact power plant to save weight [104].

Silencing was critical for avoiding detection, especially as submarine tracking methods evolved. Until the late 1950s, active sonar, which detects submarines by bouncing sound waves off their hulls, was the primary detection method. However, active sonar had limitations, including its range and the fact that submarines could detect the sonar's sound waves and realize they were being targeted. In response, the U.S. and British navies began developing passive sonar systems, which detect noises produced by the submarine itself, such as machinery and reactor coolant pumps [104].

Early nuclear submarines had loud machinery, particularly the pumps used in pressurized-water reactors, which needed to operate continuously to prevent reactor overheating. As silencing became a focus in submarine design, efforts were made to quiet these pumps and isolate the machinery from the hull using sound-absorbing mounts. These modifications increased the size and weight of the machinery, adding to construction costs. Natural-circulation reactors, which did not require noisy pumps, became more attractive for their silent operation. Additionally, submarine hulls were coated with sound-absorbing materials to further reduce noise. Such coatings also decreased the effectiveness of homing torpedoes by absorbing the sound waves they used to lock onto targets [104].

These trends in speed, depth, and silencing shaped the design of modern submarines, balancing the competing requirements of propulsion power, stealth, and structural integrity. The result was the development of submarines capable of fast, deep, and quiet operations, providing a strategic advantage during the Cold War and beyond.

Naval Reactor Utilisation

Naval reactors have predominantly been of the pressurized water type, differing from commercial reactors primarily in design to meet the unique demands of submarine and naval vessel propulsion. These reactors must deliver high power from a very compact volume, as space is limited aboard submarines and warships. Several features and modifications set naval reactors apart from their civilian counterparts [107]:

1. High Power Density and Enrichment: To achieve high power output in a confined space, naval reactors typically use highly enriched uranium (HEU). Early U.S. submarines used fuel enriched to about 97% Uranium-235, although recent designs have reduced this to around 93%. Western naval reactors generally use enrichment levels between 20% and 25%, while early Russian reactors ranged from 20% to 21%, with newer models using enrichment between 21% and 45%. For instance, India's Arihant-class submarines use around 40% enrichment. In contrast, modern French reactors have transitioned to using low-enriched uranium (LEU) with enrichment levels as low as 5%, matching civilian standards.

The use of HEU extends the operational life of the reactor core. Early cores needed refuelling every 10 years or so, while newer cores are designed to last much longer—up to 50 years for aircraft carriers and 30-40 years for many submarines. This reduces the need for refuelling and maintenance, allowing submarines to stay operational for longer periods without being taken out of service.

2. Fuel Composition and Design: Naval reactor fuel differs from the typical uranium dioxide (UO2) used in civilian reactors. It often consists of uranium-zirconium or uranium-aluminium alloys, which have better thermal conductivity and are more suitable for the high power density required in naval applications. In some cases, metal-ceramic composite fuels are used, such as the U-Al fuel in certain Russian submarines, which may be enriched to varying levels (20-45% U-235).

To support the long life of the reactor core, "burnable poisons" like gadolinium are added to the fuel. These materials absorb excess neutrons during the early stages of reactor operation, helping to control the reactivity as the fuel burns. Over time, the burnable poisons are consumed, balancing out the increasing neutron-absorbing properties of fission products accumulating in the fuel, thus maintaining stable reactor performance.

3. Reactor Pressure Vessel and Shielding: Naval reactors feature compact pressure vessels designed to withstand the stresses of both underwater operations and enemy action. The pressure vessel must include an internal neutron and gamma shield to protect the reactor structure from radiation damage. This differs from some early Soviet civilian reactors, where embrittlement due to neutron exposure was a problem.

4. Operational Flexibility and Efficiency: Naval reactors need to provide flexible power output to meet varying demands, such as rapid acceleration during evasive manoeuvres. However, space constraints limit the size of steam systems, resulting in lower thermal efficiency compared to land-based nuclear power plants. Most naval reactors also avoid using soluble boron for reactivity control, which is common in commercial reactors. Instead, they may use boron as a burnable poison directly in the fuel.

5. Durability Against Shock and Vibration: Submarines and warships face significant mechanical stress from ocean turbulence and potential enemy action. Naval reactors are built to handle these conditions, requiring designs that can withstand shock and vibration while maintaining safe and reliable operation.

Reactor Design Variations and Fuel Trends

Naval reactor designs have evolved significantly over time, adapting to various technological advances and operational requirements. For example, the latest U.S. Virginia-class submarines use S9G reactors, which are designed for a 33-year service life without refuelling, and have the capability to operate with natural convection cooling under low-power conditions. The trend in newer French naval reactors is toward using LEU fuel with lower enrichment levels (about 5-6% U-235), aligning with civilian standards and minimizing the need for military-specific enrichment facilities.

Historical Examples and Unique Reactor Types

Russia's approach to submarine propulsion has included some unique reactor types. Early Russian submarines used VM-A pressurized water reactors with 20-21% enriched uranium, providing limited operational life. The Alfa-class submarines were a notable departure, using liquid metal-cooled reactors with highly enriched uranium (90% U-235) for exceptionally high speed and depth capabilities. However, operational issues with maintaining the liquid metal coolant led to the early retirement of these vessels.

The U.S. has also experimented with different reactor types, such as the sodium-cooled reactor used in the USS Seawolf (SSN-575). This reactor offered higher thermal efficiency due to superheated steam, but problems with the sodium coolant's radioactivity and the need for continuous heating to prevent freezing led to its replacement with a conventional pressurized water reactor.

Dual- and Single-Reactor Designs

Most modern submarines use a single reactor, except for some larger ballistic missile submarines and surface ships, which employ twin-reactor setups for redundancy and increased power output. The U.S. Ohio-class SSBNs, for instance, have a single reactor requiring mid-life refuelling, while the future Columbia-class will feature reactors that do not need refuelling for their entire service life, thereby shortening maintenance periods.

Safety and Performance Standards

Naval reactors are designed to operate safely under a wide range of conditions. Their compact design and robust shielding help contain radiation, while the use of highly enriched fuel enables long intervals between refuelling. Performance standards for naval reactors prioritize reliability, longevity, and the ability to adapt to rapidly changing operational demands, reflecting the critical role of submarines and warships in military strategy.

The UK and Russia have extensive histories in naval nuclear reactor development for submarines, with unique approaches and designs that have evolved over decades [107].

UK Nuclear Submarine Reactors

The UK's nuclear submarine fleet began with the Rolls-Royce PWR1, a reactor based on the American Westinghouse S5W, provided under the 1958 US-UK Mutual Defence Agreement. The PWR1, producing approximately 78 MWt, powered the first 23 British nuclear submarines. It utilized high-enriched uranium fuel and required refuelling approximately every ten years.

For its ballistic missile submarines, the UK developed the PWR2 reactor, which powers the Vanguard-class SSBNs. The PWR2 has a thermal power output of around 145 MWt and drives two steam turbines that power a single pump-jet propulsion system delivering 20.5 MW of propulsive power. The Vanguard-class submarines, with a displacement of 15,900 dwt submerged, benefited from the increased power and longer intervals between refuelling. The Astute-class attack submarines also use the PWR2, allowing these 7,400 dwt vessels to achieve efficient propulsion and extended operational capability. The "Core H" variant of the PWR2, introduced with newer Astute-class submarines, has been designed to last for the entire 25-year service life of the vessel without refuelling, marking a significant advancement in reactor longevity [107].

The PWR3 reactor, designated for the Dreadnought-class SSBNs (which will replace the Vanguard-class), represents a new step in British naval nuclear reactor technology. It is expected to draw heavily from U.S. reactor designs, particularly the S9G reactor used in Virginia-class submarines. The PWR3 will incorporate advanced safety features, more efficient power generation, and a longer operational life compared to its predecessors. It will be more expensive to build but will reduce maintenance costs over the submarine's lifespan [107].

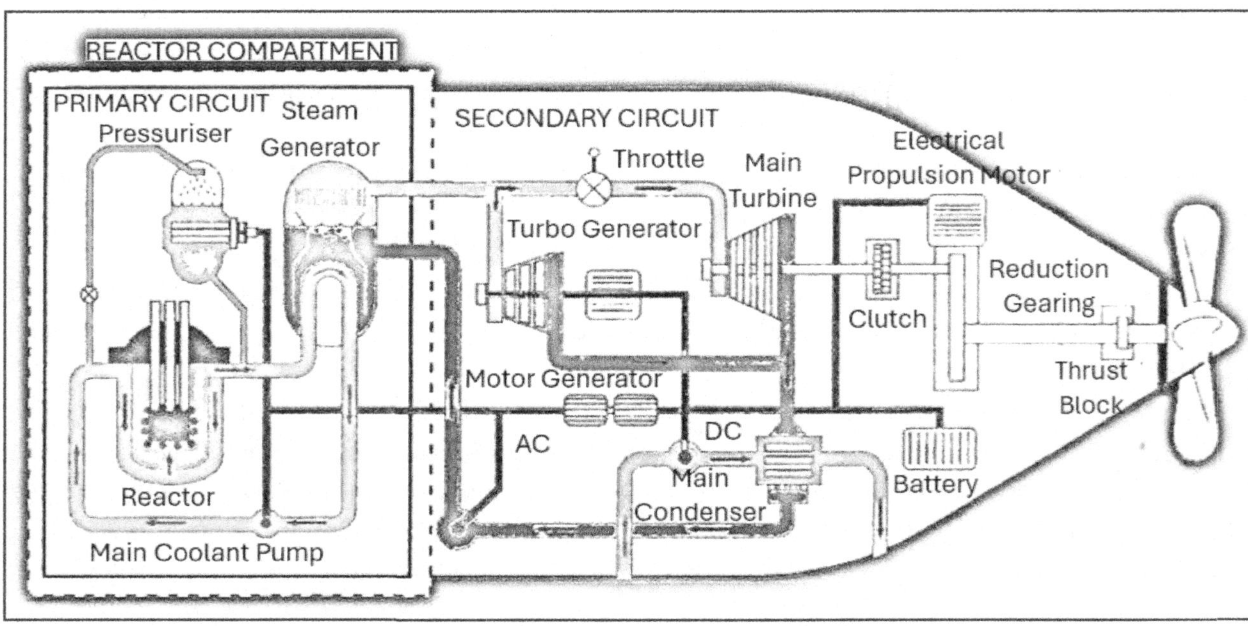

Figure 53: UK nuclear submarine layout.

Russian Naval Reactor Development

Russia has a long history of naval reactor development, starting with four generations of pressurized water reactors (PWRs) since 1959. Early reactors like the OK-150 were used in icebreakers such as the Lenin, while more advanced OK-900 series reactors powered the Arktika-class icebreakers and other civil vessels. These designs evolved independently of Russia's VVER power reactors, being specifically adapted for the marine environment by OKBM Afrikantov [107].

The KLT and RITM reactor designs followed, offering improvements in efficiency and operational safety. The RITM-200 reactor, with a thermal output of 175 MWt, powers the new LK-60 icebreakers, providing up to 60 MW of propulsive power. These reactors use low-enriched uranium (about 20%) and are designed for extended service life, with refuelling intervals of 7-10 years and a total operational life of 60 years. The RITM-400, under development for the LK-120 icebreakers, will offer even greater power output with a thermal rating of 315 MWt and a 10-year refuelling cycle [107].

Russian Submarine Reactor Systems

Russia's primary submarine reactor is the VM-5 with the OK-650 steam generator system, producing 190 MWt and using 20-45% enriched uranium. The OK-650 system is used in various Russian submarine classes, including third-generation attack submarines (SSNs) and large ballistic missile submarines (SSBNs), where two such units provide a combined 74 MW of propulsive power.

The Borei-class SSBNs, Russia's latest generation of ballistic missile submarines, are the first to utilize pump-jet propulsion and are powered by a single OK-650 reactor producing 195 MWt. Russia is also developing a fifth-generation naval reactor, reportedly a supercritical water reactor (SCWR) designed to operate for 30 years without refuelling.

The Yasen-M class SSGNs (submarine guided missile cruisers) are comparable to the U.S. Virginia-class in capabilities and will replace older Akula-class submarines. The Yasen class features the KTP-6 reactor, an advanced integral PWR based on the OK-650 system [107].

Special Cases and Historical Developments

Notably, Russia's Alfa-class submarines used liquid metal-cooled reactors, specifically the BM-40A and OK-550 fast neutron reactors, which operated with very highly enriched uranium (90%). These reactors allowed for high speed and depth capabilities but faced significant operational challenges, particularly with maintaining the lead-bismuth coolant. As a result, the Alfa class was retired early [107].

Russia also experimented with twin-reactor configurations in surface ships, such as the KN-3 reactors in Kirov-class battle cruisers, each producing 300 MWt.

International Collaborations and Unique Reactor Uses

The UK obtains its highly-enriched uranium fuel from the USA, and many reactor technologies used in British submarines have U.S. origins. Russia, on the other hand, has a more diversified history of reactor development, with contributions from multiple design bureaus [107].

China's first nuclear submarines, such as the Han-class SSNs and Xia-class SSBNs, reportedly used reactor designs based on Russia's OK-150. Over time, Chinese reactors evolved to incorporate features from both Western and Russian designs, with newer classes like the Shang and Jin using reactors with thermal outputs in the range of 150-175 MWt [107].

India's Arihant-class SSBNs feature an indigenously developed PWR with a thermal output of 100 MWt, enriched to around 40%. Brazil and Argentina have also explored nuclear submarine propulsion, though their projects remain in early stages of development.

Challenges in Dismantling and Decommissioning

The decommissioning of nuclear submarines involves removing the reactor fuel and cutting out the reactor section for disposal. In Russia, some decommissioned submarines remain stored afloat due to limited disposal facilities. The U.S. has established procedures for defueling and shipping reactor compartments to dedicated disposal sites, as demonstrated with the USS Enterprise.

Naval nuclear reactor technology has evolved significantly, driven by the need for extended operational life, safety, and performance. While the UK relies on U.S. collaboration and maintains a fleet of submarines with advanced reactors like the PWR2 and upcoming PWR3, Russia has developed a variety of reactors for both submarines and icebreakers. Each country's approach reflects unique technological and strategic priorities, with a focus on reactor longevity, refuelling intervals, and power output adaptations for specific vessel types [107].

The Functioning of PWRs

A submarine's pressurized water reactor (PWR) is a type of nuclear reactor used to generate power for propulsion and onboard systems. The PWR design is the most widely used for both naval and civilian nuclear power plants, as it provides a safe and efficient method for generating heat from nuclear fission. The basic principle behind a PWR is to use heat produced by the nuclear reaction to generate steam, which then drives turbines that produce mechanical energy for propulsion and electricity for other systems on the submarine.

In a PWR, heat generated from nuclear fission is transferred to water under high pressure in a primary loop, preventing it from boiling. The heat is then transferred to a secondary loop through steam generators, where it converts water into steam to drive turbines for propulsion and electricity generation. The primary and secondary systems are separate, ensuring that radioactive water from the primary loop does not mix with the non-radioactive water in the secondary loop.

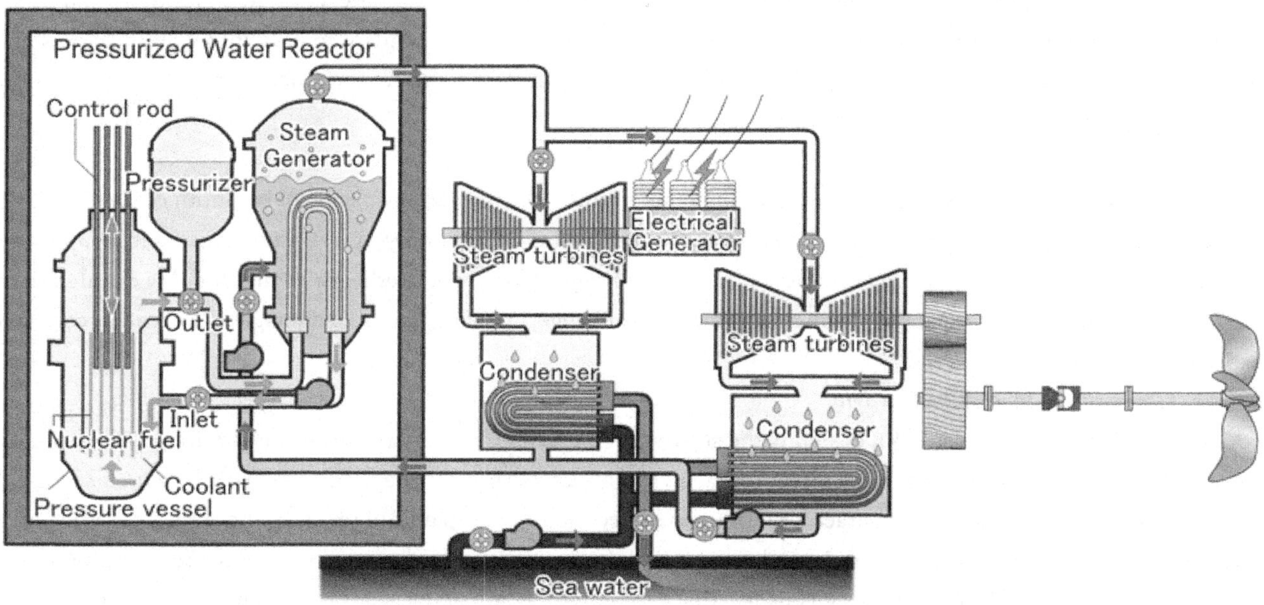

Figure 54: Pressurized Water Reactor. Tosaka, CC BY-SA 3.0, via Wikimedia Commons.

Nuclear Fission Process

At the core of the reactor lies the fuel rods, which contain enriched uranium, a radioactive material capable of sustaining a nuclear fission reaction. In this process, the nuclei of uranium atoms split into smaller parts when bombarded by neutrons, releasing a significant amount of heat energy. The reactor utilizes highly enriched uranium, often more than 20% enrichment, which enables it to generate substantial power in a relatively small space, making it suitable for use within the confines of a submarine. This level of enrichment also extends the reactor's operational life, allowing the submarine to go years without needing refuelling.

Primary Loop – Heat Transfer

The heat produced during nuclear fission is transferred to a coolant, which is ordinary water (referred to as "light water"). This water serves as both a coolant and a neutron moderator, slowing down the neutrons to sustain the fission process. The water in the primary loop is kept at high pressure, preventing it from boiling even at temperatures significantly above the normal boiling point of water. This pressurized water is circulated through a closed loop, known as the primary loop, which includes the reactor vessel, steam generators, pumps, and associated piping. This system is sealed to ensure that any radioactive water remains contained within the loop, minimizing the risk of contamination.

Steam Generation – Secondary Loop

The heated water from the primary loop flows through the steam generator, where the heat is transferred across a watertight barrier to a separate, secondary water loop. In the steam generator, the primary loop's pressurized water releases its heat to the secondary loop, causing the water in the secondary system to boil and turn into steam. Because the secondary water is at a lower pressure, it converts to steam more readily, enabling efficient energy transfer. The primary and secondary loops are physically separated, ensuring that no radioactive water from the primary loop contaminates the secondary loop.

Turbines and Energy Production

The steam generated in the secondary loop is used to drive the main propulsion turbines, which turn the submarine's propeller to provide thrust and propel the vessel through the water. In addition to propulsion, the steam also powers turbine generators that produce electricity for various onboard systems, including life support, communications, and weapons systems. After passing through the turbines, the steam is cooled and condensed back into water using a condenser. The condensed water is then pumped back into the steam generator to repeat the process, creating a continuous, closed-loop system that maximizes energy efficiency.

Cooling the Reactor

To ensure that the reactor remains at safe operating temperatures, the primary loop coolant is continuously circulated back into the reactor to absorb more heat. Even when the submarine is stationary or not actively using its propulsion system, the heat generated by the nuclear reaction must be managed to prevent overheating. This ongoing circulation of coolant ensures that the reactor stays within a safe temperature range, avoiding potential damage to the reactor core and associated systems.

Safety Features – Isolation of Radioactivity

A crucial safety feature of the PWR design is the separation between the primary and secondary loops. The primary loop contains radioactive water due to its direct contact with the reactor core, while the secondary loop remains non-radioactive because there is no direct contact with the primary coolant. The separation prevents the spread of radioactive contamination, as the heat is transferred through a watertight barrier within the steam generator, without any mixing of the water from the two systems. This design ensures that any potential radioactive leakage remains contained within the primary loop, enhancing the safety of the reactor.

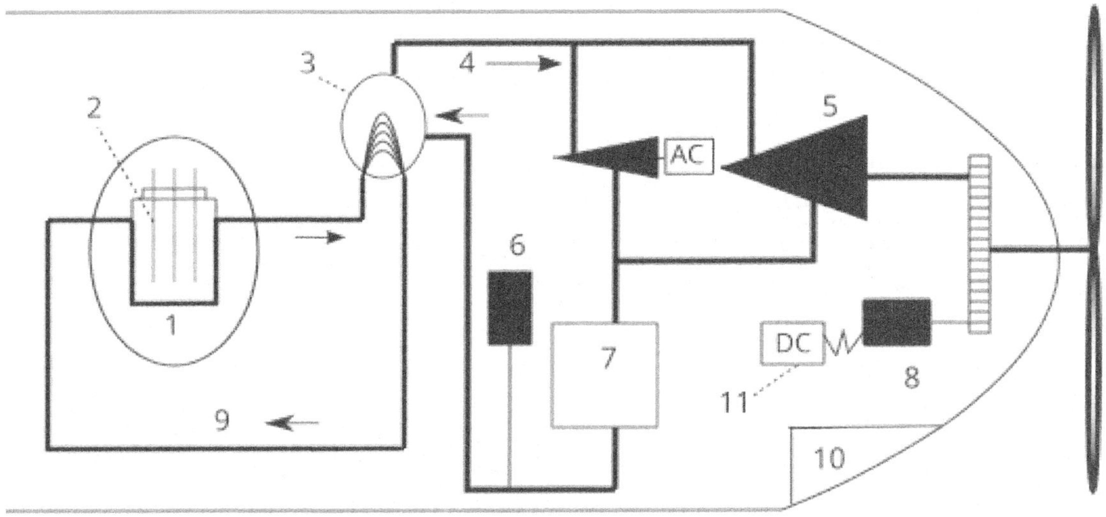

Figure 55: Submarine PWR reactor plant basic scheme. Matrek, CC BY-SA 3.0, via Wikimedia Commons.

Fuel

Pressurized Water Reactors (PWRs) utilize enriched uranium as their nuclear fuel, primarily due to the properties of light water used as a coolant and moderator in the reactor. The use of light water (ordinary water) presents a challenge because it absorbs a significant number of neutrons, which are essential for sustaining the nuclear fission chain reaction. To compensate for this neutron absorption, the uranium fuel must be enriched, which means increasing the concentration of the fissile isotope Uranium-235 [108].

Enrichment Process: Natural uranium contains only about 0.7% of Uranium-235, with the rest being mostly Uranium-238, which is non-fissile. For use in a PWR, the concentration of Uranium-235 is typically increased to around 4% through a uranium enrichment process. This higher proportion of Uranium-235 allows the reactor to achieve a sustained and controlled chain reaction, despite the neutron absorption characteristics of light water. Enrichment is crucial to enable the reactor to produce enough heat for power generation [108].

Fuel Rods and Fuel Bundles: The enriched uranium is fabricated into small cylindrical pellets, which are then packed into long, slender metal tubes known as fuel rods. Each fuel rod is filled with these pellets and sealed to prevent the release of radioactive materials. For a PWR, individual fuel rods are grouped together into fuel assemblies or bundles. Each bundle typically contains about 200-300 fuel rods, depending on the reactor design.

A large reactor core may contain between 150 and 250 fuel bundles, resulting in approximately five cubic meters of uranium, which corresponds to about 80-100 tonnes of uranium in total. The bundles are arranged vertically within the reactor core, allowing coolant (water) to flow around them, removing heat generated during the fission process [108].

Managing Fuel in the Reactor Core: As the reactor operates, nuclear fission causes the Uranium-235 atoms in the fuel to split, releasing energy in the form of heat, along with fission products and free neutrons. Over time, the density of the fuel inside the rods changes due to the formation of fission gases and other byproducts, leading to the development of small voids within the fuel rods. These voids can create pressure build-up within the rods, posing a risk of rupture.

To prevent this, the fuel rods are pre-pressurized with helium gas at around 3.4 MPa (megapascals). The helium helps absorb some of the fission gases released during operation and mitigates the stress on the cladding, which is the metal tube surrounding the fuel pellets. As the reactor operates, the pressure inside the fuel rods gradually equalizes with the high pressure within the reactor core, helping maintain the integrity of the fuel rods throughout their operational life [108].

Importance of Fuel Design in Reactor Safety: The design of fuel rods and fuel bundles is a critical aspect of reactor safety. The materials used for cladding, typically zirconium alloys, are chosen for their low neutron absorption and resistance to high temperatures. Properly maintaining the integrity of the cladding ensures that radioactive fission products remain contained within the fuel rods, preventing contamination of the reactor coolant and limiting radiation exposure.

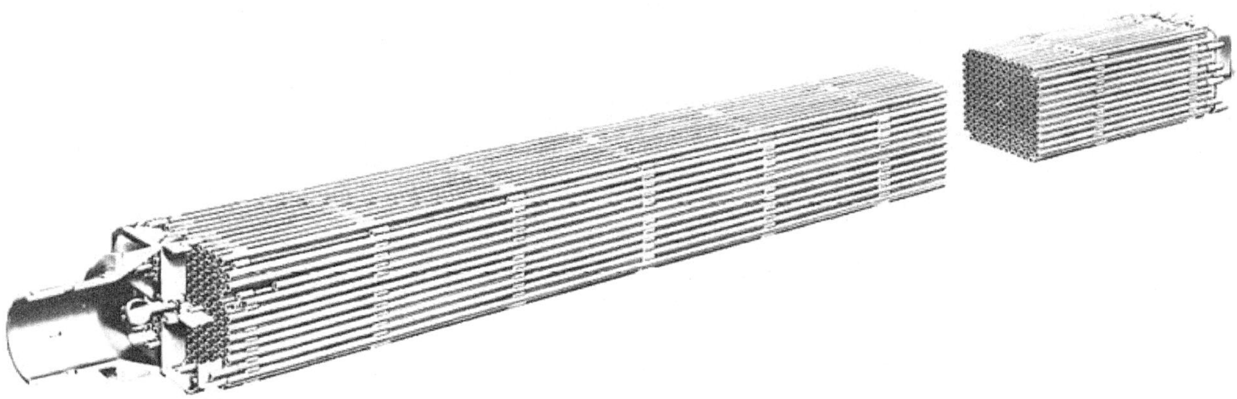

Figure 56: A nuclear fuel element for the cargo ship NS Savannah. The element contains four bundles of 41 fuel rods. The uranium oxide is enriched to 4,2 and 4,6 percent of U-235. U.S. Maritime Administration. PD-USGov, Public domain, via Wikimedia Commons.

The use of enriched uranium in PWRs enables efficient and sustained nuclear reactions, while the design of fuel rods and bundles ensures safe operation under the reactor's high-pressure and high-temperature conditions. The meticulous arrangement of fuel within the core and the management of fission by-products are essential for maintaining reactor stability and safety over the fuel's lifespan.

Coolant and Moderator

In a Pressurized Water Reactor (PWR), light water serves two essential roles: as a coolant and as a moderator. Light water, which is ordinary water consisting mainly of hydrogen and oxygen, is much more common and readily available than heavy water. Heavy water contains a higher proportion of the hydrogen isotope deuterium, but light water accounts for approximately 99.99% of all naturally occurring water [108].

The primary function of light water as a coolant in a PWR is to remove the heat generated by the nuclear fission process in the reactor core. During fission, the splitting of uranium atoms releases a substantial amount of energy in the form of heat. The coolant, which is kept under high pressure to prevent it from boiling even at elevated temperatures, circulates through the reactor core to absorb this heat. The heated water then transfers the energy to a secondary system, where it generates steam to drive turbines for electricity production [108].

In addition to its cooling role, light water acts as a moderator. A moderator is a substance that slows down fast neutrons produced during fission to thermal energies, which are more effective in sustaining the chain reaction. Slower, thermal neutrons have a higher probability of being absorbed by Uranium-235 atoms in the fuel, causing more fission reactions and sustaining the nuclear process.

While light water is not as effective a moderator as heavy water or graphite due to its relatively high neutron absorption rate, it serves an important safety function. Its dual role as both coolant and moderator contributes to the inherent safety of the PWR design.

The use of light water provides a significant safety advantage in the event of a loss of coolant accident (LOCA). In such scenarios, if the coolant is lost, the light water would no longer be present to act as a moderator, which would automatically halt the chain reaction. This self-limiting safety feature reduces the risk of a runaway reaction and potential reactor meltdown.

Furthermore, if the reactor coolant overheats and forms steam bubbles inside the core, the reduction in water density decreases the moderating effect. This phenomenon, known as a "negative void coefficient," causes the nuclear chain reaction to slow down or stop because fewer thermal neutrons are available to sustain the reaction. This characteristic helps stabilize the reactor's behaviour under abnormal operating conditions, enhancing the overall safety profile of the PWR [108].

While the use of light water provides inherent safety benefits, it also requires the reactor to use enriched uranium fuel. Since light water absorbs some neutrons, it reduces the number of neutrons available to sustain the chain reaction with natural uranium. Therefore, the uranium must be enriched, increasing the proportion of fissile Uranium-235 to ensure a sufficient number of fission reactions occur to generate power [108].

Light water plays a crucial role in PWR operation by serving as both the coolant and the moderator. Its widespread availability and the safety features associated with its moderating properties make it an effective choice for reactor designs, despite its limitations compared to heavy water or graphite moderators. The inherent safety provided by light water, particularly in accident scenarios, is a key factor in the continued use and development of PWRs for nuclear power generation.

Emerging Propulsion Innovations

Emerging submarine propulsion innovations are critical for enhancing the operational capabilities of modern naval fleets. This response synthesizes various advancements in propulsion technologies, including hybrid systems, advanced battery technologies, fuel cell technology, nuclear propulsion innovations, and novel propulsion concepts.

Hybrid Propulsion Systems are gaining traction due to their ability to combine the advantages of different energy sources. The Diesel-Electric Hybrid system integrates traditional diesel engines with electric propulsion, offering flexibility in energy use and extended underwater endurance. This configuration allows submarines to operate quietly and efficiently, especially during stealth missions [109]. Air-Independent Propulsion (AIP) technologies, such as fuel cells and Stirling engines, significantly enhance the submerged endurance of non-nuclear submarines, enabling them to remain submerged for extended periods without surfacing to recharge batteries [110]. Furthermore, the Nuclear-Diesel Hybrid system combines the long-range capabilities of nuclear propulsion with the cost-effectiveness of diesel engines, optimizing fuel efficiency for various mission profiles [109].

Advanced Battery Technologies are also pivotal in submarine propulsion. Lithium-Ion Batteries have emerged as a superior alternative to traditional lead-acid batteries, providing higher energy density, faster charging times, and longer operational life. This technology is already being implemented in next-generation submarines, such as Japan's Soryu-class [109]. Looking towards the future, Solid-State Batteries promise even greater advancements in energy density and safety, potentially allowing for longer underwater endurance and more reliable performance [109].

Figure 57: JS Ōryū. Japan Maritime Self-Defense Force (JMSDF), CC BY 4.0, via Wikimedia Commons.

The JS Ōryū, the eleventh Sōryū-class submarine of the Japan Maritime Self-Defense Force (JMSDF), represents a significant advancement in submarine technology, particularly with its incorporation of lithium-ion batteries. This transition to lithium-ion technology enhances the operational capabilities of submarines by providing greater energy density and reduced maintenance requirements compared to traditional battery systems. The increased energy density can potentially double the submerged range of submarines, thereby improving their operational agility and effectiveness in various missions [111].

The financial backing for the JS Ōryū was substantial, with a budget allocation of ¥64.3 billion under the 2015 Japanese Defence Budget, which translates to approximately US$601.3 million in 2019 [112]. This investment underscores Japan's commitment to modernizing its naval capabilities and ensuring that its fleet remains competitive in an increasingly complex security environment. The funding is part of a broader strategy to enhance the technological sophistication of Japan's defence systems, which includes the integration of advanced battery technologies that promise improved performance and safety [113].

Moreover, the integration of lithium-ion batteries into submarine design does not come without challenges. Safety concerns, particularly regarding thermal runaway and the potential for combustion, necessitate rigorous risk assessments and safety protocols [114, 115]. Studies have shown that lithium-ion batteries, while offering superior performance, pose significant fire safety risks if not properly managed [114, 116]. Therefore, the development of robust monitoring systems and fault diagnosis capabilities is essential to ensure the safe operation of these advanced battery systems in submarines [115, 117].

Fuel Cell Technology is another area of significant innovation. Proton Exchange Membrane (PEM) Fuel Cells are utilized in modern submarines to generate electricity quietly by combining hydrogen and oxygen, thus supporting extended underwater operations [110]. Additionally, Solid Oxide Fuel Cells (SOFC) offer higher efficiency and operational flexibility by utilizing various fuels, which could further enhance submarine capabilities [110].

PEM Fuel Cells operate by facilitating the electrochemical reaction between hydrogen and oxygen, producing electricity, water, and heat as byproducts. This reaction occurs within a membrane that selectively allows protons to pass while blocking electrons, which are forced to travel through an external circuit, generating electricity. The quiet operation of PEM Fuel Cells is particularly advantageous for submarines, as it allows for stealthy underwater operations without the noise associated with traditional combustion engines [118]. The efficiency and reliability of PEM technology have made it a preferred choice for various applications, including transportation and portable power systems [118].

On the other hand, Solid Oxide Fuel Cells (SOFCs) operate at significantly higher temperatures, typically between 600°C and 1000°C. This high-temperature operation allows SOFCs to achieve higher efficiencies and greater fuel flexibility compared to other fuel cell types. SOFCs can utilize a variety of fuels, including hydrogen, natural gas, and even syngas, which enhances their operational versatility [119, 120]. The electrochemical process in SOFCs involves the oxidation of the fuel at the anode and the reduction of oxygen at the cathode, with oxygen ions conducting through a solid electrolyte to complete the circuit [121, 122]. This mechanism not only allows for efficient electricity generation but also supports co-generation applications, where both electricity and heat can be produced simultaneously [123, 124].

The advancements in SOFC technology have led to improvements in materials and designs, which are critical for enhancing performance and durability. For instance, recent studies have focused on optimizing anode materials and configurations to improve the electrochemical performance and longevity of SOFCs [125, 126]. Additionally, the ability of SOFCs to operate on various fuels, including ammonia and biomass-derived gases, positions them as a promising solution for sustainable energy generation [127, 128].

In the realm of Nuclear Propulsion Innovations, technologies such as High-Temperature Gas-Cooled Reactors (HTGRs) and Small Modular Reactors (SMRs) are being explored for their potential to improve thermal efficiency and reduce cooling requirements, thereby enhancing stealth and power output. However, specific references to these technologies in the context of submarine propulsion were not found in the provided citations.

In the context of submarine propulsion, the exploration of advanced nuclear technologies such as High-Temperature Gas-Cooled Reactors (HTGRs) and Small Modular Reactors (SMRs) is gaining traction due to their potential to enhance thermal efficiency and reduce cooling requirements. These innovations are particularly relevant for military submarines, where stealth and power output are critical operational parameters. HTGRs, characterized by their ability to operate at high temperatures while maintaining safety and efficiency, could theoretically allow for more compact reactor designs that fit within the constraints of submarine hulls. This compactness could lead to reduced thermal signatures, thereby enhancing stealth capabilities, which is paramount in submarine operations [129].

SMRs, on the other hand, offer modularity and scalability, which can be advantageous in military applications. Their design allows for easier integration into existing submarine architectures, potentially leading to quicker deployment and operational readiness. The modular nature of SMRs also means that they can be manufactured off-site and transported to the submarine construction facilities, streamlining the construction process and reducing costs [129]. Furthermore, the inherent safety features of SMRs, such as passive cooling systems, could significantly mitigate the risks associated with nuclear propulsion in confined environments like submarines [105].

The operational experience accumulated since the launch of the first nuclear-powered submarines, such as the USS Nautilus, underscores the reliability of nuclear propulsion systems. With over 12,000 reactor years of operational experience, the engineering advancements in nuclear propulsion have demonstrated that these systems can be both safe and effective [106]. The integration of advanced reactor technologies like HTGRs and SMRs could further enhance this reliability, providing submarines with the capability to operate for extended periods without surfacing, thus maintaining a strategic advantage in naval warfare [105].

Moreover, the development of these technologies is not just limited to enhancing propulsion efficiency but also extends to improving overall energy management within submarines. The implementation of sophisticated energy management systems can optimize the use of power generated by nuclear reactors, ensuring that energy is allocated efficiently to various submarine systems, including propulsion, navigation, and weaponry [109]. This holistic approach to energy management is crucial for maximizing the operational capabilities of modern submarines.

Pump-Jet Propulsion Systems represent a shift towards quieter and more efficient propulsion methods. These systems are increasingly used in modern submarines due to their ability to operate efficiently at higher speeds while minimizing cavitation noise, which is crucial for stealth operations [130, 131]. Rim-Driven Thrusters (RDT), which integrate electric motors into the rim of the propeller, further enhance manoeuvrability and reduce noise, making them an attractive option for future submarine designs [130, 132, 133].

Pump-jet propulsion systems and rim-driven thrusters (RDT) represent significant advancements in submarine propulsion technology, focusing on enhancing stealth and operational efficiency. The pump-jet propulsion system operates by drawing water into a pump, which accelerates the water and expels it through a nozzle. This design minimizes cavitation noise, a critical factor for stealth operations, as cavitation can produce detectable sound signatures that compromise a submarine's stealth capabilities [134]. The hydrodynamic characteristics of pump-jet systems are heavily influenced by the boundary layer flow around the submarine, which can be optimized to improve propulsion efficiency significantly [134]. Studies indicate that the pump-jet system can achieve higher propulsion efficiency compared to traditional propeller systems by effectively managing the boundary layer flow, thus enhancing the submarine's operational capabilities at various speeds [134].

Rim-driven thrusters further enhance submarine manoeuvrability and noise reduction. These systems integrate electric motors into the rim of the propeller, allowing for a more compact design that reduces the mechanical complexity associated with traditional propulsion systems [106]. The design of rim-driven thrusters allows for a more efficient transfer of energy to the water, which not only improves propulsion efficiency but also minimizes the noise generated during operation. This is particularly advantageous for modern submarines that require high levels of stealth for tactical operations [106]. The application of computational fluid dynamics (CFD) in analysing the performance of these systems has shown promising results, indicating that rim-driven thrusters can outperform conventional systems in terms of both efficiency and noise reduction [135].

Emerging concepts such as Magnetohydrodynamic (MHD) Propulsion and Bladeless Propulsion systems are also being researched. MHD propulsion offers the potential for silent operation by using magnetic fields to accelerate seawater, although challenges related to power requirements remain. However, specific references to MHD propulsion in the context of submarines were not found in the provided citations. Bladeless propulsion concepts, inspired by fluid dynamics, aim to reduce noise associated with traditional propellers, although they are still largely experimental.

MHD propulsion operates on the principle of using magnetic fields to accelerate seawater, thereby generating thrust without the mechanical components typical of traditional propellers. This technology has the potential for silent operation, which is a crucial advantage for submarines that require stealth in military applications. However, the primary challenge lies in the substantial power requirements necessary to generate the magnetic fields and maintain effective propulsion. Current research indicates that while MHD systems could theoretically enhance the stealth capabilities of submarines, practical implementation remains hindered by these power constraints and the need for further technological advancements [106, 136].

On the other hand, bladeless propulsion systems, which are inspired by fluid dynamics, aim to minimize the noise typically associated with traditional propeller systems. These systems utilize oscillating structures to generate thrust through the interaction of fluid flow and mechanical motion, thereby reducing cavitation and noise levels. Although bladeless propulsion is still largely experimental, its potential for quieter operation aligns well with the operational requirements of modern submarines, particularly in covert missions. The development of such technologies could lead to significant advancements in underwater vehicle design, enabling more efficient and stealthy operations [137].

In addition to these emerging technologies, existing propulsion systems such as pump-jet thrusters have been widely adopted in submarines due to their efficiency and reduced noise output. These systems utilize a combination of a pump and a jet to propel the submarine, offering advantages such as high propulsion efficiency and strong anti-cavitation capabilities. For instance, the American "MK48" torpedo and various classes of submarines like the Seawolf and Virginia-class utilize pump-jet technology, demonstrating its effectiveness in current naval applications [72, 134, 138].

Figure 58: Virginia-class attack submarine USS Mississippi (SSN 782). Defense Visual Information Distribution Service, Public Domain, via NARA & DVIDS Public Domain Archive.

Finally, the integration of Artificial Intelligence (AI) in propulsion control systems is revolutionizing submarine operations. AI can optimize energy management by switching between propulsion modes for maximum efficiency and stealth, as well as improving navigation and evasion tactics in complex underwater environments. However, specific references to AI in submarine propulsion were not found in the provided citations.

Chapter 5

Submarine Power and Electrical Systems

Electrical Distribution in Submarines

Submarine Systems

Submarine systems are specially designed for the unique underwater environment in which they operate, while also sharing some similarities with surface ships. The primary systems on a submarine support essential functions like propulsion, manoeuvring, power management, and communication, and they include the propulsion system, diving and surfacing system, power generation and distribution, and various others. Each of these systems plays a vital role in ensuring the submarine's operational capabilities and mission success [139].

Power Generation and Distribution System: The primary power source on a diesel-electric submarine comes from diesel alternators, which typically consist of two sets for port and starboard. These alternators serve several key functions: recharging the batteries after submerged operations, supplying power for onboard systems, and propelling the submarine when in "snorting" condition. Battery charging is essential, as batteries supply the electric power needed for submerged propulsion. The charging time is determined by the battery's capacity and specifications rather than the diesel alternators themselves.

When a submarine operates in "snorting" mode, it uses diesel generators while staying just below the surface, with a snorkel mast raised above the water to allow air intake for the diesel engines. This reduces the risk of detection while recharging the batteries. Modern submarines often feature Air-Independent Propulsion (AIP) systems, which enable submerged operation by using stored air, further reducing the likelihood of detection [139].

Detection Systems

Submarines use a range of detection systems to monitor their surroundings and gather data for navigation and weapon control. These include periscopes, radar, and sonar systems [139].

- **Periscopes**: Submarines have two types of periscopes, search and attack. The search periscope provides navigational capabilities and enhanced optical performance, while the attack periscope is optimized for minimal detection. It is used for observing shipping traffic, estimating distances, and providing input to the fire control system. The periscope is equipped with a heave control system to maintain a consistent height above the water surface, even with the submarine's movement.

- **Radar Systems**: While surfaced or in snorting mode, submarines can use radar to detect surface objects and ships. There are two types: navigational and attack radar. The radar mast extends above the water surface, and because radar signals can reveal the submarine's position, it is used sparingly when stealth is a priority.

- **Sonar Systems**: Sonar is the primary detection tool used by submarines. Active sonar emits sound waves to detect objects underwater, while passive sonar listens for external noises. Active sonar provides detailed information but can reveal the submarine's position, so passive sonar is often used for stealth operations. The sonar hydrophones are arranged to minimize disturbances and detect signals from various directions effectively.

Diving and Surfacing System: Submarines use ballast tanks to control buoyancy, allowing them to dive, surface, and maintain neutral buoyancy. These tanks are filled with water when diving and with air when surfacing. Controlling buoyancy is crucial for managing the submarine's depth and trim, which ensures safe and efficient underwater operations [139].

Air Conditioning and Ventilation System: Maintaining air quality and temperature control is vital for crew safety and comfort. Submarines use air conditioning systems to regulate internal temperatures and ensure the air is breathable by removing excess carbon dioxide and humidity. These systems also prevent electronic equipment from overheating.

Hydraulic and Compressed Air Systems: Hydraulic systems control many of the submarine's moving parts, including steering, diving planes, and periscope raising. Compressed air systems are used for tasks such as blowing water out of the ballast tanks during surfacing, operating masts, and supplying air to crew members during certain operations.

Cooling and Lubrication Systems: Cooling systems prevent overheating of machinery, while lubrication systems reduce friction and wear on mechanical components. The water cooling system circulates seawater through heat exchangers to remove heat from the diesel engines and other equipment, while the lubrication system ensures smooth operation of moving parts [139].

Missile and Torpedo Firing System: Submarines are equipped with complex weapon systems for launching missiles and torpedoes. These systems involve storage, loading, and firing mechanisms that are designed to function in the challenging underwater environment. Precision in firing control and noise suppression is crucial for maintaining stealth during combat operations.

List and Trim Management System: Maintaining the submarine's stability and orientation underwater is achieved through a list and trim management system. This system involves transferring water between trim tanks to adjust the submarine's balance and ensure optimal control during operations.

Masts, Periscopes, and Antennas: Submarines have various masts for communication, radar, and periscope functions. These masts are pressure-resistant structures that can be raised or lowered as needed, ensuring that the submarine remains functional at various depths and in different operational conditions.

Fire-Fighting System: Safety is paramount aboard a submarine, and a fire-fighting system is in place to handle emergencies. It includes portable extinguishers, fire suppression systems, and ventilation controls to contain and extinguish fires [139].

Advanced Detection and Navigation: The depth gauging and echo-sounding system help submarines measure their depth and detect underwater features. This is essential for navigation and avoiding underwater obstacles. Sound power telephonic and intercom systems provide internal communication, ensuring coordination among the crew.

Propulsion System and Shafting: The propulsion system consists of the main engine, reduction gears, and shafting that connects to the propeller. Modern submarines may use pump-jet propulsion for reduced noise. The shafting system transmits power from the engine to the propeller, ensuring efficient movement [139].

These interconnected systems work together to enable submarines to perform complex underwater missions while maintaining stealth, safety, and operational efficiency.

Submarine Electrical Systems

Submarine electrical systems play a crucial role in supporting propulsion, lighting, weaponry, communication, life support, and various onboard electronics. These systems are specially designed to ensure dependable operation in the challenging underwater environment while maintaining stealth. Here's an in-depth look at how submarine electrical systems function:

Submarines rely on multiple power sources to guarantee continuous operation under various conditions. In nuclear-powered submarines, the primary electrical power source is the nuclear reactor, which generates heat to produce steam that drives turbines connected to electric generators. This electricity is then used for propulsion, powering onboard systems, and recharging backup batteries. On the other hand, diesel-electric submarines utilize diesel generators to charge large battery banks, which provide electrical power when the submarine is submerged.

In nuclear submarines, the reactor-generated steam powers turbine generators, whereas, in diesel-electric models, the diesel generators supply electricity for propulsion and other systems when the submarine is surfaced or in a "snorting" condition, where air intake is allowed via a snorkel.

Batteries are essential for providing electrical power during submerged operations, particularly in diesel-electric submarines. These large battery banks consist of multiple cells designed to store enough energy to support several hours or days of underwater operation, depending on power demands. In nuclear-powered submarines, batteries serve as an emergency power source in case of a reactor shutdown.

In diesel-electric submarines, the battery banks supply power to electric propulsion motors and auxiliary systems while submerged. These batteries are recharged using diesel generators when surfaced. Additionally, some modern diesel-electric submarines are equipped with air-independent propulsion (AIP) systems, which generate electricity underwater without the need to surface, thereby extending submerged endurance and minimizing the risk of detection.

The submarine's power distribution system delivers electricity from the generators and batteries to various systems and equipment onboard. It is engineered to handle different voltages and currents, ensuring each component receives the appropriate power.

A main distribution board manages and directs the generated electrical power to various circuits. Submarines employ both alternating current (AC) and direct current (DC) systems. Typically, AC power is used for lighting, heating, ventilation, and auxiliary equipment, while DC power is reserved for critical systems such as propulsion motors, weaponry, and emergency equipment.

Electrical power is vital for propulsion, especially in electric-drive submarines, where the propulsion motor is powered by the electrical system, delivering the torque necessary to turn the propeller or pump-jet. In diesel-electric submarines, electric motors drive the propeller while submerged, drawing power from the batteries. In nuclear submarines, the electric motors are powered directly by electricity generated from the reactor-driven steam turbines. The propulsion motor's speed can be adjusted to control the submarine's velocity, enabling silent running at low speeds, which is crucial for stealth.

Submarine electrical systems are designed with redundancy to ensure power availability even during emergencies. In case of a main power failure, backup batteries automatically supply power to essential systems. Submarines may also be equipped with additional small diesel generators or emergency power units to provide electricity during critical situations. Automatic switching mechanisms enable a seamless transition from primary to backup power, ensuring the continuous operation of vital systems like life support, communication, and navigation.

Efficient power management is essential, particularly on diesel-electric submarines where battery life is limited. Submarines utilize load shedding techniques to conserve battery power during extended submerged operations, shutting down non-essential systems to extend operational time. Additionally, modern submarines feature energy-efficient lighting and equipment to further reduce power consumption.

The electrical system also supports various auxiliary systems on the submarine, which include life support (such as ventilation, air purification, and oxygen generation), communication (internal systems and external equipment like sonar and satellite links), navigation and detection (sonar and radar systems), and weapon systems (powering launch mechanisms and targeting controls).

Due to the challenging marine environment, submarine electrical systems incorporate special insulation and safety measures to prevent short circuits and electrical failures. All electrical components are waterproofed, and connectors are sealed to prevent water ingress. Additionally, electrical systems are grounded to prevent electromagnetic interference and ensure safety.

Overall, submarine electrical systems are essential for powering propulsion, life support, communication, navigation, and weaponry. Through a combination of generators, batteries, effective power management, and redundancy, submarines achieve reliable and safe operations while maintaining stealth in their underwater environment.

Diesel-electric submarines primarily utilize diesel generators to produce electricity, which is then used to charge batteries and power the submarine's systems while submerged. This configuration allows for increased operational stealth and efficiency compared to traditional mechanical propulsion systems.

The core of the diesel-electric system consists of diesel generators that convert mechanical energy from diesel engines into electrical energy. These generators are crucial for charging the submarine's batteries, which serve as the primary power source when the submarine is submerged. The use of batteries, particularly lead-acid or more advanced lithium-ion batteries, significantly enhances the submerged endurance and operational range of the submarine. Studies indicate that modern lithium-ion batteries can potentially double the submerged range compared to conventional lead-acid systems, making them a favourable option for future submarine designs [111].

Energy management is another critical aspect of diesel-electric submarines. An optimal energy scheduling procedure is essential to efficiently manage the limited resources available in an isolated underwater environment. The energy management system must balance the power generated by the diesel generators with the energy consumed by various onboard systems, including propulsion, navigation, and life support [109]. This involves running the diesel generators at surface or snorkeling depths to recharge the batteries while ensuring that the load distribution between the generators and the battery systems is optimized to reduce fuel consumption and emissions [140].

The propulsion system in diesel-electric submarines typically employs DC motors, which are favoured for their efficiency and control capabilities. These motors can be configured in various ways, including using multiple motors with different power ratings to enhance performance at low speeds, thereby improving the overall efficiency of the submarine's drive system [141]. Additionally, the design of the electrical distribution network is critical, as it must handle the varying loads from propulsion systems and other electrical consumers onboard. This includes managing the power distribution to propulsion drives, auxiliary systems, and hotel loads, which are essential for crew comfort and operational capabilities [142].

Moreover, the integration of advanced technologies such as hybrid propulsion systems is gaining traction in modern submarine design. These systems can switch between diesel and electric modes, allowing for greater flexibility and efficiency in power management. The coupling mechanisms used in these systems are designed to facilitate quick transitions between power sources without complex electronic controls, thereby simplifying maintenance and enhancing reliability [143].

Diesel-Electric Main Generators and Motors and Auxiliary Generator

Modern submarines typically use either four main motors or two double-armature main motors to drive the propellers. In the case of four-motor configurations, the motors are arranged in pairs, with each pair connected to a propeller shaft via a reduction gear. For submarines equipped with double-armature motors, these motors are directly coupled to the propeller shafts, operating within the speed range needed to drive the propellers effectively [144].

The gear units used in submarines equipped with a reduction gear installation are of the single reduction, double helical type. These reduction gears are crucial for translating the high rotational speed of the main motors, typically around 1300 revolutions per minute (rpm), to a slower speed appropriate for the propeller, approximately 280 rpm. The reduction gear's design ensures smooth power transmission and minimizes mechanical noise, which is vital for maintaining the submarine's stealth [144].

The submarine's main motors can draw power from two primary sources: the main diesel engines or the storage batteries. When surfaced, the submarine's main generators, driven by the diesel engines, provide the necessary power for the motors, enabling efficient propulsion. In submerged conditions, where the submarine must maintain stealth and cannot run its diesel engines, the main motors receive power from the storage batteries. These batteries store energy generated by the main generators while surfaced, ensuring that the submarine can continue to operate underwater without needing to surface frequently [144].

The auxiliary generator plays a versatile role in the submarine's electrical system. Driven by an auxiliary diesel engine, it supplies current for all auxiliary circuits, such as lighting, life support systems, and other onboard electronics, which relieves the main batteries from supporting these loads. Additionally, the auxiliary generator can charge the main batteries at a low rate, keeping them topped up while conserving the primary power sources. For propulsion purposes, it can drive the main motors at slow speeds through the batteries, which is helpful during low-speed manoeuvring or when operating in confined spaces [144].

Control of the main propulsion machinery is managed through the main propulsion control equipment, commonly referred to as the control cubicle, located in the manoeuvring room. This equipment allows operators to manage the speed and direction of the submarine's propellers, switch between power sources, and monitor the system's performance. The control cubicle ensures that the propulsion system responds to the submarine's navigational requirements, providing the necessary flexibility for various operational scenarios [144].

The integration of these propulsion components allows a submarine to transition seamlessly between surface and submerged operations, adapt to different power demands, and maintain efficiency and stealth. The combination of reduction gears, multiple power sources, auxiliary generation capabilities, and centralized control ensures the submarine remains effective in both combat and peacetime missions [144].

Main and Auxiliary Generators

The main and auxiliary generators on a submarine are crucial components for producing electrical power, which supports a range of functions from propulsion to auxiliary systems. These generators share many similarities but also exhibit some key differences in construction and application. Overview of how these systems operate and are structured [144]:

- Main generators on submarines are typically described by several specific characteristics: they are two-wire, direct current (DC), separately excited, shunt wound, compensated multipolar, and self-ventilated machines. The generators are fully enclosed to protect the components from harsh marine environments. They are paired with the main diesel engines, using direct flexible coupling to the engine through the flanged end of the generator armature shaft. This coupling allows the generator to convert mechanical energy from the diesel engines into electrical power.

- The armature shaft in these generators varies depending on the engine type. Generators coupled with General Motors engines have bearings at both ends, while those paired with Fairbanks-Morse engines only support the commutator end with bearings. These bearings are lubricated by the engine's main lubricating system, ensuring smooth rotation and minimal wear.

- The maximum rotational speed for these generators is typically around 750 rpm for General Motors engines and 720 rpm for Fairbanks-Morse engines. The generators are rated to produce approximately 2,650 amperes at 415 volts, providing around 1,100 kilowatts of electrical power.

- The armature shaft is a key component, made from a single piece of forged steel. It includes coupling flanges, thrust collars, and oil deflectors as integral parts. The core of the armature, supported by a spider mounted on the shaft, is constructed from laminated magnetic steel punchings, which are pressed and keyed in place to ensure a tight fit. The armature windings, which consist of single-turn coils held in slots by wedges, are insulated with mica to prevent electrical leakage. These windings help convert mechanical rotation into electrical energy.

- An equalizer winding is also present in submarine generators, ensuring balanced current distribution in the various circuits within the armature. This winding is located at the commutator end of the generator, providing additional support for electrical performance.

- The commutator, an essential part of DC generators, consists of copper segments insulated from each other by mica. It ensures proper electrical contact between the rotating armature and the stationary brushes, which collect the generated current. The brush rigging assembly holds the brushes in place and provides mechanisms for adjusting their positions to optimize electrical contact. Different configurations exist across generator types, with some brushes having leading and trailing angles depending on the manufacturer.

- The main field poles are constructed from laminated steel and feature slots for compensating windings that help maintain stable electrical output under varying loads. The main field coils are wrapped around these poles and are insulated to avoid short circuits. Commutating field poles assist in improving the performance of the commutator by adjusting the magnetic field at critical times during operation.

- Compensating windings are embedded within the main field pole slots and consist of copper bars insulated with mica. These windings help balance the magnetic fields and reduce distortions that may arise from high electrical loads.

- Bearings in these generators are designed to handle high rotational speeds and load variations. They consist of split-shell housings lined with a soft metal, typically babbitt, for smooth operation. Oil seals and deflectors prevent lubricant from leaking into the generator's internal components. To maintain optimal performance, bearing housings are vented to prevent vacuum formation around the shaft, and oil is supplied under controlled pressure to avoid excessive flow.

- To prevent overheating, submarine generators use forced-air cooling systems. Hot air is passed through water-cooled cores, where it is cooled before circulating through the generator's internal components, including the armature and field coils. Different generator models have varying cooling system designs, such as the Allis-Chalmers units with integrated water tubes or other models using radiator-like cores.

- Auxiliary generators, though similar in design to main generators, serve a different purpose. Rated at 300 kilowatts, these two-wire, compensated differential compound machines are primarily used to power auxiliary circuits and provide backup charging for the main batteries. The auxiliary generator is driven by a separate diesel engine, ensuring that power can be maintained even if the main engine is unavailable.

- The auxiliary generator connects to the diesel engine via a semi-rigid coupling, with its armature shaft supported by sleeve bearings lubricated from the engine's system. Like main generators, the auxiliary versions also employ thrust collars and bearing systems to handle axial loads.

Overall, the main and auxiliary generators on submarines are highly specialized machines designed to withstand the demanding marine environment. Their construction incorporates robust bearings, cooling systems, and electrical configurations to ensure consistent power generation. These generators form the backbone of the submarine's electrical system, providing power for propulsion, onboard systems, and emergency situations, thus enabling submarines to perform their missions effectively.

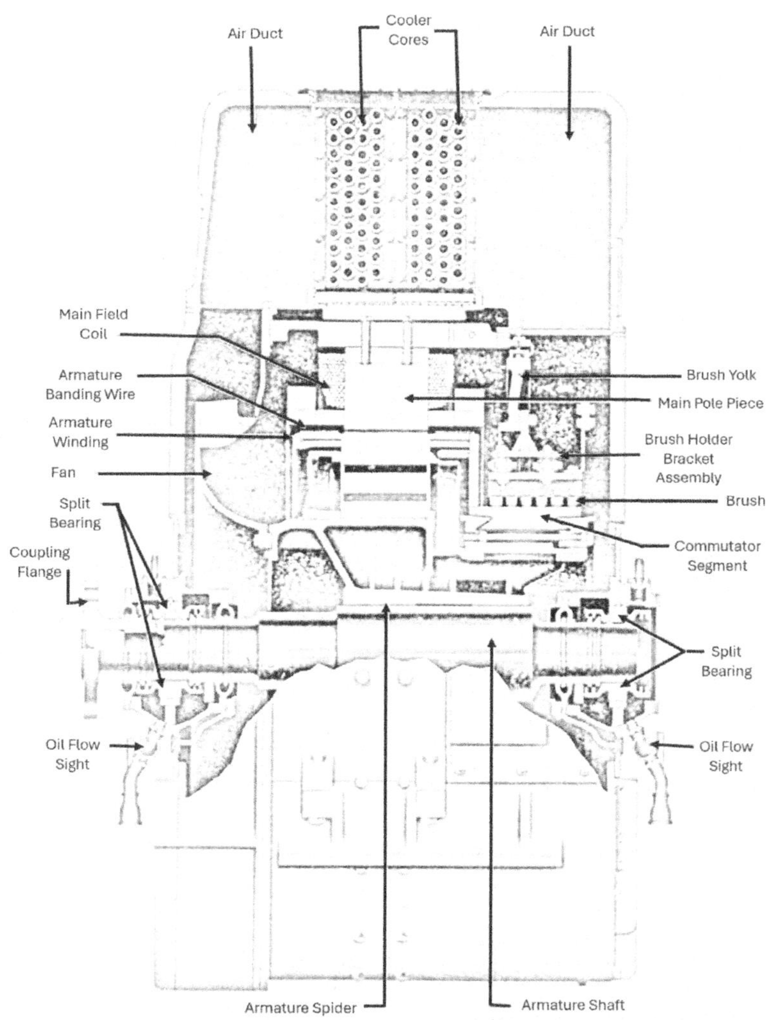

Figure 59: Cross-section of G.E. main generator [144].

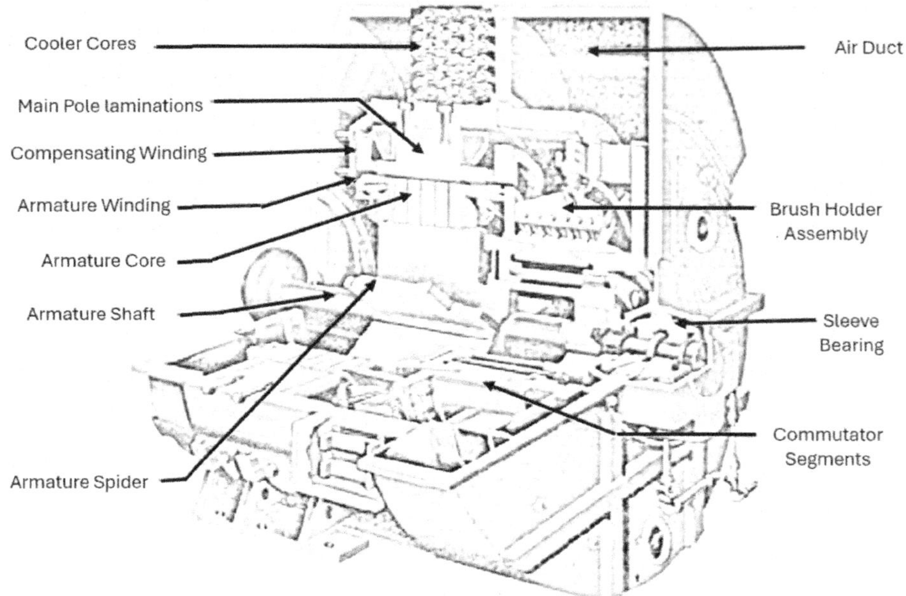

Figure 60: Cutaway of Westinghouse main generator [144].

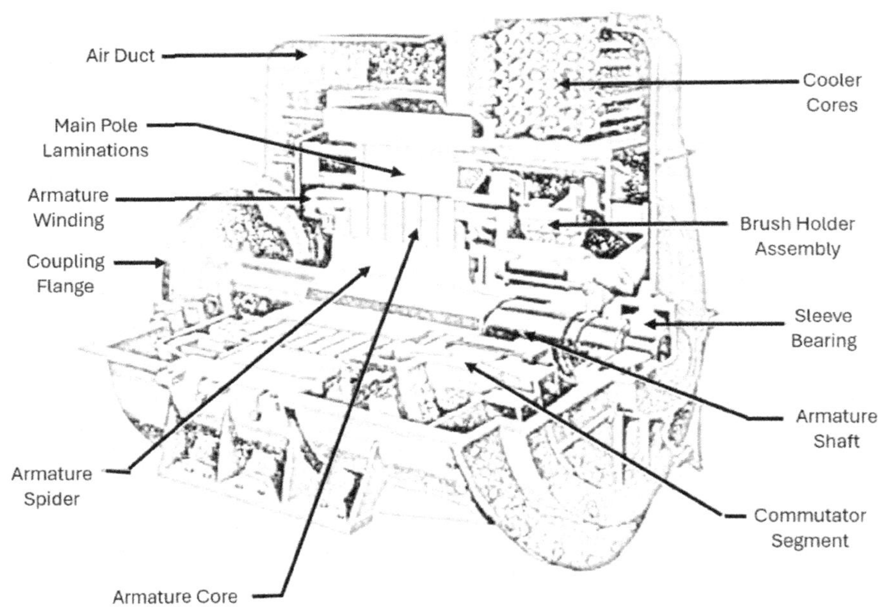

Figure 61: Cutaway of Westinghouse auxiliary generator [144].

Generator Cooling

Cooling is essential on a submarine because of the various components that generate heat during operation. The main sources of heat include the diesel engines (or nuclear reactors in some cases), the propulsion motor, and electrical equipment. Effective cooling ensures that these components operate within their designated temperature ranges, preventing damage and maintaining the submarine's operational capabilities. Here's a detailed look at what needs cooling on a submarine, why it's important, and how cooling systems are designed to meet these requirements [145]:

Components That Require Cooling

- **Diesel Generator:**
  - On diesel-electric submarines, the diesel generator is one of the primary sources of power generation. It converts the mechanical energy produced by the engine into electricity, which is then stored in batteries for use when the submarine is submerged. During operation, the diesel engine generates a significant amount of heat, which needs to be dissipated to prevent overheating. Efficient cooling is essential to keep the generator running reliably.
- **Propulsion Motor:**
  - The propulsion motor drives the submarine's propeller or pump-jet, providing thrust for movement. As it operates, the motor generates heat due to electrical losses and friction. Without adequate cooling, the motor could overheat, potentially causing a complete loss of propulsion, which would severely limit the submarine's capabilities.
- **Electrical Cabinets and Onboard Equipment:**
  - Various electrical systems on the submarine, such as weapon control systems, navigation equipment, communication systems, and life support devices, also produce heat. Cooling is necessary to keep these systems operational, especially in the confined spaces of a submarine. The risk of water contact with electrical components is mitigated by using fully welded titanium cooling components that resist corrosion and leakage.

Importance of Cooling

Cooling is crucial for maintaining the functionality and longevity of key components. If parts such as the generator, motor, or electrical cabinets exceed their maximum operating temperatures, they may sustain permanent damage or become less efficient. In emergency situations or military operations, a failure in the cooling system could be catastrophic, compromising the submarine's ability to manoeuvre, evade detection, or complete its mission. To address this, submarines are equipped with highly efficient cooling systems designed to manage heat under various operational conditions.

Cooling System Design

Submarine cooling systems typically use fin-tube heat exchangers similar to car radiators but on a larger scale. The cooling medium is water, which is distilled onboard to remove salt and other impurities, making it suitable for use as a coolant. This cooling water is circulated through the heat exchangers, where it absorbs heat from the components and then releases it into the surrounding seawater [145].

- **Cooling Units and Tube Design:**
  - Cooling units may come in single or double tube designs. Single tube units are often preferred because they are lighter, which is important for submarines where weight reduction is a priority. The cooling system must also fit within the limited space of the submarine's tube-like hull, often requiring custom-shaped coolers to maximize space utilization.

- **Material Selection:**
  - Historically, cooling units were made from copper-nickel alloys, but modern designs use titanium. Titanium offers weight advantages, excellent corrosion resistance, and durability, which are crucial in the harsh marine environment.
- **Size and Accessibility Considerations:**
  - Space on a submarine is at a premium, and every component must be as compact as possible. Additionally, replacement parts must fit through the submarine's hatches, which typically measure about 1 meter in diameter. Therefore, coolers are designed to be dismantled into smaller parts for easy transport and installation.

Redundancy and Failure Modes

In military submarines, redundancy is built into the cooling system to ensure continued operation even if part of the system fails. This involves designing for various failure modes, such as emergency shutdowns, fan failures, or reduced water flow due to pump malfunctions [145].

- **Tandem Motors and Cooling Redundancy:**
  - Submarines often use tandem motors with two coolers per motor. Each cooler is rated as a "100% cooler," meaning it can individually handle the motor's entire cooling load. This redundancy ensures that if one cooler fails, the other can maintain the required cooling capacity.
- **Contingency Measures:**
  - Additional cooling measures include air-cooling options, where doors can be opened to allow external air to help manage temperatures. The design takes into account suboptimal conditions, such as reduced cooling water flow or partial equipment failure, to maintain system functionality.

System Design Challenges

- **Weight and Space Optimization:**
  - Given the limited space and the need to minimize weight on a submarine, cooling system components are designed to be as compact and lightweight as possible. Using lighter materials like titanium and optimizing the shape of cooling units to fit within the submarine's hull helps achieve this goal.
- **Customization and Adaptation:**
  - The cooling systems are customized based on the client’s requirements, which may include specific failure modes or operational contingencies. The design process often involves collaboration with clients to ensure that the equipment meets all specifications and can withstand a wide range of operational conditions.

Cooling is a critical system on submarines that affects the performance and safety of the vessel. It ensures that essential components, including the diesel generator, propulsion motor, and electrical systems, remain operational within safe temperature limits. With carefully designed redundancies, high-quality materials like titanium, and customized layouts, submarine cooling systems are engineered to handle the unique challenges of underwater environments while optimizing space, weight, and operational effectiveness.

Main Motors

The main motors of submarines are critical components that drive the vessel's propellers, enabling both surface and submerged movement. These motors come in various designs, including geared and direct-drive configurations. They must be robust and reliable, operating in challenging underwater conditions. Here's a detailed explanation of how these motors are constructed and function [144]:

Geared Main Motors

Design and Structure

Geared main motors are two-wire, direct current (DC), compound motors equipped with several field windings, including shunt, series, commutating, and compensating windings. These windings allow the motor to operate with stability and efficiency under different conditions. The shunt field is separately excited via the excitation bus, which receives power from the submarine's batteries, providing greater control over the motor's performance. The motors are enclosed to be watertight below the field frame and waterproof above, ensuring durability in the submarine's submerged environment. Cooling is achieved through a fan attached to the armature shaft that circulates air through cores, which are cooled using circulating water.

Operation and Control

The motors have a high-speed design, capable of generating up to approximately 1370 horsepower (hp) at 415 volts and 1300 revolutions per minute (rpm). They can be connected in various series or parallel arrangements through a control system known as the main control cubicle. This flexibility allows for precise speed adjustments during both surface and submerged operations. On the surface, speed control is managed by adjusting the generator speed and shunt field, while underwater, it is controlled by varying the motor's shunt field or altering the series and parallel combinations of the motors. The ability to reverse the flow of current through the motor armature circuit also allows the submarine to change direction.

Key Components

The components of the geared main motors, such as the commutator, armature, armature windings, brush rigging, and field frame, closely resemble those of a submarine's main generators. Bearings play a crucial role in supporting the armature shaft, using split sleeve bearings housed in secure enclosures to prevent oil leakage. These bearings are lubricated by oil supplied through a pump connected to the reduction gear units.

Lubrication and Cooling

Lubrication System

The motor's bearings are lubricated by a gear-driven oil pump, which is critical for maintaining smooth operation. If the propeller shaft speed drops below 38 rpm, a standby pump is activated to provide sufficient oil pressure to the bearings and reduction gears. Oil flow is carefully monitored, with safety mechanisms such as sight glasses and overflow systems to prevent excessive pressure or flooding. The lubricating system also includes air chambers to prevent vacuum formation around the shaft, which ensures that any oil leakage is drained away from the motor's interior.

Cooling Systems

The cooling of main motors involves circulating air through water-cooled units, similar to those used for main generators. For certain motors, such as those made by Allis-Chalmers, the cooling units are divided into sections that cover most of the motor's exterior surface. This setup ensures that heat is effectively dissipated, even during continuous operation, keeping the motor within safe temperature limits. Proper cooling is essential, as overheating can lead to reduced efficiency or damage to the motor's components.

Direct Drive Motors

Double Armature Design

In newer submarine classes, traditional main motors and reduction gears have been replaced by more compact and efficient double armature motors directly connected to the propeller shafts. These motors are rated for higher power output—up to 2700 hp per motor—allowing for greater propulsion capabilities. Each motor has two armatures and is mounted directly onto the propeller shaft, one for each of the starboard and port shafts. This design eliminates the need for intermediate reduction gears, simplifying the power transmission process.

Operation and Control

Double armature motors feature a similar field winding arrangement, with shunt and series windings providing the necessary flexibility in operation. They are also equipped with compensating windings and commutating poles to ensure smooth current flow and reduce sparking. Cooling is provided by a cross-mounted water tube air cooler, and mechanical air filters ensure that circulating air remains free from contaminants. During slow-speed operations, the motor can function without additional cooling, conserving energy.

Motor Frame and Shaft Design

The frame of a double armature motor is split at an 11-degree angle, making it easier to access and remove the armature for maintenance. The frame is watertight below the split, ensuring that any condensation or liquid is contained and drained away safely. The armature is mounted on a hollow forged steel shaft, which includes bearings at both ends for smooth rotation. The forward end of the shaft is fitted with a Kingsbury thrust bearing that absorbs the propeller's thrust load, ensuring stable operation.

Bearings and Maintenance

Bearing Construction and Maintenance

The bearings used in submarine motors are made from durable materials like cast steel lined with babbitt, providing a hard-wearing surface that can withstand high-speed operation. These bearings are designed with split sleeves to allow easy replacement and maintenance. Grooves in the sleeves ensure proper oil circulation, which helps carry away heat and prevent sludge buildup. Precision tools like depth gauges and bridge gauges are used to measure bearing wear and ensure correct installation of new bearing shells.

Thrust Bearings and Shaft Stability

Kingsbury thrust bearings are used to manage the forward and reverse thrust from the propellers. They consist of a rotating collar and stationary shoes that provide even load distribution. The design allows for slight misalignments, which is critical for maintaining the integrity of the propulsion system under varying loads and conditions. These thrust bearings, along with the radial bearings, are lubricated through a motor-driven pump, ensuring smooth operation even during extended missions.

The main motors of submarines, whether geared or direct-drive, play a vital role in translating electrical energy into mechanical propulsion. Designed for efficiency, reliability, and ease of maintenance, these motors use advanced cooling, lubrication, and control systems to ensure optimal performance. With a robust structure and advanced field winding arrangements, they provide submarines with the power needed for both high-speed operations and stealthy, low-speed manoeuvres. Proper maintenance of these motors, especially in terms of lubrication and cooling, is critical for ensuring the long-term operational capability of submarines in challenging underwater environments.

Nuclear Powered Submarine Electrical Systems

The design of electrical systems in nuclear submarines is a complex and multifaceted endeavour that integrates various technologies to ensure efficient power generation, distribution, and management. Central to this design is the nuclear propulsion system, which typically employs pressurized water reactors (PWR) to generate steam for turbine propulsion. The electrical systems must accommodate the high power demands of propulsion, onboard systems, and auxiliary equipment while maintaining safety and reliability under extreme conditions.

Nuclear submarines utilize a variety of electrical distribution systems, including both alternating current (AC) and direct current (DC) configurations. The choice between these systems is influenced by factors such as efficiency, weight, and the specific operational requirements of the submarine. Studies have shown that DC systems can offer advantages in terms of reduced electrical losses and improved cable weight efficiency, which are critical in the constrained environments of submarines [146]. Furthermore, the integration of modular lead-bismuth fast reactors has been explored to enhance the safety and efficiency of nuclear propulsion systems, providing a robust power source for electrical systems [147]. Systems include:

1. Power Generation

The primary source of electrical power in a nuclear submarine is the nuclear reactor. The reactor generates heat through nuclear fission, which is used to produce steam. The steam is then directed to steam turbines that drive generators to produce electricity.

- **Nuclear Reactor**: The reactor onboard a submarine is typically a pressurized water reactor (PWR), which uses highly enriched uranium as fuel. It generates a significant amount of heat through controlled nuclear fission. This heat is transferred to a primary coolant loop containing pressurized water, which prevents boiling.

- **Steam Production**: The heat from the primary loop is transferred to a secondary loop via steam generators. In the secondary loop, the water is converted into steam, which is used to drive steam turbines.

- **Steam Turbines and Generators**: The steam turbines are connected to generators that produce electrical power. In a nuclear submarine, multiple turbines and generators may be used to ensure redundancy and provide different power outputs for various systems.

2. Electrical Distribution System

The generated electrical power is distributed throughout the submarine via a complex distribution network that supports both alternating current (AC) and direct current (DC) systems.

- **Main Distribution Board**: The power generated from the turbines is directed to a main distribution board, which acts as a control hub for distributing electricity to various systems. It manages the flow of power and directs it to different circuits, including propulsion, life support, weapons, and auxiliary systems.

- **AC and DC Power Systems**: Nuclear submarines use both AC and DC power. AC power is typically used for general systems such as lighting, heating, and ventilation, while DC power is used for critical operations like weapon systems, battery charging, and electric propulsion.

- **Redundant Power Paths**: The electrical distribution network is designed with multiple redundant power paths to ensure uninterrupted operation of essential systems. If a failure occurs in one part of the network, electricity can still be delivered via alternative routes.

3. Propulsion System Integration

Electricity is essential for the propulsion system in nuclear submarines, especially since the propulsion motors are electrically driven. The electrical design allows for precise control over the submarine's speed and manoeuvrability.

- **Electric Propulsion Motors**: The steam-generated electricity is used to power large electric propulsion motors that turn the submarine's propeller or pump-jet. These motors provide torque directly to the propulsion shaft, enabling both high-speed and low-speed operations.

- **Variable Speed Control**: The propulsion motors can be controlled to run at different speeds, allowing the submarine to maintain stealth by operating at lower speeds or to achieve high-speed manoeuvres when needed. Variable speed is achieved by adjusting the voltage and current supplied to the motors.

- **Direct Drive and Reduction Gear Systems**: Some submarines use a direct drive configuration where the propulsion motor is directly coupled to the propeller shaft. Others may use reduction gears to match the motor's high-speed output with the propeller's lower speed requirement.

4. Battery Systems and Backup Power

Even in nuclear-powered submarines, battery systems play a vital role in providing backup power in case of an emergency or reactor shutdown.

- **Battery Banks**: Large battery banks are used as backup power sources and can provide electricity to essential systems if the main power supply fails. The batteries are also used during "silent running" when the submarine needs to minimize noise by operating without the reactor.

- **Charging Mechanism**: The batteries are charged using power generated by the nuclear reactor's turbines. They are kept fully charged to ensure readiness in case of an emergency.

- **Automatic Switching**: In the event of a power loss, the electrical system can automatically switch from the main power supply to battery power, ensuring continuous operation of critical systems such as life support, communication, and navigation.

5. Auxiliary Power Systems

The electrical system also supports auxiliary power sources and equipment to ensure continuous operation under various conditions.

- **Emergency Diesel Generators**: Nuclear submarines often have emergency diesel generators that can provide electrical power during reactor maintenance or in the event of a reactor failure. These generators are also used for battery charging when the reactor is offline.

- **Portable Power Units**: Some submarines are equipped with portable power units or auxiliary power sources that can be used to power specific equipment or systems in emergency situations.

6. Power Management and Load Distribution

Effective power management is crucial in submarines to optimize electrical consumption and ensure that critical systems receive adequate power.

- **Load Shedding**: In case of a power shortage or system failure, non-essential systems can be automatically or manually shut down to prioritize power for essential functions like propulsion, life support, and weapon systems.

- **Monitoring and Control**: The electrical system includes monitoring equipment that provides real-time data on power consumption, battery status, and generator output. Operators can use this information to manage power distribution efficiently.

7. Electrical Insulation and Safety

Special considerations are made in submarine electrical systems to prevent electrical hazards and ensure safety in the underwater environment.

- **Waterproof Cabling and Connectors**: All electrical wiring and connectors are waterproof to prevent short circuits and corrosion caused by exposure to moisture. Cabling is insulated with materials that resist water ingress and pressure.

- **Grounding and Shielding**: Electrical systems are properly grounded to prevent electrical shock and electromagnetic interference. Shielding is used to minimize the risk of signal disruption from external sources.

- **Circuit Protection**: Circuit breakers, fuses, and surge protectors are employed throughout the electrical network to protect equipment from overcurrent, voltage spikes, and electrical faults.

8. Cooling and Heat Management

Electrical equipment generates heat, especially in the confined spaces of a submarine. Effective cooling is necessary to maintain optimal operating temperatures.

- **Air and Water Cooling Systems**: The electrical equipment is cooled using air and water systems. Cooling water is drawn from the ocean, distilled, and circulated to cool generators, propulsion motors, and other heat-generating components.

- **Heat Exchangers**: Heat exchangers transfer excess heat from electrical systems to the cooling water. This keeps temperatures within safe limits and prevents overheating.

- **Ventilation Systems**: Ventilation systems are also used to maintain airflow around electrical components, dissipating heat and ensuring a stable operating environment.

9. Integration with Navigation and Communication Systems

The electrical system also powers navigation and communication equipment that is crucial for submarine operations.

- **Sonar and Radar Systems**: Electrical power is used to operate sonar arrays, radar systems, and other detection equipment. These systems help the submarine detect underwater and surface objects, contributing to navigation and situational awareness.

- **Internal Communication Systems**: Submarine electrical systems support intercoms, telephones, and internal networks that allow for communication among crew members and coordination of operations.

10. Stealth Considerations

Stealth is a critical factor in submarine design, and electrical systems are engineered to minimize acoustic and electromagnetic signatures.

- **Silent Running**: During silent running, the submarine minimizes noise by reducing the speed of the electric propulsion motors and shutting down non-essential systems. The electrical system must support low-noise operation while maintaining power to essential equipment.

- **Electromagnetic Signature Reduction**: To avoid detection by enemy sensors, electrical components are designed to limit electromagnetic emissions. This is achieved through proper shielding and grounding practices.

The electrical architecture of a nuclear submarine typically includes a combination of high-voltage direct current (HVDC) systems and traditional AC systems. HVDC systems are particularly advantageous for long-distance power transmission, as they minimize losses associated with AC systems [148]. Additionally, the implementation of energy storage solutions, such as lithium-ion batteries, is being investigated to provide backup power and enhance operational flexibility. However, the thermal management of such batteries is crucial, as thermal runaway can pose significant risks in the confined spaces of a submarine [111].

The inclusion of HVDC systems in submarine electrical architecture offers significant advantages, particularly in power transmission efficiency. HVDC is more efficient for transmitting power over long distances compared to traditional AC systems. This is because HVDC minimizes the losses associated with AC power transmission, such as those caused by resistance and reactive power in the cables. For a submarine, this increased efficiency is crucial, as it allows for more effective distribution of power across the vessel without significant energy loss. HVDC can be used to power high-demand systems, including propulsion motors and key onboard equipment, ensuring that electrical energy is used as efficiently as possible.

In a nuclear submarine, where space is limited, and systems must operate under challenging conditions, the efficiency of HVDC reduces the need for large and bulky power transformers that are typically required with AC systems. This space-saving aspect allows for more compact electrical configurations and makes room for other critical equipment. Furthermore, HVDC is less prone to electromagnetic interference, which is an important consideration for the stealth operations of a submarine.

While HVDC is advantageous for certain applications, traditional AC systems still play a vital role in submarine electrical architecture. AC power is used for many standard onboard systems, such as lighting, ventilation, heating, and auxiliary equipment, as well as for supporting the integration of conventional electrical components. AC power is also often employed in systems where variable speed control is needed, such as in electric propulsion motors and certain auxiliary systems.

The combination of HVDC and AC systems allows for a versatile and adaptable electrical infrastructure. Power management systems can convert between AC and DC, depending on the specific requirements of different equipment. This hybrid approach ensures that both high-efficiency power transmission and compatibility with conventional AC-powered equipment are maintained.

Energy storage is an essential aspect of submarine electrical systems, particularly for providing backup power and enhancing the submarine's operational flexibility. Lithium-ion batteries are being increasingly investigated and implemented in modern submarines due to their higher energy density and longer lifecycle compared to traditional lead-acid batteries. These batteries can store more energy in a smaller space, which is highly beneficial in the confined environment of a submarine.

The inclusion of lithium-ion batteries allows the submarine to operate silently by using stored electrical power instead of running the reactor or generators. This silent running capability is crucial for stealth operations, as it reduces noise and minimizes the risk of detection. Additionally, these batteries can provide backup power during emergencies, ensuring that essential systems such as life support, communication, and navigation remain operational if the main power source fails.

While lithium-ion batteries offer numerous advantages, their implementation in submarines presents significant challenges related to thermal management. In a confined space such as a submarine, the risk of thermal runaway—a condition where the battery's temperature rapidly increases, leading to fire or explosion—must be carefully managed. The enclosed environment amplifies the consequences of such an event, potentially endangering the crew and the integrity of the submarine.

To mitigate these risks, advanced thermal management systems are integrated into the battery modules. These systems may include active cooling solutions, such as liquid-cooled jackets around the battery cells, or phase-change materials that absorb and dissipate heat. Additionally, battery monitoring systems continuously track temperature, voltage, and current levels to detect any early signs of thermal instability, allowing for proactive management and, if necessary, the isolation of problematic battery cells.

The integration of HVDC, AC, and advanced energy storage systems within a submarine's electrical architecture is a complex task that requires careful planning and design. The electrical control system manages the power flow between HVDC and AC networks, ensuring that each system's advantages are utilized effectively. The control system also oversees battery charging and discharging cycles, optimizing the energy storage system's performance and lifespan.

In this hybrid electrical configuration, HVDC is typically used for major power transmission tasks, such as supplying energy to the propulsion system or other high-power devices. Meanwhile, AC power is distributed to support the standard onboard systems and auxiliary functions. The submarine's electrical management system seamlessly converts power between HVDC and AC as needed, ensuring that each component receives the appropriate type of power.

Moreover, the design of electrical systems must account for the unique operational environment of submarines, including the effects of seawater and pressure on electrical components. Advanced modelling techniques are employed to simulate the interactions between various system components, including the impact of sea conditions on reactor performance and thermal control systems [136]. This multidomain simulation approach aids engineers in optimizing the design of electrical systems to ensure resilience and reliability during operation.

Factors such as seawater exposure, extreme pressure at depth, temperature variations, and the need for stealth all influence the design choices made for electrical components and system architecture. These conditions pose unique challenges that are not present in other environments, making the design of submarine electrical systems fundamentally different from those in surface ships or land-based installations.

Electrical systems on submarines must be robust enough to withstand the effects of seawater and high pressures encountered during underwater operations. As submarines operate at significant depths, the pressure on the hull and components can reach hundreds of times atmospheric pressure, leading to potential compression and deformation of materials. Electrical components, wiring, and connectors must be carefully designed to avoid failure under these extreme conditions.

Seawater, being a highly conductive medium, can cause electrical shorts and corrosion if it comes into contact with electrical components. To mitigate this, all electrical components must be sealed and insulated against seawater ingress. The use of specialized materials, such as corrosion-resistant alloys and high-durability insulators, ensures that electrical systems remain operational despite prolonged exposure to harsh underwater conditions. Connectors are often waterproofed with advanced sealing technologies, and cables are sheathed with materials that can resist the corrosive effects of saltwater.

To address the complexities involved in designing submarine electrical systems, engineers employ advanced modelling techniques to simulate the interactions between various system components. These simulations take into account multiple factors, including electrical load distribution, thermal management, pressure effects, and mechanical stress. By using a multidisciplinary approach, engineers can accurately predict how the electrical systems will perform under real-world conditions.

Multidomain simulation tools are particularly useful in assessing how different sea conditions, such as varying temperatures, currents, and salinity levels, impact reactor performance, power distribution, and cooling systems. For example, the thermal control systems of a submarine must be capable of maintaining optimal operating temperatures for electrical components despite changes in external seawater temperature. Simulations allow engineers to model heat transfer from electrical components to the cooling systems, optimizing the placement and design of cooling units to ensure efficiency and resilience.

The performance of a submarine's reactor is closely linked to the electrical system's design and its ability to manage heat. As a nuclear submarine operates, the reactor generates significant amounts of heat that must be dissipated effectively to avoid overheating and potential damage. The cooling systems must account for the varying conditions that a submarine encounters, such as changes in seawater temperature at different depths. Advanced modelling helps engineers understand how the reactor's thermal output interacts with the submarine's electrical systems and the surrounding environment.

For instance, when a submarine dives deeper, the increased pressure can affect the flow characteristics of the cooling medium used to manage reactor heat. This, in turn, can impact the efficiency of heat exchange processes and require adjustments to cooling system parameters. Multidomain simulations enable engineers to study these effects in detail, allowing them to design electrical systems that can adapt to changing conditions without compromising safety or performance.

The goal of using advanced modelling techniques and multidomain simulation in submarine electrical system design is to ensure resilience and reliability during operation. The underwater environment is unforgiving, and even minor electrical failures can have serious consequences for the submarine's safety and mission success. Therefore, the electrical systems must be designed to maintain functionality in adverse conditions and recover quickly from any disturbances.

By simulating various failure scenarios, such as power surges, thermal runaway in energy storage systems, or cooling system malfunctions, engineers can design electrical architectures that incorporate redundancies and fail-safe mechanisms. For example, systems may include backup power supplies, alternate cooling routes, and automated control systems that adjust power distribution and cooling dynamically in response to detected issues. These features enhance the resilience of the submarine's electrical systems, ensuring continued operation even in challenging circumstances.

In addition to propulsion and power distribution, the electrical systems in nuclear submarines also encompass sophisticated control systems for monitoring and managing power flows. These systems are designed to ensure that critical loads receive uninterrupted power, particularly during emergency situations. The integration of advanced sensors and monitoring technologies allows for real-time assessment of system performance, enabling proactive maintenance and minimizing the risk of failures [106].

The electrical systems in nuclear submarines are integral not just for propulsion and power distribution but also for managing and controlling power flows to various subsystems. This encompasses sophisticated control systems that are essential for monitoring, regulating, and prioritizing electrical power distribution to ensure that all critical functions remain operational, even in emergency situations. The complexity and high stakes of submarine operations demand that these electrical control systems be highly reliable, responsive, and capable of adapting to changing conditions.

A key function of submarine electrical control systems is to ensure that critical loads, such as life support, propulsion, navigation, and communication systems, receive uninterrupted power. In the event of power fluctuations, malfunctions, or emergencies, the control system is designed to prioritize the distribution of power to these vital components. This ensures the safety of the crew and the operability of the submarine during combat or other mission-critical scenarios.

For example, in the case of a sudden power loss or a fault in the electrical system, the control architecture automatically shifts power from non-essential systems to maintain the functionality of crucial equipment. This kind of load prioritization is managed by intelligent controllers that assess the current electrical demand and available power supply in real-time, allowing for immediate response to disturbances. The system may also include automated switching capabilities that can reroute power quickly to ensure continued operation of key systems.

Modern submarine electrical systems incorporate an array of advanced sensors and monitoring technologies to maintain optimal power distribution and detect potential issues before they become serious problems. These sensors are strategically placed throughout the submarine to track various parameters, such as voltage, current, temperature, and pressure. By continuously monitoring the electrical network, the control systems can identify abnormal conditions, such as overheating, overloading, or drops in voltage, and initiate corrective actions.

The data collected from these sensors is used to perform real-time assessments of the electrical system's health. For instance, thermal sensors installed near critical electrical components can detect any overheating, allowing the system to trigger cooling mechanisms or adjust power distribution to prevent damage. Similarly, current sensors can detect short circuits or faults, prompting the control system to isolate affected areas and reroute power to maintain overall system integrity. The ability to continuously monitor and respond to changing conditions helps minimize the risk of failures and enhances the submarine's operational safety.

The real-time data gathered by the monitoring systems is not only used for immediate corrective actions but also for predictive maintenance. By analysing trends in the collected data, the control system can anticipate potential failures or deteriorations in performance, enabling maintenance crews to address issues before they lead to downtime or more significant damage. For example, an increase in electrical resistance in certain circuits over time may indicate the beginning of a component's wear and tear. The control system can alert maintenance personnel to replace or repair the affected component during a scheduled maintenance period, thus avoiding unexpected failures.

Proactive maintenance based on real-time assessment also extends the lifespan of critical electrical components and improves the overall reliability of the submarine. This approach is particularly important in the constrained and hostile environment of a submarine, where repair opportunities may be limited and the consequences of system failures can be severe. Advanced diagnostics and monitoring systems contribute to a maintenance strategy that reduces operational costs and enhances mission readiness by minimizing the need for emergency repairs.

In nuclear submarines, the ability to respond effectively to emergency situations, such as a reactor shutdown or a fire in the engine room, is vital. The electrical control system plays a crucial role in such scenarios by automatically adjusting power flows to ensure that all essential systems continue to function. The system's emergency protocols include predefined responses that can isolate damaged circuits, engage backup power sources, and initiate cooling processes to manage any heat generated by electrical faults.

The integration of real-time monitoring and intelligent control mechanisms allows the submarine to remain operational under adverse conditions. For example, if an emergency situation leads to a loss of power in one section of the submarine, the control system can dynamically redistribute power from other sections, thereby maintaining critical functions like propulsion, sonar, and life support. This level of control is achieved through advanced algorithms that balance the electrical load and prioritize energy distribution based on the urgency of the situation.

The sophisticated control systems in nuclear submarines not only improve safety and resilience but also enhance overall operational efficiency. By optimizing power distribution based on the submarine's current operating conditions, the control system can reduce unnecessary energy consumption and extend the time the submarine can operate on stored power. For instance, during low-power operations, such as silent running or reconnaissance, the control system can limit power to non-essential systems, conserving energy for critical operations.

Furthermore, the seamless integration of control, monitoring, and diagnostic functions supports a more streamlined operation, where the electrical systems work in harmony to achieve optimal performance. This level of automation allows the crew to focus on mission-critical tasks without having to manually adjust power settings or monitor system status constantly.

Electrical Cabling

Electrical cabling in submarines is a critical component of the vessel's infrastructure, designed to ensure safe, reliable, and efficient transmission of power and data throughout the submarine. Given the unique underwater environment, these cables must meet stringent requirements for insulation, durability, flexibility, and resistance to harsh conditions, such as pressure, moisture, vibration, and electromagnetic interference.

Material Selection and Construction: Submarine cables are typically constructed using materials that provide high electrical insulation and mechanical strength while being resistant to corrosion and seawater. The outer jackets are often made from specialized polymers, such as cross-linked polyethylene (XLPE) or ethylene propylene rubber (EPR), which offer excellent resistance to water ingress and chemicals. Inside, the cables use conductors made from materials like copper or aluminium, which are chosen for their high conductivity and reliability.

To further protect the cable, especially in high-stress areas, the design may include layers of shielding, braiding, or armouring. These layers can provide additional mechanical protection against abrasion and crushing, and also serve to reduce the effects of electromagnetic interference by shielding the internal conductors.

Waterproofing and Pressure Resistance: Since submarines operate underwater at significant depths, the electrical cabling must be able to withstand high external pressures. To achieve this, cables are often made with multiple layers of waterproofing, which may include a combination of rubberized seals, watertight jacketing, and water-blocking tapes or compounds within the cable structure.

For high-pressure applications, such as deep-sea operations or cables that run through pressure hulls, cables are specially engineered with thicker insulation and reinforced construction to ensure they remain watertight and function reliably under compression. This pressure-resistant design prevents damage or electrical shorts due to water ingress, which could compromise the submarine's operational capabilities.

Flexibility and Durability: Submarine cabling needs to be flexible enough to accommodate the tight spaces and complex routing inside a submarine's confined environment. The cables are designed with materials and construction techniques that allow for repeated bending and movement without degradation of the insulation or conductors. For instance, cables may use fine-stranded conductors and flexible insulating materials to provide better bend radius and resilience.

Durability is another crucial factor, as cables must resist wear and tear caused by vibrations, mechanical stresses, and the movement of other components. The insulation materials used in submarine cables are selected not only for their electrical properties but also for their toughness and ability to withstand long-term exposure to oil, fuel, and other potentially damaging substances.

Fire-Resistant and Low-Smoke Materials: Fire safety is a significant concern in submarines due to the confined space and limited options for evacuation. Therefore, submarine electrical cables are often made from fire-resistant materials that can withstand high temperatures without catching fire. Additionally, they are designed to produce minimal smoke and toxic fumes in case of a fire, helping to maintain visibility and reduce the risk of harmful exposure to the crew.

Low-smoke, zero-halogen (LSZH) cables are commonly used in submarine applications because they do not emit corrosive gases when exposed to fire, thus protecting sensitive electronic equipment and ensuring the safety of the crew.

Shielding and Electromagnetic Compatibility (EMC): Submarines house numerous electronic systems and high-power equipment, which can generate electromagnetic interference (EMI). To ensure electromagnetic compatibility (EMC), electrical cabling is often shielded with metal foil, braiding, or a combination of both. This shielding minimizes the electromagnetic emissions from the cables and protects the cables from external EMI, thus maintaining the integrity of power and data transmissions.

Proper grounding and bonding of cable shielding are also important to ensure that any electromagnetic noise is safely dissipated, preventing interference with sensitive electronic systems like sonar, radar, and communication equipment.

Colour Coding and Identification: Colour coding is often used in submarine electrical cabling to help identify the purpose and voltage level of different cables. This aids in maintenance, troubleshooting, and repair work, allowing technicians to quickly identify and distinguish between power, communication, data, and control cables. In some cases, the cables are also labelled with markings or printed identifiers that indicate their function, routing, or connection points.

Routing and Installation Considerations: The routing of electrical cables inside a submarine is a complex process due to the limited space and the need to avoid interference with other equipment. Cables are typically installed in specialized cableways or conduits that provide protection and organization while allowing for easy access during maintenance. The installation must ensure that cables are adequately supported, do not create choke points, and are isolated from potential sources of heat or physical damage.

Cable clamps, ties, and guides are used to secure the cables in place and minimize vibration. In areas where cables pass through bulkheads or compartments, watertight cable glands or penetrations are used to maintain the submarine's pressure integrity.

Testing and Certification: Before installation, submarine cables undergo rigorous testing to ensure they meet all safety and performance standards. Tests may include electrical resistance, insulation integrity, pressure resistance, flexibility, and fire resistance. Cables are often certified to meet military or naval standards that specify the required characteristics for submarine applications, such as MIL-SPEC or DEF STAN in the UK.

Submarine wiring standards are often based on military or naval specifications that define the requirements for electrical components and systems used in naval vessels. These standards include guidelines for insulation, conductor materials, shielding, fire resistance, and other essential characteristics. Some of the common standards relevant to submarine wiring are:

- **MIL-SPEC (Military Specifications, United States):** MIL-SPEC standards, such as MIL-DTL-24640 and MIL-DTL-915, provide detailed requirements for submarine cables, including construction, materials, insulation, voltage ratings, and environmental resistance. These specifications ensure that submarine cables meet the rigorous demands of military use.

- **DEF STAN (Defence Standards, United Kingdom):** The UK's Defence Standards outline the requirements for submarine electrical systems, including wiring. DEF STAN 61-12, for example, addresses the specification for cables used in military vehicles and vessels, including submarines, to ensure proper insulation, fire resistance, and mechanical robustness.

- **NATO STANAGs (Standardization Agreements):** NATO countries often adhere to shared standards for interoperability, including those for submarine wiring. These agreements establish guidelines for electrical cabling that ensure compatibility and reliability across different nations' naval fleets.

- **IEC Standards (International Electrotechnical Commission):** Some aspects of submarine wiring may be guided by IEC standards, which provide international guidelines for electrical installations in ships, including submarines. IEC 60092, for instance, is a standard for electrical installations in ships and addresses cabling requirements.

MIL-DTL-24640 and MIL-DTL-915 are military specifications that set stringent standards for submarine cables to ensure their safety, reliability, and performance under challenging conditions. These standards are essential for naval vessels, including submarines, where electrical systems must operate reliably in a confined, high-pressure underwater environment.

MIL-DTL-24640 is a military detail specification that governs the requirements for lightweight, low-smoke shipboard cables used in naval vessels. These cables are designed to meet specific operational demands, including high resistance to fire, mechanical durability, and minimal smoke generation.

The lightweight nature of these cables is vital for reducing the overall weight of submarines, which directly impacts their performance and manoeuvrability [149]. The low-smoke characteristic is crucial in fire scenarios, as it minimizes smoke and toxic gas emissions, thereby preserving visibility and air quality in the confined spaces of submarines [150]. The cables are constructed from materials that resist flame spread and limit smoke emissions, typically utilizing cross-linked polyethylene (XLPE) for insulation, which is known for its flame-retardant properties [151].

The key characteristics of MIL-DTL-24640 cables are as follows:

- **Lightweight and Low-Smoke Design**:
  - MIL-DTL-24640 cables are specifically designed to be lightweight to reduce the overall weight of the submarine, which is a crucial factor in vessel design and performance.
  - The low-smoke feature ensures that, in the event of a fire, the cables emit minimal smoke and toxic gases. This is critical for maintaining visibility and air quality within the confined spaces of a submarine.
- **Material Requirements**:
  - The cables are typically constructed with materials that resist flame spread and limit smoke emissions. Insulation materials like cross-linked polyolefin or similar compounds are commonly used for their flame-retardant properties.
  - The outer jacket may also incorporate low-smoke, zero-halogen (LSZH) materials to further enhance safety by reducing toxic gas emissions when exposed to fire.
- **Environmental and Mechanical Performance**:
  - MIL-DTL-24640 cables are designed to withstand various environmental conditions, including exposure to saltwater, high humidity, vibration, and mechanical stress.
  - The cables feature a robust construction to resist abrasion, crushing, and other forms of physical damage.
- **Voltage and Temperature Ratings**:
  - These cables are available in various voltage ratings, typically ranging from 600 volts to 1000 volts.
  - They are also designed to function across a wide temperature range, accommodating both high and low-temperature environments encountered in submarine operations.
- **Applications**:
  - MIL-DTL-24640 cables are used in various shipboard applications, including power distribution, lighting, communication systems, and electronic control circuits.

Moreover, MIL-DTL-24640 cables are engineered to endure harsh environmental conditions, including exposure to saltwater, high humidity, and mechanical stress. Their robust construction allows them to resist abrasion and crushing, ensuring durability in the challenging underwater environment [152]. The cables are rated for voltages typically ranging from 600 to 1000 volts and are designed to operate across a wide temperature spectrum, accommodating the extreme conditions encountered during submarine operations [153].

MIL-DTL-915 is another military specification that addresses electrical cables and wiring used in naval shipboard applications. Unlike MIL-DTL-24640, which emphasizes lightweight, low-smoke features, MIL-DTL-915 focuses on general-purpose shipboard cables.

These cables are versatile and suitable for various applications, including power distribution, lighting, and control systems aboard naval vessels [154]. They must meet flame resistance standards to prevent fire spread and comply with fire safety requirements, although the specifications may not be as stringent as those outlined in MIL-DTL-24640 [155].

The construction of MIL-DTL-915 cables typically involves insulation materials such as rubber, cross-linked polyethylene (XLPE), or polyvinyl chloride (PVC), which provide adequate protection against physical damage and environmental exposure [156]. These cables are also designed to withstand mechanical stresses and environmental challenges, making them suitable for a wide range of naval applications [157].

Key aspects of MIL-DTL-915 include:

- **General-Purpose Design**:
  - MIL-DTL-915 cables are designed for use in a variety of general-purpose shipboard electrical applications, including power, lighting, and control systems.
  - While they may not prioritize weight reduction as much as MIL-DTL-24640 cables, they are still constructed to meet high standards of durability and performance.
- **Flame Resistance and Fire Safety**:
  - These cables are also required to meet flame-resistance standards to prevent fire spread along the cable.
  - They must comply with fire safety requirements that limit smoke and toxic gas emissions, though they may not be as stringent as MIL-DTL-24640 in this regard.
- **Construction and Insulation Materials**:
  - MIL-DTL-915 cables are typically insulated with materials such as rubber, cross-linked polyethylene (XLPE), or polyvinyl chloride (PVC), depending on the application.
  - The outer jacket materials are chosen to provide adequate protection against physical damage and environmental exposure.
- **Voltage and Application Range**:
  - These cables are available for a range of voltage applications, typically up to 600 volts.
  - They are versatile and used in various shipboard systems, including electrical power distribution, communication circuits, and control wiring.

- **Mechanical and Environmental Specifications**:
  - MIL-DTL-915 cables must meet rigorous mechanical requirements, including resistance to abrasion, impact, and crushing.
  - Environmental resistance includes the ability to function in conditions of saltwater exposure, temperature variations, and humidity, which are typical in naval environments.

Comparison and Usage in Submarines

- **MIL-DTL-24640** cables are typically used in situations where weight reduction and minimal smoke emission are priorities, such as in submarines and other areas with confined spaces. The low-smoke characteristics make these cables ideal for submarine applications where fire safety is critical.
- **MIL-DTL-915** cables are often used in more general-purpose applications where the same level of lightweight design may not be necessary. They still meet high standards for durability and fire resistance, making them suitable for a range of naval uses.

Both MIL-DTL-24640 and MIL-DTL-915 play important roles in ensuring that submarine electrical systems are reliable, safe, and able to withstand the harsh conditions encountered underwater.

DEF STAN 61-12 is a critical series of Defence Standards from the United Kingdom that delineates the requirements for electrical cables utilized in military applications, particularly in naval vessels such as submarines. These standards are essential for ensuring that the cables meet stringent specifications regarding construction, material properties, and performance requirements, thereby guaranteeing safety, reliability, and functionality in the demanding environments typical of military operations [158]. The series encompasses various parts, each addressing specific types of cables and their applications, which is crucial for the operational integrity of military systems.

DEF STAN 61-12 is a critical series of Defence Standards from the United Kingdom that delineates the requirements for electrical cables utilized in military applications, particularly in naval vessels such as submarines. These standards are essential for ensuring that the cables meet stringent specifications regarding construction, material properties, and performance requirements, thereby guaranteeing safety, reliability, and functionality in the demanding environments typical of military operations [158]. The series encompasses various parts, each addressing specific types of cables and their applications, which is crucial for the operational integrity of military systems.

The general overview of DEF STAN 61-12 highlights its role in guiding the design, selection, testing, and installation of electrical cables in military contexts. These standards are widely adopted within the UK defence sector to ensure that all electrical wiring and cabling conform to high-performance benchmarks necessary for military applications, including those in submarines [153, 158]. The standards not only facilitate compliance with safety requirements but also enhance the operational capabilities of military assets by ensuring that the electrical systems function effectively under various conditions.

Material requirements specified in DEF STAN 61-12 emphasize the use of durable and reliable materials that can withstand harsh conditions. Cables are constructed from materials designed to resist abrasion, corrosion, and other forms of physical damage, which is vital for maintaining operational integrity in challenging environments [157, 159]. Common insulation materials include cross-linked polyethylene (XLPE) and ethylene propylene rubber (EPR), which are selected for their robustness and flame-retardant properties, thereby enhancing fire safety [160, 161]. The emphasis on material properties ensures that the cables can endure the rigors of military service, particularly in underwater and high-pressure scenarios encountered by submarines.

Fire resistance and low smoke emission are critical aspects of DEF STAN 61-12, reflecting the need for cables to meet stringent flame retardancy requirements. This is particularly important in confined spaces, such as submarine interiors, where the risk of fire can have catastrophic consequences [153, 158]. The use of low-smoke, zero-halogen (LSZH) materials is prevalent, as these materials emit minimal smoke and no harmful halogen gases when exposed to fire, significantly enhancing safety for personnel onboard [153, 160]. This focus on fire safety aligns with broader military standards that prioritize the protection of both equipment and personnel.

Mechanical and environmental performance specifications within DEF STAN 61-12 address the need for cables to exhibit high tensile strength, flexibility, and resistance to various forms of mechanical stress, including vibration and impact [153]. The cables must also demonstrate resilience against environmental factors such as water, oil, chemicals, and extreme temperatures, ensuring reliable performance in the underwater and high-pressure conditions typical of submarine operations [153, 158]. These mechanical and environmental requirements are crucial for maintaining the integrity and functionality of electrical systems in submarines.

In terms of electrical properties, DEF STAN 61-12 outlines requirements for insulation resistance, dielectric strength, and voltage ratings, ensuring that cables can handle the electrical demands of military applications [153, 158]. The standards accommodate various voltage levels, from low-voltage control and instrumentation cables to medium-voltage power cables, thereby providing a comprehensive framework for electrical system design in submarines [153, 158].

Electromagnetic compatibility (EMC) is another critical aspect addressed by DEF STAN 61-12, as it specifies shielding requirements to mitigate electromagnetic interference (EMI) [153, 158]. This is particularly vital in submarines, where numerous electronic systems operate in close proximity, and effective shielding is necessary to prevent signal degradation and maintain operational effectiveness [153, 158]. The use of shielding materials, such as braided metal or foil, is essential for protecting against external electromagnetic noise.

The series is divided into multiple parts, each focusing on different types of cables. For instance, Part 4 specifies requirements for single-core and multicore flexible cables, while Part 5 addresses multicore screened cables used in instrumentation and control systems [158]. Part 6 focuses on fire-resistant cables that must remain operational during fire incidents, underscoring the comprehensive nature of DEF STAN 61-12 in addressing various cable applications [158].

Testing and quality assurance processes outlined in DEF STAN 61-12 are rigorous, ensuring that cables meet all specified criteria through mechanical, electrical, and fire testing [153, 158]. These processes are crucial for maintaining consistent performance across all manufactured cables, thereby meeting the high standards required for military applications.

Cross-Linked Polyethylene (XLPE) and Ethylene Propylene Rubber (EPR)

Cross-Linked Polyethylene (XLPE) and Ethylene Propylene Rubber (EPR) are two popular insulation materials used in electrical cabling across various fields, including marine, industrial, and military applications. Each material possesses unique properties that make it suitable for specific conditions, especially when durability, electrical insulation, and resistance to environmental factors are crucial.

Cross-Linked Polyethylene (XLPE)

XLPE is a type of polyethylene where the polymer chains are chemically cross-linked to form a three-dimensional network. This cross-linking process enhances the material's thermal and mechanical characteristics, making it highly resistant to heat, chemicals, and mechanical stress. XLPE's structure provides excellent electrical insulation properties, including high dielectric strength, which makes it suitable for high-voltage and medium-voltage applications.

The material is frequently used in power cables because of its superior electrical characteristics. It can handle higher voltages while providing strong resistance to electrical stress, which reduces the likelihood of insulation failures. Its high dielectric strength and low dielectric loss contribute to efficient power transmission, even across long distances.

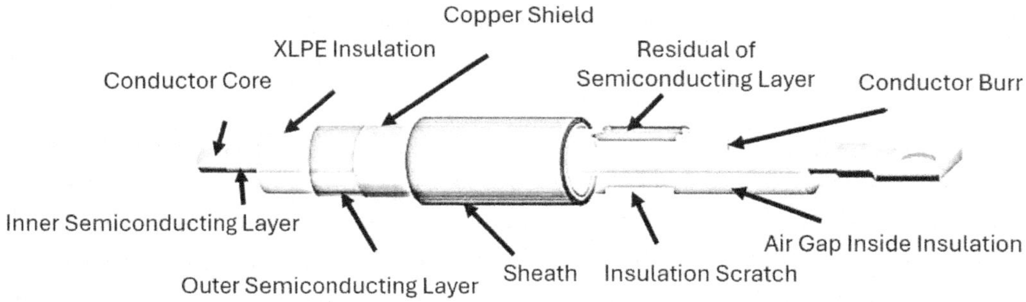

Figure 62: Typical insulation defects of cross-linked polyethylene cable.

In terms of thermal and mechanical performance, the cross-linking process grants XLPE greater thermal stability, allowing it to function continuously at temperatures up to 90°C, with short-term exposure limits reaching 250°C in case of faults. Mechanically, XLPE is robust and abrasion-resistant, making it ideal for environments where cables may experience physical wear and stress. The material's chemical resistance is also noteworthy, as it withstands exposure to chemicals, oils, and solvents, and has low water absorption, making it suitable for harsh, humid, or wet environments such as those found in marine applications.

Commonly, XLPE is used in power distribution cables for utilities, industrial installations, and marine settings, including submarines. It is particularly advantageous for high-voltage power transmission and distribution systems. However, while XLPE offers benefits like high thermal stability, electrical insulation, and resistance to chemicals, it is less flexible than some other insulation materials like rubber and can be vulnerable to UV damage without added protection.

Ethylene Propylene Rubber (EPR)

EPR is a synthetic elastomer made from ethylene and propylene monomers, resulting in a durable, flexible, and electrically insulating material. The elasticity and resilience of EPR allow it to maintain flexibility, even at low temperatures, making it ideal for applications where cables are frequently bent or flexed.

In terms of electrical insulation, EPR provides excellent properties such as high dielectric strength and low electrical conductivity, making it well-suited for medium-voltage cable insulation. The material's resistance to electrical tracking and partial discharge further helps prevent electrical faults. EPR also offers stable performance in a wide temperature range, with continuous operation possible at up to 90°C and short-term exposure tolerances of up to 130°C. This flexibility and resilience are crucial for dynamic applications where cables may face repeated mechanical stress.

Figure 63: REP-90/EPR (Ethylene Propylene Rubber) 95 mm2 cable cross section. The cable consists of three active conductors and one earth and also has 24 fibre optic strands and four 2.5 mm2 pilot cores. It is rated at 6.6kV and is used as trailing cable. Calistemon, CC BY-SA 4.0, via Wikimedia Commons.

Chemically, EPR has good resistance to water, heat, and certain chemicals, making it suitable for underwater and saltwater environments. It also has excellent resistance to ozone and UV radiation, enhancing its longevity in outdoor settings. EPR-insulated cables are commonly used in power distribution, industrial installations, and marine applications, including onboard submarine power systems and control circuits. The material's flexibility makes it a good choice for portable or mobile equipment, as well as for cables that need to be routed through tight or complex spaces.

While EPR offers advantages like high flexibility, excellent insulation properties, and resistance to environmental stressors, it does have some limitations. Its thermal stability is generally lower than that of XLPE, making it less suitable for very high-temperature applications without additional modifications.

Comparison Between XLPE and EPR

In comparing XLPE and EPR, each has specific strengths that make them suitable for different applications. XLPE is generally better suited for higher voltage applications due to its superior dielectric strength and thermal stability. EPR, on the other hand, is more flexible, making it preferable for applications that involve frequent bending or movement of the cables. Both materials have good resistance to chemicals and moisture, but XLPE holds a slight advantage in environments where exposure to chemicals is a primary concern.

Both XLPE and EPR are extensively used in electrical cabling for submarines, with each material offering distinct benefits depending on the application's requirements. While XLPE is favoured for high-voltage applications due to its excellent electrical and thermal properties, EPR is chosen for its flexibility and durability in dynamic environments. The choice between the two materials depends on factors such as voltage requirements, operating temperature, mechanical flexibility, and environmental conditions.

Battery Systems and Power Storage

Submarine battery systems and power storage are crucial components of submarine electrical architecture, providing energy for propulsion, onboard systems, and emergency power. These systems ensure the submarine can operate submerged for extended periods without reliance on external power sources. The type of battery system used in a submarine depends on the propulsion type (diesel-electric or nuclear) and specific operational requirements.

Submarine battery systems have traditionally relied on lead-acid batteries, which have been the mainstay due to their proven reliability, safety, and ability to supply high power outputs for propulsion and other submarine systems. However, advancements in battery technology are driving the adoption of newer chemistries, such as lithium-ion batteries, to meet the increasing demands for improved performance, longer submerged endurance, and reduced operational risks. Each battery type has distinct characteristics that impact its suitability for different submarine classes, making the choice of battery technology a crucial factor in submarine design.

Lead-Acid Batteries

Lead-acid batteries have long been the most widely used energy storage solution for conventional diesel-electric submarines. They provide reliable power for submarine propulsion and auxiliary systems, such as lighting and life support, when the submarine is submerged and unable to use its diesel engines. The long history of lead-acid batteries in submarines attests to their dependability and rugged nature. They can deliver large bursts of current, which is essential for rapid propulsion manoeuvres or high-power equipment operation.

Despite their advantages, lead-acid batteries have some significant limitations. Their relatively low energy density means that they require considerable space and weight to store a useful amount of energy. Submarines have limited space, so using lead-acid batteries can reduce the available room for other equipment or supplies. Additionally, the depth of discharge—how much of the battery's capacity can be used before recharging—is limited to avoid shortening the battery's lifespan. Frequent deep discharges can lead to battery degradation, requiring more frequent replacements.

Lithium-Ion Batteries

Lithium-ion batteries are emerging as a viable alternative to traditional lead-acid batteries in modern submarine designs. These batteries offer a much higher energy density, allowing them to store more energy in a smaller and lighter package. This increased energy density translates to extended submerged endurance, enabling submarines to remain underwater for longer periods without needing to surface for recharging. Lithium-ion batteries also boast faster charging capabilities, which means submarines can recharge more quickly, reducing the time spent at snorkel depth and the associated risk of detection.

The lighter weight and reduced space requirements of lithium-ion batteries are significant advantages for submarines, where every kilogram and cubic meter count. This weight reduction allows for greater flexibility in submarine design, such as the addition of more sophisticated equipment or the carrying of extra supplies. However, lithium-ion batteries come with challenges, particularly regarding thermal management. These batteries are prone to thermal runaway, a condition where excessive heat generation can lead to fire or explosion. In the confined and high-pressure environment of a submarine, such an event could be catastrophic. As a result, lithium-ion battery systems require advanced cooling mechanisms and safety measures, such as temperature monitoring and fire suppression systems, to mitigate these risks.

Several modern diesel-electric submarines, such as Japan's Soryu-class, have already adopted lithium-ion battery technology to enhance their underwater performance and increase their operational capabilities. These submarines benefit from the higher energy density and rapid recharge rates provided by lithium-ion batteries, making them more versatile and effective in various mission scenarios. As battery technology continues to advance, lithium-ion batteries are likely to become more common in submarines, potentially replacing lead-acid batteries entirely for some applications. Nonetheless, careful consideration of thermal management, safety protocols, and lifecycle costs remains essential for their widespread adoption.

Diesel-Electric Submarine Battery Systems

Diesel-electric submarines use battery systems as a vital component for their underwater operations. These batteries provide the power needed for propulsion, lighting, life support, and other electrical systems while the submarine is submerged and unable to run its diesel engines. The submarine's battery power is replenished by charging the batteries while the submarine is on the surface or operating at snorkel depth, where it can take in air to run its diesel generators. Managing the charging and discharging processes effectively is crucial for the submarine's performance and stealth capabilities.

The diesel generators onboard a diesel-electric submarine are used to recharge the batteries, typically when the submarine is on the surface or at snorkel depth. Charging involves running the diesel engines to power the generators, which convert mechanical energy into electrical energy for storage in the batteries. Since diesel engines require an air supply for combustion, the submarine must either surface or stay near the surface with a snorkel mast extended above the water to allow air intake during the charging process.

The depth of discharge—how much of the battery's stored energy is used before recharging—affects the submarine's operational efficiency and battery lifespan. A deeper depth of discharge can provide extended underwater time but may reduce the number of charging cycles the batteries can endure. Charging time is another critical factor, influencing the submarine's "indiscretion ratio," which measures the proportion of time spent at snorkel depth to charge versus the time spent submerged. A lower indiscretion ratio is preferable, as it reduces the risk of detection by minimizing the duration the submarine remains near the surface.

Submarine battery systems are typically arranged in banks, consisting of multiple cells connected in series and parallel to achieve the desired voltage and storage capacity. The configuration is designed to provide the necessary power for prolonged submerged operations, while also accommodating the specific electrical requirements of the submarine's propulsion and onboard systems. The capacity and number of battery banks can vary depending on the submarine class and its intended operational profile.

Redundancy is a key aspect of submarine battery design to ensure operational reliability. The battery system is divided into separate banks, allowing for continued operation even if one or more banks fail. This redundancy is critical for maintaining propulsion and essential electrical functions, especially during extended submerged operations. In case of a battery failure, the submarine can isolate the affected bank and continue using the remaining banks, thus ensuring that the loss of one battery segment does not compromise the entire electrical system.

Overall, the effective use and management of battery systems in diesel-electric submarines play a crucial role in balancing submerged endurance, operational stealth, and mission flexibility.

Nuclear Submarine Battery Systems

Nuclear submarines utilize battery systems differently compared to diesel-electric submarines, as their primary source of power is a nuclear reactor. The onboard batteries in nuclear submarines are designed mainly to act as an emergency power backup, ensuring the submarine's continued operation during situations where the reactor is shut down or unable to provide sufficient power. Additionally, batteries and other power storage solutions are employed to handle auxiliary power needs and manage peak power demands, enhancing the submarine's operational resilience.

In nuclear submarines, the batteries serve as a crucial emergency power source, providing electricity when the primary power system, the nuclear reactor, is offline or in case of other power-related emergencies. Should the reactor shut down, whether due to maintenance, a fault, or damage, the batteries are immediately engaged to sustain essential systems. These critical systems include life support, communication, navigation, and some propulsion capabilities, allowing the submarine to remain safe and operational while efforts are made to restore the reactor's functionality.

The battery capacity in nuclear submarines is designed to support these critical functions for a limited duration. This backup power period provides enough time for the crew to diagnose and address the issue, restore the reactor, or, if necessary, bring the submarine to the surface for repairs. While battery-powered propulsion is limited in range and speed, it allows the submarine to manoeuvre away from potentially dangerous situations or position itself to minimize detection risk while waiting for primary power to be re-established.

Figure 64: Battery well of USS Nautilus, the first nuclear-powered submarine. The batteries are for emergency use if the reactors shut down. Z22, CC BY-SA 3.0, via Wikimedia Commons.

Nuclear submarines also utilize battery systems and other energy storage solutions to supplement power during certain operations. For example, battery power can be used to manage peak electrical demands when the submarine is carrying out high-energy tasks, such as launching torpedoes or powering advanced sonar systems. This auxiliary power capability allows for smoother load management, preventing power fluctuations or strain on the primary nuclear reactor. Additionally, batteries can provide a quiet, stealthy power source for low-speed operations, reducing noise generated by the reactor's pumps and associated machinery, thus enhancing the submarine's stealth capabilities.

In modern nuclear submarines, the exploration of advanced energy storage technologies, particularly lithium-ion batteries, is gaining traction to enhance emergency power duration and performance. Lithium-ion batteries are recognized for their high energy density, which allows for more compact storage solutions compared to traditional lead-acid batteries. This characteristic is crucial for submarines, where space is at a premium and the need for reliable backup power during critical situations is paramount [162, 163]. The integration of these advanced battery systems not only improves the overall energy efficiency of the submarine but also extends the operational capabilities during emergencies, thus reinforcing the resilience of the submarine under adverse conditions.

The design of battery systems in nuclear submarines is meticulously focused on ensuring reliability and safety. Given the unique operational environment of submarines, where they may be submerged for extended periods, the battery systems must be capable of sustaining critical operations without failure. This necessitates rigorous testing and monitoring of the battery modules to ensure they can withstand the stresses of underwater operations [162]. The implementation of real-time health monitoring systems, such as those utilizing multilayer neural networks, can significantly enhance the reliability of these energy storage systems by providing early fault diagnosis capabilities [162]. Such innovations are essential for maintaining operational readiness and ensuring the safety of the crew and vessel during emergencies.

Moreover, the shift towards high-density power storage systems is not merely a matter of performance enhancement; it also aligns with broader trends in energy management and sustainability. As the demand for cleaner energy solutions grows, the adoption of lithium-ion batteries and similar technologies in military applications, including nuclear submarines, reflects a commitment to leveraging advanced technologies for improved operational efficiency and environmental responsibility [163]. This trend is indicative of a larger movement within the naval engineering community to integrate sustainable practices into the design and operation of military vessels, thereby enhancing their overall effectiveness and reducing their environmental footprint.

Air-Independent Propulsion (AIP) and Power Storage

Air-Independent Propulsion (AIP) is a technology that enhances the submerged endurance of diesel-electric submarines by providing an alternative power source that doesn't rely on atmospheric oxygen, allowing the submarine to stay underwater for extended periods. While AIP is not a battery system, it works in tandem with traditional battery storage to improve the overall underwater operational capabilities. AIP reduces the frequency of surfacing or snorkelling required to recharge batteries with diesel engines, thereby improving the submarine's stealth and reducing its exposure to detection.

Air-independent propulsion systems can work alongside battery storage to extend a submarine's submerged endurance. AIP technology typically includes mechanisms like fuel cells, Stirling engines, or closed-cycle diesel engines that generate electricity underwater without requiring external air intake. This generated electricity can be used to power the submarine's electrical systems and propulsion, which significantly reduces the rate of battery discharge during submerged operations. As a result, the batteries last longer, and the need to surface or snorkel for recharging is minimized. This capability is particularly beneficial for stealth missions, where the risk of detection increases if a submarine has to come near the surface to recharge its batteries.

While AIP provides a continuous low-power output suitable for maintaining the submarine's basic operational functions and slow-speed propulsion, battery power remains essential for certain operational demands. Batteries are used to meet peak power requirements, such as high-speed manoeuvres, rapid changes in depth, or bursts of propulsion speed. During these high-energy activities, AIP systems may not be able to supply the required power quickly enough on their own. Thus, AIP serves to complement battery usage rather than replace it entirely, allowing submarines to optimize their energy resources for various mission profiles.

By providing a continuous source of power while submerged, AIP systems significantly enhance the stealth capabilities of diesel-electric submarines. The use of AIP minimizes the need to run noisy diesel engines, which can emit detectable sounds and increase the risk of detection by adversaries. Additionally, the reduced reliance on surfacing to recharge batteries improves the submarine's ability to remain hidden for extended periods, making it more challenging for enemy forces to locate and track the vessel.

In conjunction with power storage solutions like lithium-ion or advanced lead-acid batteries, AIP-equipped submarines can better manage power distribution, switching between AIP-generated power and stored battery power based on operational requirements. This hybrid approach ensures that submarines maintain high levels of stealth and can adapt to various mission scenarios, including long-term patrols, reconnaissance, and covert operations. Ultimately, AIP and battery systems work together to provide a balanced and flexible power management solution, enhancing the overall performance and endurance of modern diesel-electric submarines.

Ongoing Development

The development of submarine battery technology is crucial for advancing the operational capabilities of both diesel-electric and nuclear submarines. As the demand for longer submerged endurance, increased safety, and better energy management grows, new innovations in battery systems continue to emerge. However, there are several challenges that need to be addressed, alongside ongoing efforts to improve energy density, reliability, and safety.

Effective thermal management is crucial for the safety and performance of lithium-ion batteries, particularly in confined environments such as submarines. These batteries are known for their high energy density but also for their propensity to generate significant heat during operation [164]. If this heat is not adequately managed, it can lead to thermal runaway, a condition characterized by uncontrolled temperature increases that can result in fires or explosions [165, 166]. This risk is exacerbated in submarines, where space constraints and safety concerns necessitate robust thermal management solutions.

Thermal runaway in lithium-ion batteries is influenced by various factors, including the battery's state of charge (SOC), the materials used in the cells, and the design of the battery module [165-167]. Research has shown that the thermal characteristics of lithium-ion batteries can vary significantly under different conditions, such as overcharging, external heating, or mechanical abuse [166, 168]. For instance, studies have demonstrated that the thermal runaway of one cell can propagate to neighbouring cells, leading to a cascading failure within the battery pack [166, 169, 170]. Therefore, it is essential to implement effective cooling systems and thermal management strategies to mitigate these risks.

Several techniques have been developed to enhance thermal management in lithium-ion batteries. Liquid cooling systems, advanced heat sinks, and thermal barriers are commonly employed to maintain battery temperatures within safe operating limits [171, 172]. For example, hybrid thermal management systems that combine phase change materials with forced-air cooling have shown promise in improving the thermal stability of battery packs [172]. Additionally, the use of thermally conductive separators with hierarchical structures has been proposed to enhance heat dissipation and overall thermal management [171]. These innovations are vital for ensuring the safe operation of batteries in high-energy applications, such as electric vehicles and submarines.

Moreover, ongoing research is focused on developing early warning systems for thermal runaway detection. Techniques such as gas characterization and monitoring of thermal behaviour under various conditions are being explored to provide timely alerts in case of potential thermal events [173, 174]. This proactive approach is essential for enhancing the safety of lithium-ion batteries, particularly in high-stakes environments like submarines, where rapid response to thermal incidents can prevent catastrophic outcomes [175].

Unlike surface vessels, submarines operate in confined, high-pressure settings where even minor failures can lead to catastrophic outcomes. This necessitates stringent safety standards and the implementation of multiple safety layers in battery design. For instance, robust monitoring systems are essential for real-time tracking of critical parameters such as temperature, voltage, and state of charge, which are vital for preventing incidents like thermal runaway [176, 177].

Recent advancements in battery chemistry, particularly the development of safer electrolyte materials and solid-state batteries, are being explored to enhance safety further. Solid-state batteries, for example, utilize non-flammable electrolytes, significantly reducing fire risks associated with traditional lithium-ion batteries [177]. Additionally, fire suppression systems, including gas-based and water mist extinguishers, are integrated into battery compartments to mitigate fire risks. Research indicates that such systems can effectively suppress lithium-ion battery fires, thereby enhancing overall safety [178, 179].

The potential for thermal runaway in lithium-ion batteries poses a significant challenge for submarine applications. Thermal runaway can occur due to various factors, including external heating, overcharging, and internal faults, leading to uncontrollable temperature increases and possible explosions [180, 181]. To address these risks, it is crucial to implement comprehensive safety strategies, including the use of fire suppression technologies and advanced monitoring systems that can detect early signs of failure [182, 183]. Furthermore, the integration of smart battery management systems that utilize machine learning algorithms for fault diagnosis can significantly enhance the reliability of battery systems in submarines [162, 163].

Hybrid power systems that integrate battery storage with various energy sources, such as air-independent propulsion (AIP) systems and supercapacitors, are increasingly being investigated to enhance energy management and operational performance in submarines and other marine applications. These hybrid configurations allow for seamless transitions between power sources based on specific operational demands, thereby optimizing energy usage and improving overall system efficiency.

The integration of supercapacitors into hybrid energy storage systems (HESS) is particularly advantageous due to their ability to deliver rapid bursts of power, which is essential for high-speed manoeuvres in submarines. Supercapacitors possess high power density and can quickly respond to load changes, thus alleviating stress on batteries during peak demands [184, 185]. This capability is crucial in military applications where submarines may require sudden acceleration or manoeuvrability without compromising the longevity of the battery systems. Moreover, the combination of batteries and supercapacitors allows for a more balanced energy management strategy, effectively reducing reliance on any single power source [186].

In addition to supercapacitors, AIP systems play a vital role in sustaining low-power demands during prolonged submerged operations. AIP technologies enable submarines to generate electric energy silently while underwater, thereby enhancing stealth capabilities [187]. The integration of AIP with battery systems allows submarines to recharge their batteries during snorkelling operations, although this method poses a risk of detection due to the submarine's exposure to the surface [187]. The hybridization of these systems not only extends the operational range of submarines but also improves their energy efficiency by allowing for the optimal use of available power sources depending on the mission profile [188]

The hybrid configurations also facilitate advanced energy management strategies that can dynamically adjust the power distribution between batteries and supercapacitors or AIP systems. This adaptability is essential for maintaining operational efficiency and reliability, especially in environments where energy demands fluctuate significantly [189]. Research indicates that such hybrid systems can lead to significant reductions in fuel consumption and emissions, aligning with modern environmental standards and operational requirements [190].

Extended Endurance

The development of submarine battery technology is important for enhancing underwater endurance, primarily through the improvement of energy density. Traditional lead-acid batteries, while reliable, have been largely outperformed by newer technologies in terms of energy storage capacity. Lithium-ion batteries have emerged as a superior alternative due to their higher energy density, faster charging capabilities, and longer lifespan, making them increasingly favoured in modern submarine applications [111, 191]. The transition from lead-acid to lithium-ion systems represents a significant advancement in submarine technology, allowing for extended submerged operations and improved overall performance [111, 114].

Research into next-generation battery technologies is ongoing, focusing on materials and chemistries that can further enhance energy density. Solid-state batteries, for instance, are being investigated for their potential to provide higher energy densities and improved safety profiles compared to conventional lithium-ion batteries [192, 193]. These batteries utilize solid electrolytes, which can mitigate risks associated with liquid electrolytes, such as leakage and flammability, thus making them particularly suitable for the demanding environments of submarines [192, 193]. Additionally, advancements in materials science, such as the exploration of two-dimensional materials like transition-metal dichalcogenides, are showing promise for developing high-capacity anodes that could significantly improve battery performance [194, 195].

The integration of artificial intelligence in battery material design is also a noteworthy trend, as it accelerates the identification and optimization of new materials for energy storage applications [196]. This approach not only enhances the efficiency of the research process but also holds the potential for breakthroughs in battery technology that could further extend the operational capabilities of submarines [196]. Furthermore, the ongoing research into sodium-ion batteries and other alternative chemistries aims to address the environmental and resource limitations associated with lithium-based systems, ensuring a sustainable future for submarine energy storage [197].

The transition towards lithium-ion batteries and other high-energy-density storage solutions is increasingly critical, particularly in applications requiring extended underwater endurance and minimized detection risks, such as in submarines. Lithium-ion batteries (LiBs) are favoured for their high energy density, lightweight nature, and low self-discharge rates, making them suitable for various applications, including military submarines and electric vehicles [162, 179]. The operational agility of future submarines could significantly improve with advancements in LiB technology, potentially doubling submerged range capabilities if packing density can be optimized [111].

Research into solid-state batteries represents a promising avenue for enhancing submarine battery systems. Solid-state batteries utilize a solid electrolyte, which mitigates risks associated with liquid electrolytes, such as leaks and thermal runaway incidents [198]. These batteries not only offer higher energy density but also improved safety features, making them particularly appealing for underwater applications where reliability is paramount [198]. The ongoing development of solid-state technology is crucial as it addresses the limitations of conventional LiBs, which are prone to safety hazards under certain operational conditions [198].

In addition to solid-state batteries, high-capacity lithium-based systems such as lithium-sulphur and lithium-air batteries are being explored to push the boundaries of energy density and overall battery performance. These technologies promise to enhance the energy storage capabilities of submarines, allowing for longer missions without the need for frequent recharging [114, 199]. The maritime sector's shift from traditional lead-acid batteries to lithium-ion systems reflects a broader trend towards adopting advanced energy storage solutions that meet the increasing demands for efficiency and safety in underwater operations [200, 201].

Furthermore, the integration of advanced energy management systems in submarines is essential for optimizing energy resources and enhancing operational efficiency. These systems can help manage the complexities of energy storage and consumption, particularly in scenarios where stealth and endurance are critical [109]. The combination of innovative battery technologies and sophisticated energy management will likely define the future of submarine operations, enabling longer missions while maintaining a low profile against detection [111, 114].

Power Management and Redundancy

Submarine power management and redundancy are essential components in ensuring that electrical systems operate reliably under the demanding conditions encountered underwater. These practices are critical for managing power distribution, sustaining essential functions, maintaining operational stealth, and ensuring the safety of the crew. Given the unique environment and mission requirements of submarines, power management and redundancy must be carefully engineered to handle routine operations, emergencies, and failure scenarios.

Power Management

Power management in submarines is a critical aspect of their operational efficiency and effectiveness. The management of electrical power, derived from various sources such as diesel generators, batteries, nuclear reactors, and air-independent propulsion (AIP) systems, is essential for ensuring that the submarine can meet the diverse power demands of its systems, including propulsion, life support, weaponry, and sensors [109]. The complexity of power management in submarines arises from the need to balance these demands dynamically, particularly during varying operational scenarios.

Dynamic Load Distribution

Dynamic load distribution is a key feature of submarine power management systems. These systems are designed to allocate electrical power based on real-time operational requirements. For example, during routine cruising, the power requirements for propulsion may be lower, allowing more power to be directed towards sensors and navigation systems. Conversely, in high-speed manoeuvres or combat situations, propulsion demands can surge, necessitating a reallocation of power to ensure critical systems remain operational [109]. This dynamic allocation is crucial for maintaining operational readiness and efficiency, particularly in the constrained environment of a submarine.

Dynamic load distribution is an essential aspect of submarine power management, enabling the allocation of electrical power to different onboard systems based on real-time operational needs. This approach is critical in the confined environment of a submarine, where the power supply must be carefully balanced to ensure optimal performance and operational readiness. Dynamic load distribution ensures that the electrical demands of various systems—such as propulsion, navigation, sensors, life support, and weaponry—are met efficiently, even as mission requirements shift.

Dynamic load distribution allows submarine power management systems to adapt to various operational scenarios, ensuring the efficient allocation of resources. During routine cruising or low-speed operations, propulsion power requirements are relatively low. This allows more electrical power to be allocated to non-propulsive systems, such as navigation equipment, sensor arrays, communication systems, and life support. The power management system continuously monitors these systems' power requirements, directing resources where they are most needed without compromising the submarine's overall efficiency.

When the submarine transitions to high-speed manoeuvres, combat situations, or emergency operations, propulsion demands can spike significantly. The power management system dynamically reallocates power from lower-priority systems to the propulsion system to meet the increased demand. For instance, non-essential loads such as air conditioning, certain lighting systems, or other auxiliary equipment may be temporarily reduced to prioritize propulsion and critical mission functions. This reallocation ensures that the submarine can achieve the necessary speed and manoeuvrability to evade threats or engage targets while maintaining operational effectiveness.

The dynamic load distribution is achieved through real-time monitoring and control systems that assess the power consumption and status of various onboard systems. The submarine's power management system uses sensors and automated controls to gather data on power usage, system loads, and potential faults. These sensors detect changes in operational conditions and adjust power distribution accordingly. For example, if a sudden surge in propulsion power is detected, the system can quickly divert power from other systems to accommodate the increased load.

This real-time capability is crucial because it allows the submarine to react swiftly to changing operational needs, which is especially important in unpredictable combat or emergency situations. Automation plays a key role in making these adjustments rapidly and efficiently, with minimal input from the crew, thereby reducing the risk of human error during critical moments.

Dynamic load distribution also contributes to the submarine's stealth capabilities by managing power usage in a way that minimizes noise and heat signatures. By controlling power flow to propulsion and auxiliary systems, the power management system can adjust the speed and operation of electric motors and pumps to reduce acoustic output. For instance, when operating in stealth mode, the submarine may reduce the speed of certain fans or pumps to lower noise levels, reallocating power to quieter systems or adjusting propulsion settings to achieve silent running.

In terms of efficiency, the system's ability to adapt to varying power demands ensures that energy is used optimally, preventing waste and conserving battery power for diesel-electric submarines. This helps extend submerged endurance and reduces the need for frequent battery recharging, which is critical for maintaining stealth and operational capability over extended missions.

A key feature of dynamic load distribution is the ability to prioritize critical systems over non-essential loads. The power management system categorizes submarine systems based on their importance to mission success and safety, ensuring that essential functions like life support, propulsion, navigation, and communication always receive sufficient power. In the event of a power shortfall or system failure, lower-priority systems can be powered down or operate in a reduced capacity to free up power for more critical applications.

Additionally, dynamic load distribution is integrated with system redundancy measures to ensure continuous operation in the face of faults or damage. Should a component of the power generation or distribution system fail, the load can be redistributed to other functioning systems, allowing the submarine to continue its mission with minimal disruption.

The ability to dynamically adjust power distribution enhances a submarine's operational readiness by ensuring that it can quickly transition between different mission modes. Whether performing routine patrols, engaging in high-speed pursuits, or evading enemy detection, the submarine's power management system adapts to changing demands in real-time, maintaining optimal performance across all systems. This flexibility allows submarines to remain effective in diverse and challenging operational environments, thus improving mission success rates.

Dynamic load distribution in submarine power management ensures that electrical resources are allocated efficiently based on real-time operational requirements. By dynamically adjusting power flows to meet shifting demands, the system enhances the submarine's efficiency, stealth, and operational readiness while maintaining critical system functionality in various scenarios. This capability is vital for maximizing performance and ensuring the submarine remains prepared for any situation.

Battery Management

Battery management is another vital component of power management, especially in diesel-electric submarines. These vessels rely heavily on large battery banks for submerged operations, making effective management of battery usage, charging, and discharge cycles essential [111]. The depth of battery discharge must be carefully monitored to prevent damage and extend battery life, while charging cycles are regulated to ensure that batteries are adequately replenished when diesel engines are operational [109]. This careful management minimizes the "indiscretion ratio," which refers to the time spent recharging, thus maximizing the submarine's submerged endurance and operational capabilities [109].

Battery management plays a critical role in the power management strategy of diesel-electric submarines, where large battery banks provide the primary source of electrical power during submerged operations. The batteries supply energy for propulsion, life support, navigation, and other essential systems when the submarine is underwater and cannot use its diesel generators. Effective battery management encompasses the careful regulation of battery usage, monitoring of discharge cycles, and optimized charging processes to ensure the submarine's operational readiness, endurance, and battery longevity.

One of the primary aspects of battery management is monitoring the depth of battery discharge during submerged operations. Discharge refers to the amount of battery capacity used to power the submarine's systems. If batteries are discharged too deeply or too frequently, it can lead to degradation of the battery cells and reduce the overall lifespan of the battery bank. Therefore, careful monitoring is essential to prevent excessive discharge and maintain the health of the batteries. Submarine battery management systems are equipped with sensors and control algorithms that continuously track the state of charge (SOC) and state of health (SOH) of the battery cells, ensuring that discharge levels remain within safe limits.

Limiting the depth of discharge also allows the batteries to maintain a higher reserve capacity, which is crucial for emergency situations. By preventing battery depletion beyond a certain threshold, the submarine can still rely on battery power for critical functions in case of unexpected events, such as system failures or sudden changes in mission requirements.

Charging cycles are another vital aspect of battery management in diesel-electric submarines. The submarine's diesel generators recharge the batteries when the vessel is surfaced or at snorkel depth, allowing air intake for the diesel engines. The charging process must be carefully managed to ensure that the batteries are adequately replenished without overcharging, which could damage the battery cells. Modern battery management systems use advanced algorithms to regulate the charging rate, monitor battery temperature, and adjust charging parameters based on the battery's SOC and SOH.

The goal is to optimize the charging cycles to restore the battery's full capacity as quickly and efficiently as possible while minimizing wear on the battery cells. Rapid charging may be necessary in situations where the submarine needs to minimize time spent at or near the surface for stealth purposes. However, controlled charging processes are used to avoid excessive heat buildup and ensure the long-term health of the batteries.

A key objective of effective battery management is to reduce the submarine's "indiscretion ratio," which is the proportion of time spent recharging the batteries at snorkel depth compared to the total time of submerged operations. A lower indiscretion ratio enhances the submarine's stealth capabilities, as spending less time near the surface reduces the risk of detection by enemy forces. Battery management strategies, therefore, focus on optimizing the charging cycles to maximize battery capacity while minimizing recharge time. This includes implementing adaptive charging techniques, which adjust charging rates based on operational conditions and mission requirements.

The management of battery usage during submerged operations also influences the indiscretion ratio. By carefully controlling the power consumption of onboard systems and implementing load-shedding strategies for non-essential equipment, the submarine can extend the battery life and prolong submerged endurance. This approach not only maximizes operational capabilities but also improves the submarine's ability to remain undetected for longer periods.

Battery management systems are designed with redundancy and safety features to ensure the submarine's operational readiness and protect against potential hazards. These systems continuously monitor battery parameters, including voltage, current, temperature, and SOC, to detect any anomalies or signs of deterioration. If a problem is detected, such as an overheating battery cell or a significant drop in voltage, the system can automatically isolate the affected battery bank to prevent further damage and maintain power to essential systems.

In addition, battery management systems are equipped with thermal management solutions to prevent overheating during both charging and discharging cycles. This is especially important for modern high-energy-density batteries, such as lithium-ion, where thermal runaway could pose a significant safety risk in the confined environment of a submarine. By integrating active cooling and thermal monitoring technologies, battery management systems help maintain safe operating temperatures and ensure reliable performance.

The advancements in battery management are also relevant to hybrid power solutions, where batteries work in conjunction with air-independent propulsion (AIP) systems. In such configurations, battery management systems help optimize the balance between battery use and AIP output, providing additional flexibility and extending the submarine's submerged endurance. By carefully managing the charging and discharging cycles, the battery management system ensures that sufficient power is available for peak demands, such as rapid propulsion or emergency manoeuvres.

Battery management is a vital component of diesel-electric submarine power management, focusing on the regulation of battery usage, monitoring of discharge cycles, and optimization of charging processes. Effective battery management reduces the indiscretion ratio, extends battery life, ensures operational readiness, and supports the integration of hybrid power solutions, thus enhancing the submarine's submerged endurance and overall mission effectiveness.

Peak Power Demands

Submarines must be equipped to handle peak power demands that arise during specific operations, such as weapon deployment or rapid acceleration. To manage these surges, submarines may integrate hybrid energy storage solutions, including supercapacitors or auxiliary batteries, which can provide additional power for short durations without overloading primary generation sources [109]. This capability not only enhances the submarine's performance but also ensures stable power delivery across all systems.

Submarines are designed to perform a range of complex operations that require sudden increases in power demand. Situations like launching torpedoes or missiles, engaging in rapid acceleration to evade threats, or conducting emergency manoeuvres can place substantial strain on the submarine's power systems. To manage these peak power demands, submarines often incorporate hybrid energy storage solutions that can quickly supply additional energy when needed, ensuring smooth and stable power delivery across all systems.

Hybrid energy storage solutions, such as supercapacitors or auxiliary battery systems, are used in conjunction with the submarine's primary power sources, such as diesel generators or nuclear reactors. These hybrid systems are specifically designed to provide high-power output over short durations, making them ideal for handling rapid power surges without overloading the primary generation units. By storing and releasing energy quickly, hybrid storage solutions can bridge the gap during peak demands, ensuring that critical systems receive the power they need without compromising the stability of the overall electrical network.

Supercapacitors are highly efficient at storing and delivering large amounts of electrical energy in a very short time. Unlike traditional batteries, which rely on chemical reactions to generate electricity, supercapacitors store energy electrostatically, allowing them to discharge almost instantaneously. This makes them particularly well-suited for submarine applications where rapid power delivery is essential, such as during weapon deployment or emergency propulsion. When integrated into a submarine's power management system, supercapacitors can quickly absorb excess energy from primary generators or auxiliary batteries and release it as needed, stabilizing voltage levels and ensuring a consistent power supply.

Auxiliary batteries, which can be separate from the main battery bank, provide another layer of support for managing peak power requirements. These batteries are typically configured to supply power for specific operations, such as high-speed propulsion or powering weapons systems. During times of peak demand, auxiliary batteries can be activated to supplement the main batteries or other power sources, providing a temporary power boost. This approach allows the submarine to maintain performance across critical systems without placing excessive strain on the primary batteries or generators. Additionally, auxiliary batteries can be recharged during periods of lower power demand, ensuring they are ready for the next surge.

The use of hybrid energy storage solutions enhances a submarine's performance by providing flexibility in power management. During operations where rapid power adjustments are needed, these systems ensure that energy is readily available, minimizing delays and maintaining operational readiness. This capability is especially important in military settings, where the ability to respond swiftly to changing situations can be the difference between mission success and failure.

Moreover, hybrid storage solutions help protect the submarine's primary power generation equipment by reducing the risk of overload. When peak demands are met using auxiliary batteries or supercapacitors, the primary generators can continue to operate within their optimal range, reducing wear and extending their service life. The ability to stabilize power delivery during high-demand situations also reduces the likelihood of power fluctuations, which can negatively impact sensitive electronic systems and navigation equipment.

Stable power delivery is crucial for maintaining the submarine's overall operational capabilities. Systems such as communication, navigation, life support, and electronic warfare require consistent power to function effectively. By incorporating hybrid energy storage solutions, submarines can prioritize power allocation based on real-time needs. For instance, during a rapid acceleration or an emergency, power can be redirected from non-essential systems to propulsion or weapons, with hybrid storage providing the necessary backup.

Incorporating hybrid energy solutions also allows for better energy management strategies. When the submarine is cruising at a steady pace with lower power requirements, the hybrid storage systems can be recharged in preparation for future demands. This ensures that the submarine is always prepared to handle unexpected power needs, enhancing its capability to perform extended missions with minimal risk of power shortfalls.

In conclusion, equipping submarines with hybrid energy storage solutions, such as supercapacitors or auxiliary batteries, is a vital strategy for managing peak power demands. These systems provide the flexibility needed to handle rapid surges in energy requirements, ensure stable power delivery across all onboard systems, and protect primary power generation sources from being overloaded. By integrating these advanced storage technologies, submarines can maintain high performance and operational readiness in a variety of challenging scenarios.

Energy Conservation Techniques

Energy conservation techniques are also employed to extend submerged endurance and reduce the risk of detection. Load shedding, where non-essential systems are temporarily powered down, is a common practice during critical operations. For instance, lighting and air conditioning may be minimized to conserve battery power during extended submerged periods [109]. The adoption of energy-efficient technologies, such as LED lighting and variable-speed motors, further aids in reducing overall power consumption, thereby enhancing the submarine's operational capabilities [109].

Energy conservation techniques are crucial for submarines, especially during extended submerged operations where the availability of power is limited. Efficient power management not only helps extend the submarine's submerged endurance but also plays a critical role in maintaining stealth by reducing the frequency of surfacing or snorkelling for battery recharge. One of the primary strategies used for energy conservation is load shedding, which involves selectively powering down non-essential systems to save energy during critical operations. This approach, combined with the use of energy-efficient technologies, significantly enhances the submarine's operational capabilities.

Load shedding is a widely used energy-saving practice in submarines. During extended submerged operations, where maintaining a low profile is essential to avoid detection, power demands must be carefully managed. By temporarily shutting down non-essential systems, such as interior lighting, ventilation, or air conditioning, energy can be conserved and directed to critical systems like propulsion, navigation, and communication. This is especially important during times when the submarine needs to remain submerged for longer durations or operate under conditions that require maximum stealth, as it minimizes the depletion of battery reserves.

Strategic load shedding allows the submarine to prioritize essential functions, ensuring that power is available for emergency operations or peak power demands. For instance, during silent running modes, where noise and energy consumption must be kept to a minimum, load shedding can significantly reduce the power draw, thereby extending the time the submarine can stay submerged without having to surface for battery recharging.

The adoption of energy-efficient technologies is another essential aspect of submarine power management. The use of modern, low-power solutions such as LED lighting and variable-speed motors helps reduce overall energy consumption. LED lights, for example, consume significantly less power than traditional incandescent bulbs, providing the same level of illumination while using a fraction of the energy. This not only conserves battery life but also generates less heat, which in turn reduces the cooling requirements, further saving energy.

Variable-speed motors used in systems like ventilation, cooling pumps, and hydraulic systems contribute to energy conservation by adjusting their power output based on actual demand. Unlike traditional motors that run at a constant speed, variable-speed motors can operate at lower speeds when full power is not necessary, thereby reducing power consumption. This flexibility in adjusting power usage according to need is particularly beneficial during low-activity periods, enabling more efficient energy management and extended submerged operations.

The implementation of load shedding and energy-efficient technologies has a direct impact on the submarine's operational capabilities. By reducing the overall power consumption, the submarine's battery life can be significantly extended, allowing for longer underwater endurance without surfacing. This extended submerged time not only enhances mission capabilities but also reduces the risk of detection by minimizing exposure at the surface. The more efficiently the submarine can manage its energy resources, the better it can perform stealth missions, evade threats, and conduct extended patrols in hostile waters.

Energy conservation also helps optimize the submarine's power management strategy. By lowering the baseline power consumption, it becomes easier to allocate energy to high-priority systems when needed. This ensures that during critical moments, such as evasive manoeuvres or combat situations, sufficient power is available for propulsion and weapons systems without depleting reserves too quickly. As a result, the submarine maintains higher operational readiness, with the flexibility to adapt to changing mission requirements.

Reducing power consumption not only conserves battery energy but also plays a role in reducing the acoustic and thermal signatures of the submarine. By minimizing the use of noisy equipment like air conditioning compressors or ventilation fans, the submarine becomes quieter, making it harder for enemy sonar to detect. Lower power consumption also means less heat is generated, which decreases the thermal footprint and reduces the chances of being detected by thermal imaging systems. Thus, energy conservation techniques not only extend the endurance of the submarine but also contribute to its stealth capabilities, enabling it to operate more discreetly.

Redundancy in Submarine Electrical Systems

Redundancy is an essential aspect of submarine design to ensure that critical systems remain operational even if a component or power source fails. The confined and isolated environment of a submarine makes it challenging to repair electrical faults during operations, so redundancy is built into the power generation, distribution, and storage systems.

- **Multiple Power Sources:** Submarines typically have multiple power sources, such as diesel generators, batteries, nuclear reactors, or AIP systems, to provide redundancy. For diesel-electric submarines, having multiple diesel generators ensures that if one generator fails, others can take over the power load. In nuclear submarines, the nuclear reactor is the primary power source, but batteries serve as a backup in case of a reactor shutdown, allowing essential systems to continue functioning [111]. This multi-source approach ensures that submarines can still operate even if one power source becomes compromised.

- **Redundant Battery Banks:** Battery redundancy is vital for diesel-electric submarines that rely heavily on battery power for submerged operations. Battery systems are typically arranged in banks, with multiple cells connected in series and parallel configurations. If one battery bank experiences a failure, the other banks can continue to supply power. This arrangement ensures that the submarine remains operational, albeit at reduced performance, while mitigating the risk of complete power loss [117].

- **Redundant Electrical Distribution Paths:** Electrical distribution systems in submarines are designed with multiple distribution paths or circuits to ensure that power can be rerouted in case of a fault or damage. This involves using dual or triple-redundant power lines, circuit breakers, and switchboards, allowing operators to isolate a faulty circuit and maintain power to critical systems [146]. The distribution paths are also segmented, ensuring that a localized fault does not affect the entire electrical grid.

- **Emergency Generators and Backup Systems:** Submarines are equipped with emergency generators or alternative power systems that can provide electrical power in the event of main power source failure [111]. These emergency systems are typically powered by auxiliary diesel engines or specialized energy storage solutions such as emergency batteries. They are designed to power life support, navigation, communication, and other essential systems, ensuring that the submarine remains operational during emergency situations [111].

- **Automatic Switching and Isolation:** Redundant power systems in submarines are often equipped with automatic switching mechanisms that detect power loss or faults and immediately switch to a backup power source. This capability minimizes downtime and ensures continuous power supply to critical systems. Similarly, fault isolation mechanisms help identify and isolate faulty components, preventing further damage and allowing the submarine to continue operations without significant interruption [202].

- **Control System Redundancy:** The control systems that manage power distribution also have redundancy to maintain functionality. Redundant sensors, controllers, and monitoring devices are used to detect faults in the electrical system and to ensure accurate data is available for power management decisions [203]. In case of a control system failure, backup controllers can take over, maintaining the ability to manage power flows and respond to emergencies.

Integrated Power Management Systems

Modern submarines are increasingly equipped with integrated power management systems that effectively combine power generation, storage, and distribution into a cohesive framework. These systems leverage advanced software and automation technologies to monitor and control power flows, optimize energy usage, and manage redundancies. The integrated approach facilitates real-time adjustments to power distribution based on operational needs, enhancing the submarine's resilience by coordinating all elements of the electrical system.

The architecture of these integrated power management systems is designed to support dynamic control of power consumption, which is crucial in the confined and demanding environments submarines operate in. For instance, the implementation of power controller systems (PCS) allows for complex multiple-input multiple-output (MIMO) power management policies that can dynamically adjust power distribution in response to varying operational demands [204]. This capability is essential for maintaining stability and efficiency, particularly in systems that utilize medium-voltage direct current (MVDC) configurations, which are increasingly being adopted in modern naval vessels [205].

Moreover, the integration of software-controlled power management techniques enables submarines to optimize energy usage effectively. By employing sophisticated algorithms, these systems can selectively manage power to different components, ensuring that energy is allocated efficiently based on real-time requirements [109]. This is particularly important in submarines, where power resources are limited and must be managed judiciously to support various onboard systems and missions.

In addition to optimizing energy usage, integrated power management systems enhance the resilience of submarines. By coordinating the various elements of the electrical system, these systems can quickly adapt to changes in operational conditions, such as shifts in power demand or unexpected failures in power generation or storage components. This adaptability is crucial for maintaining operational readiness and safety in the unpredictable underwater environment [146]. Furthermore, the use of advanced monitoring and management technologies allows for proactive maintenance and fault detection, further contributing to the reliability of submarine operations [206].

Chapter 6
Life Support Systems

Submarine Air Systems

Air systems are crucial to the functioning of a submarine, as virtually every aspect of its diving, surfacing, and overall operation depends on compressed air. These systems support a wide range of tasks that are essential for the submarine's functionality, making them one of the most versatile onboard systems. The importance of the air systems extends beyond mere ventilation and air quality management; they are integral to propulsion, weapon deployment, buoyancy control, and various mechanical operations [207].

Role in Submarine Operations

Compressed air is essential for diving and surfacing procedures. To dive, the submarine uses compressed air to adjust its buoyancy by releasing air from the ballast tanks, allowing water to fill them and making the submarine heavier. Conversely, to surface, air is blown into the ballast tanks to expel the water, increasing buoyancy and lifting the submarine back to the surface. This air management is fundamental to the submarine's ability to control its depth and remain submerged or emerge as required [207].

Beyond buoyancy control, compressed air plays a vital role in other operational systems. For example, the main hydraulic system relies on air pressure maintained in air-accumulator flasks to operate various hydraulic mechanisms. Additionally, compressed air is used to discharge torpedoes from the submarine's torpedo tubes, enabling weapon deployment during combat. Furthermore, starting the main propulsion engines often requires air pressure to initiate the combustion process. These applications highlight how essential compressed air is to both routine and critical operations aboard a submarine [207].

Air Systems for Onboard Life Support

In addition to operational tasks, compressed air is also important for maintaining life support systems on the submarine. When a submarine remains submerged for extended periods, the quality of onboard air can deteriorate due to the accumulation of carbon dioxide and other contaminants. Compressed air and oxygen are used to revitalize the air within the submarine, ensuring a breathable atmosphere for the crew. This is particularly important during long-term underwater missions where access to fresh air is limited [207].

Pressure management within the submarine is also tested using compressed air to ensure the integrity of the pressure hull. Tightness tests are conducted to detect any potential leaks in the hull, ensuring the safety of the crew and the submarine's ability to maintain internal pressure at depth. This illustrates another critical function of the air systems in monitoring and maintaining the submarine's structural integrity.

Compressed Air as a Versatile Power Source

Compressed air is a versatile power source that supports a variety of primary and secondary functions within a submarine. One key advantage is its ability to be stored easily in high-pressure air banks located throughout the vessel. This stored air can be used on demand without the need for additional energy expenditure, providing a readily available source of power. The submarine's high-pressure air compressor compresses air using a series of pistons that increase the air pressure in stages, allowing it to be stored for later use [207].

This compressed air can be regulated to provide different pressure levels, from low to high, depending on the needs of various onboard systems. Its compressibility also allows it to act as a cushion, reducing the impact of mechanical forces on equipment during operations. The ability to distribute compressed air to different parts of the submarine through an interconnected network of air lines further enhances its versatility and makes it an indispensable component of the submarine's infrastructure.

Air Systems in Emergency Situations

In emergency situations, such as a power loss or mechanical failure, the air systems provide critical backup capabilities. Compressed air can be used to operate emergency ballast blow mechanisms to bring the submarine to the surface if the primary systems fail. This ensures that the crew has a reliable means of escaping dangerous situations and returning to a safe depth. Additionally, certain air-powered tools and equipment may be utilized for repairs or emergency operations, allowing the submarine to maintain functionality even under challenging conditions [207].

Typical Air System in a Diesel-Electric Submarine

A typical air system in a diesel-electric submarine plays a critical role in various operational functions, including diving and surfacing, propulsion, onboard life support, and equipment operation. The system is designed to store, distribute, and regulate compressed air throughout the submarine, supporting both essential and auxiliary tasks. A typical air system found in a diesel-electric submarine includes:

1. High-Pressure Air Storage: The air system begins with high-pressure air storage tanks or air banks, which hold compressed air at pressures typically ranging from 200 to 300 bar (2,900 to 4,350 psi). These tanks are distributed throughout the submarine to ensure balanced weight distribution and to provide multiple sources of air for different systems. The high-pressure air is used for a variety of purposes, including diving and surfacing, engine starting, and emergency procedures.

The air storage banks are connected to the submarine's air compressor system, which replenishes the air supply when the submarine is surfaced or at snorkel depth. These compressors draw in atmospheric air, compress it to the required pressure, and then store it in the air banks. The system includes multiple compressors for redundancy, ensuring that air can be generated reliably.

2. Air Compressors and Control Systems: Diesel-electric submarines are equipped with high-pressure air compressors that refill the air banks when on the surface or while operating at snorkel depth. The compressors typically have multiple stages of compression to achieve the necessary high pressures. Control systems manage the operation of the compressors, including pressure monitoring and automatic shutdown or adjustment when the air storage reaches the desired level.

Air compressors are usually cooled using seawater or internal cooling systems to prevent overheating. The compressors may also include moisture separators to remove any water vapor from the compressed air, ensuring that only dry air is stored in the tanks. This helps to prevent corrosion in the air storage tanks and connected systems.

3. Air Distribution System: The high-pressure air stored in the air banks is distributed throughout the submarine via an interconnected network of air lines. The distribution system includes a series of valves, pressure regulators, and control mechanisms that direct air to various subsystems at the required pressure levels.

The air distribution system is typically divided into different pressure zones:

- **High-Pressure Zone:** For applications that require the full pressure of the stored air, such as torpedo launch systems or emergency ballast blow operations.
- **Medium-Pressure Zone:** For tasks like operating pneumatic tools or certain hydraulic systems, where lower air pressure is sufficient.
- **Low-Pressure Zone:** For air supply to non-essential systems or for controlling ballast tank operations during normal diving and surfacing.

The system includes manual and automated control valves to manage air flow, allowing the crew to adjust the air supply to various systems as needed. It also features safety relief valves to prevent overpressure conditions.

4. Diving and Surfacing Operations: The air system is crucial for controlling the submarine's buoyancy during diving and surfacing manoeuvres. When diving, the submarine opens vents on the ballast tanks to allow seawater to flood in, making the vessel heavier. The air from the tanks is vented to the atmosphere or the sea. When the submarine needs to surface, compressed air from the high-pressure air banks is used to blow seawater out of the ballast tanks, restoring buoyancy and enabling the submarine to rise to the surface.

The system allows for both "soft blow" operations, where air is released gradually to adjust the submarine’s buoyancy with precision, and "emergency blow" operations, which involve rapidly expelling water from the ballast tanks in a short time to bring the submarine to the surface quickly in case of emergencies.

5. Engine Starting and Air Breathing: Diesel-electric submarines rely on diesel engines to recharge the batteries while surfaced or at snorkel depth. Compressed air is used to start the diesel engines, providing the initial force required to turn over the engine. The air system also supports the submarine's snorkel operation, where a snorkel mast is raised to draw in atmospheric air for engine operation while the submarine remains just below the surface.

The air system includes an air intake arrangement to provide clean air to the engines and maintain an adequate supply for combustion. Filters may be used to remove particulates from the air before it enters the engines.

6. Torpedo and Missile Launching: Compressed air is used for weapon deployment, such as launching torpedoes or missiles. The air system provides the necessary pressure to eject the torpedo from its tube, overcoming the external water pressure and sending the weapon on its way toward the target. The air system can also be used to clear water from the torpedo tubes after a launch.

7. Life Support and Air Quality Management: Maintaining breathable air inside the submarine is critical for the crew's health and safety. The air system includes provisions for introducing compressed air and oxygen into the submarine to maintain acceptable air quality levels during extended submerged operations. The air is circulated through filters and scrubbers that remove carbon dioxide, moisture, and other contaminants, ensuring a safe and habitable environment.

In some cases, the air system can also be used to pressurize certain compartments for testing their tightness or conducting maintenance tasks.

8. Emergency Operations: In emergency scenarios, such as loss of power or system failures, the air system provides backup capabilities to ensure the submarine can surface safely or continue limited operations. Emergency air blow systems can be used to expel water from the ballast tanks and bring the submarine to the surface even if other systems are inoperative. The air system also powers emergency pneumatic tools and other essential equipment.

9. Pneumatic Tools and Systems: Various pneumatic tools and systems onboard the submarine rely on compressed air for operation, including hydraulic pumps, air tools, and other machinery. The air system can supply these tools with air at the appropriate pressure, allowing for maintenance and repair tasks to be carried out.

Example of a Diesel-Electric Air System

A submarine has five distinct air systems, each serving specific functions to support various operational requirements. These systems include the 3,000-pound (psi) high-pressure and torpedo impulse air system, the 600-pound MBT (Main Ballast Tank) blowing system, the 225-pound service air system (also referred to as ship's service air), the 10-pound MBT blowing system, and the salvage air system. Here's a detailed breakdown of these air systems [207]:

1. The 3,000-pound High-Pressure and Torpedo Impulse Air System

The 3,000-pound air system is the primary high-pressure air system onboard the submarine, designed to compress, store, and supply air at a maximum pressure of 3,000 psi (20,684 kPa). This system comprises high-pressure compressors, an air manifold, interconnected piping, valves, and compressed air storage banks.

The primary role of this system is to supply air to other lower-pressure air systems, including the 600-pound (4,137 kPa) MBT blowing system and the 225-pound (1,551 kPa) service air system. It also powers the hydraulic accumulator air-loading manifold and supplies the forward and aft torpedo tube impulse-charging manifolds through the 600-pound reducing valves. Additionally, this system features an external charging connection to allow air replenishment from external sources, if necessary.

2. The 600-pound MBT Blowing System

The 600-pound MBT blowing system's function is to expel water ballast from the main ballast tanks during surfacing operations, making the submarine lighter and allowing it to ascend to the surface. This system receives compressed air from the 3,000-pound air system via a distributing manifold.

The compressed air at 600 psi (4,137 kPa) is used to blow out the water from the ballast tanks, including fuel ballast tanks when they are used as ballast tanks. This capability allows the submarine to adjust its buoyancy, enabling it to surface effectively and with the appropriate balance.

3. The 225-pound Service Air System

The 225-pound service air system, or ship's service air, provides air at a pressure of 225 psi (1,551 kPa). It is used for various auxiliary operations within the submarine, including blowing air into specific tanks, operating pneumatic tools, and performing other services that require compressed air.

The 225-pound system consists of a service air manifold, interconnecting piping, and a series of valves that control the flow of compressed air to different locations throughout the submarine. This system is essential for maintaining air quality, performing maintenance tasks, and supporting equipment operation when submerged.

4. The 10-pound MBT Blowing System

When the submarine is on the surface, the 10-pound MBT blowing system is used to conserve compressed air stored in the air banks. This system operates at a pressure of 10 psi (69 kPa) and is intended for use only after the submarine has surfaced sufficiently to allow the opening of induction valves and hatches.

The system comprises a low-pressure blower, control manifold, and piping that directs air to the various main ballast tanks. The 10-pound system is a more economical method for maintaining ballast tank buoyancy once the submarine is on the surface, reducing the reliance on high-pressure air from the 3,000-pound system.

5. The Salvage Air System

The salvage air system is composed of three separate subsystems: the MBT external salvage, compartment external salvage, and compartment internal salvage systems. These subsystems are designed to restore buoyancy or provide emergency air in case of damage or flooding.

- **MBT External Salvage:** Allows compressed air from an external source to be used for blowing the ballast tanks to help surface the submarine.

- **Compartment External Salvage:** Provides connections to supply compressed air from an outside source to pressurize specific compartments for salvage operations.
- **Compartment Internal Salvage:** Utilizes the ship's existing compressed air supply to pressurize individual compartments for emergency salvage.

These salvage capabilities ensure that the submarine can perform emergency procedures in case of flooding or damage, providing crucial versatility for rescue operations.

The air systems on a submarine are interdependent, with the higher-pressure systems (3,000 psi) supplying air to lower-pressure systems (600 psi and 225 psi). The design also ensures that there is sufficient flexibility and redundancy for emergency and operational needs. Each system serves specialized functions, from ballast management and torpedo launching to maintenance tasks and emergency operations, making the air systems vital for submarine safety, performance, and survivability.

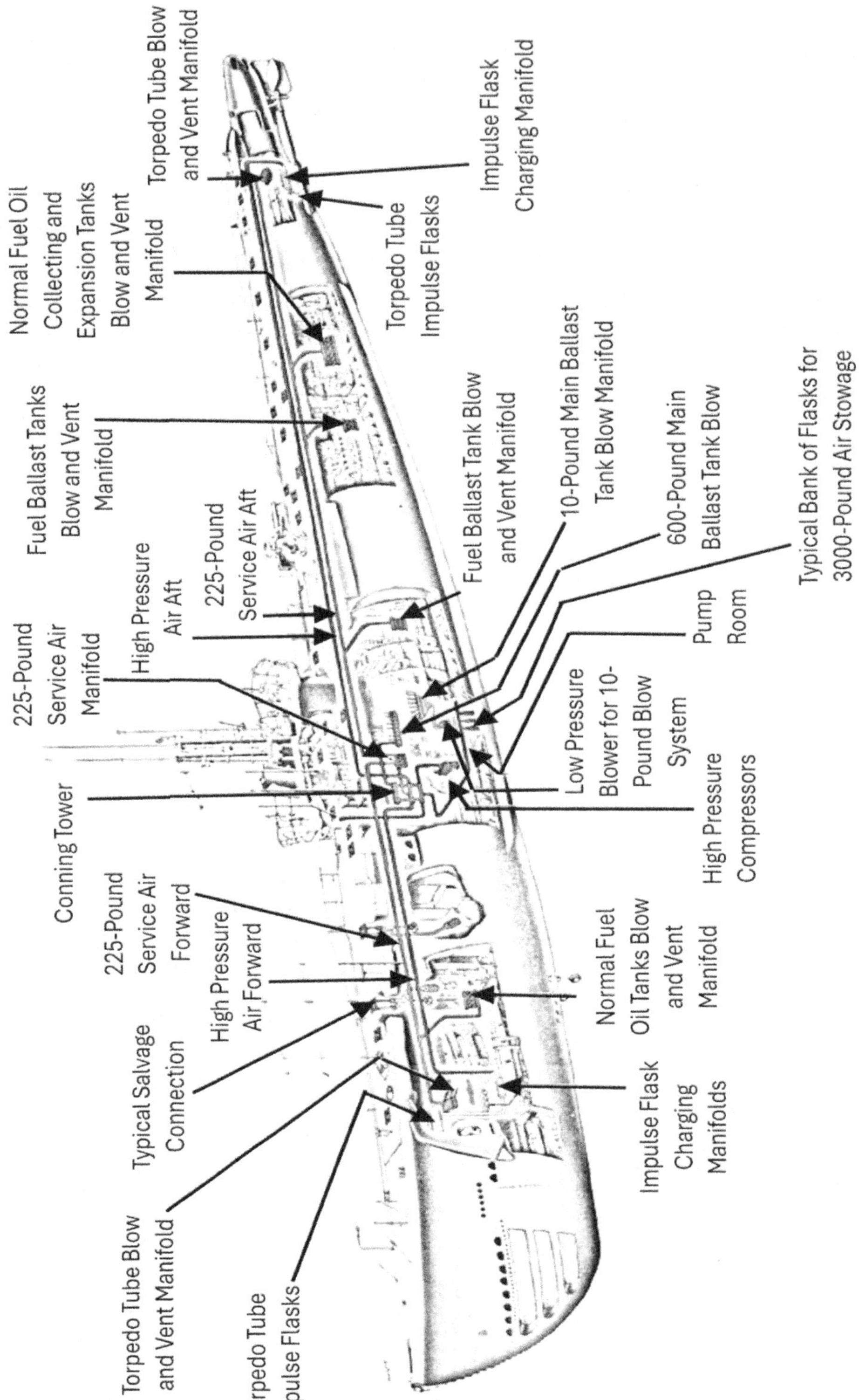

Figure 65: Example schematic of a submarine air system.

Typical Air System in a Nuclear Submarine

A typical nuclear submarine's air system includes several interconnected subsystems that provide high-pressure air, support ballast tank management, and ensure the functionality of life-support and other essential systems. Here's a detailed breakdown of the main components of a typical air system in a nuclear submarine:

1. High-Pressure Air System: The high-pressure air system (usually rated at around 3,000 psi or 20,684 kPa) is the backbone of the submarine's air supply. It consists of high-pressure compressors, air storage banks, manifolds, valves, and piping. The primary function of this system is to compress and store air for various uses throughout the submarine.

- **Air Storage Banks:** Compressed air is stored in multiple high-pressure air banks located throughout the submarine. These banks provide a reserve of air that can be used for various systems, ensuring redundancy and flexibility.
- **Compressors:** The high-pressure compressors are used to recharge the air storage banks. They can be run using power from the submarine's nuclear reactor and can replenish air while submerged or surfaced.
- **High-Pressure Manifold:** The manifold distributes compressed air from the storage banks to different subsystems and allows the controlled release of air where needed.

2. Main Ballast Tank (MBT) Blowing System: The MBT blowing system is used to control the buoyancy of the submarine by blowing air into the main ballast tanks, displacing water to make the vessel less dense and allowing it to surface. There are typically two main types of MBT blowing systems:

- **High-Pressure Blowing System:** This system uses air from the high-pressure air storage banks (3,000 psi) to blow water out of the ballast tanks when the submarine needs to surface quickly.
- **Low-Pressure Blowing System:** When the submarine is already on or near the surface, a low-pressure blower (typically operating at 10 psi or 69 kPa) is used to expel water from the ballast tanks, conserving high-pressure air.

3. Torpedo Impulse Air System

The high-pressure air system is also used for the torpedo impulse system, which provides the necessary air pressure to launch torpedoes from the submarine. The compressed air forces the torpedo out of the launch tube, ensuring it exits the submarine smoothly.

- **Impulse Tanks:** High-pressure air is stored in dedicated impulse tanks, which are connected to the torpedo tubes. The system is designed to allow rapid recharging and firing of multiple torpedoes.

4. Service Air System: The service air system operates at a lower pressure, typically around 225 psi (1,551 kPa). It is used for various auxiliary functions on the submarine, such as:

- **Pneumatic Tools:** Air-powered tools for maintenance and repairs.
- **Ventilation and Control Systems:** Assisting with opening and closing hatches or valves and regulating the airflow within the submarine.
- **Blowing Smaller Tanks:** Adjusting trim and buoyancy using variable ballast tanks.

5. Hydraulic System Support: The air system is also integral to maintaining the hydraulic systems on a nuclear submarine. The high-pressure air helps pressurize hydraulic accumulators, which in turn power various hydraulic equipment, such as steering controls, periscope raising mechanisms, and other critical mechanical systems.

6. Life Support and Emergency Systems: Compressed air plays an essential role in life support and emergency systems on a nuclear submarine. This includes:

- **Air Revitalization:** Compressed air can be used to help maintain breathable air quality by mixing with oxygen to replenish the air supply during extended submerged operations.
- **Emergency Blow System:** In case of an emergency, the high-pressure air system is used for an "emergency blow" of the main ballast tanks, rapidly expelling water to bring the submarine to the surface as quickly as possible.
- **Escape Compartment Air Supply:** Air systems can supply pressurized air to escape compartments or to support other emergency procedures, such as compartment salvage or damage control.

7. Cooling and Dehumidification: Air systems also play a role in cooling and dehumidifying the atmosphere within the submarine. This is important to prevent the build-up of moisture, which could affect sensitive equipment and the comfort of the crew.

- **Air Conditioning Integration:** The air system works in tandem with the air conditioning and ventilation systems to regulate the internal temperature and humidity, ensuring a safe and habitable environment.

8. External Air Connections: Nuclear submarines are equipped with external air connections that allow for the recharging of air systems when docked or during resupply operations. This can be done via dedicated shore connections, ensuring the submarine's air systems are fully charged before deployment.

Oxygen Generation and Carbon Dioxide Scrubbing

On a submarine, maintaining a breathable atmosphere is essential for the safety and well-being of the crew, especially during extended underwater operations. Two critical aspects of this are oxygen generation and carbon dioxide (CO_2) scrubbing. Together, these processes ensure a consistent supply of fresh air and remove harmful CO_2 from the environment, creating a liveable atmosphere inside the submarine.

Oxygen Generation

The primary methods for oxygen generation in submarines include electrolysis of water, the use of stored oxygen tanks, and chemical oxygen generation. Each method has its own advantages and limitations, making them suitable for different operational scenarios.

Electrolysis of Water is the predominant method employed in modern submarines for oxygen generation. This process involves the electrolysis of seawater or freshwater, where water is distilled to remove impurities before being subjected to an electrical current. This current splits water into its constituent elements: hydrogen and oxygen. The oxygen produced is then released into the submarine's atmosphere, while the hydrogen is typically vented overboard or stored. The efficiency of this method allows for a continuous supply of oxygen, which is crucial for long-duration submerged operations [208, 209]. Notably, advancements in electrolysis technology, such as the use of proton exchange membrane (PEM) electrolyzers, have further enhanced the reliability and efficiency of oxygen generation systems in submarines [208].

The equipment used for oxygen production on submarines is specially designed to ensure efficiency, safety, and continuous operation in the challenging underwater environment. Key components and systems involved in the electrolysis process include the electrolyzer, water purification units, power supply, gas handling mechanisms, cooling systems, control features, and safety measures.

The electrolyzer unit serves as the core component for splitting water molecules into oxygen and hydrogen using an electrical current. It consists of multiple electrolysis cells arranged in series or parallel, each equipped with an anode (positively charged) and a cathode (negatively charged) submerged in a water solution. When direct current (DC) is applied, water is split at these electrodes, with oxygen forming at the anode and hydrogen at the cathode. Some electrolyzers incorporate a proton exchange membrane (PEM) or similar separator to prevent the produced gases from mixing, which could pose safety risks.

Before electrolysis, water must undergo purification and distillation to eliminate salts, minerals, and other impurities. This is typically achieved through desalination methods such as reverse osmosis or distillation, which produce fresh water suitable for electrolysis. Additionally, deionization filters ensure that the water is ultra-pure, preventing any ionic contaminants from causing equipment fouling or corrosion and thus maintaining electrolysis efficiency.

A stable power supply is crucial for driving the electrolysis process, as the equipment requires direct current to separate water into hydrogen and oxygen. The power supply system features voltage regulators and current controllers to maintain optimal operating conditions and adjust the power output based on the submarine's oxygen production requirements.

Gas separation and handling systems are essential for managing the output gases after electrolysis. Gas separators or demisters are used to remove residual water droplets from the generated oxygen and hydrogen, ensuring that these gases are dry and free of contaminants. The oxygen is then stored in dedicated tanks or directly released into the submarine's atmosphere to maintain appropriate oxygen levels. Buffer tanks may also store excess oxygen for later use. Conversely, the hydrogen is typically vented overboard through special vent lines and safety valves, though some submarines may store hydrogen for applications such as fuel cells.

The process of electrolysis generates heat, necessitating an efficient cooling system to prevent overheating and maintain optimal conditions. Cooling is usually achieved using seawater heat exchangers, which help dissipate the heat and regulate the temperature of the electrolysis cells within safe operational limits.

To ensure safe and efficient operation, the system is equipped with a comprehensive control and monitoring setup. This includes various sensors that track parameters such as water quality, gas flow rates, pressure, temperature, and voltage. Alarms and automatic shutdown mechanisms are integrated to detect any anomalies or hazardous conditions, such as high gas pressure or overheating, and respond accordingly. An automatic control unit adjusts the power supply and water flow based on oxygen demand, ensuring smooth and efficient operation.

Given the potential hazards associated with handling gases like oxygen and hydrogen, the electrolysis equipment is designed with robust safety features. These include pressure relief valves to prevent over-pressurization, explosion-proof enclosures to avoid ignition risks, and hydrogen detectors to monitor for leaks. Together, these measures help ensure early detection and prevent dangerous situations, keeping the submarine's atmosphere safe for its crew.

Stored Oxygen Tanks serve as a backup oxygen supply in submarines. These tanks contain compressed oxygen that can be released into the atmosphere during emergencies or when the electrolysis system is undergoing maintenance. While stored oxygen tanks provide a reliable short-term solution, their limited capacity means they are not suitable for prolonged use. The integration of stored oxygen systems ensures that submarines can maintain safe oxygen levels even in the event of a failure in the primary generation system [210, 211].

Chemical Oxygen Generation is another method utilized, particularly as an emergency backup. This method typically involves the use of chemical reactions to produce oxygen from compounds such as sodium chlorate or potassium superoxide. These chemical oxygen generators can be activated in emergencies to provide a rapid supply of oxygen, although they are less common than electrolysis systems [210, 212]. The versatility of chemical oxygen generation allows submarines to have an additional layer of safety, ensuring that crew members have access to breathable air even under adverse conditions [213].

Carbon Dioxide Scrubbing

Carbon dioxide scrubbing is a critical process in submarines to maintain a safe and breathable atmosphere for the crew. As crew members exhale, carbon dioxide (CO_2) accumulates in the confined space of the submarine, leading to potentially harmful concentrations that can cause headaches, dizziness, and even unconsciousness if not managed effectively [214]. To mitigate these risks, submarines are equipped with CO_2 scrubbers, which are specialized systems designed to remove excess CO_2 from the air.

The primary mechanism of CO_2 scrubbers involves passing the submarine's air through canisters filled with chemical absorbents. Commonly used materials include monoethanolamine (MEA) and lithium hydroxide, which chemically react with CO_2 to form stable compounds, effectively removing CO_2 from the air [215]. The reaction not only cleans the air but also allows for its recirculation within the submarine, ensuring a continuous supply of breathable air. The use of MEA has been well-documented as an effective solvent for CO_2 capture in various applications, including power plants and submarines [215].

Modern submarines have begun to adopt regenerative CO_2 scrubbers, which offer enhanced operational efficiency. These systems utilize solid amine or zeolite compounds that can adsorb CO_2 under normal conditions and release it when heated, allowing for the regeneration of the scrubber without the need for constant replacement of absorbent materials [214]. This regenerative process not only reduces the logistical burden of resupplying chemicals but also improves the overall sustainability of submarine operations by minimizing waste. The ability to vent the absorbed CO_2 overboard further enhances the efficiency of these systems, making them a preferred choice for contemporary submarine designs [214].

Submarine carbon dioxide scrubbing involves several critical components and systems to ensure the removal of CO_2 from the air, maintaining breathable conditions for the crew during extended underwater operations. The primary equipment for this process is the carbon dioxide scrubber, which captures and chemically absorbs CO_2 from the air. These scrubbers typically use a chemical absorbent such as soda lime or lithium hydroxide. The absorbents, often in granular or pellet form, are housed in canisters or trays inside the scrubber. As air passes through the scrubber, the absorbent reacts with the CO_2, converting it into solid compounds like calcium carbonate or lithium carbonate, which effectively removes the CO_2. Once scrubbed, the purified air is recirculated back into the submarine's atmosphere.

Modern submarines often utilize regenerative CO_2 removal systems, which employ solid amine adsorbents. In this setup, CO_2 is captured onto a solid amine material, which is then regenerated by heating to release the captured CO_2, allowing it to be vented outside the submarine. Additionally, lithium hydroxide canisters may be used as a supplementary or emergency method for CO_2 removal. The canisters react with CO_2 to form lithium carbonate and water, providing a disposable solution for air scrubbing during critical situations.

Figure 66: Machinist's Mate 1st Class Alberto Lezama, from New York, switches the number one carbon dioxide scrubber into recirculation mode to perform maintenance on the batteries aboard the Los Angeles-class attack submarine USS Norfolk (SSN 714). (U.S. Navy photo by Mass Communication Specialist 1st Class Todd A. Schaffer/Released). U.S. Navy photo by Mass Communication Specialist 1st Class Todd A. Schaffer, Public domain, via Wikimedia Commons.

Efficient air circulation and filtration are also vital for effective CO_2 scrubbing. Fans and blowers are used to maintain a continuous flow of air through the scrubber units, ensuring thorough distribution of the scrubbed air throughout the submarine compartments. The air circulation system may also include particulate filters to remove dust, fibres, or other airborne contaminants, further purifying the air before or after it passes through the scrubber.

Monitoring and control systems play a crucial role in maintaining the air quality within the submarine. CO_2 sensors strategically placed throughout the vessel continuously monitor carbon dioxide levels, providing real-time data. The scrubbers are equipped with automatic control systems that adjust the scrubbing process based on the detected CO_2 concentration. When levels rise above a specific threshold, the system increases air circulation through the scrubbers to enhance CO_2 removal, ensuring the air remains safe for the crew.

Heat management is another important consideration in submarine CO_2 scrubbing, as some regenerative systems generate heat during the scrubbing process. Heat exchangers and cooling systems help dissipate this heat, keeping the equipment within safe operating temperatures. In certain designs, the excess heat may be recovered and used to warm other areas of the submarine, improving overall energy efficiency.

Lastly, CO_2 venting systems are essential for disposing of the captured carbon dioxide. In regenerative systems, once the CO_2 is released from the absorbent material, it is vented outside the submarine. This venting must be carefully managed to prevent detection by sonar or other enemy systems, ensuring operational stealth.

Together, these systems enable submarines to maintain breathable air for extended missions underwater, making carbon dioxide scrubbing a crucial component of life support in submarine operations.

Oxygen Level and CO_2 Concentration Monitoring

Monitoring oxygen levels and carbon dioxide (CO_2) concentrations is essential for maintaining a safe atmosphere, particularly in confined environments such as submarines. The management of these gases is critical for ensuring crew safety and optimal physiological function.

Oxygen monitors are vital in submarine systems, where maintaining oxygen levels around 21% is crucial. Automated systems continuously monitor these levels and can adjust the oxygen generation rate accordingly. For instance, when oxygen levels drop below the target threshold, these systems can increase the oxygen supply to restore balance. Research indicates that automated devices are significantly more effective than manual controls in maintaining oxygen saturation (SpO_2) within desired ranges, thereby preventing both hyperoxia and hypoxia [216-218]. The efficacy of these automated systems is particularly pronounced in clinical settings, where they can respond rapidly to fluctuations in oxygen levels, thus enhancing patient safety [219, 220].

Similarly, CO_2 levels are closely monitored, with a typical upper safety limit set at around 0.5% by volume. Elevated CO_2 concentrations can impair cognitive function and overall health, making it imperative to manage these levels effectively. Automated scrubbers are employed to increase their activity when CO_2 levels rise, thereby ensuring that the crew is not exposed to harmful concentrations. Studies have shown that automated oxygen titration systems can help mitigate the risks associated with hypercapnia, particularly in patients with conditions like obesity hypoventilation syndrome [221, 222]. These systems not only enhance safety but also reduce the workload on medical staff, allowing for more efficient management of oxygen and CO_2 levels [218, 223].

The integration of oxygen and CO_2 monitoring systems into automated frameworks is crucial for the safety of submarine environments. Automated control systems, such as closed-loop oxygen control, have been shown to reduce the incidence of hyperoxic episodes, which can occur when oxygen levels exceed safe limits [224, 225]. These systems are designed to maintain oxygen levels within a narrow range, thus minimizing the risks associated with both hypoxia and hyperoxia. Furthermore, the implementation of such technologies has been associated with improved outcomes in various clinical scenarios, including respiratory failure and chronic obstructive pulmonary disease (COPD) [220, 225, 226].

Emergency Procedures and Backup Systems

Submarine emergency procedures and backup systems for air management are vital for ensuring the crew's safety during critical situations. Redundancy is built into these systems to provide multiple layers of backup for both oxygen generation and carbon dioxide scrubbing. This approach ensures that even if primary systems fail, there are alternative methods available to maintain breathable air inside the submarine.

One critical emergency backup system is the use of oxygen candles, which can produce oxygen through a chemical reaction when ignited. These candles consist of a chemical compound that releases oxygen when burned, providing an immediate source of breathable air in emergency scenarios. Oxygen candles are stored onboard and are only used in extreme situations where the submarine's regular oxygen generation systems are inoperative. They offer a straightforward and reliable means of boosting oxygen levels in a confined environment, ensuring that the crew can continue to breathe while resolving the primary system's failure.

For carbon dioxide scrubbing, portable CO_2 scrubbers serve as a vital emergency backup. In the event of a malfunction in the primary CO_2 scrubbers, these portable units can be deployed quickly to specific compartments to remove carbon dioxide and maintain local air quality. The portable scrubbers use chemical absorbents similar to those in the main system, such as soda lime or lithium hydroxide, to chemically react with and remove CO_2 from the air. Their portability allows for flexible placement within the submarine, enabling targeted CO_2 scrubbing where it is most needed. This capability is especially important in emergency situations where a localized air quality problem may arise, or where access to certain compartments is limited due to damage or operational constraints.

Together, these emergency procedures and backup systems form a comprehensive strategy for maintaining life support on a submarine, even under adverse conditions. The inclusion of oxygen candles and portable CO_2 scrubbers provides a layer of resilience, ensuring that the submarine's air management system can continue to function effectively during critical situations. This redundancy is essential for the safety and survival of the crew, allowing for continued operation and time to address and repair primary system failures.

Challenges in Oxygen Generation and CO_2 Scrubbing

The challenges associated with oxygen generation and CO_2 scrubbing in submarine environments are multifaceted, primarily revolving around heat management and the supply of chemicals necessary for effective scrubbing.

Heat Management: Both oxygen generation and CO_2 scrubbing processes are exothermic, generating heat that can significantly increase the internal temperature of a submarine. This heat must be effectively managed to maintain operational safety and comfort for the crew. Submarine ventilation and cooling systems are crucial in dissipating this heat, ensuring that the internal environment remains stable. The integration of advanced thermal management systems can enhance the efficiency of these processes, as demonstrated in various studies that analyse the thermal performance of different CO_2 capture technologies [227, 228]. For instance, the thermal integration analysis of chemical scrubbing units highlights the importance of managing heat to optimize energy efficiency, which is critical in confined environments like submarines [227].

Chemical Supply: The availability of chemicals for CO_2 scrubbing is another significant challenge that can limit the duration of submerged operations. The reliance on chemical scrubbing methods, such as amine scrubbing, necessitates a continuous supply of absorbents, which can be logistically challenging in extended missions [229, 230]. The design of efficient scrubbers and the implementation of regenerative systems can mitigate the need for frequent chemical resupply. For example, hybrid solvent systems have been shown to enhance CO_2 absorption capacity while reducing regeneration energy, thereby improving the overall sustainability of the scrubbing process [229, 231]. Moreover, innovative approaches such as the use of metal-organic frameworks (MOFs) have been explored to improve the thermal and chemical stability of adsorbents, which could further extend the operational duration of submarines without the need for resupply [232].

Integration with Life Support Systems

Oxygen generation and carbon dioxide scrubbing are essential components of a submarine's life support system, which operates in conjunction with various subsystems to ensure a breathable and safe atmosphere for the crew. The management of air quality within a sealed underwater environment involves not only the provision of adequate oxygen and the removal of CO_2 but also the regulation of humidity, air purity, and ventilation. Effective ventilation systems are crucial for maintaining air quality, as they help to continuously monitor and manage CO_2 levels, which can accumulate rapidly in confined spaces like submarines [233, 234]. The integration of these systems is vital for creating a balanced and comfortable environment, as highlighted by findings that emphasize the importance of adequate ventilation in indoor air quality management [233].

Humidity control is another critical aspect of the submarine's life support system. High humidity levels can lead to condensation, which poses risks such as corrosion of equipment and structural components. Dehumidifiers play a key role in managing moisture content, ensuring that humidity levels remain within optimal ranges [235]. The integration of humidity control with CO_2 scrubbing systems is particularly beneficial, as it allows for the simultaneous regulation of moisture and CO_2, preventing condensation during the air recycling process [235]. This combined approach is essential for protecting sensitive electronic systems and other equipment from potential damage caused by high humidity and condensation [235].

Air purification is also a vital component of the life support system, working alongside oxygen generation and CO_2 scrubbing to maintain air quality. While CO_2 is a primary concern, other airborne contaminants, including odours, smoke, and trace chemicals, can accumulate in the submarine's atmosphere. Air purification systems, such as activated charcoal filters, are employed to remove these impurities effectively [236]. Activated charcoal filters are particularly adept at absorbing organic compounds and odours, while specialized filters target specific gases [236]. By ensuring that these additional contaminants are filtered out, the air purification system contributes to a clean and safe breathing environment for the crew during prolonged underwater missions [235].

Water Purification Systems

A submarine's water purification system is essential for providing the crew with safe drinking water and for supporting various operational needs, such as cooling systems, hygiene, and oxygen generation. Due to the submarine's isolated and underwater environment, it must have a reliable system to convert seawater into fresh water through desalination and purification processes. The main techniques used for water purification on submarines are reverse osmosis and distillation, with advanced filtration methods ensuring the water's safety and quality.

Reverse Osmosis (RO)

Reverse osmosis (RO) is a widely used and effective method for water purification on submarines, designed to meet the critical need for freshwater while operating underwater for extended periods. The process involves forcing seawater through a semi-permeable membrane that allows water molecules to pass while blocking dissolved salts, impurities, and other contaminants. This separation results in fresh, potable water that can be used for drinking, cooking, hygiene, and other onboard requirements.

The process begins with the use of a high-pressure pump that pressurizes the seawater to force it through the RO membrane. The high pressure, typically between 800-1,200 psi (5.5-8.3 MPa), is needed to overcome the natural osmotic pressure of seawater and push water molecules across the membrane while leaving salts and impurities behind. The pump is a critical component of the system, as it must maintain consistent pressure to ensure the efficiency of the RO process and produce a continuous supply of fresh water. The high-pressure pump consumes a significant amount of energy, making energy efficiency an important factor in RO system design, especially on diesel-electric submarines where power is more limited.

Figure 67: Reverse Osmosis Water Purifying Unit plant. U.S. Navy photo by Mass Communication Specialist 1st Class Drae Parker, Defense Visual Information Distribution Service, Public Domain, via Picryl.

The heart of the RO system is the semi-permeable membrane, which acts as a barrier that selectively filters out dissolved salts, bacteria, viruses, and other contaminants. The membrane consists of multiple layers, with pores small enough (typically around 0.0001 microns) to allow only water molecules to pass through while rejecting larger molecules, including dissolved salts and organic materials. The quality and condition of the membrane are crucial to the system's performance, and it needs to be periodically inspected and replaced when fouling or damage occurs to maintain optimal water quality.

Before seawater reaches the RO membrane, it undergoes pre-filtration to remove larger particles and contaminants, such as sand, silt, and debris. Pre-filtration typically involves a series of filters, including coarse filters and fine particulate filters, to ensure that the water entering the RO membrane is free of large solids that could clog or damage the membrane. Effective pre-filtration is essential to prolong the lifespan of the RO membrane and reduce the frequency of maintenance required, thereby improving the overall efficiency of the purification process.

After the seawater has passed through the RO membrane and been converted into fresh water, it often undergoes post-treatment to enhance its quality for human consumption. Post-treatment processes can include remineralization, where essential minerals such as calcium and magnesium are added back to the water. This step is important because the RO process removes not only harmful substances but also beneficial minerals, resulting in water that may taste flat or lack essential nutrients. Additionally, ultraviolet (UV) sterilization or activated carbon filtration may be used as a final step to eliminate any remaining microorganisms or trace chemicals, ensuring the water is safe and pleasant to drink.

While the RO system is highly effective at producing fresh water, it does come with some challenges, particularly related to energy consumption and maintenance. The high-pressure pump used in the system requires significant energy input, which is an important consideration on submarines where power resources are carefully managed. Moreover, the RO membrane is prone to fouling over time, which can reduce its effectiveness and necessitate regular cleaning or replacement. To address these challenges, ongoing efforts focus on improving energy efficiency, developing more durable membranes, and optimizing pre-treatment processes to extend the life of the RO system components.

Distillation

Distillation is an essential water purification method used on submarines, transforming seawater into drinkable fresh water through the process of heating, evaporating, and condensing. This method is particularly advantageous for submarines due to its ability to effectively remove dissolved salts and impurities, producing high-quality water suitable for consumption and various onboard uses. The distillation process involves several key components and steps that ensure the continuous production of freshwater, which is vital for submarine operations.

The distillation process begins with the evaporator, where seawater is heated until it reaches its boiling point, causing the water molecules to vaporize and form steam. The heating source in a nuclear submarine typically comes from excess heat generated by the submarine's nuclear reactor, which makes the process highly efficient and cost-effective. In a diesel-electric submarine, where a nuclear reactor is not available, a dedicated heating element or steam generated by diesel engines may be used as the heat source. The evaporator is specifically designed to maximize the contact between the seawater and the heat source, ensuring rapid evaporation while leaving behind the salts, minerals, and other impurities that are dissolved in the seawater.

Figure 68: One of two fresh water evaporators located in the forward engine room of the USS Pampanito (SS-383). BrokenSphere, CC BY-SA 3.0, via Wikimedia Commons.

After the steam is generated in the evaporator, it is directed to a condenser, where the vapor cools and condenses back into liquid form, producing fresh water. The condenser uses a cooling medium, typically seawater, to facilitate the rapid cooling of the steam and ensure efficient condensation. The freshwater produced by the condenser is collected in a storage tank, ready for use in various onboard applications, such as drinking, cooking, hygiene, and cooling. The distillation process ensures that the water is free from salts and other contaminants, making it safe for the crew to use for extended periods underwater.

The distillation process results in a byproduct known as brine, which is the concentrated saltwater remaining after the freshwater has evaporated. To prevent the accumulation of salt and other impurities within the system, the brine must be discharged overboard. This discharge is carefully managed to avoid detection by sonar or other enemy monitoring systems, as it may produce a thermal signature in the surrounding seawater. The brine discharge process ensures that the distillation system can continue operating effectively, with minimal buildup of unwanted substances.

Distillation is particularly efficient on nuclear submarines due to the availability of excess heat from the reactor, which can be repurposed to power the evaporator without significantly affecting the submarine's overall energy consumption. This makes distillation a natural fit for nuclear-powered vessels, as the process can be sustained continuously with little additional energy cost. In contrast, diesel-electric submarines face challenges with energy efficiency, as the heat for the distillation process requires additional energy input, typically from the diesel engines. This can impact the operational efficiency and power availability for other systems on the submarine, making it less suitable for long-term submerged operations.

Distillation offers the advantage of producing high-quality freshwater with a minimal risk of contamination. The process is effective in removing not only salts but also microorganisms and other impurities, ensuring the safety and health of the crew. The technology is also robust and can operate for extended periods, provided that the system is well-maintained. However, distillation systems require regular maintenance to prevent scaling and build-up in the evaporator and condenser, which can affect efficiency. Additionally, the energy requirements for distillation, particularly on diesel-electric submarines, may necessitate careful management of onboard power resources.

Filtration and Water Treatment

After the primary purification processes of reverse osmosis or distillation, submarine water treatment involves further steps to ensure the water is of high quality and safe for the crew's consumption and other onboard uses. These additional processes help eliminate residual impurities, improve taste, and reintroduce essential minerals to the water. The main methods include activated carbon filtration, UV sterilization, and remineralization.

Activated carbon filters play a crucial role in the final stage of water purification by removing any remaining chemicals, odours, and organic compounds that could affect the water's quality. These filters work by adsorbing impurities onto the porous surface of the carbon material, effectively trapping substances like chlorine, volatile organic compounds (VOCs), and any other contaminants that may not have been removed during the primary purification process. The activated carbon also helps improve the taste of the water, making it more palatable for drinking and suitable for cooking. This step is particularly important because even trace amounts of certain chemicals can impact water safety and crew well-being during extended submarine missions.

Ultraviolet (UV) sterilization is another vital step in ensuring the water is microbiologically safe for consumption. In this process, water is exposed to UV light, which has a wavelength that can penetrate the cell walls of microorganisms such as bacteria, viruses, and protozoa. The UV light damages their DNA, rendering them unable to reproduce and effectively inactivating them. This method is highly effective in destroying any pathogens that may have survived earlier purification stages, ensuring that the water remains free from harmful microorganisms. UV sterilization is advantageous because it does not require the use of chemicals, thus avoiding the introduction of any additional substances into the water.

Both reverse osmosis and distillation processes remove almost all dissolved minerals from the water, which can lead to "flat-tasting" water and may not be ideal for long-term consumption due to the absence of essential nutrients. To address this, a remineralization step is often included in the submarine's water treatment system. During remineralization, essential minerals such as calcium, magnesium, and sometimes potassium are added back into the purified water. These minerals not only enhance the taste but also provide health benefits by contributing to the crew's daily mineral intake, which is important during prolonged underwater missions when dietary options are limited.

The combined use of activated carbon filtration, UV sterilization, and remineralization ensures that the submarine's water supply meets stringent safety and quality standards. While reverse osmosis and distillation effectively remove a wide range of contaminants, the additional treatment steps guarantee that any remaining impurities or potential hazards are addressed. This multi-stage approach to water purification is essential in the confined and isolated environment of a submarine, where access to external water sources is unavailable and maintaining the health of the crew is a top priority.

Storage and Distribution

After purification, the water on a submarine is stored in specialized tanks and carefully managed to ensure its availability and quality throughout the mission. The water is distributed across the submarine for a variety of purposes, including drinking, hygiene, cooling, and oxygen generation. Effective storage and distribution are essential for maintaining sufficient reserves, especially during extended missions or in emergency scenarios when access to external water sources is impossible.

The purified water is stored in dedicated tanks, which are typically placed in secure areas of the submarine to optimize weight distribution and space utilization. The tanks are made from materials resistant to corrosion and contamination to maintain water quality. Once stored, the water is distributed through a network of pipes to different parts of the submarine, supplying the crew with drinking water, showers, and sinks, as well as supporting various operational systems such as cooling and oxygen production.

Water management is an ongoing process on submarines, as maintaining the right levels and reserves is crucial for long-term missions. The distribution system includes pumps and valves that control the flow of water, ensuring that each area of the submarine receives the necessary supply while prioritizing essential functions. Additionally, some water may be held in reserve to accommodate potential emergencies or equipment failures, providing a safety buffer if regular supply becomes compromised.

Given the limited capacity for water storage on submarines, conservation is a key aspect of water management. To reduce consumption, submarines are equipped with water-saving fixtures such as low-flow faucets and showers, which limit the amount of water used without compromising functionality. Additionally, water recycling systems are employed to reclaim and reuse water from showers, sinks, and other non-potable sources for applications such as flushing toilets or pre-filtering before further treatment.

These conservation measures not only help extend the available water supply but also reduce the frequency of water purification cycles, thereby conserving energy and minimizing the load on the purification systems. Effective water conservation is especially critical during extended submerged operations, where the ability to replenish water is severely limited.

The quality and safety of the submarine's water supply are continuously monitored using sensors that measure parameters such as temperature, pressure, salinity, and microbial content. These sensors are placed strategically throughout the water system, providing real-time data that helps identify any issues before they can compromise the health and safety of the crew. Regular testing and sampling are also conducted to ensure the water meets strict quality standards, with results used to adjust purification processes as needed.

Maintenance of the water purification and distribution systems is essential for reliable operation. This includes routine tasks such as cleaning reverse osmosis membranes, replacing activated carbon filters, and servicing pumps and valves. In the case of distillation-based systems, heat exchangers and condensers require regular cleaning to prevent scale build-up and ensure efficient operation. Timely maintenance helps avoid unexpected equipment failures and ensures that the submarine can sustain its water needs throughout its mission.

Integration with Life Support Systems

The integration of the water purification system with a submarine's life support systems is essential for supporting various critical functions. Purified water is not just a resource for drinking and hygiene but also plays a pivotal role in maintaining the overall functionality of the submarine, especially during prolonged underwater missions. By connecting the purified water supply to different systems within the submarine, a cohesive and efficient life support network is established, optimizing the use of limited resources.

One of the most critical integrations of the water purification system is with the submarine's oxygen generation process. Distilled water is used in the electrolysis unit to produce oxygen and hydrogen. During electrolysis, purified water is split into its constituent elements, with oxygen supplied to the submarine's atmosphere to replenish breathable air and hydrogen typically vented overboard. The use of distilled water is crucial here, as any impurities could damage the electrolyzer or interfere with the process, potentially compromising the quality of the oxygen produced. Therefore, the water purification system ensures that the water fed into the oxygen generation unit is of the highest purity, supporting the continuous and safe operation of life support.

The purified water is also an essential component of the submarine's cooling systems, which are responsible for preventing overheating of critical equipment, including the propulsion system, power electronics, and other operational machinery. Water is circulated through heat exchangers to absorb and transfer heat away from these components, ensuring that they operate within safe temperature ranges. The quality of the water used in cooling is important because impurities can cause scaling, corrosion, or clogging within the cooling circuits, potentially leading to equipment failure. The purified water from the submarine's water treatment system, free from minerals and contaminants, helps maintain the efficiency and longevity of the cooling systems.

Purified water is also supplied to the submarine's personal hygiene facilities, such as showers, sinks, and toilets, supporting the crew's daily sanitation needs. Maintaining clean and safe water for personal use is critical for the crew's health, especially during extended deployments where waterborne illnesses could pose a significant risk. The water purification system ensures that the water used for hygiene is free from harmful contaminants, contributing to a safe and comfortable living environment. Moreover, any wastewater generated from these facilities can be recycled and treated for reuse, enhancing water conservation aboard the submarine.

The integration of purified water with these systems—oxygen generation, cooling, and hygiene—demonstrates the interconnected nature of submarine life support. Each system is dependent on the quality and availability of purified water, and any compromise in water quality can affect multiple aspects of the submarine's operation. By ensuring that the water purification system is linked effectively to these functions, the submarine can sustain extended underwater operations with greater resilience and safety.

Temperature and Humidity Control

The Submarine Service is known for its rigorous demands, setting it apart as one of the most challenging sectors within the armed forces. Given the unique environment and mission requirements of submarines, personnel must meet strict physical criteria to ensure they are capable of handling the stresses and responsibilities that come with life underwater. To support the crew's well-being and maintain high operational standards, the living and working conditions on submarines are designed to be as comfortable as possible. One of the most significant aspects of ensuring these conditions is the quality and preservation of food, as well as maintaining a breathable and healthy atmosphere on board [237].

Proper food preservation is essential for sustaining a submarine crew during long deployments. Certain perishable items, including meats and dairy products, are prone to rapid spoilage if kept at ordinary temperatures, making refrigeration an essential component on submarines. By keeping food at low temperatures, refrigeration slows down the growth of bacteria, mold, and other microorganisms that contribute to spoilage. Effective food preservation techniques directly impact the submarine's mission duration, as the vessel's ability to remain at sea is partly dependent on how long essential food supplies can be kept in good condition. A well-maintained refrigeration system extends the shelf life of these perishable items, ensuring that the crew can enjoy a balanced diet over extended periods without needing to resupply frequently [237].

While ensuring that food is nutritious and safe to eat is the primary function of submarine refrigeration, the palatability of the food also plays a crucial role in crew morale. Some foods and drinks are best enjoyed when served chilled, and refrigeration makes this possible. For instance, drinking water may meet hydration needs at room temperature, but it is far more refreshing when cool, especially in warm conditions. Similarly, certain foods, such as ice cream, lose their appeal if they are not kept at the appropriate temperature. By keeping food and beverages at ideal temperatures, the submarine's refrigeration system enhances the quality of life on board, providing the crew with some comforts that contribute to overall satisfaction and morale.

The quality of the air is just as important as the quality of food on a submarine. During long underwater missions, the air inside a submarine is repeatedly recycled, making proper conditioning essential for maintaining a healthy atmosphere. Air conditioning systems help regulate the temperature on board, making the environment more comfortable for the crew, while ventilation systems ensure a continuous flow of fresh air, removing impurities and excess moisture. Effective air management is critical to the crew's well-being, as it prevents the buildup of contaminants that could lead to health issues or discomfort. The ability to maintain a controlled atmosphere also contributes to energy efficiency by keeping equipment at optimal operating conditions.

Another important aspect of maintaining onboard conditions involves protecting the submarine's sensitive equipment from moisture. High humidity levels can cause condensation to form on electrical components, which may interfere with their proper functioning or even cause damage. The refrigeration and air conditioning systems play a role in reducing humidity levels, thus preventing moisture-related issues. By keeping the air dry and cool, these systems not only ensure a comfortable living environment for the crew but also safeguard the equipment that is crucial for the submarine's operations [237].

The preservation of food, air quality management, and equipment protection are all interconnected elements that contribute to the effective functioning of a submarine. Proper refrigeration and air conditioning systems are vital to maintaining the health, morale, and operational readiness of the crew, enabling them to perform their duties effectively while submerged for long periods. These systems ensure that the submarine can remain at sea for extended missions by providing safe, nutritious food, breathable air, and well-protected equipment, all of which are essential to the success of submarine operations.

Heat

Coldness and heat are terms often used to describe temperature conditions, but they are relative rather than absolute. When we say something is cold or hot, we are comparing its temperature to a standard, which is typically the temperature of the human body, about 98.6°F (37°C). For example, if someone touches ice, they perceive it as cold because its temperature is significantly lower than the temperature of their hand. Conversely, if they drink coffee, it feels hot because its temperature is higher than their mouth. Despite these subjective perceptions, the terms cold and hot lack precise definitions. In reality, ice, while cold compared to the human body, is warmer than liquid air, just as coffee is cooler than boiling water [237].

Refrigeration is essentially the process of removing heat from a substance, whether that substance is solid, liquid, or gas. By extracting heat, the temperature of the matter decreases, making it cooler. Historically, various methods have been employed to lower temperatures, some of which are no longer in widespread use. For example, one could cool air by compressing it, removing the excess heat from the compressed air, and then allowing it to expand. Another method is evaporation, such as when water in a canvas-covered canteen cools down as the water on the surface evaporates, absorbing heat from the canteen and the water inside it. Adding certain salts to water, or mixing ice with salt, can also lower the temperature of the mixture.

The two primary large-scale refrigeration methods are the use of ice and mechanical refrigeration. Ice has been used for centuries to preserve food by circulating air around blocks of ice. As the air cools, it absorbs heat from nearby items, maintaining a continuous refrigeration cycle. However, on a submarine, ice would quickly melt, limiting its practical use. Mechanical refrigeration, which involves using liquids that boil at lower temperatures than water, has become the preferred method. In mechanical refrigeration, a refrigerant absorbs heat at low temperatures and pressures and then releases it at higher temperatures and pressures through processes like compression and expansion. This method offers significant advantages, particularly on ships where consistent refrigeration is essential [237].

Understanding heat is fundamental to operating refrigeration systems. Heat is a form of energy and is not something that can be seen or touched. It is known through its effects on matter. When heat is removed from a substance, the substance becomes cooler. Matter itself can exist in three states—solid, liquid, and gas—and can transition between these states with the addition or removal of heat. For example, water can exist as ice (solid), liquid, or steam (gas), depending on the amount of heat applied.

Heat can be measured by its intensity or by the quantity possessed by a substance. Intensity refers to temperature, which is gauged using a thermometer. In the Fahrenheit scale, water freezes at 32°F and boils at 212°F, while on the Celsius scale, the corresponding points are 0°C and 100°C. For more scientific measurements, the Kelvin scale is used, where absolute zero (the temperature at which no heat remains) is set at 0K, equivalent to -273.16°C (-459.69°F).

The heat applied to a substance can be categorized as sensible or latent. Sensible heat raises the temperature of a substance without changing its state, such as heating water to a warmer temperature. Latent heat, on the other hand, is involved in changing the state of a substance, such as turning ice into water (latent heat of fusion) or water into steam (latent heat of vaporization). The amount of latent heat varies between substances; for water, it takes 143.33 British thermal units (Btu) per pound to convert ice to liquid at 32°F and 970.4 Btu per pound to convert water at 212°F to steam [237].

Total heat refers to the sum of sensible and latent heat in a substance. It is sometimes calculated from a specific reference point, such as 32°F for water or -40°F for refrigerants like Freon. In practical applications like refrigeration, understanding the total heat content helps in evaluating the efficiency of the cooling process.

In mechanical refrigeration, a refrigerant absorbs heat at low temperatures and pressures and then releases it at higher temperatures and pressures. This process typically involves the refrigerant undergoing expansion, compression, and heat transfer with air or water as the cooling medium. Refrigerants are selected based on their ability to efficiently transfer heat and are essential to the cooling process [237].

Cooling involves managing both sensible and latent heat. For example, when cooling air, the temperature may decrease without condensation (sensible heat), but when moisture in the air condenses, latent heat is released. Effective refrigeration systems must address both to maintain desired temperatures and humidity levels.

Pressure

Atmospheric pressure is a fundamental concept affecting everything on Earth, as it results from the weight of the air above us. At sea level, when the air temperature is 32°F (0°C), this pressure is approximately 14.696 pounds per square inch (psi). For practical engineering purposes, it is often rounded to 14.7 psi. Atmospheric pressure can vary due to changes in weather conditions, but this standard measurement provides a consistent reference for various calculations [237].

Atmospheric pressure can be measured using different types of barometers. The mercury barometer is a classic instrument that consists of a vertical glass tube over 30 inches long, closed at the top and open at the bottom. The lower end is submerged in a dish of mercury. Atmospheric pressure pushes down on the mercury in the dish, causing the mercury in the tube to rise. At standard sea level pressure (14.7 psi), the mercury column stands at 29.921 inches. The upper part of the tube above the mercury is nearly a vacuum, with no air or other gases present, allowing for accurate pressure measurements.

The aneroid barometer is another device used to measure atmospheric pressure. Unlike the mercury barometer, it does not use any liquid. Instead, it consists of a small, sealed metal box with a partial vacuum inside and a flexible surface. As atmospheric pressure changes, the flexible surface moves, and this movement is translated by a lever system to a pointer on a circular scale, indicating the atmospheric pressure.

Barometer readings are typically given in inches of mercury. To convert these readings to psi, one can use the ratio of 14.696 psi for 29.921 inches of mercury. The conversion factor is approximately 0.491 psi per inch of mercury. For example, if a barometer reads 30 inches, the corresponding pressure is roughly 14.7 psi (30 x 0.491).

Atmospheric pressure influences the boiling point of water. At sea level, where the atmospheric pressure is 14.7 psi, water boils at 212°F (100°C). As altitude increases, atmospheric pressure decreases, lowering the boiling point. For instance, at 1,000 feet above sea level, the pressure drops to around 14.14 psi, and water boils at 210°F (98.9°C). This relationship between pressure and boiling point is consistent across different altitudes, with the boiling point continuing to decrease as altitude increases [237].

Changes in pressure directly affect the temperature at which substances change from one state to another, such as from liquid to vapor. This relationship can be demonstrated using a sealed container with a piston. By pushing the piston in, pressure inside the container increases, raising the temperature needed for the liquid to boil. Conversely, pulling the piston out decreases the pressure, lowering the boiling point. The exact pressure-temperature relationship for phase changes is well-documented for various substances and is essential for applications like refrigeration.

In enclosed systems like pipes or tanks, pressure is often measured using a Bourdon-tube pressure gauge. This gauge features a curved, flattened tube that straightens slightly when pressure increases. The movement is transferred through a lever system to a pointer on a scale, which shows the pressure in psi. The gauge's zero point corresponds to atmospheric pressure (14.7 psi), meaning that any reading above zero represents the pressure above atmospheric pressure, referred to as gauge pressure.

While gauge pressure indicates the pressure above atmospheric pressure, absolute pressure measures the total pressure inside a system, including atmospheric pressure. For example, if a gauge reads 6 psi, the absolute pressure would be 20.7 psi (14.7 + 6). Accurate measurements of absolute pressure are crucial for various engineering and scientific applications [237].

When pressures fall below atmospheric levels, they are termed partial vacuums. A vacuum gauge measures such pressures, typically in inches of mercury below the standard atmospheric level. A full vacuum corresponds to zero absolute pressure or approximately 30 inches of vacuum.

Compound gauges, also known as pressure-vacuum gauges, are used for systems that experience both positive pressures and vacuums. These gauges have a scale extending to the left and right of zero, with readings above atmospheric pressure in psi and below atmospheric pressure in inches of vacuum. This versatility makes compound gauges useful in applications like refrigeration, where pressures can vary significantly [237].

Vaporisation and Physical Conditions of Vapours and Liquids

There are two main forms of vaporization: ebullition and evaporation. These processes differ in speed and visibility, but both involve the transformation of a liquid into a vapor state. Understanding these types of vaporization is essential for fields like refrigeration and air conditioning, where controlling heat transfer and phase changes is crucial [237].

Ebullition, commonly known as boiling, is a rapid and visible form of vaporization. It occurs when a liquid is heated to its boiling point, causing bubbles to form within the liquid. These bubbles, consisting of vapor, rise to the surface and burst, releasing the vapor into the surrounding atmosphere. The process typically starts at the bottom and sides of the container where the liquid is in contact with the heat source. As the temperature rises, more bubbles form, leading to a vigorous boiling action. This rapid phase change from liquid to vapor makes ebullition an efficient method for quickly transferring heat away from a liquid.

In contrast to ebullition, evaporation is a much slower and usually invisible process. It occurs at the surface of a liquid rather than throughout the entire volume, and it can happen at temperatures below the liquid's boiling point. When molecules near the surface gain enough energy to overcome the liquid's surface tension, they escape into the air as vapor. For example, water gradually evaporates from the surface of lakes, rivers, and seas. Similarly, wet clothes hung out to dry will lose moisture through evaporation. This form of vaporization removes heat from the liquid, making it a cooling process. In fact, evaporation plays a role in the body's natural cooling mechanism, as sweat evaporates from the skin, carrying heat away.

Sublimation is a less common phase change where a substance transitions directly from a solid to a vapor without passing through the liquid state. This can occur under certain conditions, such as when ice or snow slowly disappears without melting, even in temperatures below the freezing point. The process is also evident when wet clothes hung out in sub-freezing temperatures first freeze stiff, then gradually dry through sublimation. While sublimation is not widely used in large-scale refrigeration, it does have practical applications, such as cooling items with dry ice, which sublimes under atmospheric pressure.

The terms "vapor" and "gas" both refer to a state of matter that is neither solid nor liquid, but there is a distinction. A vapor is close to the liquid state and can readily condense back into a liquid with slight changes in temperature or pressure. In contrast, a gas exists in a state far removed from the liquid phase, requiring significant changes in pressure or temperature to liquefy. In refrigeration and related fields, the term "gas" is often used interchangeably with "vapor," although "vapor" is the more technically accurate term when discussing substances that can easily return to a liquid state.

Vapours can exist in two conditions: saturated or superheated. A saturated vapor is in equilibrium with its liquid at a given temperature and pressure, meaning it is at the boiling point for those conditions. If the vapor contains liquid particles, it is called a wet saturated vapor. When the vapor has no liquid particles present, it is known as a dry saturated vapor. It is difficult to produce a completely dry vapor, as the boiling action typically carries some liquid particles into the vapor. Additionally, any heat loss in long piping systems can cause some condensation, introducing liquid droplets into the vapor.

A vapor that is heated beyond its saturation temperature without any liquid present is considered a superheated vapor. Its temperature exceeds the boiling point for the given pressure. The amount of superheat is the difference between the vapor's temperature and the saturation temperature at the same pressure. For example, if a vapor is 20°F above its saturation temperature, it is said to contain 20°F of superheat. Superheated vapor behaves differently than saturated vapor and is used in various industrial processes where additional energy transfer is required.

Substances, whether solids, liquids, or gases, expand when heated and contract when cooled. This is more pronounced in gases than in solids and liquids. In engineering, allowances must be made for these volume changes to avoid damage to structures and machinery. Water exhibits unique behaviour, contracting as it cools until reaching 39.2°F, then expanding as it freezes. This expansion can create powerful forces capable of cracking rocks or bursting pipes.

The specific volume of a substance is defined as the space occupied by a unit weight (such as one pound) of the substance at a specific temperature and pressure. Specific volume changes dramatically during phase transitions, such as from liquid to vapor. For example, water expands approximately 1,604 times when it vaporizes from a liquid state at atmospheric pressure, illustrating the substantial volume change associated with phase changes.

Understanding these principles of vaporization, phase transitions, and material expansion is fundamental for effectively applying refrigeration and air conditioning technologies, as well as managing heat transfer in various industrial processes.

Heat Transfer

Heat transfer is a fundamental concept in thermodynamics, and it occurs in three primary ways: radiation, convection, and conduction. Each mode involves the movement of heat energy, but the mechanisms differ significantly.

Radiation is the process by which heat travels through a vacuum or transparent medium without needing a physical substance to carry it. This is how heat from the sun reaches the Earth, as radiant energy travels through the emptiness of space. In radiation, energy is transferred in the form of electromagnetic waves, specifically infrared waves in the case of heat. When these waves encounter an object, they are absorbed, causing the object to warm up. Radiation does not heat the medium through which it travels, such as air; it only heats the objects it strikes. Everyday objects, such as stoves, light bulbs, and even human bodies, radiate some heat, although to a much lesser extent than the sun.

Convection involves the actual movement of heat through a fluid medium, such as a liquid or gas. It occurs when heated particles in a substance move from one place to another, carrying energy with them. For example, in a room, warm air rises and cooler air sinks, creating a cycle that distributes heat. This form of heat transfer is also present in natural phenomena, such as ocean currents and atmospheric circulation, where heated particles move through water or air. The human body uses convection to cool down; when we exhale, the air expelled has absorbed some of our body heat, thus removing it from our system [237].

Conduction is the transfer of heat through a material, from one molecule to the next, by direct contact. It occurs in solids, liquids, and gases, though it is most effective in solids, particularly metals. For conduction to happen, there must be physical contact between substances, and heat will always travel from an area of higher temperature to one of lower temperature. For example, when one end of a metal rod is heated in a fire, the heat moves along the rod by conduction until the entire rod becomes warm. Materials vary in their ability to conduct heat; metals are good conductors, while non-metals like wood or rubber are poor conductors.

Thermal conductance is a measure of how easily a substance allows heat to pass through it. For instance, if a copper bar and an iron bar of identical size are placed in a fire, the copper bar will conduct heat more rapidly than the iron bar. This property makes certain materials more suitable for specific applications, such as using metals for heat sinks or ceramics for thermal insulation. Materials with low thermal conductance, such as cork or wool, are effective insulators because they slow down the transfer of heat.

To maintain desired temperature conditions, insulating materials are used to reduce heat transfer. Insulation is crucial in both heating and refrigeration because it prevents heat from entering or escaping a controlled environment. Effective insulation uses materials with low thermal conductance, such as cork, wool, and certain plastics, which trap air in small pockets. These air pockets reduce heat transfer by minimizing radiation and convection within the material.

Insulating materials used in refrigeration and other low-temperature applications face additional challenges. When the air's moisture comes into contact with cold surfaces, it can condense and even freeze. This condensation can penetrate porous insulation materials, reducing their effectiveness. For this reason, materials used in low-temperature environments are specially designed to resist moisture absorption. Additionally, the insulation is often coated with a moisture-resistant barrier to maintain its insulating properties.

Cold pipes, such as those in refrigeration systems, must be thoroughly insulated to prevent heat from entering and causing the refrigerant inside to warm up. Common materials for pipe insulation include cork composition, rock wool, and mineral wool. To avoid problems such as moisture intrusion or heat transfer through metal pipe supports, it is crucial to cover all seams and joints with waterproof materials. Additionally, the insulation must be applied in a way that minimizes the potential for heat conduction through the supports [237].

In the event of damage to pipe insulation, it is essential to repair the affected area promptly to prevent heat from entering and moisture from accumulating. If the original molded sections are unavailable, temporary measures should be taken to re-establish the waterproof barrier. This ensures the continued effectiveness of the insulation in maintaining the desired temperature.

Refrigeration

Mechanical Refrigeration

Submarine refrigeration systems employ a vapor compression cycle to provide cooling, utilizing a refrigerant that alternates between liquid and vapor phases. This system is crucial in submarines due to the limited space, confined environment, and need for high safety standards. A suitable refrigerant for such systems must exhibit specific properties: it should have a low boiling point, easily transition between liquid and vapor phases, and, most importantly, be safe for use in an enclosed underwater environment. Safety concerns are particularly heightened in submarines because any leakage or malfunction can pose significant health risks to the crew in a confined space.

The vapor compression cycle in submarine refrigeration involves a refrigerant absorbing heat and vaporizing at low temperatures in the evaporator. The vaporized refrigerant, now carrying the absorbed heat, is compressed by a compressor, raising its pressure and temperature. The refrigerant then flows to a condenser, where it releases the absorbed heat to a cooling medium, such as seawater, and condenses back into a liquid. The liquid refrigerant is stored in the receiver before being regulated by a thermostatic expansion valve, which controls the flow of refrigerant back into the evaporator to continue the cycle. The system is a closed-loop, meaning the refrigerant is continuously reused, and energy input from an electric motor is required to drive the compressor and maintain the cycle.

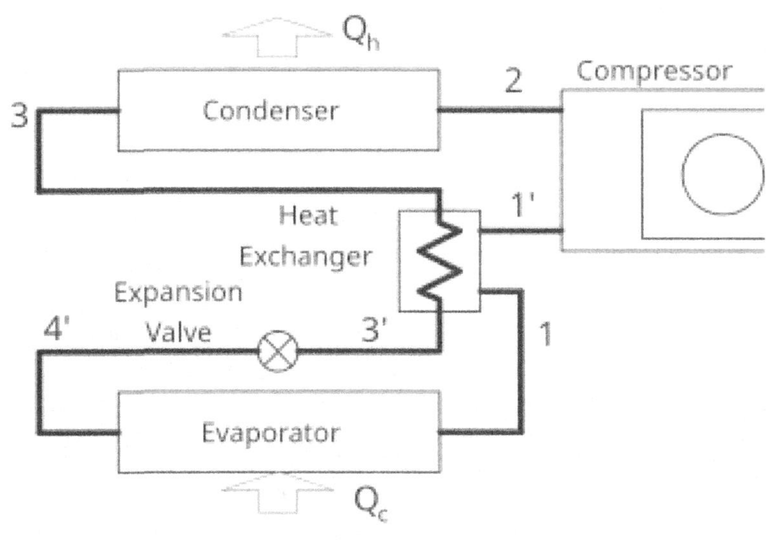

Figure 69: Refrigeration schematic. Shoji Yamauchi, CC BY-SA 4.0, via Wikimedia Commons.

The main components of the refrigeration system include the evaporator, compressor, condenser, receiver, and thermostatic expansion valve. In the evaporator, the liquid refrigerant absorbs heat from the surrounding environment or objects, causing it to evaporate. The compressor then increases the vapor's pressure, which also raises its temperature above that of the seawater used in the condenser. At the condenser, heat is transferred from the refrigerant to the seawater, allowing the refrigerant to condense into a liquid state. The receiver stores the liquid refrigerant before it passes through the thermostatic expansion valve, which controls its flow back into the evaporator, where the cycle repeats [237].

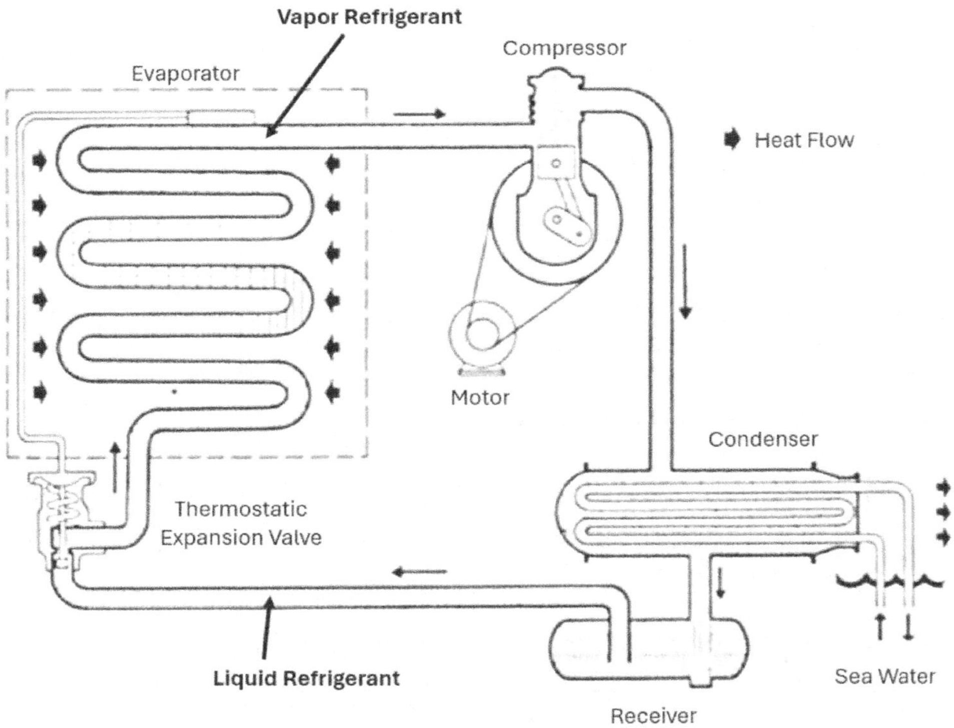

Figure 70: Mechanical refrigeration cycle.

A refrigerant is a substance used in the refrigeration cycle to absorb heat at a low temperature and release it at a higher temperature. In submarines, the primary refrigerant used is dichlorodifluoromethane (Freon-12 or F-12), which is chosen for its stability, non-flammability, and ability to operate efficiently at low temperatures. This type of refrigerant is classified as a primary refrigerant because it directly participates in the phase change processes of the refrigeration cycle. While other primary refrigerants such as ammonia, sulphur dioxide, and carbon dioxide exist, Freon-12 has been favoured in naval applications due to its safer characteristics [237].

Modern nuclear submarines typically use environmentally safer refrigerants that comply with current regulations on ozone depletion and global warming potential. The refrigerants used are chosen for their efficiency, safety, and minimal environmental impact. Here are the common refrigerants used in contemporary nuclear submarines:

1. HFCs (Hydrofluorocarbons)

- **R-134a (Tetrafluoroethane)**: This is one of the most common refrigerants used in various types of HVAC systems, including those on submarines. It has a good balance between efficiency and safety, and does not deplete the ozone layer, making it suitable for modern applications. However, it has a higher global warming potential (GWP) than some alternatives.

- **R-404A and R-407C**: These refrigerants are blends of HFCs that provide suitable properties for refrigeration and air-conditioning systems. They are used for applications requiring medium to low temperatures.

2. HFOs (Hydrofluoroolefins)

- **R-1234yf**: This HFO refrigerant has a much lower global warming potential compared to HFCs like R-134a. It is gradually being adopted in various industries, including military applications, due to its reduced environmental impact.

- **R-1234ze**: Another low-GWP HFO option, R-1234ze is used in some new refrigeration systems as an alternative to traditional HFCs. It offers efficiency and safety in line with the stringent requirements of nuclear submarines.

3. Alternative Refrigerants and Blends

- **R-513A**: A blend of R-1234yf and R-134a, R-513A has a lower GWP than R-134a while still maintaining similar performance characteristics. It is considered a drop-in replacement for existing systems designed for R-134a.

- **R-410A**: While more commonly used in air-conditioning systems on land, some naval applications may use R-410A due to its efficiency and safety profile.

4. Legacy Refrigerants (Phased Out or Being Replaced)

- **R-114 and R-12**: Older submarine systems used refrigerants like R-114 (Dichlorotetrafluoroethane) and R-12 (Dichlorodifluoromethane), which are CFCs (chlorofluorocarbons). These refrigerants have been largely phased out due to their high ozone depletion potential (ODP). Modern submarines have transitioned to refrigerants with lower environmental impact.

When selecting refrigerants for submarines, several key considerations must be taken into account to ensure compliance, safety, and performance. First, adherence to environmental regulations is crucial. Submarine refrigerants must comply with international standards such as the Montreal Protocol, which mandates the reduction of substances with high ozone depletion potential (ODP) and global warming potential (GWP). As a result, modern refrigerants are chosen based on their lower environmental impact, reducing harm to the ozone layer and limiting contributions to climate change.

Safety is another critical factor, as submarines operate in confined environments where even minor leaks could pose serious risks. Therefore, refrigerants used on submarines must be non-flammable and exhibit low toxicity to protect the health of the crew and prevent potential fire hazards.

Lastly, the efficiency of the refrigerant is essential for maintaining the submarine's internal climate under varying operational conditions. The refrigerant must perform reliably across different temperatures and pressures to ensure a stable environment for both equipment and personnel.

In modern nuclear submarines, refrigerants like R-134a, HFOs such as R-1234yf and R-1234ze, and low-GWP blends like R-513A are commonly used. These options represent a shift towards refrigerants that are more environmentally friendly while still meeting the stringent safety and performance requirements of submarine operations.

Unlike primary refrigerants, secondary refrigerants are not directly involved in the phase changes within the refrigeration cycle. Instead, they are used to transfer heat from the area to be cooled to the primary refrigerant. Typically, secondary refrigerants consist of brine solutions (saltwater) and are more common in large industrial cooling systems, such as ice production. However, secondary refrigerants are not used in modern submarines because the spaces requiring cooling are close to the refrigeration equipment, eliminating the need for an intermediate heat transfer fluid [237].

In refrigeration, a standard unit known as a "refrigeration ton" is used to quantify heat transfer rates. This measurement is based on the amount of heat required to melt one ton of ice at 32°F over 24 hours, equating to 288,000 British thermal units (Btu). When broken down, this translates to a cooling capacity of 12,000 Btu per hour. It is crucial to understand that a "refrigeration ton" refers to a rate of heat transfer rather than a physical weight of ice or refrigerant. The term helps standardize the sizing and capacity of refrigeration equipment for various applications.

Safety is paramount in submarine refrigeration due to the unique environment. The choice of refrigerant and the design of the system must ensure minimal risk of toxic exposure, fire, or mechanical failure. Submarines rely on closed-loop refrigeration systems that continuously recycle the refrigerant, minimizing the need for replenishment and reducing potential leak points. The refrigerants selected, such as Freon-12, are non-toxic, non-flammable, and capable of operating efficiently in the confined space of a submarine, ensuring the safety and well-being of the crew.

The submarine's mechanical refrigeration system relies on the vapor compression cycle, with carefully chosen refrigerants to ensure efficient and safe cooling. The design and operation of the system are tailored to the unique demands of the submarine environment, where space constraints, safety, and operational reliability are of utmost importance. By understanding the principles, components, and refrigerant types used, it is possible to appreciate the sophisticated nature of submarine refrigeration systems and their critical role in maintaining a habitable environment underwater.

Air Conditioning and Heating

Air conditioning on a submarine is a vital system that ensures a controlled and comfortable environment for the crew, equipment, and operations. Given the unique and confined nature of submarine life, maintaining the right temperature, humidity, and air quality is essential for health, safety, and operational efficiency. The submarine's air conditioning system is designed to manage heat generated by personnel, electronic equipment, and machinery while also maintaining breathable air quality.

Air conditioning on a submarine is a crucial aspect of engineering focused on creating and maintaining favourable indoor air conditions. This involves the design, construction, and operation of specialized equipment to regulate temperature, humidity, air purity, and oxygen content. The requirements for air conditioning can vary widely based on the environment, whether it's a theatre, factory, store, or, in this case, the confined quarters of a submarine, where maintaining air quality is of the utmost importance due to the prolonged underwater missions [237].

Humans are air-breathing creatures, relying on the continuous intake and expulsion of air to survive. The lungs pump air in and out of the body automatically, ensuring that oxygen is absorbed into the bloodstream, while carbon dioxide and other waste gases are expelled. Given the submarine's confined environment, maintaining air quality is crucial for the crew's health, efficiency, and morale. Air conditioning not only keeps the air breathable but also addresses issues like the removal of fumes from the galley, engine room, and battery compartments, as well as moisture and odours that accumulate in enclosed spaces [237].

Air conditioning on a submarine serves multiple purposes. Besides providing breathable air, it controls heat and moisture levels generated by the crew, machinery, and activities inside the vessel. The human body alone releases significant heat and moisture. On average, an adult emits about 500 British thermal units (Btu) of heat per hour, and a considerable amount of moisture through perspiration and respiration, which can quickly raise the temperature and humidity in a confined space. Air conditioning systems help manage these effects by cooling the air and removing excess moisture.

Additionally, air conditioning plays a critical role in protecting equipment, especially sensitive electrical systems, from moisture-induced corrosion and malfunction. Submarines generate a significant amount of moisture from various sources, including the crew, cooking, batteries, and bilges. If left unchecked, this moisture could condense on cool surfaces and damage critical systems. The air conditioning system helps extract this moisture from the air and collects it in a tank for non-potable uses, such as laundry.

The air that humans breathe is a mixture of various gases, with approximately 21% oxygen, around 78% nitrogen, and trace amounts of other gases. While breathing, only about 4% of the oxygen in inhaled air is absorbed by the body, which means the same air can be circulated and breathed multiple times before oxygen levels drop to an unsafe threshold. Submarine air conditioning systems must maintain proper oxygen levels to prevent hypoxia and other health issues while ensuring that carbon dioxide and other impurities are removed [237].

The air conditioning system on a submarine consists of multiple components that work together to control temperature, humidity, and air circulation. It includes compressors, condensers, evaporators, air handlers, ducts, and sensors. These components are integrated to form a closed-loop system where a refrigerant continuously cycles to absorb heat from inside the submarine and release it outside. The design prioritizes compactness and efficiency due to the limited space and resources available in a submarine.

Modern air conditioning engineering consists of two main components: the actual conditioning of air and ventilation. Air conditioning involves altering air temperature, humidity, purity, and oxygen levels to meet specific standards, while ventilation ensures the replacement of stale air in an enclosed space with fresh, conditioned air. This is especially critical in submarines, where the air inside the vessel can become stale due to continuous recirculation during long dives, necessitating effective systems to maintain a healthy environment for the crew [237].

The core of the air conditioning system operates using a vapor-compression refrigeration cycle. This cycle involves the refrigerant circulating through four main components: the compressor, condenser, expansion valve, and evaporator.

- **Compressor:** The refrigerant enters the compressor as a low-pressure gas. The compressor compresses the refrigerant, increasing its pressure and temperature.

- **Condenser:** The high-pressure, high-temperature refrigerant gas then moves to the condenser, where it releases heat to the seawater or outside air, causing the refrigerant to condense into a high-pressure liquid.

- **Expansion Valve:** The refrigerant passes through the expansion valve, where it expands, lowering its pressure and temperature.

- **Evaporator:** In the evaporator, the refrigerant absorbs heat from the submarine's interior air, cooling the air in the process. The refrigerant evaporates back into a low-pressure gas, and the cycle repeats.

Heat exchange in submarine air conditioning typically involves seawater cooling. The submarine's condenser transfers heat from the refrigerant to the seawater outside the hull. Seawater pumps circulate cold seawater through the heat exchanger to carry away the heat. This cooling mechanism is highly effective because of the large temperature difference between the seawater and the refrigerant.

The cooled air from the evaporator is circulated throughout the submarine via air handlers and a network of ducts. Fans distribute the conditioned air to different compartments, ensuring uniform temperature and air quality. Given the enclosed environment, humidity control is crucial to avoid condensation, which can lead to equipment malfunction and corrosion. The air conditioning system incorporates dehumidifiers to maintain optimal humidity levels by removing excess moisture from the air.

In addition to temperature and humidity control, submarine air conditioning systems must address air quality. This involves filtering the air to remove particulates and trace contaminants. The system uses activated charcoal filters and other filtration methods to eliminate odours and harmful substances. This helps maintain a healthy atmosphere and ensures that the air remains breathable during extended underwater missions.

Submarine air conditioning systems are designed with redundancies to ensure continuous operation even if some components fail. Backup systems and multiple compressors are typically installed to provide operational resilience.

The system also includes safety features like pressure sensors and alarms to monitor refrigerant levels, pressures, and temperatures, ensuring the system operates within safe parameters.

Heat and moisture generated by the crew's activities significantly impact the air conditioning system's workload. A person's body emits heat, comparable to several electric light bulbs, and releases moisture through evaporation, especially when performing physical tasks or in higher temperatures. In a submarine, with limited ventilation, this can quickly lead to uncomfortable conditions. The air conditioning system must, therefore, cool the air and regulate humidity to maintain a stable and comfortable environment for all onboard [237].

The unique environment of a submarine presents challenges for air conditioning. Space constraints, the need for quiet operation to avoid detection, and continuous exposure to seawater, which can cause corrosion, all influence the design. Maintenance and regular monitoring of the air conditioning system are essential to keep it functioning effectively under these demanding conditions.

Heating on a submarine is an essential aspect of the environmental control system that ensures the crew's comfort, maintains safe operational temperatures for equipment, and helps regulate humidity levels. Submarine heating systems are designed to function effectively in the confined underwater environment, where external temperatures can be extremely cold, especially at greater depths or in polar regions.

In the context of submarine operations, the primary heat sources utilized for heating depend significantly on the type of submarine—nuclear or diesel-electric. For nuclear-powered submarines, the nuclear reactor serves as the main heat source. The reactor generates substantial heat as it produces energy to propel the submarine and support onboard systems. This heat is effectively harnessed and distributed throughout the vessel via a sophisticated network of heat exchangers and air ducts, ensuring that the crew remains comfortable and that critical equipment does not freeze. The efficiency of this heat distribution system is crucial for maintaining operational readiness and crew welfare, especially during prolonged underwater missions [238, 239].

In contrast, diesel-electric submarines rely on the heat generated by their diesel engines during operation. These engines produce waste heat, which can be captured through heat exchangers and circulated within the submarine to provide necessary heating. However, when the submarine is submerged for extended periods and the engines are not running, alternative heating methods must be employed. In such scenarios, electric heaters powered by battery banks are utilized to maintain adequate temperatures within the vessel. This dual approach to heating—using both engine waste heat and electric heaters—ensures that diesel-electric submarines can operate effectively even when not actively using their engines [238, 240].

The operational dynamics of both types of submarines highlight the importance of efficient heat management systems. For nuclear submarines, the reliance on a reactor necessitates robust safety protocols and efficient heat exchange systems to prevent overheating and ensure crew safety. Meanwhile, diesel-electric submarines must balance the use of engine heat with battery management to sustain operations during silent running periods. The integration of these heating systems is vital for the overall functionality and safety of submarine operations [238-240].

The heat distribution system on a submarine plays a crucial role in maintaining a comfortable environment for the crew and ensuring the proper functioning of equipment. Since submarines operate in cold underwater conditions, keeping the internal temperature consistent and warm enough is essential. The heat generated from the primary sources—either the nuclear reactor in nuclear submarines or the diesel engines in diesel-electric submarines—is effectively distributed throughout the submarine using a combination of heat exchangers, forced-air heating, and radiant heating systems.

Heat exchangers are at the core of the heat distribution system. These devices transfer heat from the primary heat source to the air or a secondary fluid that circulates throughout the submarine. In nuclear submarines, heat is derived from the reactor's coolant system. The coolant absorbs heat from the reactor core and transfers it to a secondary water or air system via the heat exchangers. This secondary system is isolated from the reactor's primary coolant to prevent any potential radioactive contamination. The heated water or air from the secondary system then circulates through the submarine's heating network.

For diesel-electric submarines, heat exchangers draw waste heat from the diesel engines when they are in operation. This heat, which would otherwise be lost, is captured and transferred to the air circulating throughout the submarine. Utilizing heat exchangers in this manner maximizes energy efficiency and ensures that the crew has a reliable source of warmth.

Forced-air heating systems are used to distribute the heat from the heat exchangers to various compartments within the submarine. In these systems, fans and blowers push heated air through a network of ducts that run throughout the vessel. The heated air is directed into different sections, such as living quarters, command areas, and machinery spaces, ensuring uniform temperatures across all areas. The forced-air system also doubles as part of the ventilation system, promoting continuous air circulation and preventing air stagnation.

This circulation is essential in a submarine's enclosed environment, as it not only helps maintain consistent temperatures but also aids in the removal of excess humidity and stale air. Forced-air heating can be adjusted to direct more heat to certain compartments, depending on the crew's needs or specific mission requirements, providing flexibility in managing the submarine's internal climate.

Some submarine designs incorporate radiant heating as a supplementary or alternative heat distribution method. Radiant heating involves warming surfaces, such as floors, walls, or special heating panels, which then radiate heat to the surrounding air. The principle behind radiant heating is that heated surfaces emit infrared radiation, which directly warms objects and people in the vicinity without needing extensive ductwork or fans.

Radiant heating is particularly effective in areas where uniform warmth is desired without the noise or airflow associated with forced-air systems. It can be more energy-efficient, as the heat is directed where it is needed most, such as in living spaces or control rooms. Additionally, radiant heating helps reduce heat loss through ducting, ensuring that the heat generated is used effectively within the submarine.

To maintain optimal conditions, the heat distribution system can be divided into zones, allowing different compartments to be heated independently. This zoning approach ensures that spaces like the control room, living quarters, or storage areas receive the appropriate level of heat based on their unique requirements. The system can automatically adjust heat levels using thermostats and sensors located throughout the submarine, providing precise temperature control.

Temperature control and regulation within a submarine are critical for ensuring a comfortable living environment for the crew and protecting the functionality of sensitive equipment. The unique operating conditions of a submarine, which involve prolonged periods submerged underwater, present challenges that require a sophisticated heating and temperature management system. The heating system utilizes thermostats, sensors, automatic controls, and zoning techniques to monitor and adjust the internal temperature, ensuring that each compartment maintains optimal conditions.

Temperature regulation within the submarine begins with a network of thermostats and temperature sensors strategically placed throughout the vessel. These sensors continuously monitor the ambient temperature in various compartments, such as living quarters, control rooms, machinery spaces, and storage areas. The data collected by the sensors is transmitted to the submarine's central environmental control system, which uses this information to make real-time adjustments to the heating system.

The system automatically increases or decreases the heat output to maintain the desired temperature range based on pre-set conditions. This automation ensures that the internal environment remains stable, even as external conditions change or as the submarine transitions between different operational states, such as diving, surfacing, or entering various oceanic thermal layers. Maintaining stable temperatures is crucial for the crew's comfort and health, and also prevents the thermal expansion or contraction of materials that could impact the performance of mechanical or electronic equipment.

Given the varying heating requirements of different compartments within a submarine, the heating system is designed with a zoning capability. This means that the submarine is divided into several thermal zones, each with independent temperature control. For instance, living quarters, where the crew sleeps and rests, are typically kept at a warmer, more comfortable temperature. In contrast, areas such as storage spaces, ballast tanks, or non-critical compartments may be maintained at lower temperatures to conserve energy.

The ability to regulate each zone separately provides greater flexibility and efficiency in temperature management. If certain areas require less heat due to a reduced number of personnel or inactive machinery, the environmental control system can automatically reduce the heat output in those zones, directing the saved energy toward other areas where it is more needed. Zoning is especially beneficial during long underwater operations, as it allows for better management of the submarine's energy resources, which are limited while submerged.

To complement the thermostats and zoning features, the submarine's heating system is equipped with automatic controls that dynamically adjust the heating levels based on real-time temperature data. These controls include feedback mechanisms that detect changes in temperature and quickly modify the heat output to prevent any significant deviations from the set temperature range. For example, if a sensor detects a sudden drop in temperature due to a malfunction or a change in external water temperature, the heating system will promptly increase the heat in that zone to maintain stability.

The automatic control system is also programmed with safety protocols to ensure that the internal temperature never exceeds certain limits. Overheating can be just as problematic as insufficient heating, as it may cause crew discomfort, equipment malfunctions, or condensation buildup. The controls are therefore designed to balance the heating requirements across the submarine while avoiding temperature extremes.

Maintaining a precise level of temperature control is also essential for energy efficiency on submarines. The heating system is integrated with other environmental controls, such as ventilation and air conditioning, to optimize energy use. By utilizing sensors to detect when compartments are unoccupied or require less heat, the system can reduce heating in those areas, conserving energy for more critical operations. This efficiency is crucial, particularly on diesel-electric submarines, where the energy available while submerged is limited and must be managed carefully.

Aside from crew comfort, temperature control is vital for the protection of sensitive equipment onboard. Many electronic systems, navigation instruments, and weaponry have specific operating temperature ranges, and deviations from these ranges can lead to malfunctions or decreased accuracy. The heating system helps to keep these areas within safe temperature limits, preventing issues such as condensation or thermal stress on electronic components.

Humidity control is a critical aspect of maintaining a safe and functional environment within submarines. The interplay between heating and humidity management is essential for preventing condensation on cold surfaces, which can lead to equipment corrosion and mold growth. As the air inside a submarine is heated, its capacity to hold moisture increases, thereby reducing relative humidity levels. This reduction is vital in mitigating the risks associated with high humidity, such as the deterioration of equipment and potential health hazards among crew members [241, 242].

Heating systems in submarines are designed to not only raise the air temperature but also to facilitate humidity control. When air is heated, it can hold more moisture, which decreases the likelihood of condensation forming on surfaces that are cooler than the air temperature. This principle is particularly important in confined spaces like submarines, where the accumulation of moisture can lead to significant operational challenges. The use of dehumidifiers in conjunction with heating systems further enhances humidity control by actively removing moisture from the air. This dual approach ensures that the internal environment remains healthy and conducive to the operation of sensitive equipment [241, 243, 244].

Moreover, the integration of advanced climate control technologies, such as condensing heat recovery systems, can significantly improve energy efficiency while maintaining optimal temperature and humidity levels. These systems can reduce energy consumption by up to 30% compared to traditional heating, ventilation, and air conditioning (HVAC) systems [243]. The ability to independently manage temperature and humidity through sophisticated control strategies is crucial for submarines, where environmental conditions must be carefully monitored and adjusted to ensure the safety and comfort of the crew [244].

Submarines are equipped with backup heating systems to ensure the safety and comfort of the crew and the proper functioning of onboard equipment in the event of a primary heating system failure. These backup systems are essential because a loss of heating can lead to rapidly decreasing temperatures within the confined underwater environment, potentially causing health risks for the crew and affecting the performance of temperature-sensitive equipment. Backup heating solutions provide an extra layer of safety, particularly during extended missions when surface support is not readily available.

One of the primary backup heating solutions in submarines is the use of electric heaters. These heaters are powered by the submarine's battery banks and can be activated in emergency situations or when the primary heating system is offline. Electric heaters are typically installed in critical areas, such as the control room, navigation spaces, and crew living quarters, where maintaining a stable temperature is vital for both crew safety and the operation of essential equipment.

The heaters are designed to provide immediate heat to these compartments, compensating for the lack of warmth from the main heating system. Since submarines are equipped with substantial battery storage to power essential systems during submerged operations, these batteries can also support electric heaters for a period of time. The use of electric heaters ensures that critical areas of the submarine remain at a habitable temperature, allowing the crew to continue performing their duties and manage the submarine's operations while the primary system is being repaired.

In addition to fixed electric heaters, submarines are often equipped with portable heating units that can be deployed as needed. These portable heaters offer flexibility in addressing specific heating needs in various compartments. For example, if a certain area experiences a sudden drop in temperature or requires additional localized heating, the portable units can be quickly moved to that space. They are particularly useful in situations where temperature variations may occur, such as near open hatches, airlocks, or equipment that requires additional heat for optimal performance.

Portable heaters can be powered by the submarine's electrical systems or by portable battery packs, making them adaptable for different situations. This flexibility allows the crew to respond swiftly to temperature fluctuations in areas that are not typically covered by the main heating system. For instance, if a crew member is working in a storage compartment or maintenance area that is not regularly occupied, a portable heater can be used to provide temporary warmth without the need to activate the main heating in the entire zone.

The inclusion of backup heating systems, such as electric and portable heaters, reflects the critical importance of maintaining a stable thermal environment on submarines. The confined underwater setting poses unique challenges, as there is no access to external heat sources and ventilation options are limited. Any failure of the primary heating system can lead to rapid temperature drops, which may cause condensation on sensitive equipment, increase the risk of hypothermia, or hinder the ability of the crew to perform their duties effectively.

Backup heating systems offer a redundant solution, ensuring that even if one layer of the heating system fails, there is still a reliable source of heat available. This redundancy is especially important in life-threatening emergencies or in the case of prolonged heating system failures, as it provides the crew with time to diagnose and repair the primary system without compromising safety.

The presence of electric and portable heaters also plays a role in emergency response preparedness. In scenarios where the submarine needs to remain submerged for an extended period due to threats or environmental conditions, maintaining a comfortable and safe internal environment is crucial. Backup heating ensures that the submarine can continue its mission without compromising the well-being of the crew or the operational readiness of the equipment. It also enables the submarine to comply with safety standards and protocols that mandate adequate onboard heating capabilities, regardless of the situation.

When submarines remain submerged for extended periods, they face unique heating challenges. Being surrounded by cold ocean water, especially in deep or polar regions, leads to significant heat loss through the submarine's hull. The cold water continuously absorbs heat from the vessel, causing the internal temperature to drop. To maintain a comfortable and safe environment, the submarine's heating system must compensate for this heat loss by supplying a continuous flow of warmth throughout the vessel.

The cold environment outside the submarine results in rapid heat loss, as the metal hull conducts heat away from the interior. Without a continuous source of heat, the internal temperature could quickly reach uncomfortable or even dangerous levels. This is particularly challenging during deep-sea operations or under ice where the water temperature is close to freezing. The rate of heat loss can increase significantly in such conditions, necessitating a robust and reliable heating system to keep the crew warm and equipment functioning optimally.

For nuclear submarines, maintaining heat during extended submersion is less problematic because the onboard nuclear reactor constantly generates a significant amount of heat. This heat can be harnessed and distributed throughout the submarine's heating and ventilation system. The reactor produces not only the necessary electrical power but also a surplus of heat that can be used to keep living and working areas at a comfortable temperature. The continuous operation of the reactor ensures a steady and reliable heat source, allowing for effective climate control even during prolonged submersion.

Nuclear submarines also use heat exchangers to transfer excess heat from the reactor coolant to the air or water systems used for internal heating. This method allows the submarine to maintain a stable internal temperature, preventing heat loss from affecting the crew's comfort and the functionality of temperature-sensitive equipment. As a result, the heating demands of nuclear submarines are more easily met, making it possible to stay submerged for extended missions without compromising internal conditions.

For diesel-electric submarines, heating during extended submersion presents more significant challenges. These submarines rely on battery power when submerged, as the diesel engines cannot operate underwater. Since the batteries also power essential systems such as propulsion, life support, and navigation, energy conservation is crucial. The need to conserve battery power often limits the use of electric heating systems, which can consume significant energy. Therefore, managing energy resources becomes a key factor in maintaining heat during long dives.

To address this challenge, diesel-electric submarines may use alternative heating strategies, such as utilizing residual heat from diesel engines before submersion or using heat stored in thermal masses within the vessel. This heat can be gradually released to maintain a comfortable temperature without placing a heavy burden on battery power. Additionally, careful management of air circulation and ventilation helps distribute existing heat more effectively, reducing the need for active heating in certain areas.

For diesel-electric submarines, managing heating during prolonged submersion requires balancing energy consumption with the available battery reserves. Submarines may employ zoning techniques to focus heating efforts on critical areas, such as control rooms and living quarters, while allowing less-used compartments to cool slightly. This approach helps conserve battery power while still providing adequate warmth where it is most needed.

In some cases, the crew may use measures such as wearing thermal clothing or adjusting their activity levels to maintain body heat. If necessary, temporary heat sources like portable heaters may be employed, but these measures are typically used sparingly to avoid depleting energy reserves. Effective energy management and the ability to adapt heating strategies are essential for ensuring crew safety and comfort during extended submerged operations.

Chapter 7

Weapons and Combat Systems

Modern submarines are equipped with a wide range of sophisticated weapons and combat systems, allowing them to engage in various combat scenarios, including anti-surface warfare (ASuW), anti-submarine warfare (ASW), land attack missions, and special operations. These systems include advanced torpedoes, missiles, mines, countermeasures, and electronic warfare equipment.

Torpedo Systems

Torpedoes are the backbone of submarine weaponry, serving as the primary offensive tool for engaging enemy vessels, including other submarines and surface ships. Modern torpedo technology encompasses a range of advanced capabilities that enhance accuracy, effectiveness, and versatility in various combat scenarios. The sophisticated systems found in today's torpedoes enable submarines to deliver precise and powerful attacks, even in challenging underwater environments.

Heavyweight torpedoes are the mainstay of submarine combat, designed for long-range, high-power engagements against both submarines and surface ships. These torpedoes are typically larger and heavier than their lightweight counterparts, allowing them to carry more powerful warheads and achieve greater range. Heavyweight torpedoes often feature advanced guidance and homing systems that enable them to detect, track, and engage targets autonomously. They are equipped with a variety of sensors, including active and passive sonar, which allow the torpedo to home in on the target based on acoustic signals. Examples of heavyweight torpedoes include the American Mark 48, which can adjust its speed and depth during its approach, making it effective against both deep-diving submarines and fast-moving surface ships. The torpedo's versatility and power make it suitable for a wide range of combat situations, from open ocean engagements to littoral (coastal) warfare.

While heavyweight torpedoes are designed for long-range attacks, lightweight torpedoes are primarily used for anti-submarine warfare (ASW). These torpedoes are smaller, making them easier to deploy from various platforms, such as submarines, surface ships, helicopters, or fixed-wing aircraft. Their compact size allows for rapid launch and manoeuvrability, making them effective in scenarios where quick response times are critical. Lightweight torpedoes typically have shorter range and smaller warheads compared to heavyweight variants but are well-suited for engagements where the submarine needs to react to threats in close proximity. They are also useful in situations where multiple torpedoes may need to be deployed quickly to overwhelm a target or counter an incoming threat. The flexibility to launch lightweight torpedoes from diverse platforms extends the submarine's ASW capabilities beyond its own arsenal.

Wire-guided torpedoes are equipped with a spool of wire that remains connected to the launching submarine, allowing for real-time control of the torpedo's course. This guidance method enables the submarine crew to adjust the torpedo's trajectory based on updated target information received during its approach. If the target changes direction or speed, the wire guidance allows the torpedo's path to be corrected mid-course, significantly improving the chances of a successful hit. The wire provides a communication link between the submarine's fire control system and the torpedo, ensuring that guidance commands can be relayed even in environments where acoustic homing alone may be less effective due to noise or countermeasures. Modern wire-guided torpedoes can still operate independently if the wire connection is severed, switching to their onboard homing systems to complete the attack autonomously.

Figure 71: Submarine USS Cavalla SS 244, SSK 244, AGSS 244 forward torpedo room. Pingpaul, CC BY-SA 3.0, via Wikimedia Commons.

Modern torpedoes incorporate a range of homing technologies to improve target acquisition and engagement. Acoustic homing uses sonar signals to detect and track the target based on the sound it emits, while active sonar systems send out pulses and listen for echoes to determine the target's position. Passive sonar systems, on the other hand, rely solely on listening to sounds in the environment, making the torpedo harder to detect. Some torpedoes also utilize wake-homing, which follows the turbulent wake left behind by a moving ship. This method can be particularly effective against surface vessels, as the wake provides a reliable trail for the torpedo to follow even if the ship attempts evasive manoeuvres.

Submarines equipped with missile launch capabilities significantly enhance their versatility and effectiveness as strategic and tactical strike platforms. The ability to deploy a range of missiles, from land-attack and anti-ship variants to intercontinental ballistic missiles, makes submarines vital components of a nation's defence strategy. These capabilities enable submarines to strike targets on land, engage enemy ships, and provide nuclear deterrence, all while remaining submerged and relatively undetected.

One of the primary missile capabilities of modern submarines is the launch of land-attack cruise missiles, such as the U.S. Tomahawk or Russia's Kalibr missile. These missiles are used for precision strikes against high-value land targets, such as military installations, infrastructure, or command centres. Typically, cruise missiles have a long range of up to 1,500 kilometres (930 miles), allowing submarines to engage targets from a significant distance without surfacing. Cruise missiles follow a low-altitude, terrain-hugging flight path, which makes them difficult to detect and intercept. They are guided by various navigation and targeting systems, including GPS, inertial navigation, and terrain-matching technology, which help ensure that they strike their intended targets accurately. The ability to launch these missiles while submerged allows submarines to conduct stealthy strike missions with minimal risk of detection.

Missile Systems

Submarines are also equipped with anti-ship missile capabilities, enabling them to engage enemy surface vessels from a considerable distance. Missiles such as the Exocet, Harpoon, or Russia's P-800 Oniks can be launched from torpedo tubes or vertical launch systems, giving submarines a flexible means of striking enemy ships. These missiles typically use active radar homing or other guidance methods to locate and lock onto their targets. Anti-ship missiles are designed to fly close to the surface of the water, making them challenging to detect and intercept. Their high-speed approach and powerful warheads make them highly effective against a variety of ship classes, from frigates and destroyers to aircraft carriers. By deploying anti-ship missiles, submarines can support naval warfare operations, disrupt enemy supply lines, and threaten hostile fleets.

Figure 72: A Poseidon C3 missile is launched from the nuclear-powered strategic missile submarine USS Daniel Boone (SSBN 629) . The U.S. National Archives, Public Domain, via NARA & DVIDS Public Domain Archive.

Strategic nuclear-powered ballistic missile submarines (SSBNs) are equipped with Submarine-Launched Ballistic Missiles (SLBMs), such as the U.S. Trident II or Russia's Bulava missile. SLBMs are a key element of a nation's nuclear deterrence strategy, providing a second-strike capability that ensures retaliatory potential even if land-based nuclear forces are compromised. These missiles are capable of traveling intercontinental distances, with ranges exceeding 8,000 kilometres (5,000 miles), allowing submarines to target almost any location on Earth from a submerged position. SLBMs are equipped with multiple independently targetable re-entry vehicles (MIRVs), enabling them to carry several nuclear warheads that can be directed at different targets. This capability significantly increases the destructive power of each missile, enhancing the deterrence value of the submarine force. Launching SLBMs from underwater provides an additional layer of stealth, making detection by enemy forces difficult and enhancing the survivability of the submarine.

The integration of missile capabilities allows submarines to perform a variety of missions, from tactical strikes and naval engagements to strategic nuclear deterrence. For conventional warfare, cruise and anti-ship missiles enable submarines to support broader military objectives by attacking land targets or neutralizing enemy ships. In a strategic context, the deployment of SLBMs from SSBNs offers a powerful deterrent against potential nuclear adversaries. The ability to perform these missions while remaining submerged gives submarines a significant tactical advantage, as they can approach their targets discreetly and launch attacks with minimal warning. This flexibility makes submarines essential components in modern naval warfare and global security strategies.

Mine Laying Capability

Submarines are highly effective platforms for deploying naval mines, using stealth to lay mines in strategic locations while remaining undetected. Mine-laying capability adds a versatile dimension to submarine warfare, allowing them to establish barriers or deny access to key maritime areas, disrupt shipping routes, and protect friendly forces by creating defensive minefields. The ability to deploy mines from submarines enables naval forces to control strategic choke points, harbors, and coastal areas with minimal risk to the submarine itself.

Naval mines can be deployed from submarines using two primary methods: torpedo tubes or specialized mine-laying systems. Mines launched through torpedo tubes are typically designed to fit within the dimensions of a standard torpedo, making it easy to deploy without any significant modifications to the submarine. In some cases, mines are also housed in external compartments or specialized mine-laying tubes that can carry multiple mines for deployment. The flexibility in deployment methods allows submarines to conduct mine-laying operations discreetly and efficiently, enhancing their ability to surprise and disrupt enemy movements. Once released, the mines sink to a designated depth and anchor themselves to the seabed, becoming active and ready to engage targets.

Modern submarine-laid mines are equipped with sophisticated activation mechanisms that make them highly effective against various types of vessels. These activation triggers include magnetic, acoustic, and pressure sensors. Magnetic mines detect the magnetic signature of a passing ship, particularly those with metallic hulls, and detonate when the detected field changes. Acoustic mines are triggered by the noise generated by ship engines or propellers, making them effective against vessels emitting distinct sound patterns. Pressure mines sense the changes in water pressure caused by the displacement of a ship as it moves through the water, and detonate accordingly. The use of these multiple triggers ensures that submarine-laid mines are capable of targeting a wide range of ships, from small vessels to large warships, and can even be tuned to ignore friendly ships by recognizing specific signatures.

The deployment of naval mines by submarines offers several strategic advantages. By laying mines in vital maritime choke points, such as straits, shipping lanes, or harbor entrances, submarines can effectively block or restrict enemy access, forcing adversaries to alter their routes or face the risk of mine strikes. This capability can be used to isolate or blockade enemy ports, disrupt supply lines, and delay naval reinforcements. Additionally, laying defensive minefields in areas where friendly naval forces are operating can create protective barriers against hostile ships, limiting their maneuverability and providing an added layer of security. The threat posed by undetected mines also acts as a psychological deterrent, causing enemy vessels to exercise greater caution or avoid mined waters entirely.

Submarine-laid mines have evolved significantly, with modern designs incorporating advanced technologies to increase their lethality and versatility. Some mines are programmable and can be set to remain dormant for a specified period before becoming active, allowing for delayed activation in areas where immediate mining might reveal the submarine's position. Additionally, modern mines can be equipped with computer systems that allow for remote activation or deactivation via coded acoustic signals, providing naval forces with control over when and where the mines are operational. This level of control can be crucial for ensuring that minefields do not inadvertently endanger friendly ships or disrupt planned naval operations.

While mine-laying adds a valuable capability to submarines, it also involves certain risks and challenges. Deploying mines in hostile or heavily patrolled waters carries the risk of detection, as the act of mine-laying may reveal the submarine's position. There is also the potential for mines to inadvertently affect civilian shipping if not properly marked or controlled. Therefore, precise planning and coordination are necessary to minimize risks and ensure that the mines serve their intended military objectives.

Special Operations Support

Modern submarines have evolved to not only serve as strategic offensive and defensive weapons but also as versatile platforms capable of supporting special operations forces (SOF). These specialized capabilities allow submarines to carry out covert missions, such as reconnaissance, sabotage, and the deployment of elite military units, while remaining hidden underwater. The integration of SOF support systems, such as lock-in/lock-out chambers, dry deck shelters, and swimmer delivery vehicles, expands the submarine's mission profile and enhances its operational flexibility in diverse maritime environments.

Lock-in/Lock-out Chambers

Lock-in/lock-out chambers are specially designed compartments that enable divers or special forces personnel to enter and exit the submarine while it is submerged. These chambers are isolated from the rest of the submarine, allowing the water pressure inside to be equalized with the external underwater environment. By doing so, divers can leave the submarine discreetly, and similarly, return to the submarine without the need for surfacing, which could compromise the submarine's stealth. The lock-in/lock-out system supports various missions, such as underwater demolition, intelligence gathering, or direct action, where it is crucial for special operations personnel to deploy and recover covertly. The capability to conduct these operations while the submarine is submerged significantly reduces the risk of detection by enemy forces.

Dry Deck Shelters (DDS)

Dry Deck Shelters (DDS) are detachable, watertight chambers mounted externally on the deck of the submarine. These shelters provide an enclosed, dry space that can house special equipment, such as swimmer delivery vehicles (SDVs), inflatable boats, or specialized diving gear. The DDS is designed to be pressurized, allowing personnel to access the stored equipment and deploy it underwater. One of the primary advantages of the DDS is its ability to accommodate larger equipment and support more complex missions than the lock-in/lock-out chambers alone. For example, a submarine equipped with a DDS can transport and launch SDVs that allow SOF teams to travel significant distances underwater, thereby extending the reach of their operations.

The modular nature of the dry deck shelter also enhances a submarine's versatility. When not in use for special operations missions, the DDS can be removed, or its equipment can be reconfigured for other purposes. This flexibility allows submarines to adapt to various mission requirements without significant structural modifications.

Figure 73: A Dry Deck Shelter (DDS) equipped submarine is attached to the submarine's rear escape trunk to provide a dry environment for Navy Seals to prepare for special warfare exercises or operations. DDS is the primary supporting craft for the SDV. U.S. Navy photo by Chief Journalist Dave Fliesen File# 060206-N-1464F-005, Public Domain, via Picryl.

Swimmer Delivery Vehicles (SDVs)

Swimmer Delivery Vehicles (SDVs) are mini-submarines or underwater vehicles designed to covertly transport special operations forces from the submarine to shore or target areas. These vehicles are typically launched from the submarine's dry deck shelter or lock-in/lock-out chamber and are used to deliver SOF personnel to their mission area while minimizing the risk of detection. SDVs provide an extended operational range and allow personnel to travel at greater speeds than swimming alone, making them ideal for missions requiring stealth and rapid movement.

Figure 74: Members of the Navy Sea-Air-Land (SEAL) team emerge from a dry deck shelter aboard the submerged nuclear-powered attack submarine USS SILVERSIDES (SSN-679). The U.S. National Archives, Public Domain, via NARA & DVIDS Public Domain Archive.

The design of SDVs often includes advanced navigation and communication systems, as well as compartments to store weapons and other mission-specific equipment. Some SDVs are capable of carrying multiple operators, while others may be designed for single-person use. The versatility of these vehicles allows SOF teams to conduct a variety of missions, such as reconnaissance, mine clearance, or direct action against underwater or coastal targets. Additionally, their small size and quiet operation enable them to avoid detection by enemy sonar systems, making them a valuable asset in contested environments.

Strategic and Tactical Advantages

Equipping submarines with special operations capabilities provides a range of strategic and tactical benefits. The ability to insert and extract SOF personnel discreetly enhances a nation's capacity for unconventional warfare, allowing for the execution of high-risk missions without exposing conventional naval assets. This capability is particularly valuable for counter-terrorism, hostage rescue, and sabotage operations, where the element of surprise is critical. Furthermore, the use of submarines to support SOF missions allows forces to operate far from friendly shores, extending the reach of military influence.

The integration of SOF support systems also increases the submarine's flexibility in multi-role missions, where it may need to transition between traditional submarine warfare tasks and special operations support. This adaptability allows submarines to contribute to a broader range of mission sets, from strategic deterrence and anti-ship warfare to covert action and humanitarian assistance.

While integrating SOF support capabilities into submarine operations provides significant advantages, there are challenges and considerations. The addition of dry deck shelters and other equipment can affect the submarine's hydrodynamic properties and reduce its stealth characteristics if not properly designed and maintained. Moreover, operating lock-in/lock-out chambers and SDVs requires specialized training for both submarine crews and special operations personnel to ensure safety and mission success. Submarines equipped for SOF missions may also face limitations in terms of the number of personnel and the amount of equipment they can carry, necessitating careful planning to optimize mission logistics.

Electronic Warfare and Sensor Systems

Submarine combat effectiveness is significantly enhanced by advanced sensor and electronic warfare systems, which play a critical role in detection, tracking, and deploying countermeasures. These systems are vital for submarines to maintain situational awareness, avoid detection, and engage enemy forces effectively. The integration of sonar, electronic support measures (ESM), and visual detection technologies ensures that submarines remain formidable opponents in the underwater battlespace.

Sonar (Sound Navigation and Ranging) is the primary tool submarines use for underwater detection and navigation. Modern submarines are equipped with both passive sonar and active sonar systems. Passive sonar systems listen for sounds in the surrounding environment, such as noise generated by enemy vessels or other underwater objects. These systems are crucial for maintaining stealth, as they do not emit any signals that could reveal the submarine's position. Passive sonar arrays are usually mounted on the hull, with some extending to flank array sonar systems for better lateral coverage, and towed array sonar, which is deployed behind the submarine to detect distant threats over a wide area.

Figure 75: Four sonar technicians are shown at their stations on the BBQ-5C active-passive sonar system to spot objects outside the nuclear-powered attack submarine USS LOUISVILLE (SSN-724) during manoeuvres. The U.S. National Archives, Public Domain PDM 1.0, via NARA & DVIDS Public Domain Archive.

Active sonar, on the other hand, involves emitting sound pulses and analysing the echoes that bounce back from objects in the water. This system is particularly useful for precise detection, tracking, and range estimation of targets. However, the use of active sonar comes at the cost of potentially exposing the submarine's location, as the emitted sound waves can be detected by enemy forces. Submarines rely on a combination of hull-mounted sonar and specialized systems like flank arrays and towed arrays to enhance detection capabilities at different depths and distances. These sonar systems enable the submarine to detect and classify other submarines, surface ships, and underwater obstacles, enhancing both offensive and defensive combat operations.

Electronic Support Measures (ESM) are essential for identifying and tracking electromagnetic emissions from enemy radar and communication systems. Submarines equipped with ESM systems can detect and analyse radio frequency signals, allowing them to identify and locate enemy ships, aircraft, or other submarines that are using radar or communication systems. By passively intercepting these signals, submarines can gather valuable intelligence without revealing their own location.

ESM systems are particularly valuable in detecting threats from surface vessels or aircraft operating with radar systems. By monitoring these signals, submarines can determine the proximity of threats and decide whether to engage, evade, or deploy countermeasures. ESM also contributes to maintaining a submarine's low profile, as the submarine can gain situational awareness without the need to surface or transmit its own signals.

While traditional periscopes are still used for visual observation when submarines operate near the surface, modern submarines are increasingly equipped with optronics masts. Unlike periscopes, which physically penetrate the submarine's hull, optronics masts use digital cameras, infrared sensors, and other advanced imaging technology. These masts provide a broader range of visual and infrared capabilities, allowing the submarine to remain at a greater depth while still maintaining situational awareness above the water.

Figure 76: Lieutenant (LT) Barry Rodrigues, Officer of the Deck (OD), mans the attack periscope onboard the Los Angeles class nuclear-powered fast attack submarine. The U.S. National Archives, Public Domain PDM 1.0, via NARA & DVIDS Public Domain Archive.

Optronics masts offer several advantages over traditional periscopes. First, because they do not require a direct physical connection to the submarine's interior, they allow submarines to operate with greater stealth, as the mast can be deployed briefly and at a shallower angle. Additionally, optronics systems offer enhanced imaging capabilities, including infrared vision, which is essential for low-visibility conditions or nighttime operations. This ability to capture high-quality images and video, even at long ranges, allows submarines to gather intelligence on surface vessels, aircraft, and other threats while remaining hidden below the surface. Optronics masts also integrate with the submarine's combat and navigation systems, providing real-time data to support decision-making during combat or covert operations.

The combination of sonar, ESM, and optronics masts creates a comprehensive sensor network that gives modern submarines unparalleled combat effectiveness. These systems work together to detect, classify, and track potential threats from a range of sources, both above and below the surface. By integrating the data from sonar and ESM with visual input from optronics masts, submarines are able to form a complete picture of their surroundings, even in the complex and noisy underwater environment.

This sensor fusion allows submarines to maintain a strategic advantage, whether it's through avoiding detection, positioning themselves for an attack, or deploying countermeasures to neutralize incoming threats. The advanced detection and tracking systems also play a vital role in guiding submarine-launched weapons, such as torpedoes and missiles, ensuring that they hit their intended targets with precision.

Combat Management System (CMS)

The Combat Management System (CMS) is an integral part of modern submarines, serving as the command centre that synthesizes data from various onboard systems to provide a comprehensive tactical picture. This system is essential for processing vast amounts of information from sensors such as sonar, radar, electronic support measures (ESM), and visual inputs from periscopes or optronics masts. The CMS enhances the crew's situational awareness by consolidating and interpreting this data, enabling timely and informed decision-making in complex operational environments [245, 246].

The integration of multiple sensor data streams is a critical function of the CMS. By collecting and processing information from sonar systems for underwater detection, visual observations from periscopes, and electromagnetic emissions detected by ESM systems, the CMS creates a unified interface. This integration allows the crew to assess their surroundings with clarity and accuracy, significantly reducing the complexity associated with managing disparate information sources [247]. The ability to provide real-time updates on enemy positions and underwater threats is vital for effective submarine operations, as it aids in identifying potential threats and tracking enemy movements [248]

Target identification and tracking are among the primary functions of the CMS. When enemy vessels or aircraft are detected, the CMS analyses sensor data to classify and prioritize these targets based on the threat they pose. This capability is crucial in dynamic combat situations, where the CMS can monitor multiple targets simultaneously and continuously update their positions, speed, and heading. Such precision is essential for launching weapons like torpedoes or missiles, which require accurate targeting to ensure effectiveness [249-251]. The CMS's ability to automate tracking and provide real-time tactical recommendations significantly enhances the crew's operational effectiveness [252].

Weapon deployment is another critical aspect managed by the CMS. The system determines the appropriate weapon to be launched based on the tactical scenario and calculates optimal firing solutions by considering factors such as target speed and distance. For example, during torpedo launches, the CMS facilitates wire-guided control, allowing operators to steer the torpedo toward the target until impact. Additionally, the CMS coordinates the deployment of countermeasures, such as decoys or jamming systems, to protect the submarine from incoming threats, thereby enhancing its survivability [253-255].

Real-time decision-making is facilitated by the CMS, which provides the crew with essential tools to respond to evolving combat situations. The user-friendly interface of the CMS presents critical information clearly, allowing operators to focus on immediate threats without being overwhelmed by data. By automating various processes involved in threat detection and weapon deployment, the CMS reduces cognitive load, enabling crew members to concentrate on strategic decision-making [256-258]. For instance, if a hostile vessel is detected, the CMS can automatically suggest the best course of action, whether it be launching a torpedo or deploying countermeasures [259].

Furthermore, the CMS operates in conjunction with the submarine's navigation and propulsion systems, ensuring coordinated management of both weapon deployment and submarine movement. This integration allows the CMS to recommend optimal positioning and depth for stealth operations while maintaining an ideal course during engagements. In multi-submarine operations, the CMS can also share data with other units, facilitating coordinated attacks and joint defensive strategies, which are essential for effective naval operations [240, 260, 261].

Countermeasure Systems

Submarines are equipped with advanced countermeasure systems that are essential for avoiding detection and evading threats, thereby enhancing their survivability during combat or covert operations. These systems are critical for maintaining stealth, misleading enemy sensors, and neutralizing incoming threats such as torpedoes and aircraft attacks. The effectiveness of these countermeasures allows submarines to conduct missions undetected, protecting themselves from hostile forces [256].

One of the primary methods for evading detection involves the use of acoustic decoys and jammers. Submarines operate in a sonar-dominated environment where enemy forces utilize both active and passive sonar systems to locate underwater vessels. To counter this, submarines deploy acoustic decoys that mimic their acoustic signature, generating false signals that confuse enemy sonar [256]. These decoys are strategically released to create multiple false targets, diverting attention from the actual submarine. Jammers complement this strategy by emitting broadband noise that interferes with sonar signals, making it difficult for the enemy to pinpoint the submarine's location. This technique is particularly effective in active sonar environments, where the enemy relies on sound pulses and echoes for detection [256].

In addition to evading sonar detection, submarines face significant threats from enemy torpedoes. Modern submarines are equipped with torpedo countermeasure systems that include decoy launchers and acoustic interference devices. These systems can launch decoys that either imitate the submarine's acoustic signature or produce noise patterns designed to confuse the incoming torpedo's guidance system [256]. When an enemy torpedo is detected, the submarine's decoy launcher deploys these countermeasures to distract the torpedo from its intended path. The decoys generate acoustic signals that mimic the submarine's movements, tricking the torpedo into homing in on the wrong target, while the submarine can perform evasive manoeuvres to distance itself from the threat [256].

Moreover, submarines must also defend against aerial threats, particularly from maritime patrol aircraft and helicopters equipped with advanced sonar systems or anti-submarine warfare (ASW) weapons. Some submarines are fitted with anti-aircraft missile systems that enable them to engage low-flying aircraft attempting to detect or attack them. These short-range missile systems provide submarines with a valuable capability to defend against airborne threats, especially when operating near the surface, where they are more vulnerable to detection [256].

Nuclear and Diesel-Electric Propulsion Systems

The propulsion system of a submarine significantly influences its operational capabilities and combat endurance, shaping its strategic and tactical roles in naval operations. There are two main types of submarine propulsion systems: nuclear-powered and diesel-electric, each offering distinct advantages and limitations.

Nuclear-powered submarines, including attack submarines (SSNs) and ballistic missile submarines (SSBNs), are equipped with nuclear reactors that provide virtually unlimited energy, allowing them to remain submerged for extended periods without needing to surface. This ability gives nuclear submarines a significant operational advantage, as they can operate undetected underwater for months at a time, conducting reconnaissance, covert patrols, or strategic missions.

One of the key benefits of nuclear propulsion is the unlimited range it provides. Unlike diesel-electric submarines that rely on limited fuel supplies and require regular surfacing to recharge batteries, nuclear submarines can travel vast distances without refuelling. This capability makes them ideal for long-range missions, such as nuclear deterrence patrols (for SSBNs), where they must remain hidden in deep waters for extended durations, ready to launch submarine-launched ballistic missiles (SLBMs) if needed.

Speed is another advantage of nuclear-powered submarines. The immense power generated by the nuclear reactor enables these submarines to achieve high sustained speeds, both underwater and on the surface, giving them enhanced mobility and tactical flexibility. This speed is particularly advantageous in combat scenarios, where nuclear submarines can quickly respond to threats, pursue targets, or evade detection by enemy forces. Their speed and endurance make them critical assets for global naval operations, enabling them to project power across oceans and protect vital sea lanes.

However, nuclear-powered submarines have higher operational and maintenance costs due to the complexity of the nuclear reactor and its associated systems. Despite this, their strategic value—especially in maintaining a nuclear deterrent—is unmatched, making them a cornerstone of many nations' naval capabilities.

Diesel-electric submarines offer a different set of capabilities, primarily suited for coastal defence and regional operations. These submarines use diesel engines to travel on the surface and to recharge their batteries, which power their electric motors for underwater propulsion. While diesel-electric submarines have shorter underwater endurance than nuclear-powered ones, they are more cost-effective and require less maintenance.

A key limitation of traditional diesel-electric submarines is the need to surface periodically to recharge their batteries. This surfacing exposes the submarine to potential detection, making it vulnerable to enemy surveillance during these periods. To overcome this challenge, many modern diesel-electric submarines are now equipped with Air-Independent Propulsion (AIP) systems. AIP allows these submarines to stay submerged for longer periods without surfacing, significantly extending their underwater endurance. While AIP systems cannot provide the same endurance as nuclear reactors, they enable diesel-electric submarines to remain submerged for several weeks, enhancing their stealth capabilities.

Although diesel-electric submarines are slower and have limited range compared to their nuclear counterparts, they excel in littoral environments where manoeuvrability and silence are crucial. Their smaller size and quieter operation make them difficult to detect, especially in shallow waters or areas with high levels of background noise. These submarines are highly effective in anti-submarine warfare (ASW) and anti-surface warfare (ASuW) roles, where their stealth and ability to conduct ambushes are advantageous.

Torpedo Launch Systems and Payloads

The history and development of torpedoes dates back to the late 16th century, with the earliest known use occurring in 1585 by the Dutch, who created an explosive-packed ship. This primitive torpedo was a direct descendant of naval mines. The concept of a self-propelled underwater explosive device evolved over centuries, and during the American Revolution in 1778, barrels of gunpowder were used in the "Battle of the Kegs" as makeshift torpedoes. David Bushnell is credited with the first American use of a torpedo in 1775, where he attempted to attack a British flagship using his submarine, the Turtle, which carried a 150-pound mine. Though unsuccessful, it marked the start of torpedo experimentation in the U.S [262].

Robert Fulton, continuing Bushnell's efforts, developed floating mines during the early 19th century. These mines were anchored to the ocean floor, offering a more stable solution compared to previous designs that drifted with the tide. Fulton's designs were sold to the American Navy and set the stage for further developments in mine and torpedo warfare, which played a critical role during the American Civil War, especially for the Confederates, who used mines extensively to offset the Union Navy's superiority [262].

In 1866, Robert Whitehead revolutionized torpedo design by creating the first self-propelled "automobile" torpedo. Whitehead's torpedo introduced the idea of a weapon that would actively seek out its target, rather than passively waiting. His initial design used a two-cylinder compressed air engine, allowing the torpedo to reach speeds of up to 6.5 knots over a distance of 200 yards, becoming the foundation for all subsequent torpedo designs [262].

By 1869, the U.S. Navy established a torpedo station in Newport, Rhode Island, where Whitehead's designs were studied and replicated. However, the early versions developed at Newport failed to progress beyond the testing phase due to issues with the air flask and hull's watertight integrity. The program was briefly halted, but the Navy persisted.

In 1870, Lieutenant Commander John A. Howell developed a new torpedo design powered by a flywheel spinning at 10,000 revolutions per minute. Known as the Howell torpedo, about 50 were produced for tactical use before the U.S. Navy transitioned to the Whitehead and Bliss-Leavitt torpedoes, which became the standard until 1910. The Whitehead Mk 5 could travel up to 4,000 yards at speeds of 27 knots, representing significant advancements in torpedo technology [262].

Between 1910 and 1915, the focus shifted to improving torpedo detonators, allowing them to explode not only upon direct impact but also from glancing blows, which greatly increased their effectiveness. During World War I, torpedo development was minimal, but the introduction of the Mk 7, a steam-driven torpedo, allowed for more flexible usage across destroyers and submarines [262].

The interwar period (1918-1939) saw critical developments in modern torpedoes. The Mk 13, an aircraft-launched torpedo, and the Mk 14, used by submarines, became iconic. The Mk 14 was responsible for sinking over four million tons of Japanese shipping during World War II, while the Mk 15, deployed from destroyers, remained in service until the 1950s [262].

During World War II, the capture of German U-570, which carried the plans for an electric torpedo, led to the development of the Mk 18 torpedo. This electric-powered torpedo had two distinct advantages: it left no visible wake and required less manufacturing effort. Additionally, the Mk 24 homing torpedo, also known as "Fido", used acoustic signals to detect and follow enemy submarines, proving highly effective by sinking around 15 percent of enemy submarines from 1943 onwards [262].

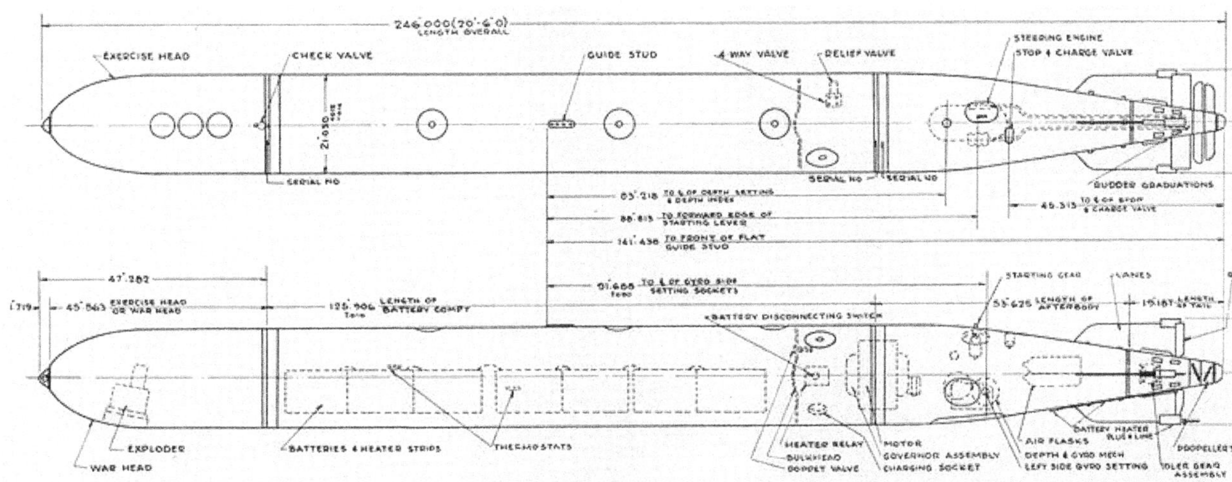

Figure 77: Mark 18 torpedo general profile, as published on the US Navy Torpedo Mark 18 (Electric) basic service manual, published April 1943. United States Navy, Public domain, via Wikimedia Commons.

The 1950s brought about more advanced torpedoes like the Mk 44 and Mk 46, which were designed to counter the growing threat of enemy submarines. The development of nuclear-powered submarines during this time also necessitated faster torpedoes, leading to the introduction of the Mk 45, capable of traveling at speeds of 40 knots with a range of 11,000 to 15,000 yards. It was later replaced by the Mk 48, which remains the primary active torpedo in the U.S. Navy today. The Mk 48 is 19 feet long, weighs approximately 3,500 pounds, and carries a 650-pound warhead. Notably, it can "re-attack" its target if it misses on the first attempt [262].

In response to the Soviet Union's introduction of the Alpha-class submarine—a fast and deep-diving adversary—the Mk 50 "Barracuda" torpedo was developed in the 1970s. This torpedo, capable of speeds over 40 knots and a range of 20,000 yards, could be launched from various platforms, including planes, helicopters, and surface ships, further diversifying its strategic use [262].

A typical submarine torpedo tube from the peak of WW II US submarine technology, operates in a manner similar to a large naval gun, with certain key differences tailored to the unique requirements of underwater combat. The torpedo tube is cylindrical, resembling a gun barrel, with a breech door at the rear (inside the submarine) and a muzzle door at the front (outside the submarine, submerged in water). While a naval gun fires projectiles using explosives, the submarine's torpedo tube launches a self-propelled torpedo using compressed air. The compressed air provides the initial force, after which the torpedo's own propulsion system takes over [263].

At the heart of the torpedo tube's functionality are the two doors: the breech door, located inside the submarine, and the muzzle door, which opens out to the sea. These doors work together in a highly interlocked system to prevent water from flooding into the submarine. The breech door is opened to load the torpedo into the tube, and the muzzle door remains closed to prevent seawater from entering. Once the torpedo is in position and the breech door is secured, the muzzle door can be opened to allow the torpedo to be fired. However, due to the high water pressure outside the submarine at depth, additional steps are needed to ensure that the muzzle door can be opened safely [263].

Figure 78: Torpedo tube of the WWII design submarine HMS Alliance (an Amphion class submarine). Geni, CC BY-SA 4.0, via Wikimedia Commons.

Before loading a new torpedo, the tube must be drained of any water that entered during the previous torpedo launch. This is done through a system of valves and drains controlled from the breech end. Once the tube is drained, the breech door can be safely opened, and the torpedo is loaded. After loading, the tube is flooded with water to equalize the pressure inside the tube with the pressure of the surrounding sea, making it possible to open the muzzle door without resistance from external water pressure. Flooding the tube from internal tanks also helps maintain the submarine's balance by ensuring the water weight is evenly distributed.

With the tube flooded and the muzzle door open, the torpedo is ready to fire. Compressed air from the submarine's high-pressure air system is released into the tube, providing the initial push to eject the torpedo. The air pressure must be sufficient to overcome the surrounding water pressure to ensure the torpedo is launched effectively. Once the torpedo exits the tube, the compressed air is vented to avoid releasing a telltale bubble of air that could reveal the submarine's position to enemy forces.

After the torpedo is fired, seawater fills the tube to compensate for the weight of the torpedo that was just launched. This is a critical step, as the sudden loss of a heavy torpedo's weight could unbalance the submarine and affect its trim (equilibrium). The water acts as ballast, maintaining the submarine's stability. Once the tube is filled with water and the muzzle door is closed, the water is drained from the tube into a tank, and air is blown into the tube to clear it for the next loading cycle.

A key safety feature of the torpedo tube is the interlocking system that prevents both the breech and muzzle doors from being open at the same time. This ensures that seawater does not flood into the submarine, which would be catastrophic. The system essentially acts as an air-lock, similar to a submarine's escape hatch, where either the inner or outer door can be opened, but never both at once.

In a simplified view, the torpedo tube could be imagined as a straightforward system of a barrel, doors, and a compressed air tank. The primary tasks involve controlling the doors, managing air and water pressure, and firing the torpedo by releasing compressed air. However, in reality, submarine torpedo tubes involve more complex mechanisms, including precise air pressure regulation, water management, and firing controls to ensure efficient and stealthy torpedo deployment [263].

This system allows submarines to launch torpedoes while submerged, maintaining operational stealth and the ability to engage enemy targets with precision. The careful management of air, water, and balance ensures that the submarine remains hidden and operationally effective throughout the firing process.

As a more recent example, the Collins Class submarines, used by the Royal Australian Navy, are equipped with the capability to discharge Mk48 torpedoes, a powerful and advanced weapon system designed to target and destroy both enemy submarines and surface ships. These torpedoes are significant in size, measuring six meters in length and weighing approximately two tonnes (2,000 kilograms). Due to their large size and weight, loading, preparing, and launching these torpedoes requires precise processes and coordination within the submarine [264].

Figure 79: A Mark 48 advanced capacity torpedo in a storage rack in the torpedo room of the nuclear-powered attack submarine USS ASHEVILLE (SSN-758). PH2 Rick Gilmore, The U.S. National Archives, Public Domain PDM 1.0, via GetArchive.

In a typical sequence, the Mk48 torpedo is loaded into one of the submarine's torpedo tubes, which serves as the chamber for launching the weapon. Before the torpedo can be fired, it must be carefully prepared for launch by the submarine's crew. This includes feeding target data into the torpedo from the submarine's combat management system. The combat system is a highly advanced piece of technology that allows the submarine to gather, analyse, and relay critical information to the torpedo, including the location and movement of the enemy vessel or target.

Once the Mk48 torpedo is discharged from the submarine, it becomes self-propelled. An onboard engine activates immediately upon exiting the torpedo tube, allowing the torpedo to travel through the water toward its intended target. What makes the Mk48 torpedo particularly effective is its guidance system, which is controlled by the submarine's combat system via an electrical guidance wire. This wire remains connected to the torpedo even after it is launched, allowing the submarine to adjust the torpedo's course in real-time based on changes in the target's position or behaviour. This wire-guidance system greatly enhances the accuracy of the torpedo, making it a highly reliable weapon in naval combat.

Figure 80: MM2 (SS) Joe Hackett inspects the MK 48 ADCAP torpedo for proper alignment as it enters the torpedo tube. The U.S. National Archives, Public Domain PDM 1.0, via Picryl.

The final stage of the torpedo's mission is to reach the optimal point beneath or near the target vessel. At this critical moment, the torpedo's explosive charge is detonated, causing catastrophic damage to the enemy vessel. The Mk48 torpedo is designed to deliver a powerful explosion, capable of disabling or sinking even large warships or submarines. The precision with which the Mk48 can be guided to its target ensures that it can strike at the most vulnerable part of the enemy ship, maximizing the chances of a successful mission.

The Collins Class submarines are undergoing upgrades to integrate new capabilities, including enhancements to their heavyweight torpedo systems and replacement combat systems. These upgrades will ensure that the Collins Class remains a potent and modern force in submarine warfare, keeping pace with technological advancements in torpedo and combat systems.

A key milestone in the upgrade process was achieved by HMAS Waller, one of the Collins Class submarines. It became the first submarine in the world to successfully fire the new Mk48 Mod 7 heavyweight torpedo. During a military exercise off the coast of Hawaii, HMAS Waller used this advanced torpedo to sink a decommissioned warship, demonstrating the effectiveness and precision of the new weapon system. This achievement highlights the ongoing evolution of the Collins Class submarines and their capabilities in modern naval warfare.

Torpedo Characteristics

The propulsion systems of early torpedoes were primarily mechanical, utilizing large flywheels and compressed air. These systems relied on air flasks, constructed from nickel steel, to withstand the high pressures necessary for effective propulsion. The release of this compressed air was meticulously controlled to drive the torpedo's propellers, allowing for a degree of precision in their movement [265]. As technology advanced, the propulsion methods evolved to incorporate combustion processes, specifically burning kerosene or alcohol in conjunction with compressed air. This innovation produced high-pressure gases that powered compact engines, significantly enhancing both the speed and range of torpedoes. However, the combustion process also introduced considerable heat, which posed challenges regarding engine durability [266]. By World War I, many torpedoes had transitioned to wet heater designs, which injected water into the combustion chamber to mitigate heat and increase gas volume, thereby improving overall efficiency [267].

The U.S. Navy standardized the use of two-stage impulse turbine engines for its torpedoes, while other nations, such as Japan and Britain, explored alternative configurations, including piston engines [268]. These advancements in propulsion technology were crucial for maintaining competitive naval capabilities during the war. Additionally, the introduction of electric torpedoes marked a significant shift in design philosophy. Powered by large battery banks, electric torpedoes offered stealth advantages due to their minimal wake, despite being slower than their combustion counterparts. Designers addressed challenges related to temperature fluctuations and battery depletion by implementing preheating mechanisms and governors to regulate speed [269]. Countries like Germany, Japan, and the United States effectively deployed electric torpedoes, particularly within their submarine fleets, highlighting the strategic importance of stealth in naval warfare.

To counteract the rotational torque produced by single propellers, which could destabilize a torpedo's trajectory, the design of counter-rotating propellers became prevalent. This innovation, attributed to the Davison design, ensured that torpedoes maintained a stable course, enhancing their effectiveness in combat scenarios [270]. The combination of these propulsion technologies—mechanical, combustion, and electric—illustrates the evolution of torpedo design and the continuous quest for improved performance and reliability in naval warfare.

Maintaining the correct depth during a torpedo's run is crucial for ensuring successful target engagement. Depth-keeping mechanisms in torpedoes utilize a combination of hydrostatic sensors and control surfaces to stabilize the device and adjust its trajectory. These systems are designed to maintain the desired depth, but they often require fine-tuning to operate effectively. Historical analyses indicate that many nations faced significant challenges with depth control in their prewar torpedo designs, leading to operational inefficiencies and missed targets [271].

For instance, American torpedoes during World War II were found to run approximately 11 feet (3.4 meters) too deep. This issue was not fully understood until comprehensive testing with target nets was conducted, spearheaded by Rear Admiral Charles Lockwood in Australia [271]. These tests revealed that the reliance on lighter practice warheads had skewed the testing data, resulting in inaccurate depth settings for the torpedoes. The findings underscored the importance of rigorous testing and calibration of depth-keeping mechanisms to ensure operational effectiveness.

Moreover, the complexity of adjusting the control surfaces in response to varying underwater conditions highlights the necessity for sophisticated control systems. The integration of hydrostatic sensors and control surfaces is essential for achieving the desired depth, but these systems can be prone to calibration errors, particularly in the context of pre-war designs. The challenges faced by the United States in this regard serve as a case study for the broader issues encountered by other nations in their torpedo development efforts [271].

Directional control was achieved through gyro guidance systems, which used gyroscopes to keep the torpedo on its intended path. These systems proved reliable in wartime, as any design flaws were immediately noticeable during limited testing. The gyroscope was typically kept spinning by a jet of compressed air, and some torpedoes could execute simple course corrections after launch. Later in the war, guided torpedoes became operational, particularly acoustic homing torpedoes that detected and followed propeller noise from enemy ships. Though wire-guided torpedoes were experimented with, they were not fully operational by the end of World War II.

A torpedo warhead is the explosive payload designed to detonate upon reaching or nearing its target. This critical component ensures the weapon's lethality, with various technologies working together to trigger the explosion precisely at the intended moment. Torpedo warheads are essential for disabling or destroying enemy submarines, surface vessels, and other maritime targets.

The functionality of a torpedo warhead depends on several key components, each playing a crucial role in ensuring its overall effectiveness and lethality.

At the heart of the torpedo warhead lies a high-energy explosive material, such as PBX (Plastic Bonded Explosives) or Torpex. These materials are chosen for their ability to generate powerful blasts upon detonation, producing shockwaves that travel effectively underwater. The force of these shockwaves maximizes damage by breaching hulls, causing structural failure, and flooding compartments, making them especially devastating against ships and submarines.

The fuse mechanism controls the precise moment when the warhead detonates. Torpedoes are equipped with different types of fuses to suit various tactical scenarios:

- **Impact Fuse**: This fuse triggers the warhead upon direct contact with the target, ensuring immediate destruction on collision.

- **Proximity Fuse**: Designed to detonate the warhead when the torpedo passes near the target, these fuses use sensors—often magnetic—to detect the ship's hull, creating a devastating shockwave under the keel.

- **Delayed Fuse**: This fuse activates only after the torpedo penetrates the target's hull, allowing the explosion to occur inside the vessel for maximum internal damage.

The warhead is housed within a durable casing, typically constructed from high-strength steel or composite materials. This casing is essential for maintaining the warhead's structural integrity during launch, travel, and impact. Additionally, it provides protection during transport and handling, ensuring the explosive remains stable and safe under various conditions.

Modern torpedoes incorporate advanced safety mechanisms to prevent accidental detonation. These mechanisms ensure that the warhead remains unarmed while it is onboard the launching vessel. Only after the torpedo has travelled a safe distance is the warhead armed, reducing the risk of unintended explosions near friendly ships or submarines. These arming mechanisms are critical for operational safety, allowing crews to handle the torpedoes without fear of premature detonation.

Together, these components—explosive charges, fuses, casing, and safety mechanisms—ensure that the warhead delivers a controlled, effective, and devastating impact. They reflect the precision required in modern naval warfare, where each element contributes to the reliability, safety, and destructive power of the torpedo.

As a torpedo approaches its target, sensors and gyroscopes play a vital role in ensuring precise alignment and maintaining the correct depth for an effective strike. Modern torpedoes are often equipped with wire-guided systems or acoustic homing technology, allowing the submarine to communicate with the torpedo during its run. These systems enable real-time course adjustments, enhancing accuracy by compensating for changes in the target's movement or environmental factors, such as currents. This ensures that the torpedo remains on a collision course with the intended target.

The warhead's fuse mechanism determines the exact moment of detonation, depending on the mission's objectives and the type of target:

- **Impact Fuse**: This fuse detonates the warhead upon direct contact with the target, causing an immediate explosion that can breach the vessel's hull.

- **Proximity Fuse**: In this case, the torpedo detonates when it passes beneath the target's keel, relying on magnetic sensors to detect the ship's structure. The explosion generates a powerful shockwave directly under the ship, often breaking its back. This method is particularly effective for sinking larger vessels, as the resulting structural damage compromises the entire integrity of the ship.

When a torpedo warhead detonates underwater, it creates a high-pressure bubble that expands rapidly, followed by a sudden collapse. This rapid pressure change generates a destructive force, capable of rupturing the hull of the target and flooding compartments. The shockwave's immense energy compromises the ship's buoyancy and stability, leading to catastrophic failure and often causing the vessel to sink quickly. This underwater blast effect makes torpedoes particularly devastating weapons in naval warfare, as the pressure dynamics are amplified in the dense aquatic environment.

The development of warheads used in modern military applications is a culmination of decades of research into explosives, detonation mechanisms, and delivery strategies. This historical evolution can be traced back to World War II, where various nations sought to enhance their destructive capabilities through innovative warhead technologies.

During World War II, the Japanese Type 97 explosive exemplified advanced warhead technology. The Japanese developed torpedoes such as the Type 93 "Long Lance," which utilized a stable mixture of TNT and hexanitrodiphenylamine. This combination was not only powerful but also highly insensitive to shock, making it safer for handling while still capable of causing significant damage to enemy vessels with warheads weighing over 490 kg [272]. In contrast, the Allies initially relied on pure TNT, which proved inadequate against heavily armoured targets. Consequently, they transitioned to Torpex, a blend of RDX, TNT, and aluminium powder, which offered 150% more explosive power than TNT alone and improved underwater shockwave propagation. This advancement made Torpex-filled warheads standard in Allied forces by the war's end, significantly contributing to their naval victories [272, 273].

PBX combines high-powered explosives such as RDX (Research Department Explosive) or HMX (High Melting Explosive) with a polymer binder. This formulation enhances both the explosive's stability and performance, offering improved handling safety. One of the key advantages of PBX is its resistance to accidental detonation, meeting Insensitive Munitions (IM) standards. This feature ensures that the explosive remains stable even under adverse conditions, such as impact or extreme temperatures, making it suitable for torpedoes and other naval munitions.

Some modern explosives, such as PBXN-109, contain powdered aluminium within their chemical composition. The addition of aluminium increases the heat output and shockwave intensity during detonation, which is especially beneficial for underwater explosions. The enhanced energy release ensures greater structural damage to targeted ships or submarines, improving the warhead's effectiveness. Aluminized explosives are now widely used for underwater weapons, as they amplify the destructive effects of pressure waves and hull breaches.

Developed during World War II, Torpex is a mixture of RDX, TNT (Trinitrotoluene), and aluminium. It was initially used in torpedoes and naval mines due to its ability to create more powerful underwater blasts compared to TNT alone. While newer explosives have since been developed, Torpex remains influential in modern warhead design. Its composition effectively amplifies blast effects, ensuring that underwater detonations deliver maximum damage by generating intense shockwaves and secondary explosions. This legacy formulation still serves as the basis for some modern explosive technologies, highlighting its effectiveness in maritime warfare applications.

Modern warheads have built upon these historical advancements, incorporating new explosive compounds to enhance power and reliability. For instance, contemporary torpedoes like the U.S. Mk 48 Mod 7 and the German DM2A4 SeaHake utilize Plastic Bonded Explosives (PBX), which blend RDX or HMX within a polymer binder. This formulation not only increases explosive power but also enhances stability, ensuring reliability under the extreme conditions of submarine warfare. Furthermore, modern warheads often employ aluminized explosives, similar to Torpex, to improve underwater shockwave effectiveness. For example, PBXN-109, used in the Mk 48, incorporates aluminium powder to amplify heat generation and blast pressure, making it particularly effective in underwater detonations [274].

Safety considerations have also led to the development of Insensitive Munitions (IM), which resist accidental detonation from shock, impact, or fire. This technology significantly reduces the risk of accidental explosions during handling, storage, and transport, marking a critical advancement in warhead safety.

In terms of warhead design and detonation mechanisms, modern torpedoes utilize sophisticated technologies to maximize their impact. Shaped charges are employed to direct explosive force into a concentrated jet, allowing penetration of even heavily armoured vessels before detonation [275]. Additionally, magnetic proximity fuses are now standard, enabling torpedoes to detect the magnetic field of a target's hull and detonate beneath the ship, creating a shockwave that ensures catastrophic structural failure [276]. Advanced torpedoes, such as the Mk 48 Mod 7, also feature programmable detonation modes, which include impact, proximity, and delayed detonation, allowing for tailored engagement strategies against various targets [277].

The continuous evolution of warhead technology has significantly impacted modern naval warfare. Today's torpedoes are designed to be more lethal, often requiring only a single hit to sink a vessel, and are capable of neutralizing submarines effectively through the use of magnetic fuses. Furthermore, the adaptability of programmable warheads allows for precision strikes against a range of targets, including surface ships and shore installations, ensuring that modern naval forces maintain a strategic advantage.

Exploders, designed to detonate the warhead upon impact or proximity to the target, posed significant reliability challenges. Most torpedoes used contact exploders, which relied on inertia to drive a weight into a firing pin when the torpedo hit its target. However, both the Germans and Americans found that their contact exploders would often jam under heavy impacts. While Germany quickly rectified the issue, the U.S. Navy took longer to address the problem, partly because they prioritized magnetic exploders designed to detonate under the keel of a ship by detecting its magnetic field. Both the U.S. and Germany ultimately found these magnetic exploders to be unreliable and reverted to using improved contact exploders.

The concept behind magnetic exploders was to detonate beneath a ship's hull, causing catastrophic damage by breaking the vessel's back. However, accurately detecting a ship's magnetic field proved difficult in practice. Differences in magnetic fields across various latitudes and the complexity of characterizing the field of a moving ship made the exploders unreliable, leading both powers to abandon the technology during World War II in favour of more dependable contact exploders.

Torpedo technology has evolved through innovations in propulsion, depth-keeping, guidance, warhead design, and exploders. Each component plays a vital role in ensuring that the torpedo can effectively reach and destroy its target, but the complexities of these systems have historically posed significant development challenges.

Submarine Infrastructure to Support Torpedo Systems

The infrastructure and technical systems supporting torpedoes on modern submarines are highly sophisticated, involving specialized equipment for loading, storage, guidance, launching, and safety management. These systems are meticulously designed to ensure that torpedoes can be deployed efficiently and safely, even in hostile or high-pressure underwater environments.

Modern submarines are equipped with multiple torpedo tubes—typically between four to six tubes at the bow and sometimes additional tubes at the stern. Each tube is designed to handle heavyweight torpedoes, such as the Mk48 used by the U.S. Navy, or advanced lightweight torpedoes for anti-submarine warfare.

The launch system operates on compressed air or water-ram systems to eject torpedoes from the tubes. Compressed air systems release a burst of air to push the torpedo out of the tube, while water-ram systems use pressurized seawater to achieve the same result with minimal noise, reducing the submarine's acoustic signature.

Figure 81: View of the right side of the torpedo room onboard the Los Angeles class nuclear-powered fast attack submarine USS HATFORD (SSN 768). The U.S. National Archives, Public Domain PDM 1.0, via Picryl.

To maintain safety and operational efficiency, torpedo tubes are fitted with interlocking systems that prevent both the inner and outer doors from being opened simultaneously, avoiding accidental flooding of the submarine. Additionally, tube-flooding and drainage systems ensure the smooth ejection of torpedoes by matching external water pressure at depth.

Submarines have limited internal space, necessitating careful planning for torpedo storage and handling. Torpedoes are stored in specially designed racks or cradles, ensuring they remain secure during manoeuvring or rough seas. Each torpedo can weigh up to two tonnes (4,400 pounds), requiring the use of automated handling equipment for loading them into the tubes.

Dedicated loading mechanisms, such as torpedo hoists or rail systems, transport torpedoes from the storage racks to the launch tubes. In addition, the loading and storage systems are climate-controlled to prevent the degradation of sensitive electronic components and warheads.

The Combat Management System (CMS) is central to the submarine's ability to deploy torpedoes effectively. The CMS integrates input from the submarine's sonar, periscopes, and other sensors, providing real-time targeting data and environmental conditions.

Before launch, the CMS feeds guidance information to the torpedo's onboard systems. This can include pre-programmed waypoints or target characteristics, ensuring the torpedo follows the optimal path toward its target. In modern wire-guided torpedoes, the CMS can provide mid-course corrections to adapt to changing conditions or target movements.

The submarine's fire control system ensures the precise launch and trajectory of torpedoes. Modern fire control systems employ wire-guidance technology, allowing the submarine's crew to maintain communication with the torpedo during its run toward the target. This connection enables the crew to adjust the torpedo's course based on updated sonar data, ensuring higher accuracy and mission success.

Acoustic homing torpedoes, which rely on onboard sensors to seek out targets, are also managed through the fire control system. The submarine ensures the right torpedo is selected based on the target's characteristics—such as enemy submarines or surface vessels—and environmental conditions.

Safety is paramount on submarines to prevent accidental detonations. Modern torpedo systems include arming mechanisms that only activate the warhead once the torpedo has traveled a safe distance from the submarine, ensuring no risk to friendly forces. These mechanisms can also delay detonation until the torpedo reaches its intended depth or proximity to the target.

Additionally, torpedo safeties and failsafe systems are integrated into the infrastructure. These include manual override options, enabling the crew to deactivate a torpedo in the event of a system malfunction. Periodic testing of the safety systems is conducted to ensure reliability under all operating conditions.

Torpedo systems on submarines require continuous maintenance and diagnostics to ensure optimal performance. Modern submarines are equipped with automated diagnostic tools that monitor the status of torpedo tubes, guidance systems, and other critical components. Routine checks ensure that compressed air or water pressure levels are maintained for effective torpedo launches.

Additionally, modular designs allow for rapid maintenance or replacement of key components, minimizing downtime during operations. Specialized maintenance crews are trained to handle and inspect sensitive components, such as warhead arming systems and guidance circuits.

Submarine torpedo systems are designed with acoustic stealth in mind. Launch systems, especially those using water-ram propulsion, are configured to minimize noise, reducing the submarine's chances of detection.

In addition, modern torpedo warheads are designed to detonate without producing large air bubbles, which could betray the submarine's location. Post-launch, the tubes are flooded with seawater to maintain the submarine's neutral buoyancy, ensuring the vessel remains in balance after firing.

Some modern submarines are designed with multi-role capabilities, meaning their torpedo tubes can launch not only torpedoes but also missiles, mines, and swimmer delivery vehicles (SDVs). This flexibility allows submarines to adapt to diverse mission profiles, from anti-submarine warfare to mine-laying and covert special operations.

In these configurations, the CMS and fire control systems need to handle multiple types of payloads, ensuring seamless transitions between different mission objectives.

Missile Systems and Deployment

Modern submarines are equipped with sophisticated missile systems that enhance their capabilities, allowing them to engage in a range of strategic and tactical missions. These systems include ballistic missiles for nuclear deterrence, cruise missiles for precision strikes, and anti-ship missiles for maritime warfare. Submarine missile systems are carefully integrated with advanced launch, guidance, and combat management systems, making them formidable tools in naval warfare.

Submarine-launched missile systems are critical components of modern naval warfare, providing strategic capabilities that enhance deterrence and operational flexibility. These systems can be categorized into three primary types: Submarine-Launched Ballistic Missiles (SLBMs), Land-Attack Cruise Missiles (LACMs), and Anti-Ship Missiles (AShMs).

Submarine-Launched Ballistic Missiles (SLBMs)

SLBMs are long-range, nuclear-capable missiles deployed by ballistic missile submarines (SSBNs), also known as boomers. These missiles form a crucial element of a nation's nuclear deterrent strategy, as they provide stealthy, second-strike capabilities. By launching from underwater, SLBMs allow submarines to remain undetected while positioning themselves in strategic locations, ready to launch multiple warheads on distant targets. SLBMs are unique because they can carry Multiple Independently Targetable Re-entry Vehicles (MIRVs)—individual nuclear warheads within a single missile, each capable of striking separate targets across different locations.

Examples of SLBMs:

- Trident II (D5):
    - Operated by the U.S. and U.K. navies, Trident II is a highly advanced SLBM with a range exceeding 12,000 km (7,500 miles).
    - It can carry up to 14 MIRVs, each independently targeted, making it one of the most effective strategic weapons in the world.
- RSM-56 Bulava:
    - Developed by Russia, the Bulava is designed for deployment on the Borei-class SSBNs.
    - It has multiple nuclear warheads and advanced countermeasures to evade missile defence systems.
- JL-3:
    - Currently in development by China, the JL-3 represents an upgrade from its predecessor, the JL-2, with extended range capabilities, potentially covering targets as far as the continental United States from Chinese waters.

SLBMs are launched using cold launch systems, a process that ensures the missile can be safely deployed from the submarine while submerged. Here's how the operation works:

- Cold Launch Sequence:
    - Compressed air or steam ejects the SLBM from its launch tube, propelling it upward through the water without igniting the missile's rocket engines.
    - This method minimizes heat and pressure inside the submarine, preventing damage and reducing noise that could reveal the submarine's position.
- Ignition on Surfacing:

- Once the missile clears the water surface, its rocket engines ignite, propelling it toward the target.
- The missile quickly climbs into the atmosphere, following a ballistic trajectory.

- Midcourse and Re-entry:
  - During the flight, the missile leaves Earth's atmosphere and enters space, where it follows a precise arc toward its target.
  - At the appropriate point, the missile releases its MIRVs, each of which independently re-enters the atmosphere and travels toward pre-designated targets.
- Terminal Phase:
  - The individual warheads are equipped with precision targeting systems, ensuring accuracy upon re-entry.
  - They travel at high speeds, evading missile defence systems, and detonate over or near their designated targets, delivering devastating nuclear strikes.

Figure 82: A Poseidon C-3 (UGM-73A) fleet ballistic missile lifts off after being launched from the submerged nuclear-powered strategic missile submarine USS ULYSSES. The U.S. National Archives, Public Domain PDM 1.0, via Picryl.

Advantages of SLBMs:

- Stealth and Mobility: Submarines can deploy SLBMs covertly, remaining undetected under the ocean surface.

- Second-Strike Capability: Even if an adversary launches a surprise attack, SSBNs carrying SLBMs ensure a retaliatory nuclear strike remains possible, deterring potential aggressors.
- Extended Range and Multiple Targets: With intercontinental ranges and MIRVs, SLBMs allow a single missile to destroy multiple high-value targets spread over large areas.
- Strategic Flexibility: Submarines equipped with SLBMs can patrol distant waters, providing the ability to respond quickly to shifting geopolitical threats.

Land-Attack Cruise Missiles (LACMs)

Land-Attack Cruise Missiles (LACMs) are submarine-launched precision weapons designed to strike land-based targets with high accuracy. These missiles are typically carried by attack submarines (SSNs) or guided-missile submarines (SSGNs) and are used for tactical and strategic strikes. Unlike ballistic missiles, LACMs follow low-altitude flight paths, hugging the terrain and making them difficult to detect by enemy radar systems. Their combination of precision, stealth, and manoeuvrability makes them a critical tool for modern naval operations.

Examples of Submarine-Launched LACMs [278-281]:

- **Tomahawk (BGM-109)**
    - A versatile, long-range cruise missile used by the U.S. Navy. It can strike targets up to 1,600 km (1,000 miles) away, making it effective for deep precision strikes.
    - The Tomahawk can carry conventional or nuclear warheads and has various versions, including Block IV, which allows in-flight retargeting.
- **Kalibr (3M-54)**
    - A Russian cruise missile with both land-attack and anti-ship capabilities. It has seen use in recent conflicts, demonstrating its effectiveness in striking strategic land targets from submarines.
    - Kalibr offers multiple variants with ranges between 300 and 2,500 km, making it suitable for both tactical and strategic missions.

Figure 83: A Tomahawk Land Attack Missile (T-LAM) rises into the air after being launched from the US Navy (USN) OHIO CLASS: Strategic Missile Submarine, USS FLORIDA. The U.S. National Archives, Public Domain PDM 1.0, via NARA & DVIDS Public Domain Archive.

Submarine-launched cruise missiles can be deployed from either torpedo tubes or Vertical Launch Systems (VLS), depending on the submarine's design and capabilities. The launch process is highly sophisticated to maintain stealth and ensure mission success:

- **Launch While Submerged**
    - Submarines can fire LACMs from a submerged position, ensuring they remain undetected.
    - When using torpedo tubes, the missile is encapsulated in a waterproof canister that is ejected from the tube. Once the canister reaches the surface, it breaks open, and the missile ignites.
    - Submarines with VLS can launch missiles directly upward through hatches on the hull, streamlining the launch process.
- **Flight Path and Guidance**
    - After launch, the missile flies at low altitudes, following a pre-programmed path to avoid radar detection.
    - Advanced guidance systems, including GPS and terrain-following radar, enable the missile to navigate complex environments and adjust its trajectory to avoid obstacles.
    - The missile can manoeuvre mid-flight, ensuring it stays on course even in adverse conditions or when navigating through enemy defences.
- **Target Engagement**
    - Upon reaching the target area, the missile can use optical or infrared sensors to identify the target with precision.
    - In some versions, the missile can loiter near the target area, waiting for confirmation before striking or redirecting to a new target.
    - When the missile reaches the target, it delivers its warhead with pinpoint accuracy, minimizing collateral damage.

Advantages of LACMs:

- Stealth and Precision: LACMs fly at low altitudes and use terrain-following technology, making them difficult for enemy radar to detect and intercept.
- Long Range and Flexibility: With ranges of up to 2,500 km, cruise missiles allow submarines to strike deep within enemy territory from a safe distance.
- Mid-Course Corrections: Some LACMs, such as the Tomahawk Block IV, can be reprogrammed in flight, giving commanders the flexibility to change targets based on real-time intelligence.
- Survivability: Launching from a submerged submarine increases the chance of a successful strike without revealing the submarine's position, preserving operational secrecy.

LACMs provide a critical first-strike capability for submarines, enabling nations to attack high-value targets such as command centres, airbases, and radar installations with minimal warning. These missiles can also be used in conventional conflicts to disable infrastructure or as part of a larger coordinated assault involving air and naval forces.

The covert launch capabilities of LACMs make them ideal for surprise attacks or missions requiring plausible deniability, such as pre-emptive strikes against enemy defences or leadership targets. Additionally, the ability to launch while submerged ensures the submarine remains undetected and can withdraw safely, ready to strike again if needed.

LACMs enhance the strategic flexibility of submarines, making them essential for modern naval warfare. Their ability to deliver precise, devastating strikes while maintaining stealth and survivability ensures they remain at the forefront of submarine-based offensive operations.

Anti-Ship Missiles (AShMs)

Submarine-launched anti-ship missiles (AShMs) are advanced weapons designed to target surface vessels from long distances. These missiles allow submarines to engage warships, cargo vessels, and enemy fleets far beyond the range of conventional torpedoes, enhancing the submarine's versatility in maritime warfare. AShMs are effective in both offensive and defensive operations, providing submarines with a powerful tool to neutralize surface threats.

Examples of Submarine-Launched AShMs:

- **Harpoon (RGM-84)**
  - The Harpoon is a versatile, long-range anti-ship missile used by the U.S. Navy and many allied navies.
  - It features active radar guidance, which allows it to autonomously track and engage enemy ships.
  - The missile has a range of over 120 km (75 miles) and flies at low altitudes, making it difficult to detect and intercept.

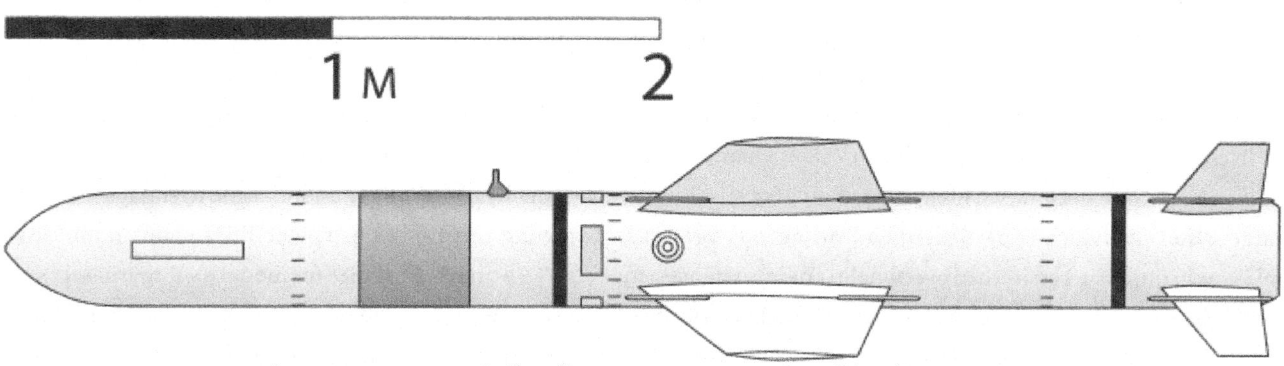

Figure 84: Harpoon missile. Allocer, CC BY-SA 3.0, via Wikimedia Commons.

- **Exocet**
  - Developed by France, the Exocet is known for its effectiveness in maritime conflicts, most notably during the Falklands War.
  - It uses active radar homing for terminal guidance and flies close to the water surface in a sea-skimming flight path, which helps it evade radar detection.
  - The Exocet is compact and can be launched from various platforms, including submarines, ships, and aircraft.

- **Kalibr (Anti-Ship Variant)**
    - This Russian missile is part of the Kalibr missile family, with specific variants designed for anti-ship roles.
    - The Kalibr missile features a sea-skimming trajectory, reducing its radar cross-section and enhancing its ability to evade enemy air defences.
    - It has a long range of up to 300 km (186 miles), making it an ideal weapon for engaging distant surface threats.

Figure 85: EXOCET MM 40 BLOCK 3. Marcomogollon, CC BY-SA 4.0, via Wikimedia Commons.

Operation of Submarine-Launched AShMs

Launch Process: Submarine-launched AShMs can be deployed using two primary methods: torpedo tubes and Vertical Launch Systems (VLS). The torpedo tube launch method involves encasing the missile in a waterproof canister, which protects it from the underwater environment. Upon surfacing, the canister opens, and the missile's rocket motor ignites, allowing for a stealthy launch while the submarine remains submerged, thus minimizing detection risk [282]. This stealth capability is crucial for maintaining operational security and surprise in naval engagements.

In contrast, submarines equipped with VLS can launch missiles directly from vertical cells, which significantly enhances launch speed and flexibility. This system allows for rapid successive launches, enabling the submarine to engage multiple targets quickly, which is particularly advantageous against fast-moving threats [283]. The ability to launch multiple missiles in quick succession can overwhelm enemy defences, making VLS a preferred choice for modern submarines [283].

Flight Path and Guidance: Once launched, AShMs utilize advanced guidance systems to ensure accuracy and evade detection. A key feature is the active radar homing system, which activates post-launch to autonomously track the target by locking onto its radar signature. This capability allows the missile to adjust its trajectory in real-time, enhancing its chances of successful engagement [283]. Additionally, AShMs are designed to follow a sea-skimming flight path, flying just above the water's surface. This low-altitude trajectory significantly reduces the missile's radar cross-section, making it difficult for enemy radar systems to detect and intercept [283]. Furthermore, many modern AShMs incorporate mid-course guidance systems that allow for updates during flight, enabling real-time adjustments based on target movements or countermeasures [283]. This adaptability is crucial in dynamic combat scenarios where targets may manoeuvre unpredictably.

Target Engagement: The final phase of AShM operation involves precise targeting and impact. As the missile approaches its target, it transitions to terminal guidance mode, locking onto the vessel's signature to ensure accuracy [283].

Many AShMs are programmed to accelerate during this final approach, striking at a shallow angle. This tactic complicates interception efforts by traditional air defence systems, as the high-speed, low-angle trajectory reduces the time available for countermeasures [283].

Upon impact, the missile's warhead detonates, creating a powerful shockwave capable of inflicting severe damage to the target vessel. The explosion can breach the hull, compromise critical systems, or ignite onboard munitions and fuel, potentially sinking the ship or rendering it inoperable [283]. The destructive capability of AShMs underscores their role as a significant threat to naval forces.

Advantages of Submarine-Launched AShMs:

- Extended Strike Range: AShMs enable submarines to engage enemy ships beyond the effective range of torpedoes, reducing the risk of counterattack.
- Stealth and Surprise: Launching from a submerged position allows the submarine to remain undetected while delivering precise strikes.
- Versatile Deployment: AShMs provide submarines with the ability to target a wide range of surface vessels, from destroyers to cargo ships.
- Sea-Skimming Flight: The missile's low-altitude flight path helps it evade enemy radar and makes interception challenging.

AShMs play a critical role in anti-surface warfare by giving submarines the ability to engage targets at long ranges and with precision. These missiles are invaluable for coastal defence, blockade enforcement, and fleet interdiction operations, allowing submarines to neutralize enemy ships before they can pose a direct threat.

Submarine-launched AShMs are often deployed in joint naval operations, working alongside surface vessels and aircraft to achieve sea control. In addition, the stealthy launch capabilities of AShMs make them well-suited for surprise attacks, disrupting enemy fleet movements or deterring enemy ships from entering strategic waters.

Missile Launch Systems on Submarines

Modern submarines rely on specialized missile launch systems to enhance their strike capabilities and operational flexibility. Two primary launch systems are used: Vertical Launch Systems (VLS) and Torpedo Tube Launch Systems. Both systems offer unique advantages, enabling submarines to deploy various missiles effectively depending on mission requirements.

Vertical Launch Systems (VLS) represent a sophisticated missile deployment technology employed by many modern submarines. These systems consist of vertical launch cells integrated directly into the submarine's hull, enhancing firepower and operational efficiency. VLS allows submarines to carry a variety of missiles in a ready-to-launch position, ensuring rapid response capabilities and reducing the need for complex reloading processes during missions.

The core functionality of VLS revolves around its design, where missiles are housed individually within dedicated launch cells. Each cell is pre-loaded with a missile, positioned vertically within the submarine's structure. This configuration enables the submarine to deploy various missile types, including land-attack cruise missiles, anti-ship missiles, and submarine-launched ballistic missiles (SLBMs). The design ensures that missiles can be launched directly from the cells, eliminating the need to pass through torpedo tubes, which streamlines the launch process and maximizes efficiency.

Figure 86: The hatches of the 12 vertical-launch Tomahawk missile tubes stand open on the bow of the nuclear-powered attack submarine USS OKLAHOMA CITY (SSN-723). The U.S. National Archives, Public Domain PDM 1.0, via Getarchive.

One significant advantage of VLS is its versatility in accommodating different missile types. For example, it allows submarines to carry and launch Tomahawk cruise missiles for precision land strikes or SLBMs for nuclear deterrence missions. This capability makes VLS-equipped submarines highly adaptable, as they can be deployed in various strategic scenarios, ranging from tactical strikes to long-range strategic missions. The ability to launch multiple missiles in quick succession further enhances a submarine's effectiveness, enabling it to respond rapidly to multiple threats or targets. This rapid response capability ensures that submarines can engage hostile forces or targets without delay, maintaining a strategic advantage.

Another key benefit of VLS is the reduction in reload time. Traditional torpedo tube launches require complex reloading procedures, which can delay operations, particularly in combat situations. In contrast, VLS eliminates this bottleneck by pre-loading missiles in individual launch cells. This allows submarines to maintain continuous operational readiness, as missiles are always ready for immediate launch. In high-stakes scenarios, where speed and precision are crucial, VLS provides a decisive advantage by minimizing delays in missile deployment.

An example of VLS in action can be seen in the Ohio-class guided missile submarines (SSGNs) of the U.S. Navy. These submarines are equipped with VLS cells that can carry dozens of Tomahawk cruise missiles, allowing them to conduct precision strikes on land targets while remaining submerged. This capability underscores the strategic value of VLS, as it enables submarines to perform covert missions, launch surprise attacks, and remain undetected in hostile environments. The integration of VLS on Ohio-class SSGNs exemplifies the importance of this technology in modern naval warfare, highlighting its role in enhancing both offensive capabilities and operational flexibility.

Torpedo tube launch systems provide an alternative method for deploying missiles, especially on submarines that do not possess Vertical Launch Systems (VLS). These systems utilize the same tubes designed for torpedo launches, offering operational versatility without requiring major structural modifications. This dual-use capability allows submarines to deploy both torpedoes and missiles from a single launch system, maximizing space and ensuring tactical flexibility.

In these systems, missiles are loaded into torpedo tubes within waterproof canisters. Upon launch, the canister travels to the surface of the water, where it opens, and the missile's rocket motor ignites. This design enables submarines to remain submerged throughout the launch process, maintaining their stealth advantage by minimizing the risk of detection. The ability to use torpedo tubes for both missiles and torpedoes enhances a submarine's mission capabilities, allowing it to switch between anti-ship, anti-submarine, and land-attack roles with ease.

A notable example of this system in use can be seen in the U.S. Navy's Los Angeles-class attack submarines. These submarines are capable of launching Tomahawk cruise missiles from their Mk 48 torpedo tubes. This ensures that even without VLS, these submarines can carry out precision strikes on both land and sea targets, making them highly effective in a variety of combat scenarios. The use of torpedo tubes for missile deployment allows older or smaller submarines to retain significant offensive capabilities without the need for structural upgrades to incorporate VLS cells.

A further example of missile launch through the torpedo tubes is the Foxtrot-class submarine. The Foxtrot-class submarine was primarily designed and built by the Soviet Union during the Cold War era, with a focus on anti-surface warfare and anti-submarine operations. As a diesel-electric attack submarine, its primary offensive weapons were torpedoes rather than modern cruise or ballistic missiles. However, it had some capabilities to fire specialized missiles under certain circumstances through its torpedo tubes.

Figure 87: Russian Foxtrot Class Submarine B-39. Robert DuHamel, CC BY-SA 3.0, via Wikimedia Commons.

Primary Weapon Systems of Foxtrot-Class Submarines:

- **Torpedoes**: The primary weapon of the Foxtrot-class was its torpedoes, launched from ten torpedo tubes (six forward and four aft). The torpedoes could be configured for use against both surface ships and submarines, providing the main offensive punch.
  - **53-65K Torpedo**: An anti-ship homing torpedo with active/passive sonar, widely used in Soviet submarines.
  - **SET-65 Torpedo**: A wire-guided, electric torpedo primarily designed for anti-submarine warfare.
  - **Type 53-38/53-39**: Older conventional straight-running torpedoes, effective against surface ships.
  - **TEST-71 Torpedo**: A more advanced wire-guided torpedo with better homing capabilities.
- **Cruise Missile Capability**: While the Foxtrot-class submarine was not originally designed to carry dedicated cruise missiles, some variants and export models were reportedly modified to launch short-range cruise missiles via torpedo tubes. These modifications, however, were not standard across the class and depended on specific operational needs and upgrades.

  - **P-15 Termit (SS-N-2 'Styx')**: In limited cases, the submarine was capable of launching this anti-ship missile from its torpedo tubes. The P-15 was a relatively short-range missile, primarily intended for coastal defence.
  - **Kalibr (Club-S 3M-54 variant)**: There are unconfirmed reports that some heavily modified or refurbished Foxtrots (after being sold to other nations) experimented with launching Kalibr cruise missiles from torpedo tubes. However, this was not part of the submarine's original configuration.
- **Mines**: The Foxtrot-class could also be used for mine-laying operations, deploying naval mines through its torpedo tubes to block strategic waterways or restrict enemy vessel movement.

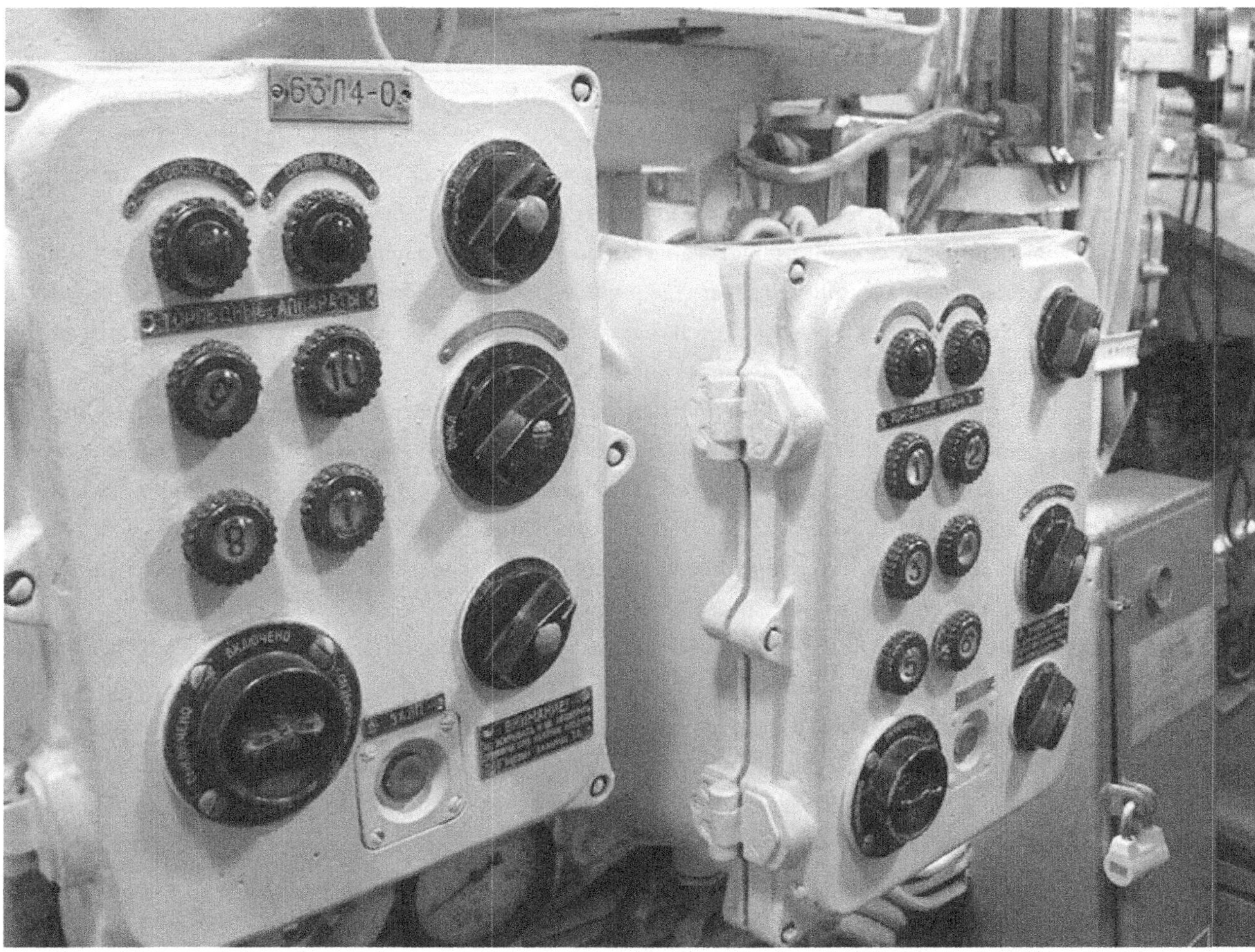

Figure 88: Torpedo launch control system on Foxtrot class submarines. Fastboy, CC BY-SA 3.0, via Wikimedia Commons.

One key advantage of torpedo tube launch systems is that they maintain missile capability without the need for specialized infrastructure. Submarines equipped with torpedo tubes can launch both torpedoes and missiles, giving commanders greater tactical flexibility. Additionally, the stealthy nature of torpedo tube launches ensures that the submarine can remain undetected throughout the process, as the missile deployment occurs from beneath the surface, preventing early warning or detection by enemy forces. This combination of versatility and stealth makes torpedo tube launch systems an essential asset for submarines that operate in complex and high-risk environments.

Guidance and Targeting Systems

Modern submarine missile systems are integral to contemporary naval warfare, leveraging advanced guidance and targeting technologies to enhance their operational effectiveness. These systems are designed to navigate complex environments, accurately strike designated targets, and adapt in real-time to changing battlefield conditions. The integration of sophisticated inertial navigation systems (INS) and Global Positioning System (GPS) technology is pivotal in ensuring that submarine-launched missiles can perform a variety of missions with high precision and reliability.

INS technology is particularly crucial for submarine-launched ballistic missiles (SLBMs), allowing them to determine their position and trajectory without reliance on external signals such as GPS. This capability is essential for strategic weapons that must operate independently, particularly in environments where GPS signals may be denied or jammed. INS utilizes gyroscopes and accelerometers to maintain accurate navigation from launch to impact, enabling SLBMs to strike intercontinental targets with exceptional accuracy [281]. The importance of INS in missile guidance is underscored by its ability to function autonomously, which is critical for maintaining the stealth and survivability of submarines during operations [284].

For cruise missiles, the incorporation of GPS guidance enhances precision by allowing these missiles to follow complex flight paths and navigate over diverse terrains. This technology is especially beneficial for land-attack missions, where missiles must avoid obstacles and accurately hit high-value targets. The ability to maintain accuracy over long distances and through varying landscapes is a significant advantage provided by GPS integration [284]. Furthermore, the use of advanced guidance algorithms, such as those based on Bézier curves, allows for optimized trajectories that improve the effectiveness of cruise missiles during terminal phases of flight [284].

In the realm of anti-ship missiles, guidance systems often employ both active and passive radar homing techniques to engage moving maritime targets effectively. Active radar homing involves the missile emitting radar waves to detect and track targets, while passive radar homing relies on reflected signals from the target itself. This dual approach enables anti-ship missiles to lock onto vessels in motion, ensuring effective strikes despite the dynamic nature of naval combat [285]. The development of radar stealth technologies further complicates the engagement landscape, as modern anti-ship missiles must contend with countermeasures designed to reduce their detection probability [286].

Moreover, data links significantly enhance the performance of modern submarine-launched missiles by facilitating mid-course corrections. These links allow missiles to receive real-time updates from the launching submarine or other assets, enabling them to adjust their course in response to target movements or changing conditions. This capability is crucial for engaging mobile or evasive targets, thereby increasing the likelihood of successful strikes [287]. The seamless integration of these guidance and targeting technologies ensures that submarine-launched missiles remain highly effective across multiple operational domains, reinforcing their role as critical components of a submarine's offensive capabilities.

Integration with Combat Management Systems (CMS)

The Combat Management System (CMS) is integral to the successful operation of missile systems on modern submarines, functioning as the central hub for managing data, coordinating weapon deployment, and ensuring mission success. This sophisticated system integrates inputs from a variety of sensors, including sonar and radar, to provide a real-time tactical overview. Through this process, the CMS enables the submarine's crew to make informed decisions regarding missile launches, target tracking, and overall mission execution.

The CMS plays a crucial role in coordinating missile launches, particularly in complex scenarios where multiple targets or threats may need to be addressed simultaneously. The system identifies and selects the appropriate missile type based on mission parameters, such as the nature of the target and environmental conditions. Before launch, each missile is programmed with specific mission data, including waypoints, trajectory paths, and target characteristics. This ensures that the missile is pre-configured to follow a precise flight path toward its objective without requiring constant manual intervention during its journey.

In many cases, the CMS offers the capability to communicate with the missile while it is in flight. Through encrypted data links, the submarine can send real-time updates to the missile, allowing it to adjust its trajectory if the target changes course or new intelligence becomes available. This mid-course correction feature is particularly valuable in fast-evolving tactical situations, as it enhances the missile's accuracy and effectiveness.

Furthermore, the CMS ensures that all missile operations are conducted efficiently and safely. By managing sensor data and integrating it with missile guidance systems, the CMS minimizes the chances of friendly fire and maximizes the likelihood of hitting the intended target. The system's ability to coordinate simultaneous missile launches from different platforms also enables the submarine to carry out complex strike missions with precision and speed. Overall, the CMS is a critical component in modern submarine warfare, ensuring seamless integration between sensors, weapons, and crew decisions to achieve tactical superiority.

Nuclear and Conventional Capabilities

Modern submarines are equipped with a diverse range of missile capabilities, allowing them to operate effectively in both nuclear and conventional warfare scenarios. This dual capability enhances the strategic flexibility of navies, providing options for both deterrence and tactical engagement. Submarines capable of launching both types of missiles play critical roles in national defence strategies, from maintaining nuclear deterrence to supporting conventional military operations.

Submarine-Launched Ballistic Missile (SLBM) platforms, such as SSBNs (nuclear-powered ballistic missile submarines), are primarily designed for nuclear missions. These vessels form the backbone of a nation's strategic deterrence, ensuring the capability to respond to a nuclear attack with a retaliatory strike. The missiles carried by SSBNs are typically long-range, intercontinental weapons with multiple independently targetable reentry vehicles (MIRVs), allowing a single missile to carry several nuclear warheads aimed at different targets. SSBNs maintain a continuous at-sea presence, ensuring that their nuclear capabilities remain operational even in the event of an enemy attack on land-based nuclear forces.

In contrast, attack submarines (SSNs) and guided-missile submarines (SSGNs) are more versatile, capable of launching a variety of conventional missiles alongside potential nuclear payloads. These submarines are often equipped with land-attack cruise missiles like the Tomahawk or Kalibr, allowing them to conduct precision strikes against military infrastructure, enemy command centers, and high-value targets onshore. The ability to launch anti-ship missiles also enables these submarines to engage enemy vessels at sea, providing tactical superiority in maritime combat scenarios.

The dual capability of these submarines offers a significant advantage to military planners. In peacetime, the presence of nuclear-armed SSBNs serves as a powerful deterrent, discouraging potential adversaries from considering nuclear or large-scale conventional aggression. In times of conflict, SSNs and SSGNs offer the flexibility to engage in conventional warfare, conducting covert attacks on enemy targets or supporting broader naval operations without escalating to nuclear confrontation. This layered approach to submarine deployment ensures that navies can respond appropriately across the spectrum of conflict, from conventional skirmishes to strategic nuclear deterrence.

Ultimately, the combination of nuclear and conventional missile capabilities in modern submarines ensures that these vessels remain relevant and effective tools in both wartime and peacetime scenarios. Their ability to adapt to various operational requirements—whether it be maintaining nuclear deterrence or supporting conventional military operations—makes them essential assets for navies worldwide, enhancing both strategic defence and tactical offensive capabilities.

Sonar and Underwater Detection Technologies

SONAR (Sound Navigation and Ranging) is a critical technology employed by submarines for underwater detection and navigation. It functions as both the "eyes" and "ears" of the submarine, particularly when submerged, as traditional methods like radar become ineffective and periscopes have limited utility. Submarines rely heavily on sonar operators to interpret sound data, especially when it is too dangerous to raise the periscope near enemy vessels. These operators provide essential information about nearby ship movements, helping the submarine navigate and avoid detection [288].

The primary purpose of SONAR (Sound Navigation and Ranging) in submarines is to detect, identify, and track objects and vessels underwater, ensuring effective navigation, surveillance, and combat capabilities. Submarines operate in environments where visual detection is limited or impossible due to water opacity and depth. Unlike radar, which relies on electromagnetic waves, SONAR uses sound waves to navigate and detect objects, as sound travels much farther and faster in water. Modern submarines rely heavily on SONAR systems to maintain stealth, avoid collisions, locate enemy vessels, and conduct search-and-rescue missions.

When a submarine is submerged, traditional instruments like periscopes or radar become less effective, making SONAR the primary tool for situational awareness. Submarines depend on these systems not only to detect other vessels but also to monitor their own noise output to remain undetected. SONAR provides real-time acoustic data, helping the crew assess threats and maintain operational safety.

All ships produce sounds as they move through the water. The most noticeable sound to a sonar operator is the noise generated by the propellers, which churn through the water, producing both sonic and supersonic frequencies. Machinery noises inside a ship also radiate through the hull and into the surrounding water, further aiding detection. Even when a vessel is stationary, subtle sounds—such as waves slapping against the hull—can be detected. The primary objective of sonar equipment on submarines is to capture and interpret these underwater sounds to assess the presence and movement of enemy vessels.

Modern SONAR systems are advanced, employing a combination of active and passive SONAR technologies to enhance detection capabilities while maintaining stealth. These systems consist of multiple components, including hydrophones, transducers, processing units, and sophisticated software that interprets sound data.

Passive SONAR listens for sound waves produced by other vessels or underwater objects without emitting any signals. This method allows the submarine to remain undetected, as it avoids generating acoustic signals that could reveal its location. Passive systems are highly effective for identifying the distinct sounds of propellers, engines, and machinery noises from enemy vessels.

The hydrophones installed on the submarine's hull pick up these underwater sounds and convert them into electrical signals. These signals are analysed using sophisticated software, which identifies the source based on the unique acoustic signature of each vessel. Passive SONAR is especially useful for submarines on stealth missions, as it allows them to gather information without alerting other vessels to their presence.

Active SONAR emits sound pulses (or "pings") into the water and listens for echoes that bounce back after hitting an object. This system provides precise information about the range, bearing, and speed of detected targets. It is particularly useful in locating enemy vessels or objects that do not produce any detectable noise.

When the submarine emits a sound wave, it travels through the water until it strikes an object. The reflected wave, or echo, returns to the submarine, where it is captured by hydrophones and processed. The time delay between the emitted sound and the returning echo allows the system to calculate the distance to the target. Active SONAR is often used during searches or manoeuvres in complex environments but is used sparingly in combat scenarios, as it can reveal the submarine's position.

The core components of sonar systems include hydrophones, cables, receiver-amplifiers, headphones, and training mechanisms. Hydrophones, the underwater microphones, are strategically placed on the submarine—both on the topside and beneath the keel. These devices convert underwater sound waves into electrical signals that travel through cables to the receiver-amplifiers, where operators can analyse the sounds through headphones. Training mechanisms allow the hydrophones to rotate, enabling operators to focus on sound sources in specific directions.

Modern SONAR systems integrate several advanced components to enhance both active and passive operations:

- **Hull-Mounted SONAR**: Arrays of hydrophones and transducers installed along the hull to detect sounds in all directions.

- **Towed Array SONAR**: A series of hydrophones deployed on a long cable, trailing behind the submarine. This system reduces self-noise interference, providing better detection capability at long ranges.

- **Flank Array SONAR**: Installed on the sides of the submarine, offering enhanced tracking capability across multiple bearings.

- **Sonobuoys**: Small, deployable sonar units that provide additional coverage. These can be released by submarines, aircraft, or ships for area surveillance.

Sonar systems can be divided into two main types: sonic and supersonic gear. Sonic sonar captures sounds within the human hearing range (below 15,000 cycles per second) and allows operators to hear natural underwater sounds. This equipment is mounted topside and is particularly useful for detecting distant noises. Operators can use sonic gear to identify the sounds of enemy machinery, propellers, and even their own submarine's noises, ensuring operational stealth. Supersonic sonar, on the other hand, detects high-frequency sounds beyond human hearing. These sounds, often used in enemy sonar systems, are converted into audible signals for the operator. Supersonic systems are essential for identifying sonar pings emitted by enemy ships searching for submarines [288].

The Weapon Control Acoustic (WCA) installation onboard a submarine manages all the supersonic sonar systems. This complex setup consists of multiple projectors and receiver-amplifiers positioned below the keel, which are controlled from the conning tower. The WCA installation also includes QB and JK/QC systems—each designed to send and receive underwater signals. These systems enable not only listening but also the emission of sonar signals to determine the range of enemy vessels. Additionally, NM gear integrated within the WCA setup helps measure the depth of water beneath the submarine's keel, ensuring safe navigation.

WCA stands for Weapon Control Acoustic, referring to a sonar-based system used in submarines for acoustic detection, tracking, and weapon guidance. It integrates supersonic listening, echo-ranging, and depth-sounding capabilities to provide critical information for navigation, surveillance, and combat operations.

WCA systems enable submarines to maintain stealth while actively or passively monitoring their surroundings. These systems help detect enemy vessels, guide weapons (like torpedoes), and provide vital data on underwater terrain and depth, supporting both tactical and navigational needs in combat and peacetime scenarios.

Sound behaves differently underwater compared to air, traveling much faster—at approximately 4,800 feet per second [288]. However, the speed and behaviour of sound waves are influenced by factors such as water temperature, pressure, and salinity. These variables can cause refraction, where sound waves bend away from their original path. Sonar operators must understand these principles to interpret signals accurately. As sound waves travel through water, they lose strength due to spreading and attenuation. Spreading occurs as sound waves radiate outward, covering more area and losing intensity. Attenuation results from obstacles like air bubbles, seaweed, or marine life, which absorb or scatter the sound waves, weakening the signal. Notably, higher-frequency supersonic sounds attenuate more quickly than lower-frequency sonic sounds, limiting their range.

Ships produce both sonic and supersonic sounds, making it necessary for submarines to employ both types of listening gear. Propellers generate a broad spectrum of frequencies detectable by either system, while machinery sounds are mostly within the sonic range. Sonic gear is crucial for monitoring both enemy vessels and the submarine's own noise emissions to maintain stealth. Meanwhile, supersonic gear excels at capturing the high-frequency sonar signals sent out by enemy escort vessels during submarine searches [288].

Figure 89: Sonar Technician (Submarines) 2nd Class Harlie Williams III, takes sonar readings aboard the Los Angeles-class attack submarine USS Newport News (SSN 750). The U.S. National Archives, Public Domain PDM 1.0, via GetArchive.

Sonic Listening

Sonic listening systems, like the JP gear used in submarines, are essential tools for detecting, identifying, and tracking sounds in the underwater environment. These systems provide the submarine crew with critical auditory information about nearby vessels, both friend and foe, and environmental sounds such as waves and marine life. The JP gear functions as an advanced listening tool, helping submarines maintain situational awareness, navigate silently, and conduct operations without being detected [288].

JP gear refers to specialized sonic listening equipment used on submarines for underwater sound detection. It serves as the submarine's primary passive sonar system, designed to detect, amplify, and analyse underwater sounds without emitting any signals, ensuring the submarine maintains stealth. The system allows the crew to gather critical information about enemy ships, sonar pings, environmental sounds, and even their own submarine's machinery noises.

JP gear is primarily used for listening to underwater sounds that occur naturally or are generated by ships, submarines, or torpedoes. Unlike active sonar, which sends out sound waves and listens for echoes, JP gear relies entirely on detecting existing sounds in the water. This passive sonar capability makes it ideal for stealth operations, enabling submarines to remain undetected while gathering information on the position and movement of other vessels.

The gear is particularly important when the submarine is operating in hostile waters, where using active sonar or periscopes could compromise its position. Instead, the crew relies on JP gear to track enemy ships silently, identify the type of vessel from its machinery noise, or detect the telltale pinging sounds from enemy sonar systems.

The hydrophone is a crucial element of the JP gear. It functions like an underwater microphone, picking up sound waves that pass through water. The hydrophone contains a long metal tube that changes size slightly when struck by sound waves. This deformation creates an electric current in the wires coiled around its wooden core. The hydrophone is designed to be sensitive to sounds from the front, while a rubber bale shields it from noise at the back, ensuring focused listening.

The training mechanism allows the operator to rotate the hydrophone in different directions. The hydrophone is mounted on a shaft that extends through the submarine's pressure hull, with a handwheel controlling its movement. This mechanism enables the operator to sweep across a full 360-degree range to detect sounds, with the target's relative bearing displayed on a circular scale.

An amplifier is used to strengthen the electrical signals generated by the hydrophone, making faint sounds more audible. The output from the amplifier can be transmitted through headphones or loudspeakers, allowing the sonar operator to interpret the sounds and report findings to the conning tower. The JP gear is powered by the submarine's batteries, ensuring silent operation and functionality even if the main power supply is compromised during combat or depth charge attacks.

The JP amplifier boosts the signals from the hydrophone through several stages. As the current passes through various amplifiers and filters, the system selectively enhances specific frequencies, allowing the operator to isolate low-frequency sounds like ship machinery or high-frequency sounds like propeller beats. Filters help the operator focus on distinct sounds by eliminating background noise, while the "magic eye" indicator visually shows the strength of detected signals, aiding in precise bearing readings.

During routine searches, the operator performs progressive sweeps with the hydrophone, scanning the area for any suspicious sounds. If a contact is detected, the operator must immediately determine the reciprocal bearing—checking if the sound comes from the front or back of the hydrophone—ensuring accurate direction reporting. This process helps the submarine crew track targets like enemy ships or torpedoes, providing vital information for navigation and attack strategies.

Submarines encounter a variety of sounds underwater. JP gear helps the operator identify sounds produced by enemy ships, such as rhythmic propeller beats or machinery hums, as well as pinging sounds from enemy sonar systems. Environmental sounds like the pounding surf or noises made by marine animals are also picked up, adding to the operator's situational awareness. Recognizing familiar sounds from the submarine's own machinery helps the crew identify unwanted noises that could compromise their stealth.

When a suspicious sound is detected, the sonar operator must quickly report the contact and adjust the amplifier controls to sharpen the target's sound. Accurate bearings are critical, requiring the operator to sweep through the sound several times to ensure precise tracking. Using filters and volume controls effectively narrows the arc of the sound, making it easier to pinpoint the source. Continuous reporting of the target's movements allows the conning officer to plan evasive actions or coordinate an attack [288].

During evasive manoeuvres, JP gear helps the crew monitor enemy movements and detect any changes in the soundscape, such as a new ship entering the area. The bass-boost filter can enhance low-frequency sounds, such as the hum of motors, aiding in detection even during silent running or when the submarine is resting on the ocean floor.

After combat or depth charge attacks, the hydrophone may need to be re-magnetized to restore its sensitivity. The operator can use the magnetizer jack and push-to-magnetize button to re-energize the hydrophone. Once the submarine surfaces, the JP gear is secured to prevent damage. The power is switched off, the hydrophone is trained to a neutral position, and the specialized headphones are carefully stored for future use, as replacements may not be available during a patrol [288].

If the JP gear detects enemy echo-ranging—the characteristic pings from active sonar systems—the operator must report the bearing and timing of the pings immediately. The time interval between pings, known as the scale, provides valuable clues about the enemy vessel's distance and intentions. Rapid reporting allows the submarine's crew to make informed decisions, whether to continue tracking the enemy or initiate evasive actions.

Supersonic Listening and WCA System in Submarines

The WCA system on submarines plays a crucial role in supersonic listening, echo-ranging, and depth-sounding. It enhances the submarine's ability to detect, locate, and track enemy vessels while remaining submerged. This system comprises several key components, including projectors, receiver-amplifiers, and remote-control units, which work together to facilitate precise underwater sound detection.

The WCA system includes two primary types of projectors: the QB projector and the JK/QC combination projector. The QB projector, mounted on the starboard side, utilizes Rochelle salt crystals, which change shape when sound waves strike them, generating an electric current. The JK/QC projector, located on the port side, features two faces: the JK face operates similarly to the QB face, while the QC face contains nickel tubes that alter their size upon exposure to sound waves, producing electrical signals. These projectors serve as the submarine's primary means of receiving underwater sounds.

The WCA system is composed of several subsystems designed to cover different aspects of sonar operations, including QB, JK/QC, and NM components. Each of these has specific roles in gathering acoustic data and maintaining situational awareness:

- **QB System**: This subsystem is responsible for passive listening. It detects enemy vessels' propeller noises, machinery sounds, and other underwater activities. It operates on supersonic frequencies, converting them into audible sound for the sonar operator. The QB projector, mounted on the starboard side, uses Rochelle salt crystals to generate electric signals when sound waves hit the projector face.

- **JK/QC System**: The JK/QC projector serves a dual function. It passively listens, like the QB system, but also actively sends out sound waves for echo-ranging operations. The JK face focuses on receiving sounds, while the QC face is used for transmitting pings to determine the range and bearing of underwater objects, including enemy vessels. This system also contributes to depth-sounding operations.

- **NM System**: The NM subsystem is used primarily for depth-sounding. It measures the depth of the water beneath the submarine's keel by sending sound waves downward and measuring the time it takes for the echo to return. This helps submarines avoid running aground or colliding with underwater obstacles.

The weak electric signals generated by the projectors are sent to receiver-amplifiers for amplification. These units convert the faint signals into audible sounds, making them suitable for analysis by sonar operators. Each projector has a dedicated receiver-amplifier located in the conning tower, with backup units available in the forward torpedo room for emergency use. Remote-control units in the conning tower control the projectors, allowing operators to rotate them in precise directions to detect sound sources. These units also display the direction in which each projector is trained, aiding in navigation and target tracking.

The projectors are mounted on shafts equipped with hydraulic mechanisms for raising and lowering them, operated by torpedomen in the forward torpedo room. These shafts also have electric motors and reduction gears for precise rotation. The training shafts are limited in rotation to protect cables, though some installations include slip rings, allowing continuous rotation without cable damage. These hoist and training mechanisms are critical for maintaining proper projector alignment during operations.

During routine operations, only one set of supersonic gear, typically the QB system, is manned. When a contact is detected, both QB and JK/QC systems are activated. The QB system focuses on tracking the primary target, while the JK/QC system scans for other vessels. This dual operation ensures comprehensive monitoring of the surrounding environment and provides essential data to support strategic decisions during combat.

The receiver-amplifier converts incoming signals from the projectors into audible sounds for analysis. It features multiple amplification stages and filters, including broad-band and peak filters, to enhance specific sound frequencies. The system also incorporates a magic eye indicator, which visually confirms the strength of the detected signal. Operators rely on the tuning controls, gain settings, and filter options to isolate target sounds and eliminate background noise effectively.

Operators use rapid and progressive search techniques to detect and track enemy vessels. In rapid search, the operator sweeps the projectors in a full circle to locate any suspicious sounds. If no contacts are detected, the operator shifts to progressive search, sweeping the projectors forward and backward in small increments to ensure comprehensive coverage. Additionally, operators conduct frequency sweeps every 15 minutes to detect enemy sonar pings, adjusting the frequency to maximize detection.

When a sound contact is made, the operator quickly determines the bearing and reports it to the conning officer. The amplifier controls are adjusted to enhance the clarity of the target sound, and the operator continuously reports bearing changes and updates on the target's movement. Operators also attempt to identify the target based on the characteristics of the propeller sounds, such as speed, rhythm, and tone, providing critical information for tactical decisions.

Accurate bearings are essential for effective target tracking. Operators achieve this by sweeping through the full arc of propeller noise and adjusting their readings as the target moves. They focus on reading bearings from the target's bow through its stern, ensuring precise tracking. Tuning controls and filters help narrow the arc of sound, while low gain settings reduce background noise, making it easier to pinpoint the target's location.

Once the submarine surfaces, operators may continue to use WCA gear at slow speeds. However, at higher speeds, the noise generated by the submarine can interfere with listening conditions, prompting the operator to secure the gear. The projector shafts are returned to their designated positions, and the equipment is powered down. Operators carefully unplug and store the headphones, which are essential for continued operations and cannot be easily replaced during a patrol.

Single-Ping Echo-Ranging in Submarines

Single-ping echo-ranging is a specialized sonar technique used by submarines to determine the precise range to a target. Unlike continuous echo-ranging, which would compromise the submarine's stealth, this method involves emitting a single burst of supersonic sound, called a "ping," and measuring the time it takes for the echo to return from the target. The returning echo provides accurate data on the target's distance. This method is usually employed when the submarine's conning officer needs a more accurate range than what can be achieved visually through the periscope [288].

When the sonar operator initiates a single-ping range, the sonar system directs a sharp burst of sound toward the target. The operator distinguishes the main echo from a series of smaller reverberations that may arise due to environmental factors. Identifying the primary echo is crucial to obtaining an accurate reading. Submarines rely on specialized sonar equipment to perform this operation. This includes projectors that emit the ping, drivers that generate the sound pulse, and range indicators that translate the time elapsed between the ping and echo into distance, displaying it directly in yards.

The drivers, located in the forward torpedo room, play a critical role in single-ping operations. These drivers send the electrical pulse to the projectors, which convert the signal into a supersonic sound wave. Two drivers are usually present: one connected to the QB projector, and the other switchable between the QC and NM projectors. While the QB and QC projectors handle sonar ranging, the NM projector focuses on echo-sounding to determine the depth beneath the submarine's keel [288].

The range indicator acts as an electric stopwatch, measuring the exact time interval between the ping and the echo. The indicator's dial is calibrated in yards, allowing the sonar operator to read the distance directly. There are various models of range indicators, such as WCA, WCA-1, and WCA-2, each with different control interfaces. Earlier models required screwdriver adjustments, while newer ones feature hand switches and run entirely on the ship's AC power.

In most situations, the QB projector is preferred for single-ping operations due to its superior sensitivity and ease of use. It can detect weaker echoes than other projectors and does not require precise tuning, unlike the QC system. However, in certain cases—specifically when the target is between 250 and 290 degrees relative—the QC projector is necessary because the QB projector's field of detection is obstructed by the JK/QC projector [288].

The sonar operator can use either the audible or visual method to interpret the echo. In the audible method, the operator listens through headphones for the return echo, adjusting the gain and tuning controls to ensure clarity. This method is most reliable for ranges beyond 1500 yards. In the visual method, typically used at closer distances, the operator relies on red flashes on the range indicator's dial. The arc of red light corresponds to the returning echo, with the trailing edge of the flash marking the precise range. While the visual method can be useful, it carries some risks, such as the possibility of confusing reverberation flashes or missing weak echoes due to insufficient power [288].

Continuous echo-ranging, often employed by surface escort ships, is avoided by submarines because it would reveal their position to the enemy. Submarines aim to maintain stealth at all times, and a continuous stream of pings would make them vulnerable to detection. Therefore, submarines rely on single-ping echo-ranging only when necessary. This method provides sufficient range data for tactical decisions while minimizing the risk of compromising the submarine's position.

Echo-Sounding in Submarines

Echo-sounding is a critical technique used by submarines to measure the depth of water beneath the keel. While the waters where submarines patrol are generally well-charted, specific operational conditions may require precise, real-time depth measurements. Echo-sounding provides this information by emitting sound waves downward to the seabed, with the ocean floor acting as the reflecting surface. This process mirrors traditional sonar echo-ranging, but instead of projecting sound horizontally, the sound is directed vertically beneath the submarine.

The primary component of the submarine's echo-sounding system is the NM projector, a steel-encased device that resembles a small coffin. This projector is securely bolted to the bottom of the submarine within the forward trim tank, allowing it to function seamlessly without disrupting the vessel's hydrodynamics. The NM projector operates using the same driver unit as the QC sonar system, ensuring efficient power management. When activated, it emits a brief burst of sound downward, which reflects off the ocean floor and returns to the submarine. The time it takes for the echo to return indicates the depth of the water beneath the submarine.

The depth measurements are displayed on an instrument called the fathometer, located in the control room. The fathometer functions similarly to the range indicator used for horizontal echo-ranging, but it is calibrated in fathoms rather than yards. Inside the fathometer is a receiver-amplifier, which processes the returning echoes and provides accurate depth readings to the crew. The fathometer's straightforward design allows the operator to monitor changing depths in real time, which is essential for safe navigation, particularly in unfamiliar or shallow waters.

To conduct an accurate depth sounding, the sonar operator must set specific switches on the system, which vary based on the estimated depth. For waters less than 300 fathoms deep, the visual mode is typically used, where depth readings appear directly on the fathometer's indicator. In deeper waters, between 300 and 1800 fathoms, the system switches between visual and audible modes, requiring the operator to listen for the echo through headphones. This method ensures that even faint echoes from greater depths can be detected. Additionally, a shoal-deep switch helps the system optimize its performance depending on the expected water depth, ensuring precise readings across varying depths.

While echo-sounding is essential for safe navigation, especially in uncharted or shallow waters, it carries inherent risks. The sound waves emitted during the process can potentially be detected by enemy vessels, compromising the submarine's stealth. Therefore, operators are trained to conduct depth soundings only when absolutely necessary. Once the depth reading is obtained, the signal interval switch must be returned to its neutral position immediately to prevent further pings from revealing the submarine's presence.

Echo-sounding offers submarines a reliable means of determining water depth, ensuring safe passage and manoeuvring, especially in unpredictable environments. However, the technique must be used cautiously, balancing the need for accurate depth information with the imperative to maintain the submarine's covert status. This combination of precision and discretion is critical to effective submarine operations.

Signal Processing and Analysis

Once the submarine's SONAR system collects sound data through its hydrophones or projectors, the raw data undergoes a sophisticated signal processing phase. This process is essential for transforming large volumes of acoustic information into actionable intelligence. Advanced algorithms and artificial intelligence (AI) play a critical role in analysing the data, helping to distinguish between meaningful signals and irrelevant noise, such as marine life, ocean currents, or ambient sounds. This filtering is crucial because underwater environments are filled with complex acoustic interference, and without precise analysis, vital information about enemy ships or threats could be missed.

The initial phase of signal processing involves noise reduction. Algorithms isolate the specific frequency patterns associated with vessels or machinery, filtering out non-relevant background noise. This phase ensures that the sounds from propellers, engines, or enemy SONAR pings stand out clearly. The system also uses frequency analysis to classify these sounds, identifying whether the source is a surface vessel, submarine, or environmental factor. Some SONAR systems employ machine learning models that have been trained with vast datasets, allowing the system to recognize known acoustic signatures, such as specific ship types or classes, with high accuracy.

After filtering and classification, the next phase of signal processing involves tracking and predicting the target's movement. Algorithms analyse the change in frequency and amplitude of the detected sounds over time, calculating the speed and direction of the target. This predictive capability is essential for navigation and evasion, giving the crew valuable insights into the trajectory of enemy vessels and allowing them to take pre-emptive actions, such as changing course or deploying countermeasures. This information also plays a critical role in guiding torpedoes and missiles, ensuring they reach their intended targets even as the target manoeuvres.

The processed data is then integrated into the submarine's Combat Management System (CMS), which serves as the central platform for coordinating the submarine's tactical operations. The CMS combines information from the SONAR system with inputs from other sensors, such as periscopes, radar, and electronic support measures (ESM), to present the crew with a comprehensive real-time tactical picture. This consolidated view allows the crew to assess the situation quickly, facilitating informed decision-making during critical moments.

The CMS displays key metrics, such as the relative position, speed, and classification of potential threats, in an intuitive interface. It provides alerts and recommendations based on the analysed data, helping the crew decide on the best course of action. Whether it involves navigating through hostile waters, evading enemy detection, or engaging with weapons, the processed SONAR data ensures that every decision is based on the most accurate and up-to-date information available.

This seamless integration of signal processing, analysis, and CMS coordination is vital for the submarine's operational efficiency. In high-pressure situations, such as enemy engagements or evasive manoeuvres, the ability to quickly interpret and act on SONAR data can mean the difference between mission success and failure. Through these advanced processes, modern submarines gain a tactical edge, enhancing their stealth, accuracy, and survivability in increasingly complex underwater environments.

SONAR Countermeasures

Modern submarines employ sophisticated countermeasures to protect themselves from enemy SONAR detection, ensuring their stealth and survivability in hostile environments. One of the primary methods involves deploying acoustic decoys, which are designed to replicate the submarine's sound signature. These decoys emit sounds that closely mimic the noise of a submarine's propellers or onboard machinery, creating false targets to confuse enemy vessels or homing torpedoes. The enemy SONAR system may struggle to differentiate between the real submarine and the decoy, allowing the actual submarine to slip away undetected or evade incoming torpedoes. This tactic is especially effective when the decoy is strategically deployed at a distance, diverting attention from the submarine's actual location.

Another crucial countermeasure used by submarines is jamming. Jammers disrupt the transmission and reception of sound waves by creating interference patterns in the acoustic spectrum. When activated, these jammers emit a barrage of noise that overwhelms the enemy's SONAR systems, making it difficult for them to interpret the incoming signals accurately. This interference can prevent enemy vessels from acquiring a reliable lock on the submarine, forcing them to rely on less accurate detection methods or abandon pursuit altogether. Jamming can also hinder active SONAR systems, which rely on sound pulses to locate objects underwater, by distorting the returning echoes and rendering them unusable for tracking or targeting.

These countermeasures work in conjunction with the submarine's evasive manoeuvres, enhancing its ability to remain undetected. For example, the submarine may deploy decoys while simultaneously changing course, further complicating the enemy's efforts to track it. Additionally, some countermeasure systems are integrated with the submarine's combat management system (CMS), ensuring that decoys and jammers are activated at precisely the right moment during engagements. This coordination allows the submarine to respond dynamically to evolving threats and maximize the effectiveness of its defences.

SONAR countermeasures not only play a critical role in torpedo defence but also help submarines maintain tactical advantages during operations. By undermining the enemy's detection capabilities, these measures allow the submarine to operate freely in contested waters, conduct reconnaissance, or launch attacks without revealing its presence. This capability is particularly valuable in modern naval warfare, where maintaining stealth is essential for both strategic missions and survival.

As technology advances, countermeasure systems are becoming more sophisticated, with modern submarines equipped to deploy decoys and jammers autonomously or in response to automated threat detection. Future developments in countermeasures are likely to incorporate artificial intelligence, enabling submarines to predict enemy detection patterns and deploy countermeasures pre-emptively. These innovations will further enhance the submarine's ability to operate covertly and avoid detection in increasingly challenging environments.

Submarine Design and Technical Infrastructure for Sonar Operations

Effective sonar operations in submarines require a complex interplay of design considerations, technical infrastructure, and advanced equipment. These systems ensure the submarine's capability to navigate, detect threats, avoid obstacles, and maintain situational awareness in the ocean's depths. Sonar operations encompass several key components, including echo-sounding, single-ping echo-ranging, supersonic listening, sonic listening, and the WCA system. Each of these capabilities demands specific design and technological features integrated within the submarine's structure.

Echo-sounding, used to measure the water depth beneath the keel, requires specialized equipment installed directly within the submarine's hull. The NM projector, essential for this task, is securely mounted flush to the hull within the forward trim tank. This placement ensures that the sound waves emitted for depth measurement are unobstructed. The NM projector operates in coordination with the fathometer, which is housed in the control room. The fathometer must be readily accessible to the crew, as it provides real-time depth data that is essential for safe navigation. To maintain stealth, the design must allow operators to control ping emissions precisely, limiting the risk of detection by enemy vessels.

Single-ping echo-ranging systems require precise alignment between projectors, drivers, and range indicators. The QB and QC projectors, housed on separate shafts, are positioned on opposite sides of the submarine's keel to ensure optimal coverage. These systems need to be mounted in a way that allows for the rotation and elevation of the projectors, operated by training motors within the forward torpedo room. The range indicators are installed within the conning tower to facilitate quick access by sonar operators during critical moments. The submarine's control systems must be designed to allow seamless switching between audible and visual modes to detect and measure distances effectively, depending on environmental and operational conditions.

Supersonic listening capabilities demand robust infrastructure, particularly through the WCA system, which manages high-frequency sonar operations. The QB and JK/QC projectors are essential components of the WCA system, positioned strategically on either side of the keel to avoid interference. Each projector's position ensures that both sides of the submarine are covered effectively, with minimal blind spots. Hydraulic mechanisms for hoisting and lowering the projectors, as well as slip rings to prevent cable tangling, are critical to maintaining operational flexibility. The remote-control units and receiver-amplifiers, typically located in the conning tower, allow operators to control the orientation and output of the projectors efficiently.

The WCA system must also provide clear channels for communication between the projectors and the control units. These systems require dedicated motor generators for power, ensuring they function reliably even during silent operations when the submarine's engines are off. In the event of emergencies, spare receiver-amplifiers and training mechanisms installed in the forward torpedo room offer redundancy to maintain listening capabilities.

Sonic listening systems rely on hydrophones installed at multiple points on the submarine's hull to capture lower-frequency sounds. These hydrophones are connected to amplifiers through shielded cables, with the amplifiers housed in the forward torpedo room or the control room. Sonic listening systems provide critical insights into enemy ship movements by detecting propeller noise, engine vibrations, and machinery sounds. The submarine's design must ensure that hydrophones are placed to minimize noise interference from the vessel's own machinery, particularly when running silently.

The sonic listening system also requires well-calibrated amplifiers, headphones, and remote control units to enable precise sound detection and analysis. Operators depend on these systems to differentiate between friendly and hostile vessels, making training and search procedures essential to their use. Additionally, the design must ensure that these components are accessible and functional even if the submarine sustains damage, with batteries providing backup power to maintain operational readiness.

Sonar operations demand careful power management to avoid generating excess noise that could reveal the submarine's position. Many sonar systems, such as the JP sonic listening gear and WCA components, draw power from the submarine's battery systems, ensuring quiet operation during stealth missions. The design of the submarine must account for sound insulation, preventing noise from internal machinery from interfering with sonar operations. Additionally, advanced filtering systems must be integrated into the sonar equipment to eliminate background noise and enhance signal clarity. This is particularly important for detecting faint or distant signals and distinguishing them from oceanic noise.

The conning tower serves as the nerve centre for sonar operations, housing critical control units and interfaces. Sonar operators stationed in the conning tower must have access to remote-control units, range indicators, and amplifiers to manage the submarine's listening and ranging activities. The arrangement of these systems within the conning tower must allow for efficient teamwork between operators, with quick access to all necessary controls. The integration of sonar operations with the submarine's combat management system ensures that sonar data feeds directly into tactical decision-making processes, providing vital information for navigation, target tracking, and weapon deployment.

Given the importance of sonar systems to submarine operations, redundancy is built into the design to ensure continued functionality under all conditions. Spare components, such as receiver-amplifiers and control units, are stored in the forward torpedo room, allowing operators to maintain sonar capabilities even if the conning tower is compromised. Hydraulic and motorized systems for training the projectors are designed to operate manually in emergencies, ensuring that the submarine can continue to detect and respond to threats.

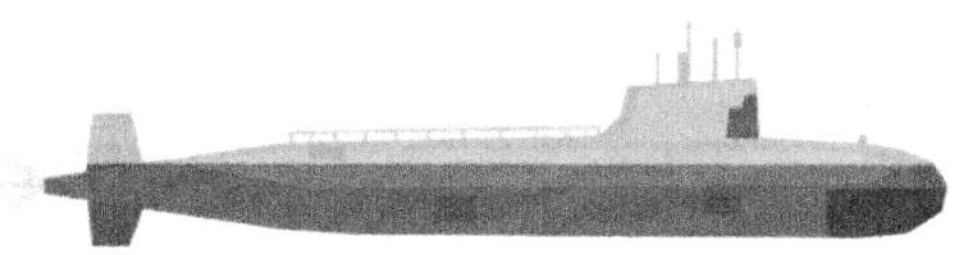

Chapter 8

Navigation and Communication Systems

Submarine Navigation Equipment

Submarine navigation underwater is a highly specialized operation that relies on a combination of advanced technologies and meticulous human expertise. Navigating beneath the ocean surface presents unique challenges that surface vessels do not encounter. Submarines operate in environments devoid of natural light and visual references, often traveling great distances at depth without surfacing. Moreover, military submarines function in stealth mode, making the use of active sonar and radar impractical due to the risk of detection by enemy forces. Submarines must therefore navigate without access to conventional visual aids or continuous satellite signals, relying instead on a sophisticated blend of internal systems and external references.

Inertial Navigation Systems (INS) play a crucial role in underwater navigation, particularly for submarines, by utilizing accelerometers and gyroscopes to measure changes in speed, heading, and movement over time. These systems operate independently of external signals, which is essential for maintaining stealth during underwater operations. The INS begins from a known position and continuously calculates the submarine's current location as it moves through the water. This self-contained nature of INS makes it particularly advantageous for military applications, where external signals could compromise operational security [289-291].

The core components of an INS include three orthogonal gyroscopes that measure angular velocity and three orthogonal accelerometers that measure linear acceleration. This configuration allows the system to provide continuous updates on the position, velocity, and orientation of the submarine [292-294]. However, while INS is highly reliable in the short term, it is susceptible to cumulative errors over extended missions due to sensor drift and biases inherent in the gyroscopes and accelerometers. This drift can lead to significant navigational inaccuracies, which necessitates periodic updates from external sources or other navigation aids to correct the accumulated errors [295-297].

The implications of navigational drift are particularly critical in underwater environments, where even minor errors can lead to substantial deviations from intended paths, especially when navigating through narrow channels or vast expanses of ocean [296, 297]. Therefore, while INS provides the necessary autonomy and stealth for submarine operations, it is often integrated with other navigation systems, such as GPS or acoustic positioning systems, to enhance accuracy and reliability

[298]. This hybrid approach helps mitigate the limitations of INS by providing occasional position updates that recalibrate the system, ensuring that submarines can operate effectively without compromising their stealth capabilities [291, 294].

Sonar navigation is an essential technology for submarines, enabling them to acoustically map their surroundings. This capability is achieved through the emission of sound waves that travel through water, reflecting off objects and the seafloor. The returning echoes provide critical information regarding underwater topography, nearby vessels, and potential hazards, which is crucial for safe navigation and tactical operations. Active sonar systems emit sound pulses and listen for echoes, allowing submarines to determine the location of objects. However, the use of active sonar can compromise a submarine's stealth, making passive sonar systems a preferred choice in many scenarios. Passive sonar detects sounds made by other vessels without emitting any sound waves, thus maintaining the submarine's covert position while still providing valuable situational awareness [299, 300].

The distinction between active and passive sonar systems is significant in military operations. Active sonar is effective for target detection and mapping but poses risks to the submarine's position if used indiscriminately. In contrast, passive sonar systems are vital for recognizing and classifying surface and subsurface targets without revealing the submarine's location [252, 299]. Passive sonar technology has evolved to include advanced signal processing techniques that enhance detection capabilities even in challenging acoustic environments, where noise and reverberation can obscure signals [300, 301]. This makes passive sonar not only a navigational tool but also an early warning system, helping submarines evade threats and maintain tactical awareness [300].

The development of sonar technology has been influenced by various research efforts aimed at improving the accuracy and reliability of underwater mapping and target detection. For instance, advancements in sonar signal classification and recognition have been explored to enhance the performance of passive sonar systems [302]. These improvements are crucial for military applications where the ability to detect and classify threats quickly can determine operational success. Additionally, the integration of sonar data with other navigational aids, such as inertial navigation systems, further enhances the situational awareness of submarines operating in complex underwater environments [303].

Submarines, when operating near the surface, can utilize satellite navigation systems, particularly the Global Positioning System (GPS), to achieve precise positional fixes. The GPS functions by triangulating signals from multiple satellites, which provides significant accuracy in determining location. This capability is critical for submarines, especially during operations where precise navigation is essential. However, the use of GPS necessitates that submarines either surface or raise an antenna mast to receive these signals, which exposes them to potential detection by enemy radar or surveillance aircraft. Consequently, submarines must judiciously limit their GPS operations to minimize risks associated with detection, often reserving such activities for moments just before diving or upon reaching designated safe areas [304, 305].

The operational constraints of submarines utilizing GPS are underscored by the inherent risks of exposure. Submarines are designed to operate stealthily, and surfacing or deploying antennas can compromise their stealth capabilities. This necessity for discretion is reflected in the selective use of GPS navigation, which is often employed in conjunction with other navigation systems to enhance reliability and safety [306, 307]. For instance, the integration of GPS with inertial navigation systems (INS) allows for continuous navigation even when GPS signals are temporarily unavailable, thus providing a fallback mechanism that enhances operational security [306, 307]. Moreover, advancements in technology, such as the development of GPS sonobuoys, have improved the ability to track submarines while maintaining a degree of stealth, as these systems can be deployed from aircraft or surface vessels without directly exposing the submarine [308, 309].

Dead reckoning is a traditional yet essential navigation technique employed by submarines to estimate their position based on known variables such as speed, heading, and time travelled. This method is particularly crucial in environments where GPS is unavailable, such as underwater or in areas with significant signal interference. The reliance on dead reckoning allows submarines to navigate effectively when other systems, including GPS, are temporarily inoperative [310].

However, the accuracy of dead reckoning is contingent upon continuous monitoring and adjustment, as external factors like ocean currents can significantly impact a submarine's course. For instance, the integration of inertial navigation systems (INS) with dead reckoning can enhance positional accuracy by compensating for these external influences. INS utilizes a combination of accelerometers and gyroscopes to track changes in velocity and orientation, thereby providing a more reliable estimate of position over time [311].

Moreover, the effectiveness of dead reckoning and INS is often augmented through the use of additional sensors and algorithms that help mitigate the inherent errors associated with inertial measurements. For example, techniques such as Kalman filtering can be employed to refine position estimates by integrating data from various sources, including inertial sensors and external references when available [312]. This approach not only enhances the robustness of the navigation system but also addresses the drift that can occur over time due to the accumulation of errors in inertial measurements.

Submarines employ echo-sounding systems for precise depth measurement, a technology that sends sound waves directly downward to ascertain the water depth beneath the keel. The fundamental principle behind echo-sounding involves measuring the time taken for sound waves to travel to the seafloor and return, which is then displayed on a fathometer. This method is crucial for safe navigation, particularly in uncharted or shallow waters, as it provides accurate depth readings essential for avoiding underwater hazards [313].

The evolution of echo-sounding technology has seen significant advancements, transitioning from single-beam systems in the 1950s to more sophisticated multibeam systems in the 1980s, which allow for comprehensive mapping of the seafloor [314]. Despite these advancements, challenges remain in achieving global seafloor topography due to the vastness of the oceans and the limitations of traditional methods [313]. The critical role of echo-sounding in navigation cannot be overstated, as it directly impacts the operational safety of submarines, especially in environments where underwater terrain is poorly mapped [313].

However, the use of echo-sounding is not without its risks. Similar to active sonar systems, echo-sounding can potentially expose submarines to detection by adversaries, necessitating careful management of its operational use [315]. The balance between obtaining accurate navigational data and maintaining stealth is a significant consideration for submarine operations, particularly in military contexts [315]. Thus, while echo-sounding is indispensable for safe navigation, its deployment must be strategically managed to mitigate the risks associated with detection [315].

Each of these navigation methods operates in tandem with the submarine's combat management system (CMS), integrating sensor data to provide the crew with a comprehensive tactical picture. Human expertise is vital throughout the process, as sonar operators interpret sound data, navigators plot courses, and commanders make critical decisions based on the information provided by these systems. Together, these technologies and techniques ensure that submarines can navigate with precision and stealth, even in the most challenging underwater environments.

Periscopes, Masts, and Optical Sensors

Periscopes, masts, and optical sensors play essential roles in submarine navigation, particularly when the vessel operates at or near periscope depth. These instruments provide critical visual information for navigation, reconnaissance, and situational awareness while minimizing the submarine's exposure to detection. Although submarines spend most of their time submerged, moments of visual confirmation remain necessary, whether for course corrections, identification of nearby vessels, or alignment with navigational landmarks. These tools, though subtle and limited in use, form an integral part of a submarine's navigational and operational framework.

The periscope is the submarine's most recognizable optical tool, allowing the crew to observe the surface from a submerged position. Modern periscopes extend through the pressure hull to provide a clear line of sight above the waterline. The periscope is equipped with a series of lenses and mirrors, offering magnified views of the horizon for identifying ships, shorelines, or navigational markers. While traditional periscopes required physical penetration through the submarine's hull, newer optronic masts perform the same function without compromising the submarine's structural integrity. These masts incorporate digital cameras and infrared sensors, transmitting real-time visual data to the control room. The ability to switch between optical and infrared imaging ensures effective operation during day and night or in adverse weather conditions.

Figure 90: Lieutenant (LT) Barry Rodrigues, Officer of the Deck (OD), takes observations through the No. 2 periscope onboard the Los Angeles class nuclear-powered fast attack submarine USS HARTFORD (SSN 768). The U.S. National Archives, Public Domain PDM 1.0, via NARA & DVIDS Public Domain Archive.

Masts serve multiple purposes beyond visual observation. In addition to hosting optronic systems, various specialized masts carry antennas, communication equipment, and navigational instruments. For example, GPS antennas mounted on masts allow submarines to obtain precise location data during brief moments at periscope depth. Radio masts facilitate encrypted communication with command centres or other vessels, while radar masts, though seldom used due to the risk of detection, can provide additional situational awareness when entering or exiting ports. Some submarines also carry electronic warfare masts designed to intercept enemy signals or jam communications. These masts enable essential interactions with the external environment while the submarine remains concealed just below the surface.

Figure 91: The periscopes and antennas of the nuclear-powered attack submarine USS ALEXANDRIA (SSN-757) protrude above the water as the submarine operates off Andros Island, Bahamas, during post-commissioning trials. The U.S. National Archives, Public Domain PDM 1.0, via Picryl.

Optical sensors enhance a submarine's visual capabilities by providing high-resolution imagery under various environmental conditions. These sensors include both visible-light cameras and infrared systems that detect heat signatures. Infrared sensors are especially valuable for detecting ships or aircraft in low-light conditions, offering a tactical advantage during covert operations. Optical systems are integrated with the submarine's combat management system (CMS), allowing the crew to analyse and respond to visual data efficiently. These sensors help confirm targets, monitor surface activity, and validate information gathered from other navigational instruments, such as sonar or GPS.

Figure 92: USS Simon Bolivar (SSBN-641) surfaced with snorkel and masts up. T6985wsx, CC BY-SA 4.0, via Wikimedia Commons.

Submarine optical sensors provide critical visual information by capturing and processing images and light signals from the surface and surrounding environment. These sensors, which are embedded in optronic masts or advanced periscope systems, allow the submarine crew to observe the outside world while remaining at or near periscope depth. Modern optical sensors enhance operational effectiveness, offering both visible and infrared imaging capabilities to ensure clarity in diverse conditions, from daylight to complete darkness.

At the core of these optical systems are high-resolution cameras, capable of capturing detailed images in the visible spectrum. These cameras function much like those in modern surveillance equipment, recording live feeds and transmitting them to display screens within the control room. The images collected allow the crew to monitor surface activities, confirm targets, navigate safely, and identify ships, landmarks, or navigational aids. Unlike traditional periscopes, which required a direct optical path from the lens to the observer's eye, modern optical systems use digital technology to transmit visual data, eliminating the need for long, physical tubes penetrating the hull.

In addition to visible-light cameras, infrared sensors play a crucial role in optical systems by detecting heat signatures. These sensors are especially effective during low-visibility conditions, such as at night or in foggy weather, where traditional visual observation is limited. Infrared technology captures the thermal radiation emitted by objects, including vessels, aircraft, or surface features, providing a detailed heat map of the environment. This capability allows submarines to detect potential threats and identify targets even when visual cues are obscured. The combination of infrared and visible imaging ensures all-weather observation, enhancing the submarine's ability to operate covertly in diverse conditions.

Optical sensors in modern submarines are integrated with stabilization systems, ensuring clear images even when the vessel is moving or affected by waves. Gyroscopes and motion compensators correct for vibrations and tilt, maintaining steady visual feeds. These systems are particularly important during high-speed manoeuvres or turbulent conditions, where uncorrected images might become blurry or distorted. The stabilized view enhances situational awareness, allowing the crew to make informed decisions with minimal visual errors.

Advanced optronic systems also feature zoom capabilities, enabling the crew to magnify distant objects for detailed observation. This zoom functionality can be controlled remotely, allowing operators within the submarine to adjust focus and magnification to gather precise visual information without surfacing. Some systems incorporate laser rangefinders to measure the distance to specific objects accurately, further enhancing navigation and target tracking capabilities.

The data collected by optical sensors is seamlessly integrated into the submarine's Combat Management System (CMS), where it is processed alongside information from sonar, radar, and other navigational tools. This integration provides a comprehensive tactical picture, allowing the crew to correlate visual data with other sensory inputs. For example, visual confirmation of a ship detected by sonar can validate its identity and position, reducing the chances of false alarms. The processed imagery is displayed on digital screens, where operators can monitor, analyse, and react to real-time events.

Despite their advanced capabilities, the use of optical sensors must be carefully managed to maintain the submarine's stealth. When masts or periscopes equipped with sensors are raised, they create slight disturbances on the water's surface that can be detected by enemy forces. Moreover, prolonged exposure of these devices increases the risk of detection by radar or aerial reconnaissance. As a result, submarines employ these sensors only briefly, with skilled operators working efficiently to capture the necessary data and retract the mast or periscope quickly.

Despite their importance, the use of periscopes, masts, and optical sensors must be carefully managed to avoid exposing the submarine's presence. Periscopes and masts, when raised, create minimal but detectable disruptions on the water's surface. In areas of heavy surveillance, such as contested maritime zones, these disturbances can be detected by enemy radar or aerial reconnaissance. Therefore, submarines typically deploy these instruments for only brief periods, minimizing the risk of detection. Skilled operators ensure that the periscope or mast is raised and retracted swiftly, capturing the necessary information within seconds.

Communication Systems in Submarine Environments

Communication systems on submarines are designed to support secure, reliable, and covert communication with other naval units, command centres, and allied vessels while minimizing the risk of detection. Given the submarine's submerged environment, where radio waves and electromagnetic signals are severely limited by water, specialized technologies are employed to maintain effective communication. These systems combine underwater, near-surface, and surface-level solutions to exchange data, voice transmissions, and encrypted messages across multiple domains.

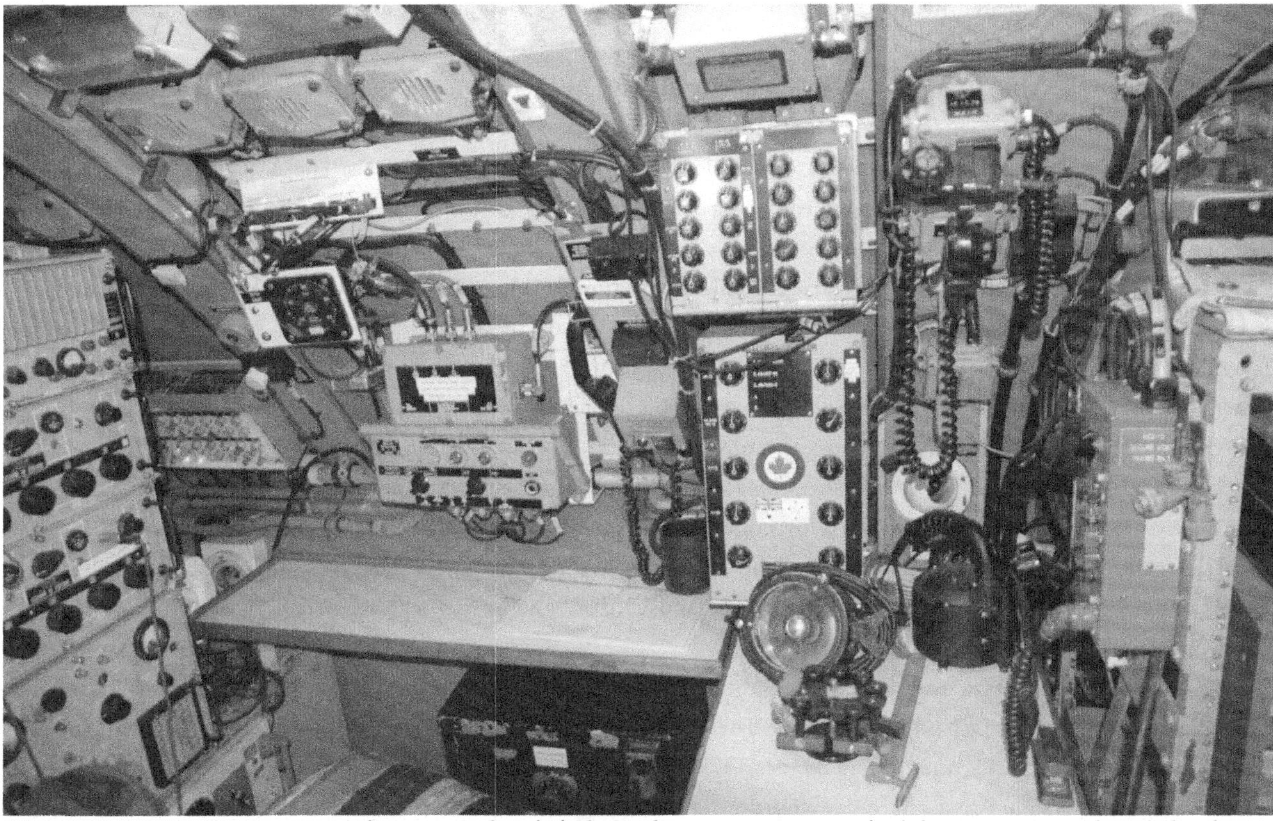

Figure 93: Communications room of HMAS Onslow, 'O' Class Submarine, 2,030 tons, built by Scotts, Greenock, completed 1969, decommissioned 1999, at the Australian National Maritime Museum, Sydney, 13 April 2016. Hugh Llewelyn, CC BY-SA 2.0, via Flickr.

Radio signals, particularly at high frequencies, do not penetrate water effectively, limiting communication options when the submarine is deeply submerged. Additionally, stealth is paramount for submarines operating in contested environments, meaning communication must be brief and carefully managed to avoid detection by enemy forces. As such, communication systems are optimized to balance connectivity, security, and operational secrecy.

Communication Methods

1. Very Low Frequency (VLF) and Extremely Low Frequency (ELF) Systems: VLF and ELF radio waves are used to transmit signals to submarines when they are submerged. These low-frequency waves can penetrate several meters into seawater, allowing submarines to receive messages while remaining at a shallow depth. However, the data transfer rate is slow, limiting communication to simple, short messages or coded instructions. VLF antennas are typically integrated into the submarine's structure or towed behind the vessel using an antenna wire.

ELF systems, used historically by the U.S. and Russia, transmit signals at even lower frequencies, enabling submarines to receive communications at greater depths. However, these signals are limited to one-way transmission, meaning the submarine can receive but cannot send messages through ELF frequencies.

2. Satellite Communication Systems (SATCOM): When at or near the surface, submarines can extend antenna masts to connect to satellite networks, enabling higher-speed data and voice communication. Satellite systems like the U.S. Navy's MUOS (Mobile User Objective System) or Russian GLONASS provide reliable links for exchanging encrypted messages, navigation data, and situational reports. Satellite communication is typically limited to brief periods to minimize exposure, as raising an antenna can compromise the submarine's stealth by making it visible to radar and aerial surveillance.

3. High-Frequency (HF) and Ultra-High-Frequency (UHF) Radios: HF and UHF radio systems are used primarily when the submarine is at the surface or periscope depth. HF radios provide long-range communication by reflecting radio waves off the ionosphere, while UHF radios are suitable for shorter-range transmissions. These systems are essential for maintaining contact with nearby vessels, coordinating fleet operations, or communicating with aircraft. However, HF and UHF signals are easily intercepted, so they are typically used with encryption to secure transmissions.

4. Acoustic Communication (Underwater Telephone Systems): When fully submerged, submarines use underwater telephone systems to communicate with other submerged vessels or surface ships. These systems transmit sound waves through the water using transducers, similar to sonar technology. Acoustic communication allows submarines to exchange basic information while remaining submerged, though it is limited by range, data speed, and susceptibility to interference from background noise and ocean conditions.

5. Buoy Communication Systems: In situations where the submarine must remain deeply submerged, it can deploy a communication buoy to the surface. The buoy, connected to the submarine via a cable, contains antennas for VHF, UHF, or satellite communication. This method allows the submarine to send and receive messages without surfacing, reducing the risk of detection. Some buoys are designed to be disposable or to self-destruct after use to avoid compromising the submarine's position.

6. Communication Masts and Antennas: Submarines are equipped with a range of communication masts that can be raised above the surface when necessary. These masts contain antennas for various frequency bands, including satellite communication, GPS, and radar. Advanced submarines often feature optronic masts that integrate cameras and sensors, further enhancing situational awareness during communication operations. Retractable masts allow submarines to establish brief communication windows while minimizing exposure.

Figure 94: SENIOR CHIEF Torpedoman (TMCS) Jeffery Leonard mans the communication console in the control room of the nuclear-powered attack submarine USS NORFOLK (SSN-714) as one of the crew operates the sound-powered telephone link. The U.S. National Archives, Public Domain PDM 1.0, via NARA & DVIDS Public Domain Archive.

Encryption and Security Measures

Given the sensitive nature of submarine missions, secure communication is paramount to prevent interception by enemy forces. Submarines employ advanced cryptographic systems and secure communication protocols to ensure that transmitted data remains confidential. Military-grade encryption is a critical component of these systems, providing robust protection against potential eavesdropping. For instance, encryption techniques such as the Advanced Encryption Standard (AES) are often utilized to secure communication channels, ensuring that even if messages are intercepted, they remain unreadable without the appropriate decryption keys [316]. The implementation of these encryption protocols is essential for maintaining operational security in military contexts, particularly in submarine operations.

In addition to encryption, submarines utilize frequency-hopping techniques to enhance the security of their communications. Frequency-hopping spread spectrum (FHSS) is a method where the carrier frequency of the transmitted signal rapidly changes according to a predetermined sequence. This technique not only helps in avoiding jamming but also makes it difficult for adversaries to detect and intercept communications [317]. Research indicates that frequency hopping can significantly reduce the probability of detection and interception, thereby enhancing the overall security of military communications [318]. Furthermore, the use of both multiplicative and additive groups in constructing optimal frequency hopping sequences has been shown to improve the robustness of these systems against jamming and eavesdropping [319].

The integration of advanced technologies in submarine communication systems is also noteworthy. Emerging methods such as quantum key distribution (QKD) are being explored for their potential to provide secure communication channels that are theoretically immune to eavesdropping [320]. Additionally, the application of self-adaptive MIMO-OFDM systems in frequency-hopping communication has demonstrated improved anti-jamming capabilities, which is particularly beneficial in the complex electromagnetic environments that submarines often operate in [321]. The combination of these technologies not only enhances the security of submarine communications but also ensures that data can be transmitted efficiently and reliably, even in hostile conditions.

Integration with Combat Management Systems (CMS)

The integration of communication systems with a submarine's Combat Management System (CMS) is critical for ensuring effective operational capabilities. This integration facilitates seamless coordination among navigation, weapon systems, and mission planning, thereby enhancing the overall tactical decision-making process. The CMS processes real-time data from various communication channels, which is essential for maintaining situational awareness and executing tactical manoeuvres effectively.

Research indicates that the Officer of the Watch (OOW) plays a pivotal role in this integrated environment. The OOW frequently communicates with the Submarine Warfare Federated Tactical Systems (SWFTS) to confirm the tactical picture and make informed decisions based on the data received from different sensors [322]. This communication is crucial during manoeuvres, as it ensures that all team members are synchronized and aware of the current operational status, thus supporting effective mission execution [248]. The integration of these systems allows submarines to remain aligned with fleet operations, receive updated mission orders, and relay intelligence data back to command centres, which is vital for mission success [246].

Furthermore, the integration of communication systems within the CMS enables submarines to assimilate data from multiple sources in real-time. This capability is essential for forming an accurate tactical picture, which is necessary for the OOW or Commanding Officer (CO) to make timely and effective decisions [246]. The ability to process and analyse data rapidly enhances the submarine's responsiveness to changing operational conditions and threats, thereby improving its tactical effectiveness [322].

Moreover, the SWFTS provides a common framework across various submarine fleets, which standardizes operations and enhances interoperability among allied forces [248]. This standardization is particularly important in joint operations where coordination and communication between different platforms are crucial for mission success. The integration of communication systems with the CMS not only supports tactical decision-making but also ensures that submarines can effectively collaborate with other units in the fleet, thereby enhancing operational effectiveness [246].

Communication Procedures and Protocols

Submarines operate under strict communication protocols to maintain stealth and minimize the risk of detection. The crew adheres to silent running procedures, where radio silence is maintained unless communication is absolutely necessary. Messages are often pre-prepared and transmitted quickly during brief communication windows. In emergencies, submarines can transmit distress signals or request assistance via designated emergency frequencies.

Emergency Communication Systems

In the event of a catastrophic failure or loss of communication, submarines carry emergency communication systems, such as satellite distress beacons or signal buoys. These systems are activated only when the submarine's crew is in immediate danger, such as after an accident or if the submarine is incapacitated. These signals are designed to alert allied forces to the submarine's position for rescue or recovery operations.

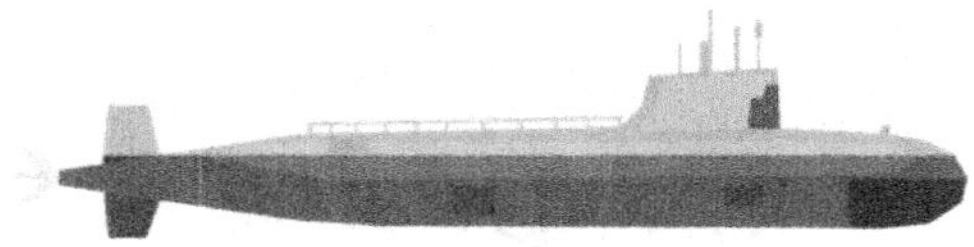

Chapter 9

Submarine Safety and Damage Control

Hydraulic Systems and System Redundancy

Hydraulic systems play a crucial role in submarine operations, providing the precise movement and control needed for essential components, particularly in environments where electrical systems might be less practical or vulnerable. These systems ensure smooth and reliable operations by generating force through the pressurization of hydraulic fluid, which is then used to actuate various mechanical elements throughout the vessel.

In the confined and demanding environment of a submarine, hydraulic systems are especially well-suited due to their compact size, high force output, and silent operation, which enhances the vessel's stealth. The hydraulic fluid flows through a closed circuit that includes pumps, valves, actuators, and reservoirs, offering precise control over the submarine's mechanical operations, even under immense underwater pressure.

Hydraulic systems are critical for controlling the submarine's diving planes and rudders, which are essential for manoeuvring. The diving planes adjust the submarine's vertical position, while the rudder governs horizontal movement. Hydraulic power enables these surfaces to respond quickly and accurately to the crew's commands, allowing the submarine to alter its depth and course smoothly.

Torpedo tubes and vertical launch systems (VLS) also depend heavily on hydraulics. These systems require substantial force to open doors, load and eject torpedoes, and launch missiles. Hydraulic mechanisms ensure the reliable operation of these components, maintaining the watertight integrity of the submarine's compartments and supporting rapid and precise weapon deployment.

The smooth and controlled motion required to raise and lower masts, such as periscopes, antennas, and snorkels, is achieved using hydraulic systems. These systems enable the quick extension and retraction of masts with minimal noise, crucial for maintaining stealth. This capability is essential during reconnaissance or communication operations at periscope depth, where exposure time must be minimized to avoid detection.

Hydraulics also operate the submarine's heavy watertight hatches, ensuring these remain sealed against high water pressure at extreme depths. In emergencies, hydraulically powered hatches provide fast access to escape trunks, enhancing crew safety. Steering and trim control are further supported by hydraulic systems, which manage the submarine's ballast and trim tanks. These adjustments maintain stability and buoyancy, allowing the vessel to remain level and agile during diving, surfacing, or cruising at different depths.

Submarine hydraulic systems consist of several vital components, including pumps that generate the required pressure to move fluid through the system. Reservoirs store hydraulic fluid, ensuring there is enough to maintain system pressure and compensate for any leaks. Valves direct the flow of fluid to specific actuators, which convert hydraulic pressure into mechanical force to move essential components such as rudders or diving planes. Filters eliminate contaminants from the fluid, preventing system failures, while accumulators store pressurized fluid, providing additional power when needed.

Hydraulic systems offer numerous advantages in submarine operations. Their compact and robust design generates high force, enabling efficient operation of critical systems within the limited space of a submarine. Since they are less susceptible to electrical interference, hydraulic systems function reliably in harsh underwater conditions. Their quiet operation also supports stealth, minimizing the risk of detection by enemy forces.

Given the importance of hydraulic systems, regular maintenance is essential to ensure optimal performance. Submarine crews routinely inspect the systems for leaks, monitor fluid levels, and replace filters to maintain peak operational efficiency. Safety protocols are in place to prevent hydraulic fluid contamination, which could compromise the system. Redundant systems and manual backups ensure critical components remain operational in the event of a hydraulic failure, maintaining the submarine's navigational and combat capabilities.

Hydraulic systems are indispensable to submarine operations, powering everything from steering and manoeuvring to weapon deployment and onboard mechanisms. Their reliability, precision, and quiet operation make them ideally suited for the challenging underwater environment. These systems enable submarines to operate effectively, ensuring mission success while maintaining safety and stealth.

Submarine hydraulic systems incorporate multiple redundancies to ensure continuous operation under demanding conditions. Given the critical role of these systems in steering, diving, weapon deployment, and hatch operation, redundancy is essential to safeguard the crew and mission. The design of these redundancies ensures that, even in emergencies or in the event of component failures, vital hydraulic functions remain operational.

The first layer of redundancy involves multiple independent pumps. Submarines are typically equipped with both electrically powered and manually operated hydraulic pumps. If the main electric pumps fail due to electrical malfunctions or damage, crew members can use manual pumps to maintain hydraulic pressure and control critical systems like rudders and diving planes. Some submarines also feature backup pumps powered by batteries, ensuring that systems remain functional even if the main electrical grid is compromised.

Multiple hydraulic circuits further enhance system reliability. These circuits operate independently to reduce the likelihood of a single-point failure disabling the entire hydraulic network. For example, one circuit may control the torpedo launch system while another manages the diving planes and rudders. If one circuit fails, the other remains operational, maintaining partial functionality of key components.

To prevent pressure loss from leaks or ruptures, submarines employ isolation valves at strategic points within the hydraulic network. These valves allow operators to isolate sections of the system where leaks occur, preserving hydraulic pressure in unaffected areas. This containment strategy ensures that critical subsystems continue to function even if part of the system is compromised.

Submarines also integrate accumulators into their hydraulic systems to store pressurized fluid. Accumulators act as emergency reservoirs, releasing stored hydraulic pressure when needed to power essential components. This feature is particularly useful during combat situations, where rapid responses are required, or in emergencies when pumps may be offline. Accumulators ensure that the crew can still operate crucial systems like steering and hatches for a limited time, even if the hydraulic pump system is inactive.

Redundant control systems add another layer of reliability. Hydraulic controls are often duplicated, with backup manual or mechanical control mechanisms available. For example, critical controls for diving planes and rudders can be overridden and operated manually if automatic or remote systems fail. This dual control capability ensures that the crew can maintain navigational and combat functions under any circumstances.

To protect against contamination, filters are installed throughout the hydraulic system. These filters prevent debris or particles from clogging valves or damaging actuators. Redundant filter systems allow for continuous operation; if one filter becomes clogged, the system can switch to a secondary filter without interrupting hydraulic flow.

Monitoring systems provide real-time data on hydraulic fluid levels, pressure, and temperature. Alarms alert the crew to any abnormalities, such as drops in pressure or fluid contamination. This allows for quick intervention, minimizing the risk of system failure.

In addition to these redundancies, emergency bypass systems can be activated if regular hydraulic pathways are blocked or damaged. These bypasses maintain the flow of hydraulic fluid to essential components, ensuring the submarine remains operable during emergencies.

Hydraulic Principles

Hydraulic systems are integral to modern submarine operations, representing a significant advancement in the development of submarines from pre-war classes. Unlike earlier submarines, which relied on hand power, electricity, or compressed air, modern submarines increasingly utilize hydraulic systems to perform critical functions. This shift toward hydraulic power stems from the unique advantages it offers: compact size, reliable performance, and quiet operation, all essential for the stealth and functionality of underwater vessels [323].

In early submarines, power was often generated manually or through electrical systems. Manual operation had limitations since it could not sustain high power demands over long periods and required continuous crew involvement. Electric systems, though widely used in submarines for propulsion and auxiliary functions, presented challenges when moving heavy components like rudders or diving planes. Electric motors can generate noise through their mechanical parts and struggle with precise, instantaneous stopping. Pneumatic power was another option; however, compressed air systems introduce risks of leakage, inefficiency, and increased maintenance requirements due to pressure losses [323].

Hydraulic systems have become the optimal solution for many onboard operations because they overcome these limitations. Hydraulics offer a balance of power and precision. The ability to generate high force from compact components and the capacity to start and stop instantaneously allow hydraulic systems to control sensitive mechanisms, such as diving planes and rudders, with exceptional precision. Furthermore, hydraulic power systems are inherently quieter than their electric and pneumatic counterparts, a vital characteristic for submarines that rely on stealth to evade enemy detection.

One of the primary applications of hydraulic systems in submarines is controlling essential steering components, such as rudders and diving planes. These systems enable the submarine to adjust its depth and direction swiftly and accurately, ensuring it can navigate efficiently in challenging underwater environments. Hydraulic systems are also employed in weapon systems, such as torpedo tubes and vertical launch systems (VLS). They allow for the smooth operation of hatches and the controlled launch of torpedoes and missiles, ensuring both safety and reliability during combat [323].

Hydraulics are essential for raising and lowering masts, such as periscopes and antennas, which must be deployed and retracted with precision and speed. The hydraulic system ensures these sensitive instruments function smoothly, minimizing the submarine's exposure to detection. Additionally, hydraulic systems operate watertight hatches, maintaining the vessel's structural integrity at various depths by keeping compartments sealed from water pressure.

A submarine's hydraulic infrastructure consists of several core components. Hydraulic pumps generate the necessary pressure to circulate fluid through the system, ensuring consistent performance. Valves control the flow of fluid, directing it to actuators responsible for moving mechanical parts like rudders, hatches, and diving planes. Hydraulic fluid, typically a lightweight oil, serves as both the power medium and lubricant for the system, ensuring smooth operation. Filters and accumulators help maintain system health, removing impurities from the fluid and providing extra pressure when needed.

Maintaining hydraulic systems is essential for submarine safety and reliability. Submarine crews regularly inspect for leaks, monitor fluid levels, and replace filters to ensure the hydraulic system remains in optimal condition. Redundant manual controls are in place to operate critical systems in the event of hydraulic failure, ensuring the submarine can continue functioning safely during emergencies.

The Structure of a Simple Hydraulic System

A basic hydraulic system, such as one used to open and close a door, demonstrates how hydraulics operate on a fundamental level. Key components include a reservoir, which stores the hydraulic fluid, a pump to pressurize the fluid, an actuating cylinder to convert fluid pressure into mechanical motion, and valves to control fluid flow. The reservoir, typically placed at a higher level to assist gravity-fed fluid flow, ensures the system remains filled and compensates for any leaks. The pump, often a hand-operated piston type in simple systems, provides the pressure needed to move the hydraulic fluid to the actuating cylinder.

In a typical system, when the pump is engaged, hydraulic fluid is drawn from the reservoir and forced into the cylinder through a pressure line. A cut-out valve prevents the pressurized fluid from escaping, holding the door in its desired position. A check valve in the return line ensures fluid only flows in one direction, preventing backflow and maintaining system integrity. When the cut-out valve is opened, the pressurized fluid returns to the reservoir, and the door moves back to its original position.

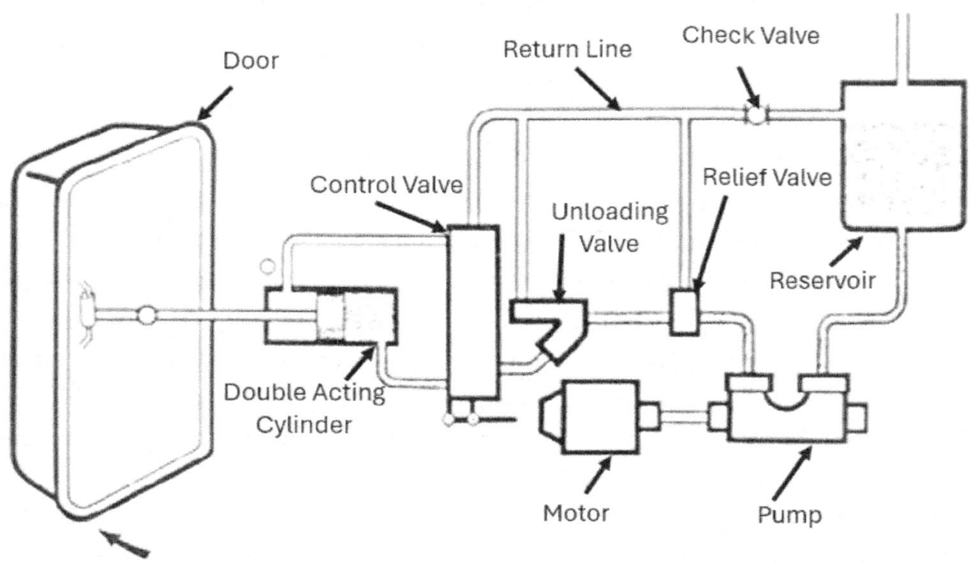

Figure 95: Power-driven hydraulic system.

While simple hand-operated systems are effective for small tasks, submarines require more powerful, automated systems to handle their complex operations. Power-driven hydraulic systems incorporate motorized pumps, double-acting cylinders, and control valves to provide rapid and precise movement. In these systems, fluid can flow to either side of the actuating cylinder to open or close a mechanism, with control valves regulating the flow direction.

Continuous operation of the hydraulic pump ensures immediate pressure availability, while relief valves protect the system from excessive pressure by redirecting excess fluid back to the reservoir. Bypass valves further enhance system efficiency by minimizing unnecessary circulation when components are not in use, reducing friction and heat build-up in the fluid.

Hydraulic systems are not without challenges. As fluid flows through pipelines and components, friction and turbulence generate heat, leading to thermal expansion of the fluid. This heat build-up can reduce system efficiency, necessitating careful design to minimize power losses. Engineers must account for friction losses by selecting appropriate pump pressures, cylinder sizes, and pipe diameters to ensure sufficient force for all operations.

In practice, submarines employ multiple hydraulic systems to control critical functions. These include steering the rudder, tilting bow and stern diving planes, and operating flood valves, torpedo tube doors, periscope hoists, and antenna masts. The main hydraulic system serves as the backbone for numerous operations, from managing engine exhaust valves to powering emergency steering systems. The integration of hydraulic power ensures the submarine can perform complex manoeuvres with precision and maintain operational readiness under various conditions.

These systems rely on the generation of hydraulic power through electric motor-driven pumps. Hydraulic motors, particularly actuating cylinders, convert the hydraulic pressure generated by pumps into mechanical energy, allowing precise control over equipment. The pumps serve as the central source of hydraulic power, maintaining consistent pressure throughout the system to enable smooth and reliable operations.

The IMO pump, a rotary pump used in submarines, is a key component of the hydraulic system. IMO stands for Ingenjörsfirman Malmström och Olson, the name of the Swedish engineering company that originally developed and patented the design for these rotary screw pumps. These pumps are renowned for their efficiency, reliability, and quiet operation, which makes them well-suited for use in demanding environments like submarines.

This pump features a cylindrical casing that houses three threaded rotors. These rotors operate similarly to worm gears, with their intermeshed surfaces creating a continuous seal as they rotate. Oil enters through the suction port and is propelled through the pump, eventually exiting via the discharge port. The precise design of the rotors ensures minimal leakage, with only a negligible amount of oil escaping back along the flow path [323].

The rotors are housed within a removable sleeve, which allows for easy maintenance and replacement. The sleeve, consisting of two bolted sections, fits snugly within the casing, providing stability during operations. The pump's design incorporates a guide bushing and packing gland to prevent leakage around the rotor shaft, with any escaping oil collected in a drip cup [323].

As hydraulic fluid is forced through the pump, it encounters resistance within the system, generating pressure. In a submarine's main hydraulic system, this pressure can range from 600 to 700 pounds per square inch. The force generated by the pressure acts against the rotors, creating an axial thrust that could increase friction and lead to premature wear. To counter this, a balancing connection is incorporated into the pump.

This balancing connection routes a portion of the discharged oil to the suction side of the pump, equalizing the pressure on both ends of the rotors. As a result, the rotors are effectively "floated" between equal oil pressures, preventing excessive axial thrust and reducing friction. This ensures the pump operates quietly, efficiently, and with minimal maintenance over extended periods, making it ideal for submarine operations [323].

The main hydraulic system is the backbone of a submarine's hydraulic infrastructure, powering numerous critical components. It distributes pressurized fluid to various systems, such as ballast tanks, air induction valves, torpedo tube doors, and bow plane rigging. It also supports ancillary systems, including forward windlass-and-capstan units and the sonar equipment (sound heads). On modern submarines, hydraulic power is also used to operate periscopes and antenna masts, which were previously electrically driven.

In emergency scenarios, the main hydraulic system provides backup power to the steering system and the bow and stern plane tilting mechanisms. Although these systems typically have independent power sources, the integration of the main hydraulic system ensures redundancy and operational reliability [323].

The main hydraulic system consists of several interdependent components. The power-generating system, which includes pumps like the IMO pump, builds the necessary hydraulic pressure. Flood and vent valves connected to ballast tanks control buoyancy, while the periscope and radio mast hoists use hydraulic pressure to extend and retract critical equipment. Forward and aft service lines distribute pressurized fluid to various parts of the submarine, and emergency systems ensure operational continuity in case of primary system failure.

The power generating system on submarines is an integral part of the hydraulic infrastructure, providing the necessary force for numerous onboard operations. This system consists of several interconnected components that generate, store, and manage hydraulic power, ensuring the smooth operation of critical submarine functions. At the heart of this system are the IMO pumps, responsible for creating the pressure needed to drive the various hydraulic mechanisms.

Two IMO pumps serve as the primary sources of hydraulic power. Each pump operates through a power rotor directly connected to an 18-horsepower electric motor, running at approximately 1,750 revolutions per minute. These pumps can function independently or in tandem, depending on the hydraulic demand. In standard conditions, one pump suffices to meet the system's needs, but both can be activated when greater hydraulic output is necessary. The pumps are equipped with manual or automatic controls to manage their operation effectively.

Oil, the working fluid in the hydraulic system, is supplied from the main supply tank, which has a capacity of 50 gallons. Typically, 35 gallons of oil are maintained within the tank, allowing room for expansion and the discharge of excess oil from the accumulator. The tank is designed to ensure a continuous flow of hydraulic fluid through the system, as the oil returns to the tank for straining and re-circulation. Sight gauges on the tank's side enable monitoring of oil levels, and a drain valve at the bottom facilitates the removal of any water that might accumulate [323].

The accumulator is another essential component, designed to store and distribute hydraulic fluid under pressure. It functions similarly to a storage battery in an electrical system, receiving oil from the pumps and maintaining it under pressure until needed. The accumulator consists of an oil cylinder, an air cylinder, and a plunger that moves between them. Compressed air exerts force on the inner surface of the plunger, while oil pressure acts on the outer surface. This configuration allows the plunger to move according to the balance between air and oil pressures, ensuring consistent pressure in the hydraulic system.

To prevent excessive axial thrust on the pump rotors, a balancing mechanism is employed. This includes a small pipe connecting the discharge side of the pump to an equalizing channel at the suction side. The channel ensures that equal oil pressure is applied to both ends of the rotors, preventing unnecessary friction and wear. This balancing feature is crucial for the longevity and efficiency of the pumps, allowing them to operate continuously with minimal maintenance [323].

The system also utilizes supply and return manifolds, which distribute hydraulic fluid to various onboard mechanisms and collect it for recirculation. The supply manifold features multiple valves to control the flow of fluid to different service lines, including emergency systems for steering and plane tilting. A quick-throw cut-out valve on the manifold allows the system to be isolated rapidly if needed, and a relief valve ensures that pressure does not exceed safe operating limits, typically set between 600 and 750 pounds per square inch [323].

The pilot valve plays a vital role in managing the automatic bypass valve, which regulates the flow of hydraulic oil based on the accumulator's status. When the accumulator is fully charged, the pilot valve directs oil to open the bypass valve, redirecting the flow back to the pump's suction side. As the accumulator discharges, the pilot valve closes the bypass, allowing the pumps to recharge the accumulator. This mechanism ensures efficient energy use and prevents overloading the system.

Safety and sound isolation are also key considerations in the power generating system. Nonreturn valves are installed to prevent backflow, protecting against oil loss in case of pipeline failure. Additionally, flexible rubber tubing connects the pumps to the rest of the system to minimize noise transmission, reducing the risk of detection by enemy forces. These hoses are supplemented by additional nonreturn valves to safeguard against accidental oil discharge in the event of a rupture.

The operation of the hydraulic power system requires careful management. Before activation, the entire system must be filled with oil, and the accumulator charged with compressed air. Quick-throw cut-out valves are opened to enable fluid flow, and the pumps are started to build up hydraulic pressure. During operation, the accumulator stores oil under pressure, ready to power various mechanisms. The system's design ensures that both pumps can operate simultaneously or alternate their use, ensuring equal wear and consistent performance [323].

The main hydraulic system exemplifies the precision and reliability necessary for submarine operations. It powers essential components such as ballast tank valves, torpedo tube doors, periscope hoists, and steering mechanisms, ensuring the submarine can perform critical functions safely and efficiently. The combination of well-coordinated components, redundancy, and safety features ensures that the hydraulic system operates smoothly, providing the submarine with the power it needs for both routine operations and emergency scenarios.

Flood and Vent Control System

The flood and vent control system is essential for a submarine's ability to manage buoyancy, allowing it to submerge, surface, and maintain depth. This system operates through a network of tanks, each with specific functions. By controlling the influx and expulsion of seawater and air, the submarine can achieve neutral, positive, or negative buoyancy as needed. The tanks surrounding the pressure hull are designed to take in seawater for submersion and expel it with compressed air to regain buoyancy. This system's precision is critical for both regular operations and emergency situations [323].

The tanks within the flood and vent control system serve different purposes. Main ballast tanks contain air when the submarine is surfaced and seawater when submerged, providing primary buoyancy control. Fuel ballast tanks initially store diesel fuel for propulsion, but once depleted, these tanks can be repurposed as ballast tanks. Special-purpose tanks such as the negative tank and safety tank play key roles during rapid manoeuvres and emergencies. The negative tank, located beneath the control room, is designed to fill with seawater to quickly increase the submarine's weight, facilitating rapid descent. Conversely, the safety tank, positioned amidships, compensates for potential buoyancy loss, such as if the conning tower is flooded, ensuring the vessel remains afloat in emergencies.

The bow buoyancy tank, located at the front of the vessel, plays a vital role during dives and surfacing manoeuvres. When diving, this tank is flooded first, tilting the submarine nose-down for controlled descent. Upon surfacing, the bow tank is the first to be emptied, raising the front of the vessel to aid in stabilization.

The operation of these tanks relies on precise control through flood and vent valves. Main ballast tanks have flood ports, and the fuel ballast tanks are equipped with hand-operated flood valves. Hydraulically operated vent valves manage the release and retention of air, facilitating the controlled filling or emptying of these tanks. The safety tank and negative tank flood valves are also hydraulically operated, though manual operation is available as a backup. The negative tank vent valve, unique in being hand-operated and vented internally, allows for precise adjustment during rapid descents.

When surfaced, the submarine's tanks are filled with air pressurized to about 10 pounds per square inch, preventing seawater from entering through the flood ports, which are always submerged. To submerge, the vent valves are opened, releasing the air and allowing seawater to flood the tanks. For resurfacing, the vents are closed, and compressed air is forced into the tanks, displacing the seawater through the flood ports, restoring positive buoyancy.

Two main control manifolds manage the flood and vent valves: the six-valve and three-valve manifolds. Located in the control room, these manifolds centralize the operation of the system. The six-valve manifold controls the bow buoyancy tank, several ballast tanks, and the safety tank. Depending on the submarine's design, the manifold may contain either six or seven valves, with each valve responsible for a specific tank or set of tanks. The valves have four operational modes—CLOSE, OPEN, HAND, and EMERGENCY—ensuring flexibility and safety in various scenarios. The EMERGENCY position is crucial, isolating the local hydraulic circuit to prevent oil loss in case of a hydraulic line failure.

The three-valve manifold operates key systems related to safety and ventilation. It manages the main engine air induction, the negative tank flood valve, and the safety tank flood valve. Each lever on this manifold is uniquely shaped for easy identification, even in low-visibility conditions. Additional safety mechanisms, such as spring-loaded pins, ensure that the air induction valve cannot be accidentally opened. This manifold, like the six-valve counterpart, features locking devices to secure the valves in their intended positions.

Hydraulic unit cylinders and associated linkages connect the control valves to the vent valves on the tanks. These hydraulic actuators enable the remote operation of vent valves, ensuring smooth and rapid adjustments to the submarine's buoyancy. In case of hydraulic failure, manual levers allow for direct control, ensuring redundancy in critical operations [323].

Steering and Plane Tilting Operations

The steering of a submarine is a complex process involving precise control over its movement in both horizontal and vertical planes. Unlike surface vessels that primarily navigate on a two-dimensional plane, submarines must manage movement in three dimensions. This requires advanced systems that coordinate the submarine's rudders and diving planes. The integration of hydraulics plays a critical role, providing the power needed for precise, reliable, and quiet operations essential for underwater navigation.

The horizontal steering of a submarine is achieved using a rudder mounted at the stern. Similar to a ship's rudder, it controls the vessel's direction by redirecting the flow of water. When the rudder is turned, water flows against it, creating hydrodynamic forces that cause the submarine to turn to port or starboard. The magnitude of the turn depends on the angle of the rudder, speed of the submarine, and water conditions. Precise control of the rudder is vital, especially when the submarine operates at high speeds or in confined spaces like harbors or undersea canyons.

Vertical control—managing the submarine's depth and pitch—is achieved through bow and stern diving planes. These planes act like wings or fins, tilting up or down to alter the vessel's angle and help it ascend or descend. By adjusting these planes in coordination, the submarine can remain level, climb to the surface, or dive deeper. The stern planes, located at the rear, are primarily responsible for controlling the submarine's trim and maintaining a stable orientation, while the bow planes, positioned at the front, are used to control ascent and descent during dives.

Hydraulics play an essential role in both steering and depth control by powering the movement of the rudder and diving planes. In a submarine, where space is limited and electrical interference must be minimized, hydraulic systems are ideal. They provide powerful and precise control while operating quietly, which is crucial for maintaining stealth. The hydraulic system includes actuators connected to the rudder and planes, allowing them to move with high precision based on commands from the control room.

When the helm or control levers are moved by the crew, the hydraulic system responds by directing pressurized fluid to the appropriate actuators. These actuators translate hydraulic pressure into mechanical force, moving the rudder or diving planes to the desired angle. The system also features feedback mechanisms that relay the position of the rudder and planes back to the control room, ensuring accurate adjustments and preventing over-correction. This feedback loop is essential for fine-tuning manoeuvres and maintaining a stable course.

The hydraulic system powering the steering and planes is designed with multiple redundancies to ensure reliability. Backup hydraulic circuits are in place, along with manual controls, to maintain steering and depth control in case of a hydraulic failure. For instance, in an emergency, the planes and rudder can be operated manually, giving the crew the ability to stabilize the vessel and maintain control. This redundancy is critical in preventing accidents and ensuring the submarine can surface safely if technical issues arise.

The hydraulic steering system is also equipped with isolation valves and non-return valves to manage pressure and prevent fluid leaks. Isolation valves allow specific sections of the system to be shut off for maintenance or in case of damage, while non-return valves ensure that the hydraulic fluid flows in only one direction, maintaining the pressure needed for smooth operation.

In addition to steering, the hydraulic system assists in trim adjustments, ensuring the submarine remains balanced throughout its journey. By moving water between trim tanks located along the length of the vessel, the hydraulic system can shift the submarine's centre of gravity and maintain proper alignment. This is essential for long missions, as changes in onboard weight distribution—such as fuel consumption or crew movement—can affect the vessel's stability.

Key Components of the Hydraulic Steering and Plane Tilting System

Hydraulic actuators are essential components of submarine steering and plane tilting systems. These actuators, typically double-acting cylinders, convert hydraulic pressure into mechanical motion to control the rudder's movement and the tilting of bow and stern planes. The use of double-acting cylinders allows hydraulic fluid to enter from either side of the piston, making it possible to achieve controlled movements in both directions. For precise tilting operations, especially for bow and stern planes, these actuators are connected to mechanical linkages, which effectively transmit the generated force to the planes. This configuration ensures that the submarine can achieve smooth, efficient movement both laterally and vertically.

Hydraulic pumps are responsible for supplying pressurized hydraulic fluid to the entire system. To ensure continuous operation, each hydraulic circuit typically includes two pumps—one active and one on standby, ready to activate if needed. These pumps are powered by electric motors, which provide a steady flow of hydraulic fluid even during extended operations. The operating pressure within these systems usually ranges between 600 and 1,000 psi (pounds per square inch), providing the necessary force to move large control surfaces like rudders and planes with precision and speed.

Control valves play a crucial role in directing the flow of pressurized fluid to the appropriate side of the hydraulic actuators, enabling the precise movement of the rudder or diving planes. In many cases, servo valves are integrated into the system to allow for proportional control, where the degree of valve opening corresponds to the desired extent of actuator movement. This proportional control ensures fine adjustments and smooth transitions. Additionally, check valves are used to prevent any backflow within the system, maintaining consistent pressure. Relief valves are also essential, protecting the system from over-pressurization by venting excess fluid to the return line when necessary, ensuring safety and reliability during operations.

Manifolds serve as critical distribution points within the hydraulic system, ensuring that fluid is directed to the correct components controlling the rudder, bow planes, and stern planes. The hydraulic system is divided into separate circuits for these different functions, with each circuit connected to its designated manifold. These manifolds contain multiple valves that enable operators to isolate specific parts of the system for maintenance or troubleshooting without affecting the entire operation. This modularity improves the system’s flexibility and reduces downtime during repairs or inspections.

Feedback sensors and control units provide real-time monitoring of the position of the rudder and diving planes, ensuring that the system responds accurately to operator commands. These sensors send continuous data to the control station, allowing for precise adjustments during manoeuvres. The control system is integrated with the submarine’s Combat Management System (CMS) or autopilot, which allows for seamless transitions between manual and automated control modes. This integration ensures that the submarine can be navigated with precision in complex underwater environments, whether by crew input or through pre-programmed courses.

The reservoir and accumulators are vital components that ensure the hydraulic system remains operational under all conditions. The reservoir stores hydraulic fluid and provides a stable supply to the pumps. It is designed to accommodate fluctuations in fluid volume due to expansion and contraction, which occur during temperature changes or varying system demands. Accumulators complement the reservoir by storing pressurized fluid, providing an instant source of hydraulic power for emergency manoeuvres. In the event of a pump failure, the stored energy within the accumulators ensures that the rudder and planes can continue functioning temporarily, allowing the submarine to maintain control and safely manoeuvre.

This integrated hydraulic system, with its combination of actuators, pumps, valves, manifolds, sensors, reservoirs, and accumulators, is essential for the safe and precise operation of a submarine. The coordination between these components ensures that the vessel can navigate efficiently and maintain stability at varying depths and conditions. The system's redundancy and real-time feedback capabilities further enhance its reliability, making it a critical aspect of submarine operations.

System Operation: Rudder and Plane Movement

The hydraulic system of a submarine plays a pivotal role in controlling both horizontal and vertical movement through precise operation of the rudder and diving planes. This system ensures that the submarine maintains its course, depth, and stability under varying underwater conditions, with real-time adjustments facilitated by hydraulic actuators, control valves, and feedback sensors.

In horizontal steering, the rudder is the key component responsible for turning the submarine to the desired heading. When the helmsman inputs a command to turn the vessel to port or starboard, the control valve responds by directing hydraulic fluid to the appropriate side of the actuator connected to the rudder. As the pressurized fluid enters the actuator, it causes the rudder to rotate to the specified angle, redirecting water flow and creating the necessary force to steer the submarine in the intended direction. To ensure precision, feedback sensors continuously monitor the rudder's position. If the rudder deviates from the input command, the control system detects the discrepancy and initiates corrective adjustments, aligning the rudder with the helmsman's command to maintain the desired course.

For vertical control, the bow and stern planes work together to manage the submarine's depth and angle of ascent or descent. When the control system receives an input to dive, hydraulic fluid is directed to tilt the planes downward, generating downward force that causes the submarine to descend. Conversely, when surfacing, the planes tilt upward, generating lift at the bow or stern, which helps the vessel rise toward the surface. The stern planes play a crucial role in maintaining the submarine's trim, ensuring that it remains level during operations by balancing the vessel's weight distribution. Meanwhile, the bow planes control the pitch of the submarine, preventing it from pitching too steeply during ascent or descent. This coordination between the bow and stern planes ensures smooth and controlled movement, enhancing the vessel's stability and manoeuvrability.

The hydraulic system also incorporates redundancy and emergency protocols to ensure safe operation in case of failure. The system includes redundant hydraulic circuits, ensuring that if one circuit fails, another can take over and maintain operational continuity. Additionally, emergency accumulators store pressurized hydraulic fluid, providing immediate power to the rudder and diving planes in the event of pump failure. This ensures that the submarine retains control over its movement, even under unexpected circumstances. In scenarios where hydraulic control is completely lost, manual override options are available. Operators can use handwheels or levers to manually adjust the rudder and planes, allowing the submarine to be maneuvered safely to the surface. This manual control capability provides an essential layer of safety, ensuring that the submarine can still navigate and surface even in emergencies.

Hydraulic Circuit Design for Steering and Planes

The design of hydraulic circuits for steering and planes in submarines is a critical aspect of ensuring operational reliability and manoeuvrability in challenging underwater environments. The hydraulic systems employed for these control surfaces are characterized by redundancy, precision, and noise reduction, which are essential for maintaining stealth and operational effectiveness.

One of the fundamental principles in submarine hydraulic circuit design is the implementation of separate hydraulic circuits for the rudder, bow planes, and stern planes. This separation is a crucial safety measure that enhances reliability by preventing a single point of failure from incapacitating all movement controls. For instance, if the hydraulic circuit for the rudder fails, the bow and stern planes can still operate independently, allowing the submarine to maintain depth control and trim until repairs can be made. This redundancy is vital for operational flexibility and risk mitigation in critical situations, as highlighted by the importance of maintaining functional hydraulic systems under various operational conditions [324-326].

The hydraulic fluid pathway within these systems is meticulously designed to ensure continuous circulation and consistent pressure. Fluid is pumped from a reservoir through supply lines by hydraulic pumps to directional control valves, which direct the fluid to the appropriate actuators based on commands from the control systems. This closed-loop system is essential for ensuring that the hydraulic pumps have a constant supply of fluid, thereby preventing interruptions during critical manoeuvres [327-329]. The design of these pathways must also consider the properties of the hydraulic fluid, as factors such as viscosity and thermal conductivity play significant roles in the overall performance of the hydraulic system [330, 331].

Isolation valves are strategically integrated into the hydraulic circuit to facilitate maintenance and manage potential malfunctions. These valves allow for the isolation of specific sections of the circuit without disrupting the entire system, which is crucial during routine maintenance or in the event of a malfunction. This modular approach enhances operational reliability and minimizes downtime, ensuring that the submarine can remain functional even when certain components require attention [332-334]. The ability to isolate sections of the hydraulic circuit not only aids in maintenance but also helps contain issues, preventing them from affecting other critical systems.

In addition to the structural design of the hydraulic circuits, the incorporation of flexible connections, such as rubber hoses or flexible tubing, is essential for noise reduction. Noise is a significant concern for submarines, as sound waves can be transmitted through rigid pipes, increasing the risk of detection by enemy sonar systems. The use of flexible components serves as a buffer, dampening vibrations and preventing the transmission of noise along the hydraulic lines, which is crucial for maintaining stealth during operations [329, 335, 336].

Hydraulic steering and plane systems in submarines offer several critical advantages that enhance their manoeuvrability, reliability, and stealth in underwater operations. These benefits make hydraulic systems the preferred choice for controlling rudders and diving planes, where precision and operational efficiency are paramount.

One of the primary advantages of hydraulic systems is their ability to provide precision control over rudders and planes. This fine control is essential during delicate manoeuvres, such as maintaining a specific depth, hovering, or navigating through narrow underwater passages. Precise adjustments to the rudder or diving planes allow the crew to respond to environmental changes or tactical needs quickly and accurately. This is particularly important during covert missions, where maintaining exact positioning is necessary to avoid detection or obstacles.

Hydraulic systems also offer high power density, meaning they can generate significant force despite being compact in size. This feature makes hydraulic actuators ideal for submarines, where space is limited, and the need to move heavy control surfaces, such as rudders and planes, is constant. The compactness of hydraulic actuators ensures that they fit seamlessly into the submarine's tight confines without sacrificing power or effectiveness. This combination of strength and space efficiency allows the submarine to operate smoothly, even under challenging conditions such as high-speed manoeuvres or heavy currents.

Another essential advantage is the quiet operation of hydraulic systems, which is crucial for maintaining the submarine's stealth. Noise reduction is a top priority for submarines, as unwanted sounds can compromise the vessel's ability to avoid enemy detection. Hydraulic systems generate minimal noise compared to other mechanical systems, such as electric motors, which can produce detectable vibrations. The inclusion of flexible hoses in the hydraulic circuits further dampens vibrations, preventing noise from traveling through the hull and being picked up by enemy sonar systems.

Reliability and redundancy are also significant strengths of hydraulic systems in submarines. These systems are designed with multiple layers of redundancy to ensure they remain functional under all circumstances. Backup pumps and accumulators provide alternative sources of hydraulic power in case the primary pump fails, allowing the submarine to maintain control over its rudders and planes. Additionally, manual overrides are available, giving the crew the ability to operate the control surfaces even if the hydraulic system experiences a complete failure. This redundancy ensures that the submarine remains manoeuvrable, improving safety and mission success in both routine operations and emergency situations.

Fire, Flooding, and Emergency Equipment and Procedures

Fire

Firefighting, suppression, and prevention on submarines are critical elements of onboard safety due to the confined spaces, limited oxygen supply, and the presence of flammable materials. The closed environment makes fire a particularly dangerous threat, as it can rapidly consume available oxygen, produce toxic smoke, and incapacitate personnel. Therefore, modern submarines are equipped with specialized firefighting systems and follow strict protocols for fire prevention, suppression, and response.

Submarines are outfitted with a range of firefighting tools to combat potential fires efficiently. Portable fire extinguishers, such as carbon dioxide (CO_2), dry chemical, and foam-based units, are strategically placed throughout the vessel. CO_2 extinguishers are used for electrical fires, as they do not conduct electricity and leave no residue. Dry chemical extinguishers are effective on combustible liquids and gases, while foam extinguishers smother fuel-based fires by cutting off the oxygen supply.

A fixed firefighting system is also installed in critical areas such as engine compartments, battery spaces, and torpedo rooms. This system often includes gas-based suppression systems like Halon or newer alternatives, which release fire-suppressing gases to displace oxygen and extinguish fires without leaving residue. These systems are activated manually or automatically in response to heat sensors or smoke detectors, minimizing the time it takes to respond to a fire.

Submarines also have fire hoses connected to the onboard water supply, allowing teams to extinguish larger fires. These hoses are usually stored in compact reels to conserve space and are fitted with specialized nozzles to adjust water flow. In situations where water use is limited to prevent electrical damage, mist or fog nozzles are employed to reduce heat without excessive water usage.

To ensure early detection, submarines are equipped with smoke detectors and heat sensors distributed throughout the vessel. These sensors are part of a centralized monitoring system that alerts the crew to any abnormal heat or smoke levels.

When a fire is detected, an alarm sounds throughout the submarine, and the control room is notified to coordinate the firefighting response.

In addition to fire alarms, ventilation shutdown systems play a crucial role. The ventilation system can be isolated in affected compartments to prevent the spread of smoke and fire. Submarines may also use pressure-tight doors to compartmentalize sections of the vessel, containing the fire and preventing it from spreading beyond the initial outbreak area.

A significant part of fire suppression involves emergency breathing apparatus (EBA). These units provide breathable air to crew members working in smoke-filled compartments, allowing them to perform firefighting duties without being overcome by smoke or toxic fumes. The submarine also carries oxygen masks and escape sets for non-essential personnel to evacuate from dangerous areas safely.

Fire prevention is a priority on submarines, where space is limited, and electrical equipment operates continuously. Routine inspections are conducted to identify potential fire hazards such as faulty wiring, overheated machinery, and fuel leaks. All flammable materials are securely stored in designated lockers to reduce the risk of accidental ignition.

Personnel are trained extensively in fire prevention protocols, including maintaining clean work areas, properly using electrical equipment, and following procedures for handling flammable substances. Batteries, a common source of heat and electrical discharge, are monitored closely to prevent overheating or hydrogen gas buildup, which can create an explosive atmosphere.

Preventive measures extend to maintenance practices, where lubrication and cooling systems are regularly inspected to prevent machinery from overheating. Ventilation systems are cleaned to remove dust and debris that could fuel a fire, and electrical systems are checked to ensure connections are secure and insulation is intact.

In the event of a fire, submarines follow a coordinated fire response plan. The control room takes command of firefighting efforts, directing personnel to the affected area while monitoring the situation through the onboard communication system. Fire teams don protective gear, including fire-resistant suits and breathing apparatus, and proceed to the scene with extinguishers, hoses, and other necessary equipment.

The crew isolates the fire by sealing pressure-tight doors, preventing smoke and flames from spreading. If the fire is near critical systems, non-essential personnel are evacuated to safer areas. The affected area may be vented to remove smoke after the fire is extinguished, and any remaining hot spots are cooled to prevent re-ignition.

Decontamination procedures follow firefighting efforts to ensure no residual smoke, fumes, or extinguishing agents remain in the air. Essential systems are inspected for damage, and repairs are initiated to restore full operational capability. Logs are updated with details of the incident, and a debriefing is conducted to review the response and identify areas for improvement.

Flooding

Flooding on a submarine presents one of the most critical emergencies due to the vessel's underwater environment and limited structural redundancy. A breach in the hull or the failure of internal systems can quickly lead to the ingress of water, endangering the crew and compromising the submarine's buoyancy and stability. To address these risks, submarines are equipped with specialized flood control equipment and follow strict procedures to detect, isolate, and manage flooding effectively.

Flood detection systems are critical for ensuring the safety and operational integrity of submarines. These systems utilize a variety of sensors to monitor water levels in critical compartments, triggering alarms when abnormal levels are detected. This early detection is vital for initiating rapid response protocols to mitigate flooding risks. For instance, submarines are equipped with flood detection sensors that continuously monitor for the presence of water, and when an anomaly is detected, alarms are activated in the control room, allowing for immediate action by the crew [337].

In addition to these sensors, visual inspection ports and gauges are strategically placed in areas susceptible to condensation or minor leaks, providing an additional layer of monitoring. The bilge systems, which are designed to collect water that accumulates during normal operations, are also closely monitored to identify unexpected surges that could indicate flooding [84]. The integration of flood detection into the submarine's control system enables operators to quickly pinpoint the affected area, facilitating an efficient response to deploy flood suppression measures and isolate the problem before it spreads to other compartments [84, 337].

The ability to compartmentalize flooding is essential for maintaining the submarine's buoyancy and preventing water ingress from spreading throughout the vessel. Pressure-tight doors are installed between compartments, allowing for rapid isolation in the event of flooding. The crew is trained to close these doors immediately upon receiving a flooding alarm, effectively preventing further water ingress [337]. Furthermore, strategically placed valves throughout the submarine's piping systems can shut off water flow to and from compromised areas, such as damaged tanks or failed piping sections. Remote valve control systems enhance safety by allowing the crew to operate critical valves from the control room without needing to enter flooded spaces, thereby reducing the risk to personnel [84].

During flooding scenarios, the submarine's control team must balance internal water distribution to maintain trim and stability. This involves shifting water between ballast tanks and adjusting buoyancy tank levels to counteract the effects of flooding. Such operational protocols are crucial for ensuring that the submarine remains stable and can navigate effectively even in compromised conditions [84, 337].

The flood suppression systems in submarines are critical for maintaining operational integrity during flooding incidents. These systems primarily utilize high-capacity emergency pumps designed to quickly evacuate water from affected compartments and bilges. The pumps are connected to bilge manifolds, which direct the water to designated discharge outlets or storage tanks. In scenarios where the primary pumping systems fail, submarines are equipped with backup or manual pumps to ensure continued operation, reflecting the redundancy built into submarine design for safety and reliability [50, 200, 338].

In cases of severe flooding, submarines deploy submersible pumps that can be placed directly into flooded compartments. These portable pumps are essential for efficiently removing water from areas that may be hard to reach with standard pumping systems. Additionally, ventilation fans play a crucial role in expelling water vapor from affected compartments, thereby preventing humidity buildup that could potentially damage sensitive equipment [50, 338, 339]. The importance of these systems is underscored by the need for submarines to maintain operational capabilities even under adverse conditions, as highlighted in various studies on submarine engineering and safety protocols [200, 338, 339].

Moreover, special precautions are taken in machinery areas to mitigate the risks of electrical fires or short circuits caused by water intrusion. The crew is trained to shut down electrical equipment in flooded areas and utilize waterproof switches to prevent accidental electrocution. This operational protocol is vital for ensuring crew safety and maintaining the functionality of critical systems during emergencies [50, 338, 339]. The integration of these flood suppression measures illustrates the complexity and foresight involved in submarine design, which prioritizes both safety and operational effectiveness in challenging underwater environments [200, 338, 339].

To manage more severe flooding caused by hull breaches, submarines carry specialized emergency repair kits. These kits contain materials such as wooden wedges, epoxy compounds, and inflatable patches that can temporarily seal small leaks. Crew members trained in damage control can quickly apply these patches to minimize water ingress until more permanent repairs are possible.

Inflatable bladders, stored onboard, are another tool used to seal breaches. These bladders are deployed inside the affected compartment and expanded to press against the hull breach, forming a temporary barrier against incoming water. This method is particularly effective in areas where external pressure from the surrounding water makes repairs challenging.

In extreme cases where flooding cannot be controlled, submarines are equipped with escape hatches and life-saving appliances. Escape suits provide thermal protection and buoyancy, allowing crew members to evacuate the submarine safely in an emergency. Escape trunks, located in strategic areas, serve as pressure chambers to allow personnel to exit the submarine without exposing the entire vessel to the external water pressure.

If flooding compromises the submarine's buoyancy, emergency ballast tank blow systems are activated. These systems release compressed air into the ballast tanks, expelling water rapidly and restoring positive buoyancy. This enables the submarine to surface quickly, reducing the risk to the crew and providing a safer environment for damage control efforts.

Damage Control Systems and Procedures

Damage control on a submarine is a critical function aimed at preserving the integrity of the vessel, ensuring the safety of the crew, and maintaining operational capability in the event of emergencies such as flooding, fire, or structural damage. Given the submarine's isolated environment underwater, the crew must act swiftly and efficiently to mitigate damage. Damage control systems are comprehensive, integrating specialized equipment, personnel training, and procedural frameworks that address a range of potential threats.

The integrated damage control system provides centralized monitoring and coordination for emergency situations. From the control room, the crew can monitor vital systems such as ballast tanks, the hull, electrical networks, and ventilation. Alarms are triggered automatically by sensors, which detect abnormal conditions like water ingress, fire, pressure drops, or toxic gas leaks. This allows the control team to assess the situation quickly and implement appropriate responses.

Submarines rely on automation where possible to support damage control efforts. Critical components, such as valve closures, water pumps, and ventilation systems, can be controlled remotely from the central station. This remote control capability minimizes the need for crew members to enter dangerous areas during emergencies, helping to reduce casualties while enabling precise responses.

Figure 96: Ten Commandments of Damage Control Aboard The Nautilus. Andrew Kalat, CC BY-ND 2.0, via Flickr.

Maintaining the pressure hull's integrity is essential to the submarine's survival underwater. In the event of a hull breach, the primary objective is to isolate the damaged compartment. The submarine is divided into multiple sections separated by pressure-tight doors, which the crew can seal to contain flooding. If necessary, mechanical patches and inflatable bladders are deployed to temporarily seal breaches, slowing or halting water ingress. These repairs may allow the submarine to remain operational long enough to surface and reach a safer environment.

Specialized emergency repair kits are carried onboard, including epoxy adhesives, wood plugs, and rubber sheets. These materials are applied by trained damage control personnel using underwater tools where necessary. In more severe cases, manual control valves are used to close off piping systems that might be contributing to flooding. Maintaining internal trim and stability is critical, and this is achieved by adjusting ballast tanks to offset water intrusion.

Figure 97: A midshipman attempts to stop flooding during a damage control simulation at Submarine Training Center Pacific. U.S. NAVY, Public Domain, via Picryl.

Fire is one of the most dangerous scenarios onboard a submarine, given the enclosed environment and the limited oxygen supply. Automatic fire suppression systems are installed in key compartments, particularly in areas housing electrical equipment, batteries, and engines. These systems release inert gases like carbon dioxide or Halon, which extinguish flames by displacing oxygen or interrupting chemical reactions.

Portable fire extinguishers are strategically placed throughout the submarine to deal with localized fires. The crew undergoes rigorous training to ensure that every member knows how to operate these systems effectively. Breathing apparatuses and protective suits are also available to allow personnel to fight fires in areas where smoke or toxic fumes are present. The ventilation system is designed to shut off airflow to sections where fire is detected, preventing flames from spreading.

Figure 98: Senior Chief Machinist's Mate Shane Jones, from Wadley, Ga., directs damage control students in the Trident Training Facility fire trainer at Naval Submarine Base Kings Bay. U.S. NAVY, Public Domain, via GetArchive.

If a fire compromises electrical circuits or vital equipment, the damage control team immediately shuts down the affected systems to prevent further risk. Redundant power and control lines are then activated, ensuring the submarine can maintain essential functions. Post-fire inspections are conducted to assess equipment damage and restore the compartment to full operational status.

Flood management procedures focus on containing water ingress, pumping out accumulated water, and maintaining the submarine's buoyancy and stability. Bilge pumps and high-capacity emergency pumps are activated immediately upon detection of flooding. Submarines also have portable submersible pumps that can be deployed directly into flooded areas to remove water. To prevent backflow and maintain pumping efficiency, non-return valves are installed along critical pipelines.

If the flooding cannot be controlled, the crew may initiate an emergency blow of the ballast tanks. This involves releasing compressed air into the tanks to force water out, rapidly increasing buoyancy and allowing the submarine to surface. Emergency blow systems are carefully monitored to prevent over-pressurization, which could damage the tanks or valves.

During flood events, crew members follow strict compartmentalization protocols. Pressure doors are sealed, and the affected section is isolated to prevent water from spreading. The damage control team balances the submarine by adjusting trim tanks, ensuring that the vessel remains stable throughout the operation. These actions are coordinated through the control room, where real-time data from flood sensors helps guide decision-making.

Toxic Gas and Smoke Management

In the context of submarine operations, the management of air quality is paramount, particularly in scenarios where toxic gases or smoke may accumulate. The ventilation systems within submarines are designed to effectively purge hazardous air, utilizing a network of fans and scrubbers specifically engineered to remove harmful gases such as carbon dioxide (CO_2) and carbon monoxide (CO). These systems are critical for maintaining a safe environment for the crew, as elevated levels of CO can lead to severe health risks, including impaired cognitive function and decision-making abilities [340, 341]. The importance of effective ventilation is underscored by research indicating that even low concentrations of CO can adversely affect human performance, making the role of ventilation systems in submarines a vital area of focus [340, 341].

Submarines are equipped with mechanisms that allow the crew to isolate affected sections and reroute airflows to contain contamination. This capability is essential for minimizing exposure to toxic environments and ensuring crew safety during emergencies. Portable gas detectors are employed to identify leaks, providing real-time data that informs the crew's actions [342]. In addition, oxygen masks are available to ensure that crew members can breathe safely in the event of gas or smoke exposure, further enhancing the safety protocols within the submarine [342].

In terms of emergency preparedness, submarines are also fitted with emergency breathing devices that supply fresh air during prolonged operations in toxic environments. This feature is crucial for maintaining crew survivability in the event of a catastrophic failure of the primary ventilation system. Furthermore, escape trunks equipped with air-tight seals provide additional refuge for crew members, allowing them to remain safe until the situation is resolved [343, 344]. The design and functionality of these systems reflect a comprehensive approach to ensuring the habitability and safety of submarines, particularly in emergency scenarios where air quality may be compromised [344].

Emergency Power and Control Systems

In the event of damage to the primary electrical system, submarines are designed with multiple redundant power sources to ensure the continuity of essential functions. These systems typically include battery banks and backup generators that provide power to critical equipment such as pumps, communication systems, and life-support systems. The redundancy in power supply is crucial for maintaining operational capabilities during emergencies, as highlighted by research on submarine systems which emphasizes the importance of reliable power generation and storage mechanisms onboard submarines [200]. The reliance on battery banks and backup generators is a standard practice in marine engineering to mitigate the risks associated with power failures [345].

Additionally, hydraulic systems serve as an independent layer of redundancy, operating separately from electrical circuits. These systems are vital for controlling rudders, diving planes, and ballast tanks, which are essential for the submarine's manoeuvrability and stability. The integration of hydraulic systems into submarine design allows for continued control even when electrical systems fail, thus ensuring operational integrity [346]. This redundancy is particularly important in scenarios where electrical failures could compromise the submarine's ability to navigate or maintain buoyancy.

Manual overrides are another critical feature incorporated into essential systems aboard submarines. These manual systems allow crew members to operate key equipment in the event of a power failure, ensuring that critical functions can still be performed. Hand-operated valves and mechanical linkages are designed to provide functionality for rudders, planes, and ballast systems, even when automated systems are offline [345]. This design philosophy reflects a broader trend in engineering towards enhancing reliability and safety in critical systems, particularly in environments where failure can have severe consequences.

Emergency lighting is a crucial safety feature that guides crew members during blackouts, facilitating safe movement throughout the submarine. The presence of emergency lighting systems is essential for maintaining crew safety and operational efficiency during power outages, as it allows personnel to navigate effectively in low-visibility conditions. The combination of these redundant systems—battery banks, hydraulic controls, manual overrides, and emergency lighting—demonstrates a comprehensive approach to risk management in submarine design, ensuring that essential functions can be maintained under adverse conditions.

Crew Training and Drills

Successful damage control relies heavily on the crew's ability to respond effectively under pressure. Submarine personnel undergo rigorous training that includes classroom instruction, hands-on practice, and simulated emergency drills. These exercises cover a range of scenarios, including flooding, fire, hull breaches, and toxic gas leaks. Drills are designed to test the crew's coordination, speed, and decision-making abilities, ensuring they are well-prepared to manage real emergencies.

Damage control teams are assigned specific roles during drills, with clear responsibilities for containment, repair, communication, and safety. After each drill, the crew conducts debriefings to identify areas for improvement and refine procedures. Regular training ensures that all crew members remain familiar with emergency protocols, equipment, and escape procedures.

Preventive Maintenance and Inspections

Prevention plays a vital role in damage control, and submarines follow strict maintenance schedules to ensure that all systems are in optimal condition. Regular inspections are conducted on the hull, piping networks, electrical systems, and fire suppression equipment. Valves, pumps, and seals are tested routinely to detect wear or potential failure points.

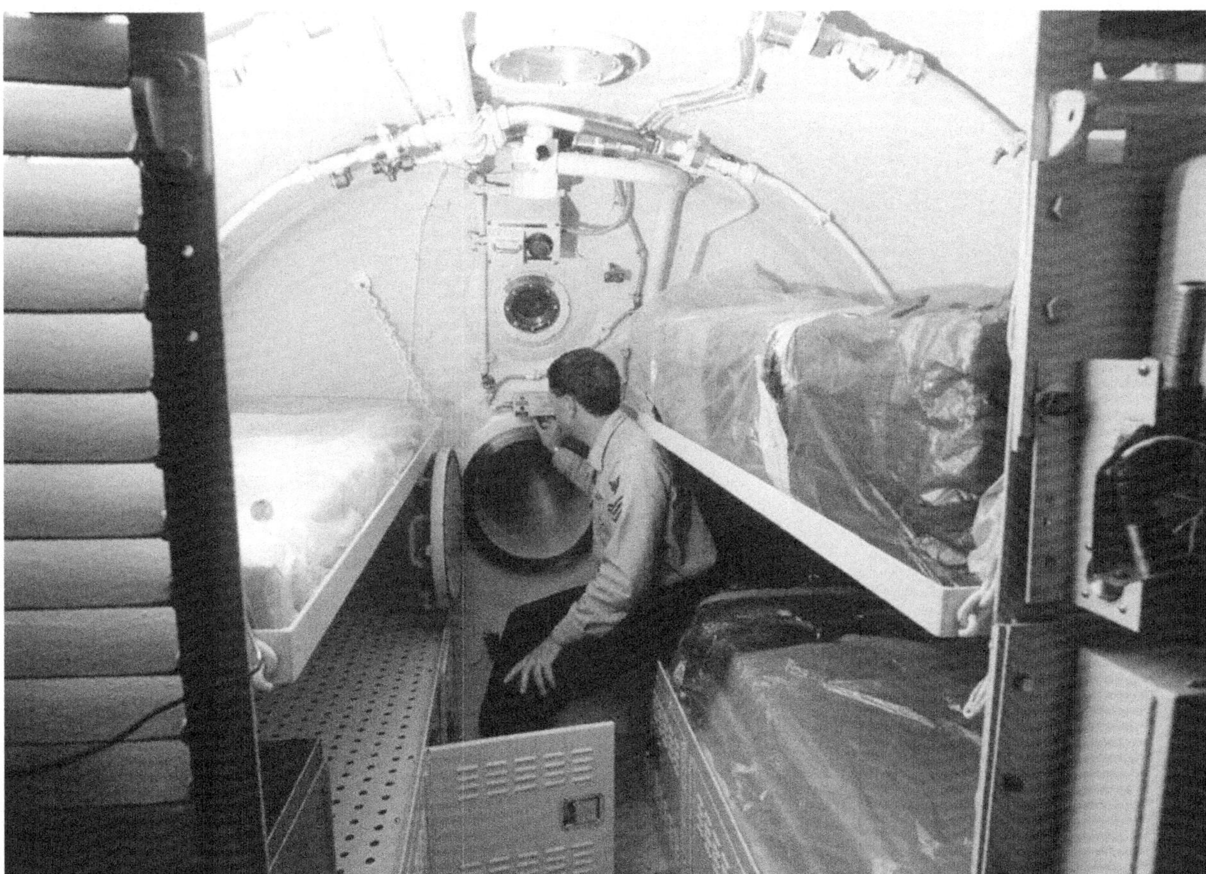

Figure 99: A crew member performs maintenance on the air lock through which supplies are passed to divers living in pressure chambers aboard the submarine rescue. The U.S. National Archives, Public Domain, via Picryl.

Corrosion prevention measures are implemented to protect the submarine's metal components from saltwater exposure. Critical systems, such as ballast tanks and escape trunks, undergo pressure testing to verify their integrity. Any issues identified during inspections are addressed promptly to prevent minor problems from escalating into emergencies.

Chapter 10

Materials and Manufacturing in Submarine Construction

Materials for Submarine Hulls

Raw Materials Used in Submarines

The primary material used in the construction of nuclear submarines is steel, chosen for its strength, durability, and corrosion resistance. Steel forms both the inner and outer hulls of the submarine, providing the structural integrity needed to endure the immense pressure encountered at extreme depths. The inner hull, also known as the pressure hull, is designed to maintain a liveable environment for the crew, housing essential systems and equipment. The outer hull serves both as a protective shell and as part of the ballast system, creating a hydrodynamic shape and reducing drag underwater. Between the two hulls are ballast tanks, which take in water to increase the submarine's weight for submersion or eject water to achieve positive buoyancy and surface [347].

Submarine architecture focuses on optimizing hull design, material selection, and overall system integration to enhance the vessel's performance, payload capacity, and stealth while also making construction more efficient and cost-effective. Advances in this field aim to achieve higher speeds with the same power by reducing drag, improve stealth by minimizing both acoustic and non-acoustic signatures, and streamline manufacturing processes through innovative structural designs and advanced materials. A holistic approach is essential, as modifications to individual components affect the hydrodynamic performance and operational capabilities of the submarine. For example, changes in hull geometry or material properties influence both manoeuvrability and stealth, as well as long-term maintenance requirements [348].

Submarine architecture requires a comprehensive approach, where geometry, material selection, and structural layout are carefully harmonized. Various hull geometries and materials have been explored to enhance hydrodynamics, reduce the submarine's target signature, and create additional surface area for embedded sensors or specialized equipment. This integrated design not only improves manoeuvrability and stealth but also supports future technological upgrades. Refining the shape of the sail—the tower-like structure on a submarine's dorsal side—can reduce drag and make maintenance easier, leading to lower life-cycle costs [348].

The use of steel remains a cornerstone of submarine construction, as it forms both the inner and outer hulls. The inner hull houses the crew, systems, and machinery, while the outer hull offers additional protection and hydrodynamic benefits. Between these two layers lie the ballast tanks, which take in or expel water to control the submarine's buoyancy, enabling it to dive or surface as needed [348].

Different types of submarines use varying hull designs depending on size and operational requirements. Smaller, modern submersibles and early submarines are typically constructed with a single hull, offering simplicity and compactness. In contrast, larger submarines, such as those developed by Soviet designers, use double hull structures, while American submarines tend to employ a single-hull approach with some light sections in the bow and stern for ballast tanks and hydrodynamic shaping. Although these sections provide streamlined movement underwater, the main hull in American designs typically consists of a single cylindrical layer, balancing efficiency and performance [348].

Double-hulled submarines feature two primary components: the light hull and the pressure hull. The light hull, sometimes referred to as casing, serves as the outer, non-watertight layer, providing an optimized shape for reducing drag. This layer protects the more critical internal pressure hull, which is designed to withstand the immense pressure at depth and maintain the structural integrity of the submarine [348].

The double hull design offers multiple benefits in terms of space utilization, structural efficiency, and damage tolerance. By placing ring stiffeners and longitudinals between the light and pressure hulls, designers free up valuable interior space within the pressure hull, reducing its overall size and weight. This approach is advantageous because the pressure hull, being thicker and heavier to endure deep-sea pressure, contributes significantly to the submarine's weight. Minimizing its dimensions improves the vessel's buoyancy and overall performance [348].

Additionally, the light hull provides an essential layer of protection. In case of impact or damage, the outer hull can absorb some of the force, potentially preventing structural breaches in the pressure hull. As long as the pressure hull remains intact, the submarine can maintain its integrity and continue operating, enhancing the vessel's survivability in combat or emergency situations. This layered approach ensures that even if external damage occurs, the critical inner workings and safety of the crew are preserved [348].

In addition to steel, submarines incorporate other metals, including copper, aluminium, and brass, each serving specific roles. Copper is often used in electrical wiring due to its excellent conductivity and resistance to corrosion. Brass, an alloy of copper and zinc, finds application in components that require a combination of corrosion resistance and mechanical strength, such as valves and fittings in seawater systems. Aluminium, known for being lightweight yet strong, is employed in non-structural components to reduce overall weight without compromising functionality [347].

Beyond metals, submarines rely on various non-metallic materials to meet the diverse requirements of their design. Plastics and composite materials are used in components that must be lightweight, corrosion-resistant, or flexible. These materials appear in internal fittings, insulation, and gaskets, where they can withstand the harsh marine environment without adding unnecessary weight. Glass is used in specialized equipment, such as sonar domes or periscopes, where optical clarity is essential. Advanced polymers and rubber-like materials are utilized for seals and acoustic insulation to reduce noise, enhancing the submarine's stealth [347].

The electronic systems onboard submarines depend on semiconductors, primarily silicon and germanium. These semiconductors form the backbone of communication systems, navigation equipment, sensors, and computing platforms. Silicon, known for its versatility and efficiency, is used in microprocessors and power electronics, while germanium plays a role in high-frequency circuits and specialized transistors. These semiconductors are crucial for the seamless functioning of critical systems, ensuring reliable performance in the demanding conditions of underwater operations [347].

The most distinctive feature of nuclear submarines is their nuclear reactor, which provides the vessel with a virtually unlimited range and endurance. The reactor core contains uranium or other radioactive materials, which undergo fission to produce heat. This heat generates steam, which drives turbines to produce electricity and power the propulsion system.

The choice of uranium, typically enriched to increase its concentration of fissile isotopes, ensures that the submarine can operate for extended periods without the need for refuelling.

Each of these materials plays a vital role in the complex design and operation of a nuclear submarine. Steel provides strength and resilience, non-ferrous metals contribute to electrical and mechanical systems, and advanced plastics and semiconductors ensure reliability in harsh underwater environments. Together, these raw materials enable the submarine to function as a self-contained, stealthy, and highly capable platform, designed to perform under the most demanding conditions while supporting the safety and effectiveness of the crew.

Material Requirements for Submarine Hulls

The choice of materials for a submarine's pressure hull is critical in balancing strength, weight, and cost. Using materials with higher yield strength allows for a thinner pressure hull, reducing overall weight and improving performance. However, high-strength materials are more expensive, requiring a trade-off between cost and efficiency. Achieving the right balance is essential for the structural integrity and operational effectiveness of the submarine, particularly given the extreme conditions it faces underwater [348].

Depth is a primary design criterion that dictates the structural requirements of the pressure hull. The hull must withstand the immense hydrostatic pressure exerted at operational depths. Engineers calculate the collapse depth—the point at which the hull fails catastrophically—by multiplying the maximum operational depth (MOD) with a factor of safety. This ensures that the hull remains structurally sound even at unexpected pressures. The calculated hydrostatic pressure at the collapse depth becomes the design pressure, guiding the selection of materials and construction standards for the hull [348].

The pressure hull must also be designed to resist the forces generated by underwater detonations, such as those caused by mines or gas bubbles. The shockwaves from explosions transmit both direct pressure and vibratory loads along the hull. These vibrations can shorten the fatigue life of the structure, and if they reach resonant frequencies, they can lead to catastrophic failure. Therefore, the materials chosen must have sufficient strength and resilience to absorb these forces without compromising the submarine's safety or structural integrity.

The pressure hull is typically constructed from thick, high-strength steel that provides a high reserve of strength. This steel is designed to maintain structural integrity even under extreme pressures and stresses encountered at depth. The hull is divided into several watertight compartments by bulkheads, further enhancing the vessel's ability to manage pressure and maintain buoyancy in case of localized damage. Precision is paramount in constructing the pressure hull, as even small deviations from design specifications can lead to a significant reduction in strength [348].

The hull's cross-section must remain perfectly circular to effectively distribute compressive forces. A deviation of just one inch (25 mm) from roundness can decrease the hull's ability to withstand hydrostatic loads by over 30%. Stiffener rings—reinforcing circular frames—are used to maintain the hull's shape and resist deformation under pressure. These rings distribute the millions of tons of compressive force exerted along the length of the hull, preventing structural failure and ensuring even force distribution.

Submarine hulls demand exceptional precision and flawless craftsmanship. Minor imperfections in the structure or welding can compromise the entire hull's ability to withstand pressure. To meet these stringent standards, construction employs a combination of stiffener rings and longitudinal supports, although transverse stiffeners are typically spaced more frequently than longitudinals due to their superior resistance to compressive loads.

All components of the pressure hull are subject to rigorous quality control, with every weld inspected multiple times using various non-destructive testing methods such as X-rays, ultrasonic testing, and dye penetrant inspection. This meticulous inspection ensures that no defects go undetected, as even a small flaw can jeopardize the submarine's ability to withstand the enormous pressures at depth [348].

Stealth Materials Technology in Submarines

Stealth technologies are critical in modern military applications, particularly for submarines, which require advanced methods to evade detection. Recent advancements have focused on innovative hull designs, noise-cancelling systems, and radar-absorbent materials that enhance the stealth capabilities of these underwater vessels.

One of the significant developments in stealth technology is the use of metamaterials to create sonar-absorbing tiles. Researchers have designed systems that can analyse enemy sonar frequencies and generate opposing sound waves, effectively masking submarines by making them indistinguishable from water. This technology can produce low-frequency sounds capable of cancelling out powerful active sonar systems employed by the US Navy and its allies [349]. Traditional passive sound-absorbing materials, while effective against higher frequency sonar, often fall short against lower frequencies due to their physical limitations [349]. The new metamaterial tiles can be strategically placed across a submarine's hull to counteract sonar beams from various angles, thus enhancing stealth capabilities significantly [349].

In addition to sound-absorbing technologies, the reduction of hydrodynamic noise is crucial for submarine stealth. Hydrodynamic noise, which becomes a dominant noise source at higher speeds, can be mitigated through innovative hull designs and noise control strategies [60]. Research indicates that the design of submarine enclosures can significantly influence hydrodynamic noise levels, suggesting that optimizing hull shapes could lead to quieter operations [60]. Furthermore, advancements in active noise control systems have shown promise in reducing unwanted sounds generated by submarine operations, thus improving stealth performance [350, 351]. Such systems utilize algorithms to adaptively cancel out noise, thereby enhancing the overall stealth profile of submarines [350, 351].

Moreover, radar-absorbent materials (RAM) are essential for minimizing radar signatures. Recent studies have explored the characteristics of various RAM, emphasizing the importance of material thickness and composition in absorbing electromagnetic waves [352, 353]. The development of new composite materials that exhibit superior radar absorption properties is ongoing, with research indicating that multi-layered structures can significantly enhance stealth capabilities [353]. These materials not only serve submarines but can also be adapted for use in aircraft and surface vessels, indicating a broad applicability of stealth technologies across different military platforms [349, 353].

Stealth materials technology plays a vital role in reducing the detectability of submarines by radar, sonar, and mine sensors, enhancing the vessel's ability to operate covertly. This involves a combination of acoustic, vibroacoustic, and radar stealth techniques, each addressing a unique aspect of detection avoidance. Below is a detailed explanation of these methods and the materials used [347].

Acoustic Stealth

Acoustic stealth focuses on reducing the noise emitted by the submarine, which could otherwise be detected by passive sonar systems. Passive sonar relies on detecting radiated noise from vessels, while active sonar identifies objects by transmitting sound waves and listening for echoes. Submarines combat these threats using specially developed acoustic rubber tiles.

These tiles consist of multiple layers, with an outer protective layer and an inner sound-absorbing layer. The inner layer contains cavities strategically designed to absorb sound waves across a range of frequencies. Variables such as the tile's thickness, cavity distribution, and rubber matrix properties are fine-tuned to optimize acoustic performance. A key feature of these tiles is their ability to withstand hydrostatic pressure up to 4 MPa, maintaining their performance even under extreme underwater conditions.

Vibroacoustic Tile

Vibroacoustic tiles isolate sonar transducers from structural vibrations and acoustic noise within the submarine's sonar dome. The tiles feature a multi-layered structure with perforated layers bonded together by specialty adhesives. This design ensures that vibrations from machinery or the hull do not interfere with sonar operations, improving the accuracy of underwater detection systems.

Additionally, vibration damping coatings (VDC) have been introduced in areas like engine rooms and decks to reduce transmitted vibrations. Epoxy-based adhesives and putty are used to secure these tiles to the submarine's hull. These adhesives ensure a strong bond between metal and rubber surfaces, maintaining effectiveness for over 20 years while preventing water seepage with polyurethane sealants.

Radar Stealth

Radar stealth is achieved through the development of radar-absorbing materials (RAM). These materials, designed using ferrite-based compounds, significantly reduce the radar cross-section (RCS) of the submarine, especially in the X-band frequency range (8–12 GHz). RAMs are incorporated into coatings, sheets, and composite structures, such as radar-absorbing paint (RAP) or radar-absorbing sheets (RASH) [347].

These coatings provide more than 10 dB attenuation, effectively reducing the detectability of the submarine's exposed components, such as the snort mast and periscope. The technology has also been applied to the superstructures of naval ships, ensuring maritime platforms maintain a low radar profile.

Advanced Materials Technology

Advanced Materials Technology in submarines focuses on improving performance, durability, and stealth while minimizing weight and reducing maintenance requirements. Composite materials play a central role by offering significant weight savings, enhanced corrosion resistance, and superior durability. These composites make submarines lighter, more manoeuvrable, and better suited for deep dives. Composite propellers further enhance efficiency by reducing vibration, which lowers acoustic signatures and contributes to stealth.

Friction Stir Welding (FSW) is another crucial innovation, used for joining aluminium alloys such as Al5083, Al6061, and Al2014. This solid-state process creates strong, reliable joints without compromising the integrity of the material, which is vital for constructing lightweight submarine structures. High Vibration Damping Aluminium Metal Matrix Composites, produced through molten flux-assisted synthesis, contain graphite and ceramic reinforcements. These composites help minimize vibrations in critical components, ensuring smoother and more reliable operations.

Rheo-casting technology offers a unique blend of forging precision and casting complexity, enabling the production of thick and thin-walled aluminium components. This method ensures the mechanical strength required for complex submarine structures while maintaining intricate design features. Bulk Metallic Glasses (BMGs), based on iron and copper alloys, are applied as coatings through high-velocity oxygen fuel (HVOF) spraying. These coatings provide exceptional hardness, up to 1000 Vickers Hardness Number (VHN), along with excellent corrosion resistance, enhancing the durability of steel and nickel-aluminium bronze components.

The HVOF spraying technique itself is essential for extending the lifespan of submarine components. By applying advanced materials like BMGs, it provides robust surface protection against corrosion and wear in harsh marine environments. Another vital advancement involves piezocomposite acoustic sensors, which are fabricated using the dice-and-fill method. These sensors enhance the sensitivity of sonar systems, enabling precise underwater threat detection and maintaining durability under high-pressure conditions.

Together, these advanced materials technologies significantly improve the performance, stealth, and reliability of submarines. They are essential for modern naval operations, ensuring that submarines remain efficient, agile, and undetectable in challenging underwater environments.

Friction Stir Welding (FSW)

FSW is employed to join aluminium alloys, including Al5083, Al6061, and Al2014. This technique uses a rotating tool to generate frictional heat, creating solid-state joints. FSW ensures strong, reliable connections, particularly in applications where traditional welding methods might compromise material integrity. It is especially valuable in manufacturing lightweight components with enhanced strength, such as those used in naval vessels [347].

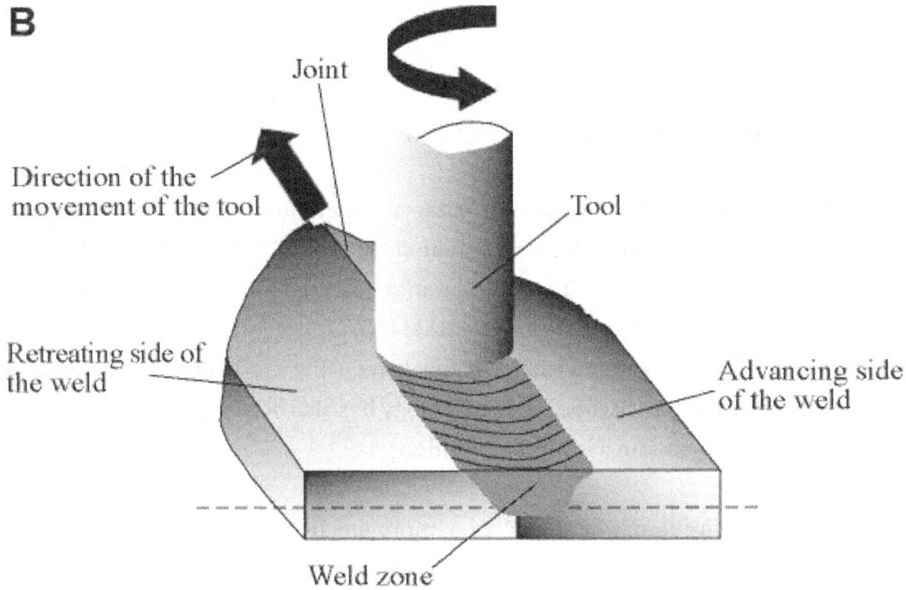

Figure 100: The Friction Stir Welding process. Anandwiki at English Wikipedia, CC BY 3.0, via Wikimedia Commons.

Friction Stir Welding (FSW) is a specialized welding technique used to join aluminium alloys, including commonly used grades such as Al5083, Al6061, and Al2014. Unlike traditional welding methods, which rely on melting the materials to create a joint, FSW operates in a solid-state process. This is achieved by using a rotating cylindrical tool that generates frictional heat as it moves along the seam between two workpieces. The heat softens the metal without liquefying it, allowing the materials to merge and form a strong bond under the combined effects of heat and mechanical pressure.

Aluminium alloys like Al5083, Al6061, and Al2014 are popular in industries requiring lightweight yet strong materials, including maritime, aerospace, and military applications. These alloys differ in composition and properties, making each suitable for specific uses. These alloys include:

Al5083: Marine-Grade Alloy

Al5083 is a non-heat-treatable alloy primarily known for its exceptional corrosion resistance, particularly in marine environments. It contains magnesium, manganese, and chromium, which enhance its resistance to seawater and industrial chemicals.

- **Properties:** Al5083 offers high strength and excellent fatigue resistance. Its mechanical properties remain consistent across a wide range of temperatures, making it well-suited for underwater applications.
- **Applications:** This alloy is widely used in the construction of ship hulls, submarines, pressure vessels, and cryogenic tanks. Its lightweight yet durable nature makes it ideal for naval vessels and other applications that demand structural integrity without compromising corrosion resistance.
- **Limitations:** Al5083 cannot be heat-treated to increase strength and is relatively soft compared to other high-strength aluminium alloys, making it less suitable for aerospace or high-impact applications.

Al6061: Versatile Structural Alloy

Al6061 is a heat-treatable alloy that contains magnesium and silicon as its principal alloying elements. It is known for being versatile, with a good balance of strength, corrosion resistance, and machinability.

- **Properties:** Al6061 offers medium-to-high strength, excellent corrosion resistance, and good weldability. It can be further strengthened through heat treatment processes such as T6 tempering, making it an excellent material for load-bearing structures.

- **Applications:** Al6061 is used across many industries, including aerospace, automotive, naval, and construction. In submarine applications, it is often employed in structural components, such as frames, supports, and piping systems, due to its high strength and ease of fabrication.

- **Limitations:** Although Al6061 is versatile, it does not provide the highest strength among aluminium alloys. In applications where extreme strength or fatigue resistance is required, other alloys may be preferred.

Al2014: Aerospace-Grade Alloy

Al2014 is a high-strength, heat-treatable alloy with copper as the primary alloying element, which gives it exceptional mechanical properties. This alloy is designed for applications requiring maximum strength but is less resistant to corrosion compared to Al5083 or Al6061.

- **Properties:** Al2014 offers high fatigue resistance, excellent machinability, and the ability to maintain strength under high-stress conditions. However, it has reduced corrosion resistance and typically requires surface treatments to protect it from environmental exposure.

- **Applications:** Due to its strength and fatigue resistance, Al2014 is extensively used in the aerospace and defence industries for components like aircraft structures, missile parts, and submarine equipment exposed to heavy loads. It is ideal for dynamic environments where weight and structural integrity are critical.

- **Limitations:** The copper content makes Al2014 more susceptible to corrosion, especially in saltwater environments, so it is often used with protective coatings or anodizing treatments to mitigate corrosion issues.

FSW is particularly advantageous for applications that require lightweight structures with enhanced strength, such as in naval vessels and submarines. Aluminium alloys, known for their favourable strength-to-weight ratios and corrosion resistance, are widely used in marine environments. However, traditional fusion welding methods can weaken aluminium by introducing thermal distortions, porosity, and defects at the joint. FSW overcomes these challenges by maintaining the original properties of the aluminium alloys, ensuring the welds are as robust as the base materials.

The process of FSW also minimizes the formation of defects like cracks or voids, which are common in high-temperature welding processes. It produces a uniform, high-strength joint with excellent fatigue resistance, crucial for components exposed to dynamic loads, such as those in ships and submarines. Additionally, since the material remains in the solid state throughout the process, FSW reduces residual stress and eliminates the need for filler materials, further enhancing the structural integrity and reliability of the joints.

FSW is highly efficient for the production of complex geometries and large structures, as it allows for precise control over the welding process. Its ability to produce consistent, high-quality welds with minimal post-processing requirements makes it a preferred choice for critical applications in the maritime industry. As a result, components manufactured using FSW contribute to the overall durability and performance of naval vessels, supporting both operational safety and long-term service life in demanding environments.

High Vibration Damping Aluminium Metal Matrix Composites

Molten flux-assisted reaction synthesis is used to create aluminium metal matrix composites with enhanced vibration damping properties. These composites contain ultrafine reinforcements of graphite and ceramics, making them suitable for use in submarine components where minimizing vibration is critical [347].

High vibration damping aluminium metal matrix composites (AMMCs) are engineered materials designed to mitigate mechanical vibrations, which is particularly vital in applications such as submarines, where structural integrity and stealth are paramount. The unique properties of these composites are achieved through various manufacturing techniques, including the incorporation of reinforcements like graphite and ceramics into a molten aluminium matrix. These reinforcements enhance the vibration-damping capabilities of the composites beyond what conventional aluminium alloys can offer [354, 355].

The synthesis process involves the introduction of ultrafine reinforcements into the molten aluminium matrix. Graphite particles are particularly effective in enhancing damping properties by reducing vibration transmission, while ceramic reinforcements contribute to the mechanical strength, hardness, and thermal resistance of the composite [356]. This combination of materials results in AMMCs that possess superior vibration-damping characteristics, which are essential for the operational efficiency of submarines.

In submarine design, the lightweight nature of AMMCs, combined with their vibration-damping capabilities, addresses critical challenges. Underwater operations generate vibrations from machinery such as engines and pumps, which can propagate through the vessel's structure. If these vibrations are not adequately controlled, they can lead to reduced operational lifespan of components and compromise the submarine's stealth, making it more detectable by sonar systems [357]. By utilizing AMMCs with high vibration-damping properties, submarines can operate more quietly and efficiently, minimizing noise transmission and mechanical stress [356].

Moreover, AMMCs exhibit excellent performance in harsh marine environments, demonstrating resistance to corrosion and fatigue over extended periods. This durability is crucial as submarine components are subjected to immense hydrostatic pressure and repeated mechanical loads. Additionally, AMMCs provide superior dimensional stability compared to pure aluminium, ensuring that critical components maintain their integrity under varying temperature and pressure conditions [354, 355].

In terms of application, high vibration damping AMMCs are utilized in key structural and mechanical components of submarines, including engine mounts, propulsion systems, and deck plates. These materials ensure that vibration-sensitive equipment, such as sensors and sonar systems, remain isolated from disturbances, thereby preserving the submarine's operational capabilities and stealth profile [356]. The combination of lightweight construction, enhanced strength, and superior vibration damping makes AMMCs invaluable for modern submarine design. Their ability to withstand demanding environmental conditions while reducing noise and mechanical stress ensures both performance optimization and increased longevity of submarine systems [354, 357].

New Rheo-Casting

This casting technique combines the advantages of traditional casting and forging, producing both thick and thin-walled aluminium alloy components with high precision. Rheo-casting is essential for creating intricate structures while maintaining mechanical properties comparable to forged materials.

Rheo-casting is an innovative manufacturing technique that integrates the advantages of traditional casting with the mechanical strength typically associated with forging. This method enables the production of both thick and thin-walled components from aluminium alloys with remarkable precision, making it particularly advantageous for industries requiring high-performance materials. The primary benefit of rheo-casting lies in its capacity to create complex and intricate structures that are often challenging to achieve with conventional casting methods while ensuring superior mechanical properties akin to those of forged materials [358, 359].

During the rheo-casting process, aluminium alloys are brought to a semi-solid state, characterized by a mixture of solid and liquid phases. This semi-solid condition imparts unique flow properties to the material, allowing it to fill moulds with enhanced accuracy. The improved flow characteristics reduce common defects associated with traditional casting techniques, such as porosity and shrinkage, resulting in components that are both highly detailed and dimensionally stable. Consequently, the need for extensive post-casting machining and finishing is minimized, which can lead to significant cost savings in manufacturing [358, 360].

The mechanical performance of components produced via rheo-casting is notably improved due to the minimization of internal defects. This process yields materials that exhibit excellent tensile strength, fatigue resistance, and overall durability, making rheo-cast parts comparable to those produced through forging. Unlike forging, however, rheo-casting accommodates complex geometries and thin-walled sections, broadening its applicability across various sectors, including aerospace, automotive, and naval engineering [359, 361].

Moreover, rheo-casting supports high-volume production while maintaining consistent quality, which is crucial for large-scale manufacturing. The combination of casting precision and the mechanical advantages of forging positions rheo-casting as a cutting-edge solution for industries that demand high-performance materials. This technique ensures that intricate, lightweight aluminium alloy components not only meet but exceed the stringent requirements of modern engineering applications [358, 359].

Bulk Metallic Glasses

Submarines also use Fe- and Cu-based bulk metallic glasses (BMGs), which exhibit excellent hardness and corrosion resistance. These materials are applied as coatings on steel and nickel-aluminium bronze components through high-velocity oxygen fuel (HVOF) spraying. BMG coatings enhance the durability of critical submarine components by providing surface hardness up to 1000 VHN (Vickers Hardness Number) [347].

Bulk metallic glasses (BMGs) are an innovative class of materials used in submarines to enhance the performance and longevity of critical components. Composed of iron (Fe) and copper (Cu)-based alloys, BMGs possess a unique amorphous structure, which means that the atoms within the material are arranged randomly, unlike the orderly crystalline structure seen in traditional metals. This lack of crystallinity imparts BMGs with exceptional mechanical properties, including superior hardness, high strength, and excellent corrosion resistance—all of which are critical in the harsh marine environment.

In submarine construction, BMGs are primarily used as protective coatings on components made from steel and nickel-aluminium bronze (NAB). These coatings are applied through High-Velocity Oxygen Fuel (HVOF) spraying, a technique that propels fine particles of the BMG material at extremely high speeds onto the surface of the component. The high velocity and impact during this process ensure that the coating bonds firmly to the substrate, creating a uniform and durable protective layer.

The HVOF process utilizes a combustion chamber where a mixture of oxygen and fuel gases (such as hydrogen, propane, or kerosene) is ignited, producing a high-temperature flame. The flame travels through a specialized nozzle, creating a supersonic jet of gases. Fine particles of the coating material, in powder form, are injected into this jet, where they are rapidly heated. The particles become semi-molten and are propelled at extremely high velocities—typically between 600 to 1000 meters per second—onto the target surface.

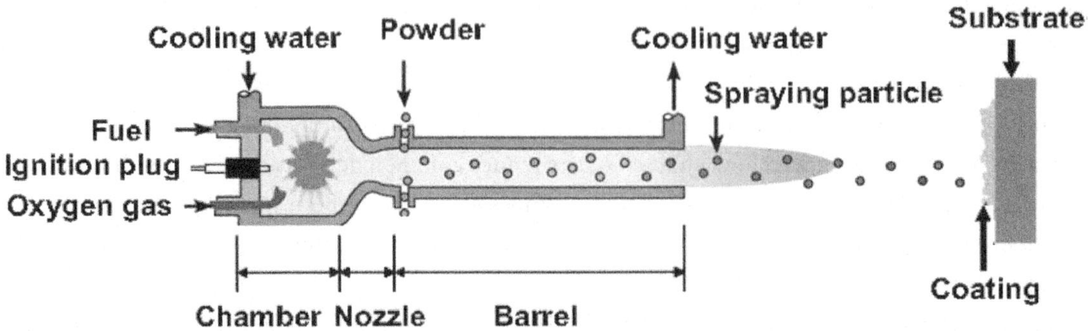

Figure 101: Schematic of high velocity oxy-fuel spraying. S. Kuroda et al., CC BY 4.0, via Wikimedia Commons.

When these particles strike the surface at such high speeds, they flatten and solidify instantly, forming a dense, tightly bonded layer. The coating adheres mechanically to the substrate, creating a strong interface that resists peeling or separation. The rapid solidification minimizes the formation of voids or defects, resulting in a dense, durable coating with excellent mechanical properties.

The key advantages of HVOF spraying include:

1. High bond strength: The extreme velocity of the particles ensures a strong mechanical bond between the coating and the underlying material, improving reliability even under stress.

2. Low porosity: The rapid cooling prevents porosity in the coating, making it highly resistant to corrosion and water ingress—critical for submarines operating in seawater environments.

3. Minimal thermal impact: HVOF spraying maintains relatively low substrate temperatures, preventing thermal distortion or weakening of the base material. This is especially important when applying coatings to precision components such as shafts and bearings.

In marine applications, such as submarine construction and maintenance, HVOF spraying is invaluable. Coatings made from Fe- and Cu-based bulk metallic glasses (BMGs) are applied using this process to enhance the hardness and corrosion resistance of key components. The HVOF-applied coatings provide exceptional wear resistance and extend the lifespan of components that would otherwise degrade rapidly due to exposure to seawater and mechanical stress.

This technique is also used to restore worn surfaces. Instead of replacing expensive components, worn parts can be recoated and repaired using HVOF spraying, reducing downtime and costs. The versatility of HVOF allows the use of a wide range of coating materials, making it suitable for diverse applications, from propellers and rudders to hydraulic cylinders and bearings.

The use of BMG coatings provides multiple advantages for submarine systems. One of the most important properties of these coatings is their extremely high hardness, with values reaching up to 1000 Vickers Hardness Number (VHN). This hardness improves the wear resistance of the coated components, ensuring they can withstand mechanical stress, abrasion, and other forms of surface degradation that occur in demanding underwater operations.

Moreover, BMGs offer excellent resistance to corrosion. Submarine components are constantly exposed to seawater, which can corrode traditional metals over time, compromising structural integrity and necessitating frequent maintenance. BMG coatings protect these components from corrosive elements, enhancing their longevity and reducing maintenance costs. This corrosion resistance is particularly valuable for parts like propellers, shaft bearings, and other critical mechanical elements, which must perform reliably without frequent replacement or repair.

The amorphous nature of BMGs also makes them less susceptible to microstructural defects that can cause fatigue over time. Traditional metals with crystalline structures are prone to crack formation under stress, especially in marine environments where fluctuating pressures and mechanical loads are common. BMG coatings mitigate this risk by providing a durable, defect-free surface, improving the overall reliability and safety of the submarine.

The use of Fe- and Cu-based BMG coatings in submarines is a significant advancement in materials engineering, ensuring that components remain functional and efficient under extreme conditions. By applying BMG coatings through HVOF spraying, manufacturers can enhance the performance, durability, and corrosion resistance of submarine components, making these vessels more reliable and cost-effective to operate over their lifecycle.

Piezocomposite Acoustic Sensing Materials

Innovative piezocomposite materials are used for advanced acoustic sensing. These materials are fabricated using the dice-and-fill method, enhancing sensitivity in sonar systems. Piezocomposites allow for precise acoustic detection, crucial for submarine operations in environments where detecting underwater threats and obstacles is essential [347].

Innovative piezocomposite materials are playing a pivotal role in enhancing acoustic sensing capabilities in submarines, addressing the need for precise underwater detection and reliable sonar operations. These advanced materials utilize the piezoelectric effect, a phenomenon in which mechanical stress generates an electric charge. This property allows the piezocomposites to detect acoustic waves in the water with exceptional sensitivity, making them essential for modern sonar systems tasked with identifying underwater threats, obstacles, and environmental changes.

Innovative piezocomposite materials are increasingly recognized for their significant role in enhancing acoustic sensing capabilities in submarines. These materials leverage the piezoelectric effect, where mechanical stress induces an electric charge, facilitating the detection of acoustic waves with exceptional sensitivity. Piezocomposites, particularly those designed with 1-3 connectivity, exhibit superior performance compared to traditional piezoelectric ceramics due to their unique combination of flexibility, durability, and enhanced signal quality, which are critical for reliable sonar operations in challenging underwater environments [362].

1-3 connectivity is a concept used in piezocomposite materials to describe a specific structural configuration that enhances their performance for applications such as sonar systems. In piezocomposites, the term "connectivity" refers to the way the piezoelectric ceramic phase and the polymer matrix are interconnected across different spatial dimensions. The 1-3 connectivity structure offers an effective blend of mechanical flexibility, sensitivity, and signal transmission, making it particularly suitable for acoustic sensing applications.

In a 1-3 piezocomposite, piezoelectric ceramic elements, typically in the form of rods or fibres, are aligned along a single dimension (hence the "1" in 1-3). These elements are embedded within a polymer matrix that provides continuity across three dimensions (represented by the "3"). The ceramic rods run parallel to each other and are evenly spaced within the matrix, without touching one another laterally. This arrangement allows the ceramic rods to provide a direct pathway for mechanical stress and electrical charge generation along their axis, while the polymer matrix surrounding them ensures structural integrity, flexibility, and mechanical isolation.

The 1-3 configuration overcomes many limitations associated with conventional piezoelectric materials, such as brittleness and lack of flexibility. By embedding the ceramic rods within a polymer matrix, the composite gains several key advantages. The polymer matrix provides mechanical flexibility, allowing the material to withstand vibrations and mechanical stresses without cracking or deteriorating. Furthermore, the matrix has a lower acoustic impedance than the ceramic rods, reducing signal loss at material interfaces and improving the performance of the piezocomposite in underwater environments.

Each ceramic rod operates independently within the matrix, minimizing interference and noise, which enhances the clarity and resolution of acoustic signals. This isolation also helps with mechanical decoupling, as the polymer matrix minimizes the transfer of structural vibrations, ensuring that only authentic acoustic signals are detected by the ceramic elements. Additionally, the geometry, density, and size of the rods can be customized to meet specific frequency ranges and acoustic needs, increasing the material's versatility for various applications.

In submarine sonar systems, 1-3 piezocomposites offer significant improvements in performance. Their enhanced sensitivity and reduced acoustic noise allow for more precise detection of underwater objects and threats. The durable and flexible nature of the 1-3 configuration ensures reliable operation in challenging marine environments, where exposure to water pressure, temperature fluctuations, and mechanical vibrations is constant. This makes the material ideal for long-term use, ensuring that sonar systems remain effective in diverse operational scenarios.

The fabrication of piezocomposite materials typically employs the dice-and-fill method, a precise manufacturing technique that involves segmenting piezoelectric ceramic elements into small sections, or "dices," which are then embedded in a polymer matrix. This structure minimizes internal mechanical stress and enhances mechanical coupling with the surrounding water, thereby improving the sensitivity and clarity of the acoustic signals detected [363-365]. By reducing cross-coupling effects between ceramic elements, piezocomposites can achieve higher sensitivity, which is vital for sonar systems that rely on the detection of subtle acoustic waves [366, 367].

The advantages of piezocomposites extend to their application in sonar systems, where they significantly improve the detection and identification of underwater objects, including vessels, mines, and geographical features. In environments where stealth is paramount, submarines utilize active and passive sonar arrays equipped with piezocomposites to maintain real-time situational awareness. The enhanced sensitivity of these materials ensures that even faint acoustic signals can be detected and analysed, allowing operators to identify potential threats at greater distances [368, 369].

Moreover, piezocomposites offer increased durability and longevity, essential for components operating under extreme underwater pressures. The polymer matrix within these composites provides flexibility, enabling sonar arrays to withstand mechanical stresses and fluctuations in water pressure during prolonged deployments. This combination of high sensitivity, reliability, and resilience positions piezocomposites as crucial components in modern submarine sonar systems, ensuring effective performance across various underwater operations [370, 371].

Stealth materials technology is a cornerstone of submarine operations, enabling these vessels to remain undetected in hostile environments. Acoustic rubber tiles, vibroacoustic insulation, and radar-absorbing coatings work together to minimize noise and radar signatures. Advanced materials such as metal matrix composites, bulk metallic glasses, and friction stir-welded components ensure that the submarine's structure is both lightweight and durable. With these technologies, submarines achieve the stealth, strength, and precision required for modern naval warfare.

Use of Composite Materials

One of the most transformative advancements in submarine technology is the introduction of advanced composites for hull construction and other critical components. These materials offer substantial weight savings, making submarines lighter and enhancing their manoeuvrability. Traditional materials like steel, while strong, add considerable mass, but advanced composites allow designers to achieve comparable or superior strength with much lower weight. This shift not only improves the submarine's agility but also enables it to carry more equipment or payload without sacrificing performance [348].

Composites are also more resistant to corrosion than conventional metal alloys, a crucial property given the harsh marine environment in which submarines operate. Corrosion resistance translates to improved longevity of the hull and significantly reduces maintenance requirements, as there is less need for repairs or anti-corrosion treatments. Furthermore, advanced composites possess greater durability, enabling them to withstand the intense pressures experienced during deep dives and high-stress conditions. The use of these materials ensures that submarines remain operational for longer periods without structural fatigue or failure, increasing their reliability during extended missions [348].

Russia is at the forefront of incorporating advanced composites in its new generation of nuclear-powered submarines, marking a significant leap forward in submarine technology. These submarines will utilize composite materials not only in their hulls but also in other crucial components, delivering substantial improvements in both stealth and endurance. The reduced weight and enhanced durability contribute to quieter operations, making the vessels harder to detect through sonar or other acoustic systems. This development positions Russia's submarines at the cutting edge of maritime stealth technology [348].

One of the most noteworthy innovations is Russia's testing of composite propellers, which could revolutionize submarine propulsion. Propellers are among the most critical elements influencing a submarine's performance, and their design must minimize noise and vibrations to maintain stealth. The use of composites in propeller blades reduces vibration and improves efficiency, resulting in a lower acoustic signature. This makes the submarine less detectable by sonar, offering a strategic advantage in naval operations. Enhanced efficiency also allows for quieter propulsion, further boosting the submarine's ability to remain undetected during covert missions.

While Russia leads in integrating composites into next-generation submarines, other nations, including the United States, are also exploring these materials [348]. As the benefits of composites—such as weight reduction, increased durability, and stealth enhancement—become more apparent, it is expected that more countries will adopt these technologies in their submarine fleets. Russia's pioneering use of composites will likely influence global trends, prompting advancements in submarine construction around the world as nations seek to maintain competitive advantages in naval technology [348].

Welding and Fabrication Techniques

Manufacturing Processes of Submarines

The manufacturing process of a submarine is intricate, requiring both manual craftsmanship and automated precision. Each step ensures that the vessel meets the highest standards of structural integrity, functionality, and safety, given the demanding conditions it will encounter underwater.

Hull Construction

The primary material used in submarine construction is steel, with plates typically 2 to 3 inches (5.1–7.6 cm) thick. These steel plates are initially cut to precise dimensions using acetylene torches. After cutting, the plates are passed through large metal rollers under immense pressure. These rollers, measuring about 28 inches (71.1 cm) in diameter and 15 feet (4.6 meters) in length, curve the plates by rolling them back and forth until the required curvature is achieved [347].

Once curved, the plates are assembled around wooden templates that define the hull's shape. These plates are hand-welded together to form individual sections of the hull. The sections are then aligned using cranes and slowly rotated beneath an automatic welder, ensuring the seams are sealed with exceptional strength. Multiple passes beneath the automatic welder reinforce these seams. To further enhance the structure, T-shaped ribs of steel are welded to the hull sections. These ribs are formed by heating steel bars to a pliable state and bending them to match the hull's curvature.

Inner and Outer Hull Assembly

The submarine's hull comprises two layers: the inner and outer hulls. The inner hull encloses the submarine's interior, containing the crew and critical equipment, while the outer hull provides an external barrier. Both hulls are fabricated by welding multiple sections together using the same process. Steel ribs are positioned between the two hulls to create space for ballast tanks, which control the vessel's buoyancy. These ribs not only add strength but also separate the inner and outer hulls.

Inside the inner hull, additional steel plates are welded in place to divide the submarine into watertight compartments. Steel decks and bulkheads are installed to further reinforce the structure. Exterior seams on the hull are polished using high-speed grinding wheels to achieve a smooth finish, which reduces drag and prepares the surface for painting. Once smoothed, multiple layers of protective coatings are applied to shield the hull from corrosion and enhance its durability.

Throughout the manufacturing process, scaffolding is erected around the hull, ensuring workers can access all areas efficiently. Inspections and quality control measures are rigorously conducted at every stage. For example, welded seams undergo x-ray testing to detect imperfections, while pipes are filled with helium to identify leaks. This meticulous approach ensures the highest safety standards, making the submarine's nuclear reactors among the safest in the world [347].

Figure 102: The forward section and sail unit of US Navy (USN) Virginia Class Submarine Pre-Commissioning Unit (PCU) NORTH CAROLINA (SSN 777), under construction at Northrop Grumman Newport News Shipyard, Virginia (VA). The U.S. National Archives, Public Domain, via Picryl.

Exterior Finishing

After the hull is assembled, external components such as rudders, propellers, and sonar equipment are manufactured and installed. Many metal components are produced using sand casting, a process that involves creating a wooden or plastic model of the desired part. This model is placed in a mould filled with hardened sand. Once the model is removed, molten metal is poured into the resulting cavity and allowed to cool, forming the final component [347].

The completed components are either welded or otherwise attached to the hull. To minimize underwater drag, sonar equipment is mounted onto the hull and covered with smooth steel sheets. Scaffolding remains in place during this phase to provide workers with easy access to all areas of the submarine.

Figure 103: Astute Class submarine Audacious under construction at Barrow in Furness shipyard in Cumbria. Defence Imagery, CC BY-SA 2.0, via Flickr.

Interior Finishing and Fitting-Out

During the construction of the inner hull, large equipment such as engines and machinery is installed. Once the inner hull is sealed, smaller equipment is brought inside. However, many essential components are not installed until after the submarine is launched. Following the ceremonial launch, the submarine is towed to a fitting-out dock, where the remaining interior work takes place.

At the dock, critical systems such as periscopes, snorkels, electronic equipment, and propulsion systems are installed. In addition to operational components, equipment to support crew comfort, such as refrigerators, electric stoves, air conditioning units, and washing machines, is also fitted. This comprehensive outfitting ensures that the submarine is fully prepared for long deployments [347].

Alloys Used in Submarine Construction

Submarine manufacturing involves the use of various alloys chosen for their specific properties. Materials used in seawater systems must be resistant to corrosion, easy to fabricate, and cost-effective while maintaining high strength-to-weight ratios. These alloys include both wrought and cast forms of copper, nickel, titanium, and aluminium, depending on the component's function.

- **Copper-based alloys** are often used for components exposed to seawater due to their excellent corrosion resistance.
- **Nickel-based alloys** are employed where both strength and resistance to high-pressure environments are required.
- **Titanium alloys** offer exceptional strength while being lightweight, making them ideal for critical structural elements.
- **Aluminium alloys** are used for lightweight components where corrosion resistance is essential, particularly in the superstructure or internal fittings.

The combination of these materials ensures that submarines are not only robust but also efficient, withstanding the harsh underwater environment while maintaining operational performance. The careful selection of materials, combined with precise manufacturing processes, results in submarines that meet the rigorous demands of modern naval operations.

Applicable Welding techniques

Welding plays a critical role in submarine construction, ensuring structural integrity, pressure resistance, and durability under extreme marine conditions. Due to the unique requirements of submarines, various advanced welding techniques are employed to join components, particularly for the hull, which must withstand significant hydrostatic pressure at depth. Submarine welding demands high precision to prevent defects, as any failure in the welded seams could compromise the vessel's integrity. Below are some of the key welding techniques used in submarine manufacturing.

Friction Stir Welding (FSW)

Friction Stir Welding is widely used in submarine construction, especially for joining aluminium alloys like Al5083 and Al6061, which are often used for lightweight internal components. FSW is a solid-state welding technique that involves a rotating tool generating frictional heat to soften and join metal surfaces without melting them. This process ensures a strong bond without introducing heat-affected zones that could weaken the structure. FSW is essential in submarine applications where weight savings are critical, and the joints need to maintain strength and corrosion resistance, such as in non-structural parts, tanks, and lightweight bulkheads.

Friction Stir Welding (FSW) is a solid-state joining process that fuses two metal workpieces without melting them, using a rotating, non-consumable tool to generate frictional heat. This process is particularly advantageous for applications that demand high weld strength and durability, such as submarine hulls, aerospace structures, and automotive components. By maintaining the material in a plasticized state rather than a molten one, FSW avoids many common welding defects and ensures robust joint quality across a variety of metals, including aluminium, titanium, magnesium, and steel.

The process begins by placing two workpieces together along a joint. The FSW tool, consisting of a profiled pin and a shoulder, is plunged into the material along the seam until the pin penetrates below the surface, while the shoulder rests atop the material. The friction from the rotating tool, combined with the pressure applied, generates heat, causing the material around the pin to soften without melting. The softened material flows around the pin and consolidates as the tool advances along the joint line. This movement, along with the applied mechanical pressure, forges the materials together, forming a strong, defect-free weld.

The unique characteristics of FSW produce distinct microstructural zones in the welded material. These include the stir zone (or dynamically recrystallized zone), the thermo-mechanically affected zone (TMAZ), and the heat-affected zone (HAZ). The stir zone, where the material undergoes severe deformation and mixing, exhibits small, equiaxed grains. This zone is often marked by an "onion-ring" pattern, a result of complex material flow. Surrounding the stir zone is the TMAZ, where the original microstructure is deformed but still recognizable. Beyond this region lies the HAZ, which experiences thermal cycles without deformation, potentially affecting mechanical properties, especially in age-hardened aluminium alloys.

FSW offers several advantages over traditional fusion welding techniques. Since the material remains in the solid state, the risks of porosity, cracking, and other solidification defects are significantly reduced. FSW welds also exhibit excellent mechanical properties, such as high strength and fatigue resistance, making the technique suitable for critical applications like submarine hull construction. Furthermore, FSW produces minimal toxic fumes and requires no filler material or shielding gas, contributing to a safer and more cost-effective welding process.

However, FSW also has limitations. The tool leaves an exit hole where it is withdrawn from the weld, and the process requires significant downward force, necessitating robust clamping systems. Additionally, FSW is less flexible than manual welding methods, making it more challenging to accommodate variations in material thickness or perform non-linear welds.

Several parameters influence the success of an FSW operation, including tool design, rotation speed, traverse speed, and plunge depth. The tool must be made from materials capable of withstanding high temperatures and abrasion, such as hot-work tool steel. Rotation and traverse speeds are carefully balanced to ensure adequate heat generation without overheating, which could lead to defects. The tool's plunge depth and tilt angle also play a critical role in ensuring proper material flow and joint consolidation. Tools designed with features like tapered pins or re-entrant flutes enhance material movement and improve weld quality.

FSW has found applications across multiple industries, including shipbuilding, aerospace, and automotive manufacturing. It is particularly valuable for joining lightweight alloys, such as aluminium, where traditional welding techniques may cause distortion or compromise strength. The material flow during FSW follows an intricate pattern. Some material rotates around the pin, forming arc-shaped features, while other material flows around the tool's sides, ensuring thorough mixing and bonding.

Figure 104: Friction Stir Welding. AstridMitch, CC BY-SA 4.0, via Wikimedia Commons.

Heat generation during FSW results from both friction and material deformation. Controlling the heat input is critical to maintaining weld quality. Excessive heat can degrade mechanical properties, while insufficient heat can lead to defects. Modern FSW systems incorporate sophisticated control mechanisms to maintain optimal tool position, load, and heat input, ensuring consistent welds even under challenging conditions. Advanced process models and simulations further aid in optimizing parameters and predicting residual stresses, contributing to the overall reliability and efficiency of the process.

Friction Stir Welding (FSW) involves specialized equipment to ensure precise control over the welding process. Below is a detailed overview of the tools and machinery required to perform FSW effectively.

1. FSW Machine or Platform

FSW machines are robust systems designed to withstand high forces and maintain precise movements during welding. These machines come in various configurations, such as gantry, column-and-boom, or robotic systems. Key features include:

- **Adjustable rotational speed** to control heat generation.
- **Traversing mechanisms** for smooth linear movement along the joint line.
- **Load control systems** to maintain appropriate downward pressure during welding.Some traditional milling machines can also be adapted for small-scale FSW, but dedicated FSW machines offer better control and pre-

cision.

2. FSW Tool

The FSW tool is essential to the welding process. It consists of two main parts:

- **Shoulder:** A cylindrical section that contacts the surface of the material. It applies downward pressure and generates frictional heat.
- **Pin (or Probe):** A smaller-diameter, profiled section extending from the shoulder, penetrating into the material. The pin stirs and forges the material as it rotates.

Tool Materials:

- **H13 Tool Steel**: Common for aluminium welding.
- **Tungsten Carbide**: Used for welding higher-temperature metals like steel and titanium.
- **PCBN (Polycrystalline Cubic Boron Nitride):** For demanding applications such as aerospace-grade welding.

The tool design (e.g., Triflute or Whorl) impacts material flow and welding quality, especially in challenging materials.

3. Workpiece Clamping System

Proper clamping is crucial to avoid material shifting during welding. Heavy-duty clamps or a **hydraulic clamping system** ensures the workpieces remain securely in place.

- **Customized fixtures** are often used for complex geometries or non-linear welds.
- **Backing plates** (usually made from copper) are placed under the workpieces to prevent material loss and provide support during welding.

4. Cooling System

As FSW generates significant heat, a cooling system prevents damage to the tool and machine.

- **Air or Water Cooling**: A coolant is directed onto the tool and the weld area to maintain optimal temperatures.
- **Internal Tool Cooling**: Some tools feature channels for circulating coolant through the tool itself.

5. Control System and Software

Modern FSW machines use **computerized control systems** for precision welding. The software regulates:

- **Rotational speed** of the tool.
- **Traverse speed** along the joint line.
- **Plunge depth** to maintain the correct welding depth.
- **Tilt angle** of the tool, typically between 2-4 degrees, to enhance material forging.

The control software can also log process data for quality control and traceability.

6. Force and Torque Sensors

Integrated force sensors monitor the **downward force**, **torque**, and other forces acting on the tool during welding. These sensors ensure that:

- The weld remains consistent.
- The tool operates within safe force limits to prevent breakage.

7. Temperature Monitoring System

Thermocouples or infrared cameras track the **temperature** in the weld zone to ensure optimal heat input. Precise temperature control helps avoid flaws such as voids or under-penetration.

8. Backup Power and Redundancy Systems

FSW machines used in critical applications, such as submarines, may include:

- **Emergency power backup systems** to prevent disruptions.
- **Redundant motors or control systems** to ensure welding can continue smoothly in case of component failure.

9. Non-destructive Testing (NDT) Equipment

After welding, NDT tools ensure the integrity of the welds. Common methods include:

- **Ultrasonic Testing (UT)**: Detects internal defects like voids or cracks.
- **X-ray Inspection**: Reveals subsurface flaws in the weld.

10. Personal Protective Equipment (PPE)

Operators need PPE, even though FSW is safer than traditional welding. Required PPE includes:

- **Heat-resistant gloves** for handling materials.
- **Safety glasses** to protect from metal debris.
- **Protective clothing** to guard against accidental contact with hot surfaces.

*Figure 105: Fillet joint friction stir welding (FSW) equipment for welding the fillet joint of an electric vehicle frame. Hwanjin Kim,1 Kwangjin Lee,2Jaewoong Kim,3 Changyeon Lee,4 Yoonchul Jung5 and Sungwook Kang,6*1 Department of Mechanical Engineering, Gyeongsang National University, Jinju 52828, Korea2 Carbon Materials Application R&D Group, Korea Institute of Industrial Technology, Jeonju 54853, Korea3 Smart Mobility Materials and Components R&D Group, Korea Institute of Industrial Technology, Gwangju 61012, Korea4 Technical Center: Daejoo Kores Co., LTD., Wanju 55316, Korea5 Dongnam Division, Korea Institute of Industrial Technology, Busan 46938, Korea6 Precision Mechanical Process and Control R&D Group, Korea Institute of Industrial Technology, Jinju 52845, Korea* Author to whom correspondence should be addressed., CC BY 4.0, via Wikimedia Commons.*

The following provides a step-by-step breakdown of the FSW process:

Step 1: Setup and Preparation of the Workpieces

1. **Material Selection:** Ensure the materials to be welded are compatible with FSW. Common materials include aluminium alloys, copper, magnesium, and titanium.

2. **Edge Preparation:** The edges of the workpieces must be clean and aligned along the joint line to ensure proper

welding. Contaminants like grease, dust, or oxides are removed.

3. **Clamping the Workpieces:** Place the workpieces securely on the welding platform using heavy-duty clamps. The clamping system must prevent any movement during welding to maintain the joint's integrity.

4. **Tool Inspection:** Check the FSW tool for wear. The tool consists of a cylindrical shoulder with a profiled pin, both critical for producing a consistent weld.

Step 2: Positioning the Tool

1. **Align the Tool:** Position the tool's pin above the joint line where the two workpieces meet. Ensure that the shoulder is slightly above the surface of the material.

2. **Adjust the Plunge Depth:** Set the plunge depth so that the pin penetrates just below the full thickness of the material, ensuring complete welding through the joint.

Step 3: Dwell Time – Preheating the Material

1. **Tool Insertion:** Lower the tool until the pin penetrates the joint and the shoulder makes contact with the surface.

2. **Dwell Time:** Allow the tool to rotate in place without advancing. This generates frictional heat to soften the material without melting it, preparing it for plastic flow.

Step 4: Initiating the Welding Process

1. **Begin Traversing the Tool:** Once the desired temperature is achieved, move the rotating tool along the joint line at a pre-determined welding speed.

2. **Material Flow:** As the tool advances, the softened material from the leading edge flows around the tool pin and consolidates behind it, creating the weld joint. The shoulder applies downward pressure, ensuring proper forging of the material.

Step 5: Maintaining Optimal Parameters

1. **Monitor Tool Rotation and Speed:** Ensure the rotation speed is consistent and the traverse speed matches the desired weld quality. Higher rotation speeds increase heat, while higher traverse speeds reduce heat.

2. **Tool Tilt Angle:** Maintain a slight tilt (2-4 degrees) to assist in forging the material as the tool moves forward. This tilt ensures better bonding and weld consistency.

Step 6: Exiting the Weld

1. **Slow the Tool:** As the tool approaches the end of the joint, gradually reduce the traverse speed.

2. **Withdraw the Tool:** Lift the tool smoothly from the material, leaving a small exit hole at the point where the pin is removed. This hole can later be filled or machined as needed.

Step 7: Post-Weld Inspection and Finishing

1. **Visual Inspection:** Check the weld for uniformity and surface quality. A good weld should have minimal surface deformation.

2. **Non-destructive Testing (Optional):** Use X-ray or ultrasonic testing to detect any internal defects that may not be visible on the surface.

3. **Finishing:** If necessary, perform light machining or polishing to smooth out the exit hole or remove any excess material.

Step 8: Final Quality Check

1. **Mechanical Testing (Optional):** Conduct tensile, fatigue, or hardness tests to verify the weld's strength and durability.

2. **Performance Validation:** Ensure that the welded component meets the operational requirements of the application, such as underwater use in submarines or aerospace structures.

Friction Stir Welding is a precise, controlled process that produces high-strength joints in a variety of metals. Following each step carefully ensures a reliable, defect-free weld. This process is especially valuable for applications like submarine hulls, where structural integrity is critical. With proper setup, monitoring, and inspection, FSW offers superior weld quality, minimal environmental impact, and enhanced mechanical performance.

Friction Stir Welding (FSW) is governed by various international standards to ensure the production of high-quality welds, particularly in industries where safety and reliability are essential, such as aerospace, marine, and automotive sectors. These standards address a range of elements, including process specifications, equipment requirements, testing protocols, and quality control measures to guarantee performance under demanding conditions.

ISO 25239 provides comprehensive guidelines for FSW when working with aluminium alloys. This standard is divided into five parts, covering essential aspects such as terminology, equipment specifications, operator qualifications, welding procedure specifications (WPS), and quality requirements, including inspections and testing. Widely adopted in industries like aerospace and marine, ISO 25239 ensures consistency and reliability in weld quality when joining aluminium alloys, which are commonly used in these sectors.

AWS D17.3/D17.3M, issued by the American Welding Society, focuses on the use of FSW in aerospace applications. This standard outlines precise guidelines for welding aerospace components, addressing welding procedures, qualification requirements, and non-destructive testing (NDT) methods. It also emphasizes performance testing to verify the mechanical properties of welds, ensuring they meet the stringent demands of dynamic aerospace environments.

BS EN 14127 is a European standard that provides procedures for non-destructive testing specific to FSW joints. It focuses primarily on ultrasonic testing to detect defects such as voids, inclusions, or incomplete penetration in the welds. The standard ensures that critical components maintain their structural integrity through rigorous inspection, which is vital in high-performance applications.

ISO 15614-12 extends FSW standards to include metals beyond aluminium, such as copper and titanium alloys. This part of the ISO 15614 standard ensures that FSW procedures meet mechanical performance requirements necessary for challenging environments, including marine and defense sectors, where reliability is critical.

The American Society of Mechanical Engineers (ASME) Section IX of the Boiler and Pressure Vessel Code provides guidelines for the qualification of welding procedures and operators for critical applications, including FSW. Although traditionally focused on fusion welding methods, the ASME standard can be adapted to incorporate FSW for specific project requirements, ensuring compliance with safety and quality benchmarks.

In aerospace manufacturing, NADCAP accreditation is often required for organizations performing FSW. The National Aerospace and Defense Contractors Accreditation Program (NADCAP) ensures compliance with aerospace industry standards through detailed audits and quality control, guaranteeing the reliability of welded components.

ISO 3834, although initially designed for fusion welding, provides a quality management framework applicable to FSW. This standard emphasizes the importance of documentation, process control, and inspection procedures, which are essential for maintaining consistent quality in manufacturing environments.

In submarine construction, FSW is gaining prominence for joining lightweight materials such as aluminium alloys and advanced composites. Standards like ISO 25239 and AWS D17.3 ensure that welded components meet the necessary strength, durability, and corrosion resistance requirements for submarine operations. Non-destructive testing procedures outlined in BS EN 14127 play a critical role in verifying weld integrity, helping to ensure the safety and performance of submarines operating under high-pressure conditions at depth. These standards collectively ensure that FSW processes in submarine construction adhere to the rigorous demands of marine environments, providing reliable and high-quality welds for critical applications.

Manual Metal Arc (MMA) Welding

Manual Metal Arc welding (MMA), also referred to as Shielded Metal Arc Welding (SMAW), is a widely utilized welding technique, particularly in the construction of submarines where high-strength steel is a primary material. This method is characterized by the formation of an electric arc between a consumable electrode and the base material, which allows for effective fusion of metals. The simplicity and versatility of MMA make it especially advantageous for applications in hard-to-reach areas, which is often the case in submarine construction [372].

The MMA process is particularly favoured for its ability to weld thick steel plates, which are essential for the structural integrity of submarine hulls. The technique is capable of producing strong welds that meet the demanding requirements of underwater environments. Research has shown that MMA welding can achieve good mechanical strength in high-strength low-alloy (HSLA) steels, although challenges remain in ensuring toughness and fatigue strength in multi-run welds [373, 374]. The ability to perform precise repairs during both construction and maintenance phases is a critical aspect of MMA welding, especially when addressing the pressure hull of submarines, where structural reliability is paramount [374].

Moreover, the use of MMA welding in submarine construction is supported by its effectiveness in managing residual stresses and ensuring the integrity of welded joints. Studies indicate that the choice of welding parameters and the chemical composition of the base materials significantly influence the mechanical properties of the welds produced [375, 376]. This is particularly relevant for high-strength steels, where the welding process must be carefully controlled to prevent issues such as cracking and reduced toughness [377]). The versatility of MMA welding allows for its application in various welding positions and conditions, making it an essential technique in the maritime industry [372, 373].

Manual Metal Arc Welding (MMAW), also known as Shielded Metal Arc Welding (SMAW), plays a crucial role in the construction, repair, and maintenance of submarines. This welding technique is especially valuable in areas requiring precise manual control. SMAW's versatility makes it suitable for both structural and non-structural components within submarine hulls and internal frameworks, ensuring reliable and durable welds.

SMAW is extensively used in hull fabrication and repairs, particularly for welding the inner and outer hull sections. In locations where automated welding systems cannot access, such as tight spaces within ballast tanks, SMAW allows for manual intervention. Reinforcing pressure hull seams and performing structural repairs in confined spaces is essential to maintaining the submarine's integrity under high pressure. This welding technique provides the level of control necessary to meet the high standards required for submarine construction.

Inside the submarine, SMAW is also applied to structural frameworks, such as bulkheads, decks, and ribs, which support the pressure hull and enable the vessel to withstand hydrostatic pressure and internal stresses. Smaller internal components like brackets, frames, and mounting points, which secure equipment and pipelines, are joined using SMAW, ensuring these critical attachments remain stable during operation. The adaptability of SMAW makes it an effective choice for such diverse tasks.

The submarine's extensive piping and plumbing systems rely heavily on SMAW. This includes the welding of pipes that carry coolant, fuel, hydraulic fluid, and air. SMAW provides strong, reliable welds necessary for these high-pressure systems. It is especially advantageous for on-site modifications and repairs where access may be limited, ensuring that critical systems remain functional.

During operations or at sea, SMAW is indispensable for emergency repairs and maintenance. Its portability allows it to be used for essential tasks such as patching leaks, repairing damaged hull sections, and re-welding deteriorated joints. In confined areas where automated welding equipment cannot be deployed, SMAW enables manual precision, allowing vital repairs to be completed promptly to ensure the submarine's operational safety.

The suitability of SMAW for submarine construction lies in its portability and accessibility. The lightweight equipment makes it ideal for working in tight spaces or areas with limited access, such as ballast tanks or piping systems. Its simplicity allows for effective use even during in-service maintenance at sea or in remote locations. SMAW is also versatile, as it can be used with various electrodes to join different metals commonly found in submarines, including high-strength steel, stainless steel, and nickel alloys. This adaptability ensures the technique meets the requirements for welding diverse materials found in the hull, piping, and internal components.

SMAW produces strong, durable welds capable of withstanding mechanical stresses and pressure variations during submarine operations, including deep dives. The excellent joint penetration achieved with SMAW ensures that the welds remain secure even under extreme conditions. Moreover, the technique's ability to deliver precise results in confined or irregular spaces ensures that critical joints are welded accurately, preventing failures in essential components.

However, SMAW does present some challenges. Skilled welders are required to ensure high-quality welds, especially for areas that are critical to the submarine's structural integrity, such as the pressure hull. Post-weld inspections are necessary to identify and rectify any potential defects, particularly in high-pressure systems or critical components. Additionally, SMAW can produce slag and spatter, which must be cleaned thoroughly to maintain the submarine's performance and corrosion resistance. Proper cleaning ensures that the vessel remains in optimal condition, minimizing the risk of degradation over time.

Figure 106: Shielded Metal Arc Welding. Weldscientist, CC BY-SA 4.0, via Wikimedia Commons.

Manual Metal Arc Welding (MMAW), also known as Shielded Metal Arc Welding (SMAW), requires a variety of tools and equipment to ensure high-quality welds, particularly in demanding environments like submarine construction. Each component plays a critical role in the process, enabling effective control over the arc, heat, and the overall welding operation.

The primary equipment used for MMAW includes the welding power supply, which generates the electrical current necessary to create and maintain the welding arc. This power supply can provide either alternating current (AC) or direct current (DC), depending on the specific requirements of the welding task and the materials being joined. DC power is often preferred for applications involving critical submarine components due to its stable arc and superior control.

The welding electrode, also known as the welding rod, is another crucial component. These electrodes are metal rods coated with flux, which serves to protect the weld pool from atmospheric contamination. When the electrode melts, the flux forms a protective gas shield and slag layer, preventing oxidation and ensuring a clean, strong weld. Different types of electrodes are used for various metals, such as steel, stainless steel, and nickel alloys, which are commonly found in submarine construction.

A welding holder or electrode holder is used to grip the electrode securely and connect it to the power source. This holder allows the welder to control the electrode's movement with precision. Connected to the power source by a cable, the electrode holder ensures consistent current flow, which is critical for maintaining a stable arc during the welding process.

Ground clamps are essential components that complete the welding circuit. These clamps connect the workpiece to the power supply, ensuring that the current flows efficiently through the electrode, workpiece, and back to the power source. A secure connection between the ground clamp and the metal being welded is necessary for smooth welding operations.

Personal protective equipment (PPE) is indispensable for ensuring the safety of welders during MMAW. This includes welding helmets with auto-darkening lenses to protect the welder's eyes from the intense arc light, as well as gloves, leather aprons, and flame-resistant clothing to shield the welder from sparks and heat. Proper ventilation or fume extraction equipment is also required to ensure a safe working environment by removing hazardous fumes generated during welding.

In addition to the core equipment, supporting tools such as chipping hammers, wire brushes, and angle grinders are used to prepare the workpiece and remove slag after welding. These tools help ensure a clean surface, which is crucial for achieving high-quality welds. Welders also use clamps, magnets, and jigs to hold workpieces in place, providing stability and precision throughout the welding process.

Welding machines used for MMAW are often portable, making them suitable for applications in tight or confined spaces, such as those found inside submarines. Some welding power supplies include advanced features like variable current settings and arc force control, providing welders with better control over the welding arc, especially in challenging environments or when working with complex materials.

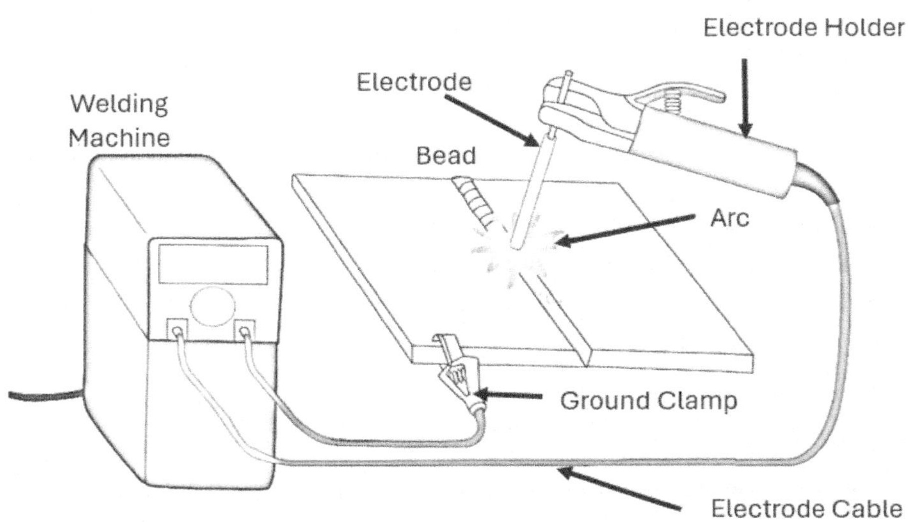

Figure 107: Schematic Diagram of MMAW Welding System.

The following provides a step-by-step breakdown of the MMAW process:

Step 1: Preparation of the Work Area and Equipment

In submarine construction, welding is often carried out in confined spaces such as the hull, ballast tanks, or engine rooms, making safety and workspace preparation critical. Start by ensuring the welding area is well-ventilated or equipped with a fume extraction system to reduce toxic gas build-up. All necessary equipment, including the welding machine, electrodes, grounding cables, and personal protective equipment (PPE), should be inspected and set up for the task.

Checklist for preparation:

- Confirm the integrity of the electrode holder and ground clamp connections.

- Choose appropriate electrodes for the metal being welded (e.g., stainless steel or nickel alloy).
- Prepare the workpieces by cleaning surfaces of rust, oil, and debris using wire brushes or grinders.
- Set up scaffolding if welding takes place at elevated points or within ballast tanks.

Step 2: Selection of Electrode and Power Settings

Choosing the right electrode is essential for achieving high-quality welds in submarines, where steel alloys and other materials like stainless steel are commonly used. Electrode selection depends on the base metal, the position of the weld (horizontal, vertical, or overhead), and the operating conditions. For example:

- **E6010** electrodes are ideal for root passes and out-of-position welding.
- **E7018** electrodes are commonly used for high-strength steel joints.

Set the welding machine to the required current and polarity based on the electrode type. Submarine construction often uses DC positive polarity for more stable arcs and better penetration.

Step 3: Positioning and Clamping the Workpieces

In submarine construction, the alignment of hull sections and structural elements is critical. Use clamps, jigs, or magnets to position and secure the workpieces. For hull repair, welding in a confined area such as a ballast tank might require specialized fixtures or braces to keep the sections aligned during welding.

- Ensure the gap between the parts matches the electrode size to allow proper penetration.
- Use tack welds (small temporary welds) to hold components in place before the full welding pass.

Step 4: Striking the Arc and Starting the Weld

To begin welding, the welder strikes the electrode tip against the workpiece, creating an electrical arc. This arc generates the heat needed to melt the electrode and the workpiece, forming the weld pool. The key to success in submarine construction is maintaining precise control over the arc length, which ensures consistent penetration without overheating the metal.

- Hold the electrode at an angle of 10-15 degrees relative to the direction of the weld.
- Keep the arc length short, roughly equivalent to the diameter of the electrode, to ensure good penetration and arc stability.

Step 5: Running the Bead

As the arc melts the electrode, both the electrode and flux coating are consumed. The molten flux creates a gas shield and slag layer that protect the weld pool from contamination. Move the electrode steadily along the weld line, maintaining a uniform speed to ensure even distribution of the molten metal.

- Use stringer beads (straight weld passes) for confined areas where access is limited.
- Employ a weaving motion for wider welds, especially for joining thicker hull sections.
- Monitor the pool size to avoid excessive heat build-up, which can warp thin plates or compromise the submarine's structural integrity.

Step 6: Slag Removal and Inspection Between Passes

Each pass leaves a layer of slag that must be removed before applying the next weld layer. Use a chipping hammer and wire brush to clean the slag. For submarine components, where multiple weld passes are often required (e.g., in pressure hulls), it is essential to remove all slag to prevent defects between layers.

- Inspect the weld visually for uniformity, cracks, or other imperfections.
- Use non-destructive testing methods, such as ultrasonic testing or dye penetrant inspection, for critical components like the pressure hull or high-pressure piping systems.

Step 7: Multi-Pass Welding for Thick Sections

Submarine construction often involves welding thick steel sections, requiring multiple passes to ensure strength and durability. For thick joints, use root passes followed by filler passes and a cap pass. Each layer must fuse properly with the previous one to prevent voids and cracking under pressure.

- Apply root passes to ensure deep penetration and solid bonding at the base of the joint.
- Use intermediate filler passes to build up the weld, ensuring the joint remains strong.
- Finish with a cap pass, creating a smooth surface that enhances structural integrity.

Step 8: Cooling and Post-Weld Treatment

Allow the weld to cool naturally to avoid introducing thermal stress. In submarine applications, where materials are subjected to extreme pressure and vibration, post-weld heat treatment (PWHT) may be required to relieve residual stresses.

- PWHT ensures the weld maintains its mechanical properties, particularly in high-pressure hull sections.
- Inspect for defects using radiographic or ultrasonic testing methods, especially in areas subject to critical stress during submarine operations.

Step 9: Final Inspection and Testing

In submarine construction, welds undergo rigorous testing to ensure structural integrity and compliance with safety standards. Both visual inspections and non-destructive testing (NDT) methods are employed to detect any defects, such as cracks, porosity, or lack of fusion, which could compromise the submarine's performance under high-pressure conditions.

- **Ultrasonic Testing (UT)** is used to detect internal flaws within the weld.
- **Radiographic Testing (X-ray)** helps identify subsurface defects, particularly in high-pressure joints like the hull.
- **Magnetic Particle Inspection (MPI)** may be applied to ferromagnetic materials to detect surface cracks.
- For areas exposed to critical stress, **pressure testing** is conducted by subjecting the welded component to high-pressure water or air to ensure it can withstand operational conditions without leaks.

Step 10: Maintenance and Repair Using SMAW

SMAW remains essential not only for initial construction but also for repairs and ongoing maintenance. Submarines operate in harsh marine environments, making regular maintenance crucial to prevent corrosion and mechanical wear. SMAW is especially useful for patching hulls, fixing pipelines, and reinforcing structural components during service or between missions.

During emergency repairs at sea, SMAW's portability makes it indispensable. Welders can access confined spaces inside the submarine, such as ballast tanks or internal bulkheads, where automated welding equipment would be impractical. These repairs are crucial to ensure the submarine's seaworthiness until it can reach a dry dock for comprehensive maintenance.

Manual Metal Arc (MMA) Welding is regulated by several international standards to ensure the safety, quality, and integrity of welded structures. These standards play a crucial role across industries such as marine, construction, energy, and defence, where the reliability of welds is essential. They provide guidelines for welding procedures, operator qualifications, material compatibility, inspection, and quality assurance, ensuring consistent performance in demanding environments.

ISO 9606-1 outlines the requirements for welder qualification testing when welding steels, ensuring that welders have the necessary skills to produce high-quality welds using MMA welding. It specifies joint configurations, welding positions, and material groupings relevant to the testing process, verifying that welders can competently handle the materials and scenarios they will encounter. Another important standard, ISO 15614, specifically Part 1, defines welding procedure specifications (WPS) and qualification requirements for arc welding of metallic materials. It ensures that the chosen welding methods meet the mechanical and quality standards necessary for structural reliability, covering material compatibility, welding parameters, and inspection techniques.

In Australia and New Zealand, AS/NZS 1554 governs structural steel welding, providing comprehensive guidelines for the design, fabrication, and inspection of steel structures welded using MMA welding techniques. This standard ensures that the welding processes meet the requirements for both general and specialized industries, including bridges and offshore platforms. The American Welding Society's AWS D1.1 is another prominent standard, widely recognized for structural welding. It applies to the design, fabrication, and inspection of steel structures, ensuring that welded joints meet strength and durability standards essential in shipbuilding, submarines, and offshore projects.

European standards EN 1011-1 and EN 1011-2 offer further guidance for welding metallic materials, with a focus on arc welding of ferritic steels. These standards address crucial aspects such as joint preparation, preheating, and post-weld heat treatment, ensuring that MMA welds meet performance requirements in sectors like marine engineering and energy production. ISO 3834, which defines quality requirements for fusion welding processes, emphasizes process control, documentation, and inspection to maintain high-quality welds. It is especially important in industries like shipbuilding and defence, where the reliability of welded joints is critical.

ASME Section IX focuses on the qualification of welding procedures and personnel, particularly in high-stakes industries such as nuclear power and pressure vessel manufacturing. This standard applies rigorous testing to ensure that the welds meet performance and safety requirements. ISO 5817, which specifies acceptance levels for imperfections in fusion-welded joints, provides detailed guidelines on defect sizes, ensuring that welds meet performance standards. This standard is essential in fields such as shipbuilding and offshore engineering, where even minor defects could compromise structural integrity.

ISO 10863 outlines the requirements for ultrasonic testing of welds, including those created with MMA welding. This non-destructive testing standard ensures that potential defects, such as cracks or incomplete fusion, are detected, safeguarding the integrity of welded components. It is especially valuable in submarine construction, where high-pressure environments demand flawless welds.

In submarine construction, MMA welding plays a vital role in joining structural components of the inner and outer hulls, as well as welding high-pressure piping systems. Standards like ISO 9606-1, ISO 15614, and ISO 5817 ensure that welders are properly qualified and that welding procedures are meticulously controlled to meet stringent requirements. Non-destructive testing, following ISO 10863, verifies the structural integrity of critical joints, which is essential for maintaining safety under extreme underwater pressure. Adhering to these standards guarantees that welded structures in submarines remain safe, reliable, and capable of withstanding harsh marine conditions.

Tungsten Inert Gas (TIG) Welding

Tungsten Inert Gas (TIG) welding, also known as Gas Tungsten Arc Welding (GTAW), is a critical welding technique employed in submarine construction due to its ability to produce high-quality, precise welds. This process utilizes a non-consumable tungsten electrode and an inert shielding gas, typically argon, to protect the weld area from oxidation and contamination. The importance of TIG welding in submarine construction is significant, as it is particularly effective for welding critical components such as pipes, valves, and interior systems where precision is paramount [378-380].

The advantages of TIG welding include its capability to create clean, defect-free welds with minimal distortion, which is essential for maintaining the hydrodynamic shape and structural integrity of submarines [381, 382]. The process allows for excellent control over the heat input, resulting in superior metallurgical properties of the welds. Studies have shown that TIG welding can achieve high-quality weld metal deposits, characterized by great precision and excellent strength, making it suitable for applications in demanding environments such as nuclear and marine industries [382, 383]. Furthermore, the use of argon as a shielding gas effectively prevents oxidation of the molten metal, ensuring that the welds maintain their integrity under high-stress conditions [384, 385].

Figure 108: TIG Welding, Airman 1st Class Caleb Wren, 4th Equipment Maintenance. Defense Visual Information Distribution Service, Public Domain, NARA & DVIDS Public Domain Archive Public Domain Search.

In addition to its mechanical advantages, TIG welding is also valued for its versatility in joining various metals and alloys, including stainless steels and aluminium [379, 386]. The process is adaptable to different welding parameters, which can be optimized to achieve desired weld characteristics, such as depth of penetration and heat-affected zone size [379, 387]. This adaptability is particularly beneficial in submarine construction, where components may vary significantly in thickness and material properties.

Moreover, the precision of TIG welding is critical in applications where the aesthetic quality of the weld is as important as its mechanical properties. The ability to produce aesthetically pleasing welds without the need for extensive post-weld finishing processes is a significant advantage in the construction of submarines, where visual inspection is often part of quality assurance protocols [381, 386]. Overall, the combination of high-quality welds, minimal distortion, and adaptability makes TIG welding an indispensable technique in the construction of submarines.

TIG welding is primarily used in the construction of high-pressure piping systems that circulate fuel, hydraulic fluid, coolant, and air throughout the submarine. These piping networks require strong, leak-proof joints to maintain operational integrity under high pressure. Since the pipes are often made of corrosion-resistant metals like stainless steel, nickel alloys, and titanium, TIG welding's precise heat control ensures that the welds are durable and maintain the material's corrosion-resistant properties.

This welding method is also widely employed in the assembly and maintenance of the submarine's pressure hull. The hull, composed of high-strength alloys, needs welds that can withstand the immense hydrostatic pressure encountered at great depths. TIG welding is used for sections requiring precise welds, particularly in complex or tight areas, such as around bulkheads or access points, where automated welding systems are impractical.

TIG welding is critical in the fabrication of seawater systems, exhaust ducts, and ballast tanks. These components are continuously exposed to corrosive seawater, making the quality and cleanliness of welds essential to prevent leaks and corrosion over time. The TIG process produces smooth, consistent welds with minimal porosity, reducing the risk of defects that could compromise the submarine's integrity during extended deployments.

Additionally, TIG welding is used to assemble smaller, intricate components within the submarine, such as electronic equipment enclosures, control panels, and valve systems. These parts demand precision to ensure proper functionality, and TIG welding offers excellent control over heat input, preventing warping or damage to sensitive components.

Tungsten Inert Gas (TIG) welding, also known as Gas Tungsten Arc Welding (GTAW), relies on specialized tools and equipment to achieve high-quality welds. Known for its precision and control, TIG welding is indispensable for critical applications such as submarine construction, where structural integrity and reliability are paramount. The primary component of TIG welding is the welding machine, which provides the necessary electrical current. These machines offer both alternating current (AC) and direct current (DC) modes. AC mode is used for materials like aluminium and magnesium, while DC mode is preferred for ferrous metals, including stainless steel and nickel alloys, which are integral to submarine construction. The welding machine also includes an amperage control feature, allowing welders to fine-tune the heat to suit different material thicknesses.

The TIG torch, or welding gun, serves as the main tool for directing the arc onto the workpiece. This torch comprises a tungsten electrode holder, a ceramic nozzle to channel the shielding gas, and a flexible hose to deliver the gas and cooling water in water-cooled systems. Some TIG torches incorporate control switches, enabling the welder to activate or stop the arc without returning to the power source. The tungsten electrode, held by the torch, plays a crucial role in generating the arc. Non-consumable in nature, these electrodes vary in type based on the material being welded. For instance, pure tungsten is used for AC welding of aluminium, while thoriated, lanthanated, or ceriated tungsten electrodes are employed for DC welding of steel and alloys. The electrode's size and shape influence arc stability and weld penetration.

The shielding gas supply system is essential in TIG welding to protect the weld pool from atmospheric contamination. Common gases include argon, known for its stability, helium, which increases heat input for thick materials, and argon-helium mixtures for specialized applications. A gas regulator and flow meter are used to control the flow rate, ensuring optimal shielding throughout the welding process. Filler rods are also a vital part of TIG welding, adding material to the joint to enhance strength and corrosion resistance. Different filler rods are chosen based on the material being welded, such as ER70S-2 or ER316L for steel and 4045 or 5356 for aluminium alloys.

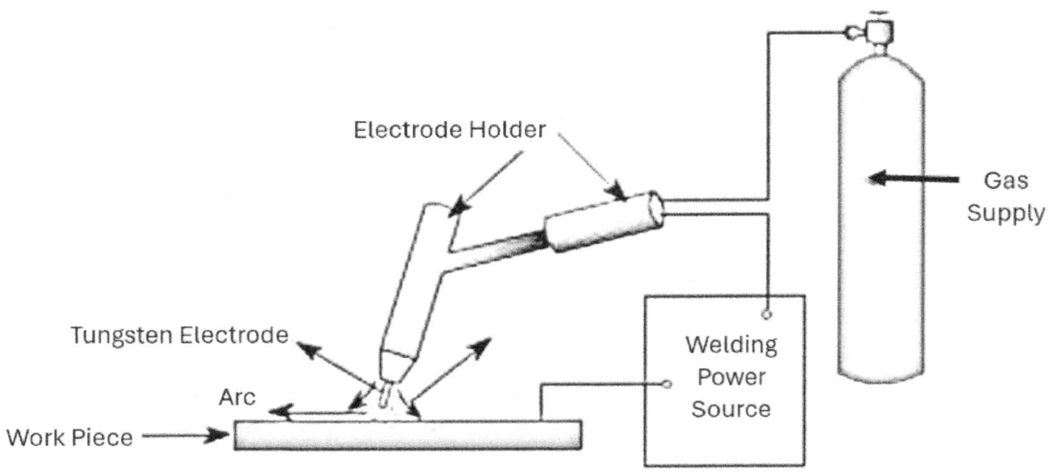

Figure 109: Schematic Diagram of TIG Welding System.

Precision in heat control is achieved through the use of a foot pedal or hand control, which allows welders to adjust the current in real time. This feature is especially crucial when working on complex submarine components, where excessive heat can lead to distortion. In high-amperage applications, water coolers are integrated to prevent the torch from overheating, ensuring that it remains functional and durable throughout the welding process. Given the confined spaces in submarine construction, welders must wear protective gear, including auto-darkening helmets, gloves, leather aprons, and respiratory equipment. This safety gear protects them from arc flashes, heat, and harmful fumes.

Clamps, jigs, and fixtures are used to secure components during welding, maintaining alignment and stability to ensure precise joints. Cleanliness is also paramount; wire brushes, grinding wheels, and solvent cleaners are employed to remove contaminants from surfaces before welding. Gas purging equipment is sometimes required when welding closed sections or pipes to shield the interior of the joint from atmospheric exposure, which is critical for preventing corrosion in submarine piping systems.

To meet the stringent standards of submarine construction, weld inspection tools such as magnifying lenses, weld gauges, and non-destructive testing (NDT) equipment are employed. These tools detect potential defects, including cracks and incomplete fusion, ensuring that the welds meet safety and performance requirements.

The following provides a step-by-step breakdown of the TIG Welding process:

Step 1: Preparation of Materials and Work Area

Before starting the welding process, all metal surfaces must be thoroughly cleaned. In submarine construction, contaminants like dirt, rust, oil, or paint can compromise the weld's strength and corrosion resistance. Use wire brushes, grinding wheels, or solvent cleaners to prepare the base metal and filler rods. Ensure that the work area is well-ventilated, especially when welding in confined spaces, to prevent gas build-up and exposure to harmful fumes.

Step 2: Setting Up the TIG Welding Equipment

Install the appropriate tungsten electrode in the TIG torch based on the material. For submarine construction, thoriated or lanthanated tungsten electrodes are often used when welding high-strength steels or nickel alloys, while pure tungsten is used for aluminium sections. Adjust the electrode to extend about 1/8 inch (3 mm) from the gas nozzle to focus the arc effectively.

Connect the shielding gas, typically argon, to the welding machine. Use a gas regulator and flow meter to ensure the correct flow rate (around 15-20 cubic feet per hour) to maintain a protective atmosphere. Select either alternating current (AC) for aluminium or direct current (DC) for steel and nickel alloys, based on the material being welded.

Step 3: Securing the Workpiece with Clamps and Fixtures

Use clamps, jigs, or other fixtures to hold the workpieces securely in place, ensuring precise alignment of the components. Submarine construction often involves working in confined spaces or with complex geometries, making it critical that all parts remain stable during the welding process. If welding pipes or enclosed spaces, gas purging equipment may be used to shield the inner surface from oxidation.

Step 4: Striking the Arc and Establishing the Weld Pool

Turn on the welding machine and adjust the amperage according to the thickness of the material. Initiate the arc by holding the tungsten electrode near the joint and pressing the foot pedal or torch switch. The arc will create a molten pool at the point of contact. In submarine welding, it is crucial to maintain a controlled pool to avoid overheating, which could lead to distortion or compromised weld quality.

Step 5: Adding Filler Rods for Strong Welds

If additional material is needed, introduce the appropriate filler rod into the molten pool at a steady pace, keeping it within the protective gas shield. For submarine hulls and internal frameworks, use stainless steel or nickel-alloy filler rods that match the base material to maintain corrosion resistance and mechanical strength. Ensure that the rod does not touch the tungsten electrode to prevent contamination.

Step 6: Managing Heat Input and Arc Length

Control the heat by adjusting the foot pedal or hand switch throughout the welding process. In submarine construction, where the risk of distortion is high due to the precision requirements, managing heat input is critical. Maintain a consistent arc length of about 1/8 inch (3 mm) to ensure uniform penetration and avoid incomplete fusion.

Step 7: Completing the Weld Pass

Move the torch steadily along the joint, guiding the arc and filler rod as needed. In submarine construction, multiple passes may be required for thicker sections, with each pass carefully controlled to ensure even fusion. Allow the weld to cool slightly between passes to prevent warping, especially when welding pressure-critical components like hull sections or piping systems.

Step 8: Inspecting the Weld

Once the weld is complete, inspect it for defects such as cracks, porosity, or incomplete fusion. In submarine construction, non-destructive testing (NDT) methods like ultrasonic or radiographic testing are commonly used to ensure weld integrity, particularly for high-pressure systems and critical joints.

Step 9: Cleaning the Weld

Remove any oxide build-up or slag with a wire brush or grinding wheel. Cleaning ensures a smooth, corrosion-resistant surface, which is essential for components exposed to seawater in submarines. If additional passes are required, re-clean the surface before starting the next pass.

Step 10: Post-Weld Heat Treatment (If Required)

For some materials used in submarines, post-weld heat treatment (PWHT) may be necessary to relieve residual stresses and restore material properties. This is especially important for pressure hull sections and high-strength piping systems.

The ISO 9606-1 standard defines qualification testing for welders using TIG welding on steel. This ensures that welders possess the required skills to handle a range of joint configurations, welding positions, and materials relevant to submarine construction. Only qualified welders can perform critical welds on submarine pressure hulls and structural components, helping to maintain the integrity of these vital elements. Similarly, the ISO 15614 Series addresses the specification and qualification of welding procedures. Part 1 of this series focuses on ensuring that TIG welding procedures meet the required mechanical properties for high-strength steel, stainless steel, and nickel alloys used in submarines. The guidelines cover essential welding parameters, joint preparation, and inspection protocols to ensure precise and durable welds.

Compliance with ASME Section IX, part of the Boiler and Pressure Vessel Code, is crucial for submarine construction, especially when welding high-pressure vessels and piping systems. This standard governs the qualification of welding procedures and personnel involved in critical applications, ensuring that all welds meet stringent performance requirements. The ISO 3834 standard, meanwhile, provides a framework for managing the quality of fusion welding processes, including TIG welding. It emphasizes process control, documentation, and inspection, ensuring consistency in both new construction and repairs.

TIG welding also plays an important role in submarine piping and internal frameworks, where stainless steel is commonly used. The AWS D1.6 Structural Welding Code provides guidelines specifically for welding stainless steel, ensuring that welded joints meet the required strength, durability, and corrosion resistance standards. Additionally, BS EN ISO 5817 defines acceptance levels for imperfections in fusion-welded joints, such as those produced by TIG welding. This ensures that submarine components, including pressure hull sections and piping systems, meet safety and performance criteria by maintaining the structural integrity of welded joints.

The ISO 10863 standard outlines procedures for ultrasonic testing (UT), a non-destructive testing method essential for verifying weld quality. UT is particularly useful in submarine construction for inspecting welds in pressure hulls, piping systems, and other critical components, identifying any defects that could compromise integrity. In Australia and New Zealand, AS/NZS 1554 provides structural steel welding guidelines applicable to TIG welding. This standard ensures the quality of steel components used in submarines, such as internal frameworks and external hull sections, by setting requirements for joint design, welding parameters, and inspections.

Submerged Arc Welding (SAW)

Submerged Arc Welding (SAW) is a highly effective welding process frequently utilized in the construction of submarines, particularly for creating long, continuous welds on both the outer and inner hulls. This method involves generating an electric arc beneath a layer of granular flux, which serves to shield the molten weld pool from atmospheric contaminants, thereby enhancing the quality of the weld. The flux not only protects the weld but also contributes to the stability of the arc and the overall welding process, making SAW particularly suitable for automated systems that require consistent and high-quality seams [388, 389].

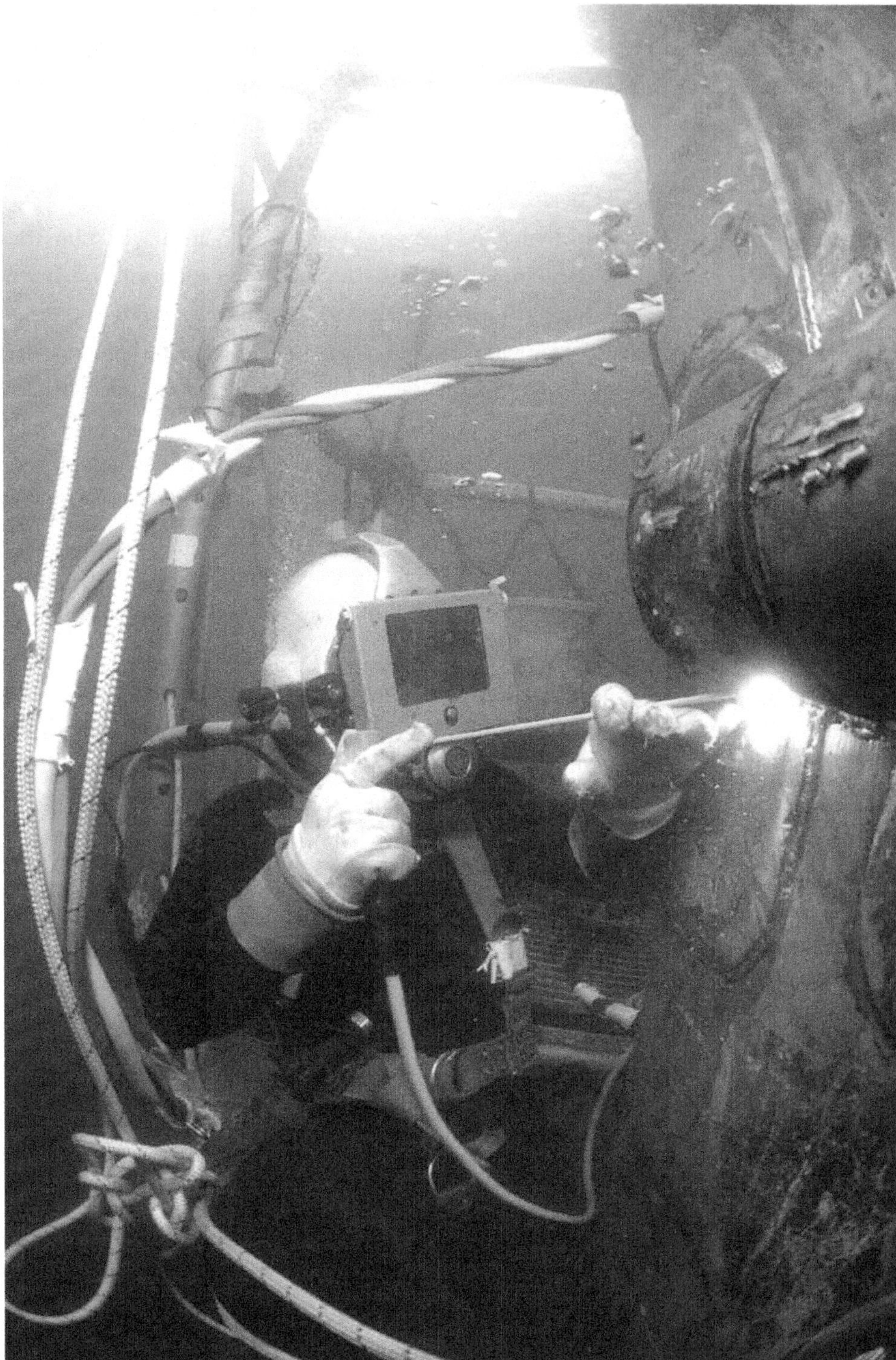

Figure 110: Submerged Arc Welding. United States Navy, Public Domain, via Getarchive.

One of the primary advantages of SAW is its ability to join thick steel plates, which is essential for the pressure hull of submarines. The process is characterized by deep penetration capabilities, allowing for effective fusion of thick materials, which is crucial for withstanding the extreme pressures encountered in deep-sea environments [390]. The mechanical properties of the welds produced by SAW are typically superior, providing the necessary strength and durability required in submarine construction. Research indicates that the welds exhibit excellent mechanical properties, which are vital for ensuring the structural integrity of submarines under high-pressure conditions [391, 392].

Moreover, the efficiency of SAW is enhanced by its high deposition rates and the ability to utilize multiple wires simultaneously, which significantly increases productivity in manufacturing settings [393, 394]. The process parameters, such as welding speed, current, and flux composition, can be optimized to achieve desired weld characteristics, including bead shape and penetration depth [395, 396]. This adaptability makes SAW a preferred choice in industries where high-quality welds are essential, such as shipbuilding and submarine construction [397].

Submerged Arc Welding (SAW) is widely used in submarine construction due to its ability to produce high-quality, deep-penetration welds with minimal defects. SAW is an automated or semi-automated welding process that uses a continuously fed wire electrode, submerged under a blanket of granular flux, to protect the weld from atmospheric contamination. This process is particularly advantageous in applications requiring thick steel plates, making it suitable for critical structural components in submarines.

SAW is extensively employed in the fabrication of pressure hulls. Submarine hulls are made from thick, high-strength steel to withstand immense hydrostatic pressure at depth. The process's ability to weld thick materials in a single pass or a few passes makes it ideal for the construction of these hulls. The high deposition rates and deep penetration ensure that the hull sections are securely welded, minimizing the risk of defects that could compromise the submarine's ability to maintain integrity under pressure.

The outer hull of a submarine also benefits from SAW. The smooth and uniform weld beads produced by this process contribute to the streamlined surface of the submarine, reducing hydrodynamic drag during underwater travel. Furthermore, the durability of the welds ensures that the outer hull can resist corrosion and withstand the mechanical stresses experienced during operations, including changes in pressure as the submarine dives and resurfaces.

SAW is also utilized in the construction of structural supports and bulkheads inside the submarine. These internal components must be robust enough to support the equipment and systems housed within the vessel. The efficiency and precision of SAW allow for the quick and reliable joining of structural elements, ensuring that the submarine can endure both internal stresses and external forces encountered during missions.

The process plays an essential role in the assembly of ballast tanks, which control the buoyancy of the submarine. Ballast tanks are situated between the inner and outer hulls, and the integrity of their welds is crucial for ensuring the submarine can submerge and surface effectively. SAW ensures that these tanks are welded to high standards, eliminating the possibility of leaks that could disrupt the buoyancy control systems.

SAW's automation capabilities make it particularly valuable for submarine construction, where precision and consistency are paramount. The use of automated SAW systems allows for the continuous welding of long joints, such as those found in hull sections, without interruptions. This reduces the potential for weld defects and ensures that the welds meet strict quality control standards required for submarines operating under extreme conditions.

In addition to new construction, SAW is also used for repairs and retrofitting during submarine maintenance. The process's high efficiency and ability to produce defect-free welds make it ideal for restoring structural integrity in damaged areas. Repair operations may include re-welding critical sections of the hull or replacing corroded structural elements with new, securely welded components.

Submerged Arc Welding (SAW) is an automated or semi-automated welding process that relies on specialized tools and equipment to achieve smooth operation, high-quality welds, and overall efficiency. A crucial component in various industries, including submarine construction, the success of SAW depends on the precise functioning and integration of its key elements.

The welding machine, or power source, plays a critical role by providing a constant voltage or current necessary for the welding process. This power source can deliver either alternating current (AC) or direct current (DC), with each mode suited for specific tasks. DC power is preferred for achieving deeper penetration, while AC power helps reduce arc blow when welding magnetic materials like steel. This versatility allows operators to match the power output to the thickness and material properties of the components being welded.

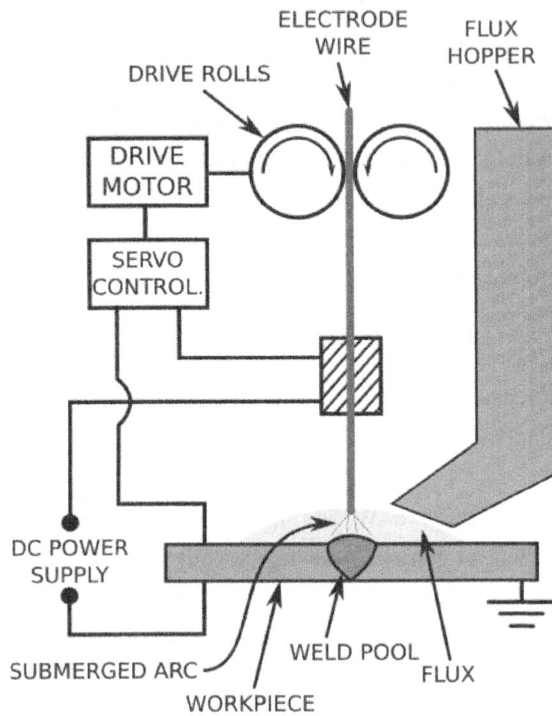

Figure 111: Schematic of submerged arc welding. Wizard191, CC BY-SA 3.0, via Wikimedia Commons.

The wire feeder unit continuously supplies electrode wire to the arc, ensuring stable operation throughout the welding process. In SAW, the electrode wire is generally a solid wire, selected based on the materials being joined. For instance, carbon steel wire is used for welding steel structures and submarine hulls, while stainless steel wire is required for corrosion-resistant components within the submarine. The wire's diameter is carefully chosen according to the thickness of the workpiece, with larger diameters reserved for welding thick sections such as those found in pressure hulls.

The flux hopper stores and delivers granular flux, a material used to shield the weld pool from atmospheric contamination, prevent oxidation, and stabilize the arc. Flux delivery is precisely controlled to ensure a consistent protective layer throughout the welding process. Two common types of flux are used in SAW: fused flux, which provides uniform properties, and agglomerated flux, which offers additional advantages such as improved slag removal. These properties help maintain the weld's quality and integrity in harsh underwater environments.

The welding head and nozzle form the focal point of the SAW process, as they direct both the electrode wire and flux to the workpiece. The welding head precisely controls the position of the electrode to ensure accurate feeding into the joint. It may also contain a contact tip that transfers current to the wire, enabling the arc to generate sufficient heat for effective welding.

A travel mechanism, often automated using a carriage or track system, moves the welding head along the joint at a steady speed, ensuring smooth and continuous welding. The travel speed and positioning are programmable, allowing the system to achieve precise welds along long seams. This level of control is essential in submarine construction, where even slight deviations can compromise the structural integrity of the hull.

The control panel serves as the command centre for adjusting key parameters such as voltage, current, wire feed speed, and travel speed. Many modern SAW systems include real-time monitoring tools to track welding performance, helping operators identify and address any inconsistencies. This is particularly important in submarine applications, where high-quality welds are vital to the vessel's safety and operational performance.

Flux recovery systems are integral to the SAW process, capturing unused flux after welding and recycling it for future use. These systems help reduce waste while ensuring consistent welding results over time. Proper clamping and fixtures are equally essential, securing workpieces during welding to prevent movement or misalignment. In submarine construction, these tools stabilize large hull sections and thick steel plates, ensuring precise welding without distortion.

Even though SAW reduces visible arc radiation due to the flux covering, personal protective equipment (PPE) remains essential for the safety of welders. Helmets or face shields protect against heat and sparks, while gloves and flame-resistant clothing guard against burns. Respiratory equipment is particularly important when welding in confined spaces, such as inside ballast tanks, to protect against fumes.

Given the critical nature of submarine construction, non-destructive testing (NDT) equipment, such as ultrasonic testing (UT) machines, is used to inspect welds for hidden defects. NDT methods help ensure the integrity of welded joints, particularly in pressure-critical sections of the hull, ensuring that the submarine can withstand extreme underwater pressures. Through the precise use of these tools and equipment, SAW enables the production of high-quality, reliable welds essential for the construction and maintenance of submarines.

Submerged Arc Welding (SAW) requires a range of specialized tools and equipment to ensure high-quality welds, smooth operation, and efficiency. This welding process, known for its automation and capability to join thick materials, relies on precise control of various components to achieve strong, reliable welds essential in demanding industries like submarine construction.

The power source, or welding machine, serves as the foundation of the SAW process. It provides either constant voltage or constant current to maintain the stability of the arc throughout the operation. Depending on the materials being welded, either alternating current (AC) or direct current (DC) power is used. DC power is typically applied for deep penetration in thick materials, while AC power is preferred for reducing arc blow in magnetic materials like steel. The power source allows operators to adjust current and voltage settings to align with the thickness of the metal and the welding speed, ensuring optimal performance.

A wire feeder unit plays a crucial role by continuously supplying the electrode wire into the arc during welding. This unit ensures a stable and consistent arc, which is necessary for achieving high-quality welds. The electrode wire used in SAW is usually a solid wire and must match the material being joined. For example, carbon steel wire is ideal for structural components in submarine hulls, while stainless steel wire is employed for sections requiring enhanced corrosion resistance. The wire diameter varies based on the thickness of the workpiece, with larger wires used for thick steel plates typical in submarine construction.

The flux hopper and flux delivery system are essential for protecting the weld pool from atmospheric contamination and stabilizing the arc. The hopper stores granular flux, which is automatically dispensed over the welding arc during the operation. Flux acts as a shield, preventing oxidation and ensuring a clean weld. There are two primary types of flux: fused flux, which offers uniform properties, and agglomerated flux, which improves slag removal. Consistent flux delivery is essential to maintain the quality of the weld and avoid contamination during the process.

The welding head and nozzle are integral to directing the electrode wire and flux toward the workpiece. The welding head ensures precise positioning of the electrode wire along the joint, while the nozzle channels the flux to form a protective covering over the arc. The head may also include a contact tip, which transfers current from the power source to the wire, generating the arc needed to melt the metal. These components ensure smooth and consistent welding by maintaining the correct alignment and current flow.

A travel mechanism, often automated with a carriage or track system, ensures the welding head moves smoothly along the joint at a constant speed. This mechanism allows for uniform welds, especially over long seams typical in submarine construction. The automated system can be programmed to control the speed, length, and positioning of the weld, ensuring precise results and minimizing operator error.

The control panel and monitoring system provide operators with the ability to adjust key parameters during the welding process, including voltage, current, wire feed speed, and travel speed. Real-time monitoring is often incorporated into modern SAW systems to track the welding operation and ensure compliance with required specifications. This is particularly crucial in submarine construction, where weld integrity is essential for the vessel's safety and performance.

Flux recovery systems are an important aspect of SAW, capturing unused flux after welding and recycling it for future use. This system helps reduce waste and ensures consistent welding performance over time. Proper flux recovery also contributes to cost efficiency, as it minimizes the amount of new flux required for subsequent operations.

Clamps and fixtures are necessary to secure the workpieces during welding, ensuring they remain properly aligned throughout the process. In submarine construction, these tools stabilize large sections of the hull and other heavy components, preventing movement or distortion that could compromise weld quality.

Personal protective equipment (PPE) is essential for the safety of welders, even though the arc is submerged under flux and produces minimal visible radiation. Welders must wear helmets or face shields to protect against sparks and heat, as well as flame-resistant clothing and gloves to guard against burns. Respiratory protection is also required when working in confined spaces, such as inside submarine ballast tanks, to prevent exposure to fumes.

The following provides a step-by-step breakdown of the SAW process:

Step 1: Preparation of Materials

Before welding, it is essential to prepare the workpieces by cleaning the surfaces to remove rust, dirt, oil, and contaminants. This ensures that the weld will bond properly without defects. In submarine construction, this step is crucial since structural components like the pressure hull must meet high standards for strength and reliability. Wire brushes, grinding tools, or solvent cleaners are often used for this preparation.

Step 2: Setting Up the Equipment

The SAW machine must be configured with the appropriate settings, including voltage, current, and wire feed speed. The settings will depend on the type and thickness of the material being welded. In submarine construction, these parameters are often adjusted for high-strength steel, stainless steel, or nickel alloys. Operators must also install the correct electrode wire—such as carbon steel or stainless steel wire—and fill the flux hopper with the appropriate flux type.

Step 3: Aligning the Workpieces and Securing Fixtures

The workpieces must be aligned and secured using clamps or fixtures to prevent movement during the welding process. In submarine construction, alignment is especially important for pressure hull sections and bulkheads to maintain structural integrity. The use of fixtures ensures that the joint stays stable throughout welding, reducing the risk of warping or misalignment.

Step 4: Setting the Travel Mechanism

For long seams, the welding head is mounted on a mechanized carriage or track system. This travel mechanism ensures that the welding head moves smoothly and at a consistent speed along the joint. In submarine construction, this automation is critical for maintaining uniform weld quality across extensive sections of the hull.

Step 5: Positioning the Welding Head

The welding head is adjusted to ensure that the electrode wire and flux are directed accurately onto the joint. In submarine construction, precise positioning is crucial, especially in pressure-critical sections of the hull and piping systems. The contact tip must be aligned with the joint to maintain arc stability and ensure even heat distribution.

Step 6: Initiating the Welding Process

The welding operation begins by starting the arc. The electrode wire feeds into the arc as the granular flux is dispensed over the arc and weld pool. The flux melts partially to form slag, which protects the weld from contamination and stabilizes the arc. In submarine welding, maintaining a consistent arc is essential for deep penetration and defect-free welds, especially for thick steel plates.

Step 7: Monitoring the Weld in Progress

Throughout the welding process, the operator monitors parameters such as current, voltage, and wire feed speed. The flux delivery system ensures that a consistent layer of flux shields the weld pool. In submarine applications, real-time monitoring systems are often used to ensure compliance with stringent welding standards and to detect any deviations that might compromise weld quality.

Step 8: Managing Flux and Recycling

As the welding progresses, unused flux is collected by a recovery system. The flux is filtered and recycled for reuse, which ensures cost-efficiency and consistency. In submarine construction, this recycling is particularly important for long welds on large sections of the hull, minimizing material waste.

Step 9: Inspecting the Weld and Removing Slag

After welding, the slag formed from the melted flux is removed using tools such as chipping hammers or wire brushes. This step ensures a smooth finish and prevents slag inclusions that could weaken the joint. In submarine construction, welds undergo rigorous inspection to ensure they meet the necessary quality standards.

Step 10: Non-Destructive Testing (NDT)

For critical components, such as pressure hulls and high-pressure piping, non-destructive testing (NDT) methods are used to inspect the welds. Ultrasonic testing (UT) and radiographic testing (RT) are commonly employed to detect internal defects, such as cracks or porosity. Submarine welds must pass these inspections to ensure they are capable of withstanding extreme underwater pressure.

Step 11: Post-Weld Treatments

In some cases, post-weld treatments, such as heat treatment or stress relief, are applied to ensure that the welds meet mechanical property requirements. This is especially important in submarine construction, where components must endure dynamic loads and pressure fluctuations.

Step 12: Final Quality Checks

Once the welding and inspection processes are complete, a final quality check is performed. Welds are evaluated for dimensional accuracy, strength, and adherence to specifications. In submarine construction, meeting these criteria is essential for ensuring the vessel's safety and operational reliability in harsh marine environments.

Electron Beam Welding (EBW)

Electron Beam Welding (EBW) is a highly specialized technique that has gained prominence in submarine construction, particularly for applications that demand extreme precision and strength. The process involves the use of a focused beam of electrons to generate heat, which melts the metal and forms a weld in a vacuum environment. This unique setting minimizes contamination and allows for the creation of exceptionally strong joints with narrow heat-affected zones (HAZ), making EBW particularly suitable for complex and high-stress components, such as nuclear reactor assemblies and sensitive sonar equipment mounts.

The vacuum environment in which EBW operates is crucial for its effectiveness. It prevents oxidation and contamination of the weld, which is a significant advantage over other welding methods. Studies have shown that the electron beam's high energy density (ranging from 10^7 to 10^9 W/cm²) allows for deep penetration and precise control over the welding process, resulting in minimal distortion and a narrow HAZ [398-400]. This characteristic is particularly beneficial for submarine components that require both precision and strength, as it reduces the risk of structural weaknesses that could arise from excessive heat exposure during welding.

The ability of EBW to produce deep, narrow welds is advantageous for critical submarine parts. The process is capable of welding thick materials in a single pass, which enhances the efficiency of the operation and improves the mechanical properties of the welded joints. Research indicates that EBW can achieve high joint efficiency and superior mechanical properties, making it ideal for demanding applications in the aerospace and marine industries [401-403]. For instance, the use of EBW in welding titanium alloys, which are commonly used in submarine construction, has been shown to yield joints with excellent ductility and strength [404, 405].

Furthermore, advancements in EBW technology, such as dynamic beam positioning and multi-beam techniques, have further enhanced its capabilities. These innovations allow for better control over the welding parameters, leading to improved weld quality and performance [406, 407]. The ability to manipulate the beam's focus and oscillation can optimize the weld geometry and mechanical properties, which is particularly important in applications where precision is paramount [408, 409].

EBW offers deep penetration, minimal distortion, and high weld strength, making it ideal for critical applications in submarine construction, where reliability and precision are paramount. Below is a detailed description of how and where EBW is used in submarine construction.

Applications of EBW in Submarine Construction:

1. Pressure Hull Fabrication In submarine construction, the pressure hull must withstand extreme underwater pressures. EBW is used to weld thick sections of high-strength steel, titanium, and nickel alloys, ensuring precise, high-strength joints with minimal distortion. The deep penetration achieved by EBW allows the creation of strong, continuous seams that are crucial for maintaining hull integrity.

2. Welding of Complex Geometries and Internal Structures EBW is used for intricate welds on internal structural components such as bulkheads, ribs, and mounting brackets. These internal elements support the pressure hull and distribute stress evenly. EBW's precision allows these components to be welded without compromising structural integrity, especially in confined spaces where conventional welding methods would be challenging.

3. Assembly of Nuclear Reactor Components In nuclear-powered submarines, EBW plays a vital role in the assembly of reactor components. The process is used to weld heat exchangers, pressure vessels, and coolant piping systems, ensuring leak-proof joints. The precision and cleanliness of EBW minimize contamination risks and ensure the reliability of components operating under extreme conditions.

4. Sealing of Piping Systems and High-Pressure Vessels Submarines contain extensive piping systems that transport air, water, hydraulic fluid, and coolant. EBW is used to seal high-pressure piping and vessels with welds that are free from porosity or contamination. This is essential for ensuring the durability and safety of systems that operate under high pressure and fluctuating temperatures.

5. Welding of Exotic Materials and Dissimilar Metals Submarine construction often involves materials like titanium, stainless steel, and advanced composites. EBW's high energy density enables it to join these exotic materials effectively. Additionally, the process is used to weld dissimilar metals, such as nickel alloys to stainless steel, which are often employed in submarine components for corrosion resistance.

6. Manufacture of Propulsion and Control System Components Precision is critical in the manufacture of propulsion systems, including propellers, turbines, and rudders. EBW is used to assemble these components with minimal heat-affected zones, ensuring high performance and durability. Welds in control systems such as bow and stern planes benefit from EBW's accuracy, maintaining smooth operation and precise alignment.

EBW offers several advantages that make it particularly useful in submarine construction. The process allows for deep penetration and narrow welds, which are essential for thick sections such as the pressure hull. EBW's precision minimizes distortion and stress, preserving the integrity of components that must endure the rigors of underwater environments. The use of a vacuum environment during EBW ensures that welds are free from contamination, essential for ensuring the safety and reliability of critical submarine systems.

In addition, EBW's ability to weld complex geometries and dissimilar materials makes it ideal for joining components with intricate shapes or varied material properties. This versatility is crucial in submarine construction, where different materials are often used together to optimize strength, weight, and corrosion resistance.

While EBW offers many advantages, it also presents challenges. The process requires a vacuum chamber, limiting the size of components that can be welded in a single operation. This constraint means that large submarine sections may need to be disassembled for welding and reassembled afterward. Additionally, the precision of EBW requires skilled operators and careful quality control to ensure defect-free welds. Non-destructive testing methods, such as ultrasonic testing (UT) and radiographic testing (RT), are often used to inspect EBW joints to guarantee their integrity.

Electron Beam Welding (EBW) is a precision welding process that uses a focused beam of high-velocity electrons to join materials. The beam generates intense heat as it strikes the weld joint, causing the metals to fuse together, typically within a sealed vacuum chamber to prevent contamination. This process is highly efficient and capable of producing deep, narrow welds with minimal heat-affected zones and low distortion, making it ideal for critical applications like submarine construction [410].

The electron beam used in EBW is generated by an electron gun, powered by a high-voltage supply. This energy source heats an incandescent cathode, which emits high-speed electrons. The emitted electron beam is then accelerated and precisely directed through a series of anodes and focusing coils, which concentrate the beam using electromagnetic fields. This focused beam delivers energy densities 100 to 1000 times greater than conventional arc welding, vaporizing a narrow path through the metal and enabling deep, high-strength welds. Due to its efficiency, EBW can operate at speeds 10 to 50 times faster than arc welding while keeping heat-affected zones minimal [410].

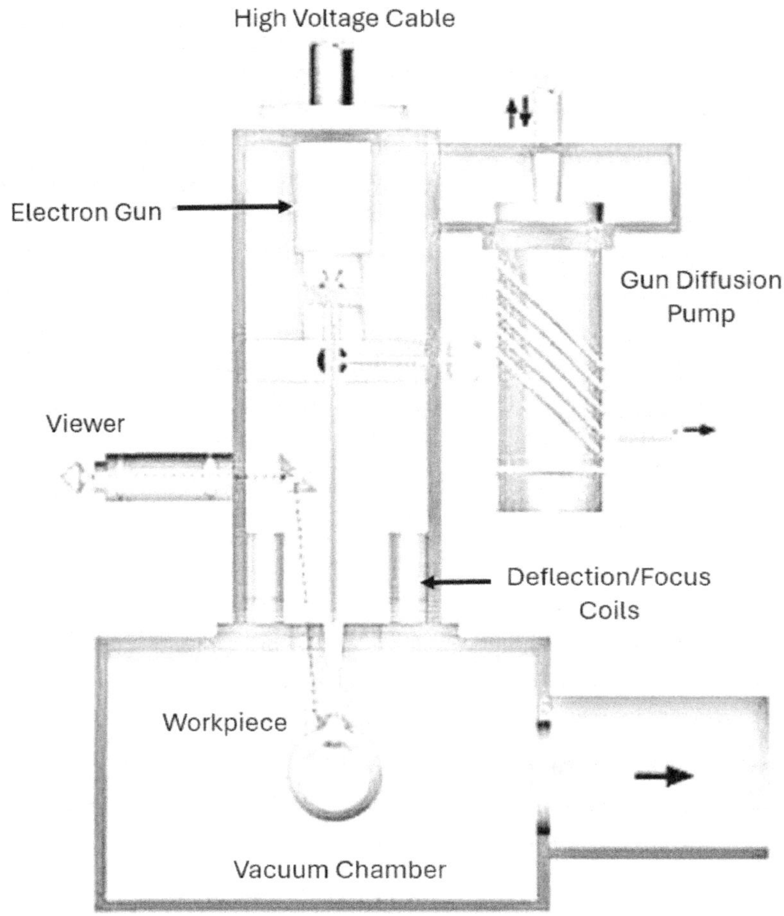

Figure 112: Schematic of Electron Beam Welding process.

During the welding process, the electron beam typically remains fixed while the components to be welded are moved beneath it using rotary or linear movement systems controlled by CNC (Computer Numerical Control) systems. These systems ensure precision in aligning the joint and can make real-time adjustments to the welding parameters during the process to maintain weld consistency [410].

The EBW process begins with a thorough cleaning of the components to ensure all contaminants are removed. If the materials are ferrous-based, demagnetization may also be necessary to avoid interference with the electron beam. Proper preparation is crucial to avoid impurities that could compromise the weld quality.

Once cleaned, the parts are secured in place using fixtures and mounted onto the CNC-controlled work movement system. The CNC system manoeuvres the parts into the optimal position under the electron beam and can make dynamic adjustments during welding. After securing the components, the vacuum chamber is sealed, and air is evacuated to create the necessary vacuum environment. Materials like titanium require a higher vacuum level to prevent oxidation during welding.

Once the vacuum is established, the electron beam is aligned with the weld joint. The power and parameters of the beam are adjusted to suit the material and joint design, ensuring consistent heat input throughout the weld cycle. These adjustments can be programmed into the CNC system to maintain optimal weld conditions.

With the beam properly configured, the welding cycle begins. The intense energy of the electron beam melts the material along the joint line, creating a precise, deep fusion. After the weld is complete, the chamber is re-pressurized to allow access to the welded components. The parts are then removed from the fixtures for post-weld inspection [410].

After welding, the components undergo a thorough inspection to ensure the integrity of the joint. Non-destructive testing (NDT) methods, such as fluorescent penetrant testing, are commonly used to detect surface cracks and defects. Radiographic testing may also be employed to examine the weld internally. These quality control measures are essential in submarine construction, where welds must meet stringent safety and performance standards due to the extreme operating conditions.

To achieve successful results with EBW, several conditions must be met. The process is compatible with a wide range of metals, including high- and low-carbon steels, stainless steel, nickel, copper alloys, and titanium. It can also join dissimilar metals effectively, although certain combinations, like aluminium with refractory metals, may yield variable results. As EBW does not typically use filler material, joint design becomes crucial. Welds with planetary or circumferential configurations are generally most effective, as they allow for even heat distribution and solid joint fusion. Close attention to joint fit during the design phase ensures that the components align perfectly, promoting a high-quality weld [410].

Since EBW is a machine-based process, it often requires sample tests to validate the welding parameters before full-scale production. These sample welds establish baseline settings that can be reused for future production to ensure consistency.

The equipment involved in EBW must function seamlessly to generate and control the electron beam, create and maintain a vacuum environment, and ensure precise movement and alignment of the components being welded. Below is a detailed explanation of the key tools and equipment required for EBW.

The electron gun is the central component of the EBW system. It generates the high-velocity electron beam used to fuse the metals. The electron gun contains an incandescent cathode, which emits electrons when heated. These electrons are accelerated using a high-voltage power supply, typically ranging from 60 kV to 200 kV. The electrons pass through anode plates and focusing coils, which shape and direct the beam. This precise control of the beam ensures that the energy is concentrated at the desired location on the weld joint.

A vacuum chamber is essential to the EBW process, preventing atmospheric gases from interfering with the electron beam and contaminating the weld. The chamber is sealed, and air is removed to create a vacuum environment. Some materials, such as titanium, require a high vacuum level to ensure a clean weld. The size of the vacuum chamber can vary depending on the components being welded. Larger chambers are used for structural submarine parts, while smaller chambers are suitable for precision work on smaller components.

The work holding and fixture system is critical for maintaining the precise alignment of the components during welding. Fixtures securely hold the parts in place within the vacuum chamber, ensuring they do not shift during the welding process. These fixtures are often mounted on CNC-controlled movement systems that allow for precise linear and rotary movement, ensuring the components are positioned correctly under the beam. The CNC system can also be programmed to adjust the positioning during welding for complex joint geometries.

A power supply unit provides the necessary electrical energy to the electron gun. It must maintain a stable and adjustable voltage to ensure consistent beam intensity throughout the welding process. This stability is crucial for producing uniform welds, especially in submarine applications where structural integrity is paramount. Some systems are equipped with feedback mechanisms that monitor and adjust the power output in real-time, ensuring optimal beam performance.

Beam monitoring and control systems are essential for maintaining the quality of the weld. These systems allow the operator to adjust beam parameters such as focus, intensity, and alignment during the welding process. Many modern

EBW systems include real-time monitoring to detect any deviations from the desired welding parameters, ensuring that corrective actions can be taken immediately to avoid defects.

A cooling system is required to manage the heat generated by the electron beam and the components within the vacuum chamber. Heat exchangers and cooling fluids, such as water or specialized coolants, are used to maintain stable operating temperatures and prevent overheating. The cooling system ensures that the electron gun, power supply, and welded components remain within safe temperature limits.

Non-destructive testing (NDT) equipment is crucial for post-weld inspection, especially in submarine construction, where weld quality is critical for safety and performance. Fluorescent penetrant testing is commonly used to detect surface defects such as cracks or voids. Radiographic inspection techniques may also be employed to identify subsurface flaws that could compromise the weld's integrity.

Operators and technicians working with EBW must wear personal protective equipment (PPE). Although the electron beam itself is contained within the vacuum chamber, there are risks associated with high-voltage equipment and potential radiation exposure. Proper PPE includes protective eyewear, insulated gloves, and flame-resistant clothing.

The flux recovery system is not typically required for EBW, as the process does not involve consumable materials like traditional arc welding. However, certain specialized EBW systems may include mechanisms for capturing and recycling vaporized material generated during the welding process to maintain a clean environment inside the vacuum chamber.

Below is a detailed step-by-step guide to performing EBW, with a specific focus on its application in submarine construction.

Step 1: Preparation and Cleaning of Components

The first step involves thoroughly cleaning the components to remove any contaminants, such as oils, grease, or dirt, that could compromise the quality of the weld. In submarine construction, where welds must resist corrosion and withstand high pressure, this step is crucial. For ferrous metals, components may also need to be demagnetized to prevent interference with the electron beam during welding.

Cleaning methods can include chemical cleaning with solvents, ultrasonic baths, or mechanical cleaning using abrasives. Careful inspection ensures the surfaces are free from debris before welding.

Step 2: Assembly and Fixturing

The components to be welded are securely clamped and aligned within specialized fixtures. These fixtures are mounted on CNC-controlled systems that allow for precise movement during welding. In submarine construction, large and heavy sections such as pressure hulls and structural frames are aligned with high precision to ensure the joint fits perfectly.

Proper fixturing prevents the components from shifting during the welding process. CNC control systems can also be programmed to adjust the orientation of the parts for complex weld geometries, ensuring the electron beam follows the desired weld path accurately.

Step 3: Vacuum Chamber Sealing and Evacuation

Once the components are loaded and positioned, the vacuum chamber is sealed, and air is removed to create a vacuum environment. The vacuum prevents atmospheric gases from interacting with the electron beam and ensures a clean weld. The level of vacuum required depends on the material; for example, titanium components need a higher vacuum to avoid contamination.

Vacuum evacuation can take several minutes to hours, depending on the size of the chamber. A consistent vacuum environment is critical for maintaining the integrity of the electron beam and ensuring defect-free welds.

Step 4: Alignment and Beam Adjustment

The electron beam is aligned with the joint, and the welding parameters are adjusted according to the material and joint type. The power level, beam focus, and travel speed are set based on the thickness of the materials being welded.

For submarine components, careful control of these parameters ensures the weld penetrates deeply while minimizing the heat-affected zone.

Real-time monitoring systems allow operators to adjust the beam's focus and alignment if necessary during the welding cycle, ensuring precision throughout the process.

Step 5: Welding Process Initiation

The welding cycle begins once the electron beam is correctly aligned and all parameters are set. The electron beam is fired at the joint, vaporizing a small portion of the material and forming a keyhole. As the beam moves along the joint, molten metal flows around the keyhole and solidifies behind it, creating a deep, narrow weld.

The CNC system controls the movement of the parts, guiding them under the fixed beam or adjusting the beam's position for intricate welds. In submarine construction, this step is critical for ensuring uniform and defect-free welds, particularly in pressure hulls where structural integrity is paramount.

Step 6: Re-Pressurizing the Chamber and Part Removal

After welding is complete, the chamber is slowly re-pressurized by introducing air back into the system. The welded part is then removed from the fixtures for inspection. For large submarine components, automated systems assist with the removal and transportation of the welded sections to the inspection area.

Step 7: Post-Weld Inspection and Testing

Submarine welds undergo rigorous inspection to ensure they meet quality and safety standards. Non-destructive testing (NDT) methods such as fluorescent penetrant inspection, radiography, or ultrasonic testing are used to detect any cracks, voids, or incomplete fusion in the weld.

Special attention is paid to pressure hull welds, as these joints must withstand extreme underwater pressure without failure. In submarine piping and propulsion systems, inspections ensure that welds are leak-proof and resistant to corrosion.

Step 8: Finishing and Quality Assurance

If the inspection reveals any imperfections, repairs or re-welding may be required. Once the welds pass inspection, the welded components are cleaned and prepared for further assembly or finishing processes. Quality assurance documentation is prepared to confirm that the welds meet all applicable standards and specifications required for submarine construction.

Laser Beam Welding (LBW)

Laser Beam Welding (LBW) is an advanced welding technique that has gained prominence in various industries, including submarine systems, due to its ability to produce high-quality joints with minimal distortion. This technique utilizes a concentrated laser beam to melt the metal surfaces, resulting in a strong joint while minimizing heat input, which is crucial for applications where precision is paramount, such as in submarine systems and sensitive mechanical assemblies.

The precision of LBW is particularly beneficial for joining thin metal sheets and intricate components. Research indicates that LBW is characterized by a high energy density, which allows for deep and narrow weld seams. This feature significantly reduces thermal distortion compared to conventional welding methods, making it ideal for applications requiring tight tolerances and minimal heat-affected zones (HAZ) [411-413]. For instance, studies have shown that LBW can effectively minimize distortions in thin sheet structures, which is critical in maintaining the integrity of complex assemblies in electronic systems [414, 415].

Moreover, the high-speed operation of LBW enhances manufacturing efficiency and consistency. The ability to automate the welding process further contributes to its appeal in industrial applications. Automation not only improves the speed of production but also ensures repeatability and reliability in weld quality [411, 416]. The low thermal input associated with LBW allows for rapid cooling of the weld, which can lead to improved mechanical properties and reduced residual stresses in the welded joints [412, 417]. This is particularly advantageous in the aerospace and defence sectors, where material performance is critical [411, 412].

In addition, LBW is noted for its versatility in joining various materials, including high-strength steels and aluminium alloys, which are commonly used in submarine construction [418, 419]. The technique's adaptability to different materials and configurations makes it a preferred choice for modern manufacturing processes, where the demand for lightweight and strong components is ever-increasing [420].

Laser Beam Welding (LBW) plays a crucial role in submarine construction, offering high precision, deep weld penetration, and minimal heat distortion. This advanced welding technique uses a concentrated laser beam to fuse materials, making it ideal for applications where precision and strength are essential. LBW is particularly valuable in submarine manufacturing due to the strict requirements for structural integrity, weight reduction, and corrosion resistance. Below is a detailed overview of how and where LBW is applied in submarine construction.

Laser beam welding is frequently employed in constructing the inner and outer hulls of submarines. These hulls must withstand extreme underwater pressure and resist corrosion over prolonged periods. LBW is preferred for certain sections of the pressure hull because it produces narrow, deep welds with minimal heat-affected zones (HAZ). This precision reduces the risk of distortion, which is critical in maintaining the cylindrical shape of the pressure hull, essential for withstanding hydrostatic pressure.

The process is especially advantageous in joining lightweight or specialized materials like titanium, high-strength steels, and nickel alloys. These materials are commonly used in submarine construction due to their excellent strength-to-weight ratios and corrosion resistance, and LBW ensures defect-free joints between these metals.

Submarines contain complex piping systems that transport fluids such as coolants, hydraulic fluids, and air. LBW is used to weld high-pressure pipes, especially in hard-to-reach areas or for joints that require minimal internal disruption. The precision of laser welding ensures leak-proof joints, reducing the chances of system failure under high pressure.

Because LBW generates little spatter and minimal internal burrs, it ensures smooth internal surfaces, critical for fluid dynamics and preventing corrosion. This characteristic makes it the preferred choice for welding stainless steel and titanium pipes used in hydraulic and cooling systems.

Submarine propulsion systems include components such as propeller shafts, turbine assemblies, and engine housings, which require precise, high-strength welds. LBW is employed to join intricate components, ensuring minimal distortion while maintaining structural integrity. The deep penetration capability of the laser beam is particularly useful for welding thick metal sections used in propulsion shafts, ensuring the strength required to withstand high torque loads.

Laser welding is also applied to produce lightweight and complex assemblies from advanced alloys, such as titanium-based components used in stealth propulsion systems. These precise welds contribute to the overall efficiency and noise reduction of the propulsion system, enhancing the submarine's stealth profile.

In modern submarines, lightweight materials and advanced sensor systems are crucial for enhancing performance. LBW is used to assemble delicate components such as sensors, electronic housings, and connectors. Its non-contact nature allows for precise joining without causing thermal damage to sensitive parts, making it ideal for welding intricate electronic assemblies.

The low heat input of LBW is also beneficial when welding structural elements that integrate sensors and wiring. This ensures that embedded systems remain functional and the surrounding materials maintain their mechanical properties, contributing to the submarine's operational reliability.

Laser beam welding offers several advantages in submarine construction. It allows for high-speed welding, reducing production time and costs. The minimal heat input lowers the chances of warping and reduces the need for post-weld treatments. Additionally, the ability to automate LBW makes it suitable for large-scale submarine manufacturing, ensuring consistency and precision across multiple welds.

The non-contact nature of LBW also allows welding in confined spaces where other welding techniques, such as arc welding, might be difficult to implement. This versatility makes it a valuable tool for repairs and modifications during both construction and in-service maintenance.

Laser Beam Welding (LBW) requires a range of specialized tools and equipment to deliver precise, high-quality welds. This process involves the use of concentrated laser energy to join materials, necessitating meticulous control over the welding environment, laser optics, and workpieces to ensure success. Below is a detailed description of the essential tools and equipment used for LBW, particularly in demanding applications like submarine construction.

The laser source generates the intense beam of light used to weld materials. In LBW, common laser sources include CO_2 lasers, fibre lasers, and solid-state Nd lasers, each suited for specific applications. CO_2 lasers are often used for large-scale welding of metals, while fibre lasers and Nd lasers provide high precision and are ideal for welding complex or delicate components. These lasers produce concentrated energy that allows for deep penetration and narrow heat-affected zones, minimizing distortion.

The optical system is critical for focusing and directing the laser beam. This system includes mirrors, lenses, and focusing heads that concentrate the beam onto the weld joint. In some setups, fibre optic cables transmit the laser beam from the source to the welding head, offering flexibility in accessing tight or confined areas. Precise control over the beam focus ensures the energy is concentrated where needed, which is essential when welding intricate or high-strength submarine components.

The welding head holds the focusing lens and nozzle, directing the laser beam onto the workpiece. The nozzle also supplies shielding gas to prevent contamination of the weld pool and protect the laser optics from splatter. In submarine applications, where precision is paramount, the welding head is often mounted on robotic arms or gantry systems, allowing for consistent weld quality across complex geometries.

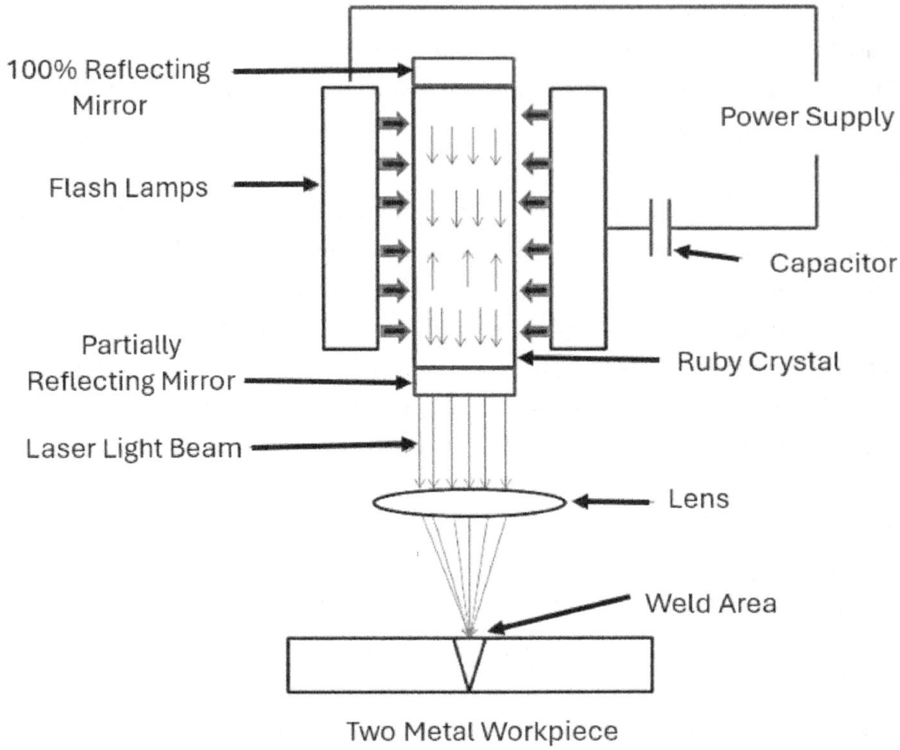

Figure 113: Schematic of Laser Beam Welding process.

Shielding gases such as argon, helium, or nitrogen are used to protect the weld pool from atmospheric contamination. The choice of gas depends on the material being welded. For example, argon is commonly used for stainless steel, while helium provides deeper penetration for thick materials. In submarine construction, shielding gas plays a crucial role in preventing oxidation and ensuring the welds are corrosion-resistant.

Proper alignment of the components being welded is critical for achieving high-quality joints. Fixtures and clamping systems are used to hold workpieces in place, preventing movement during the welding process. In automated setups, CNC-controlled positioning systems or robotic arms manoeuvre the components under the laser beam. These systems are particularly important in submarine construction, where precise alignment ensures the integrity of pressure hulls and structural components.

The laser and optical components generate significant heat during operation, requiring efficient cooling to maintain performance and prevent damage. Water-cooling systems are commonly used for high-power lasers to regulate temperature. The cooling system also prevents the overheating of the welding head and optics, ensuring consistent beam quality throughout the welding process.

The control panel allows operators to set and monitor welding parameters, such as laser power, pulse duration, and welding speed. Modern LBW systems are equipped with real-time monitoring features, enabling adjustments during the welding process to maintain quality. For submarine construction, precise control over parameters is essential to avoid defects and ensure that welds meet stringent safety standards.

LBW generates fumes and particulates that must be removed from the work area to ensure a safe environment. Fume extraction systems capture and filter these emissions, protecting operators and preventing contamination of sensitive components. In submarine construction, where welding often takes place in confined spaces, fume extraction is critical to maintaining air quality.

Although LBW does not produce the intense arc light associated with traditional welding, it still requires the use of appropriate PPE. Operators wear laser safety glasses rated for the specific wavelength of the laser being used. In addition, gloves, flame-resistant clothing, and respiratory protection are essential when working in submarine environments, where welding is often conducted in enclosed spaces.

Welds produced by LBW must undergo rigorous inspection to ensure they meet the required quality standards. Visual inspection tools, such as magnifiers, are used to identify surface defects. Non-destructive testing (NDT) methods, including ultrasonic testing and radiography, detect internal flaws. In submarine construction, these inspections are crucial to verify that welds in pressure hulls, piping systems, and structural frameworks can withstand the harsh underwater environment.

The laser system requires a stable power supply to function efficiently. High-power lasers demand significant electrical input, and specialized power units regulate voltage and current to maintain consistent laser output. In submarine construction, reliable power supply units ensure uninterrupted welding, minimizing the risk of defects caused by fluctuations in power.

Many modern LBW systems incorporate automation to enhance precision and efficiency. CNC software controls robotic arms and positioning systems, ensuring that welds are accurate and consistent. Automated welding is particularly useful for large submarine components, such as hull sections, where manual welding would be labour-intensive and time-consuming. Programmable systems also allow for repeatable welds, reducing the risk of human error.

Some LBW processes, especially those involving highly reactive metals like titanium, are conducted in vacuum chambers to prevent oxidation. While not always necessary, vacuum chambers are sometimes used in submarine construction when welding critical components that must be free from contamination. These chambers provide an inert environment, ensuring the highest quality welds for sensitive parts.

The following provides a step-by-step process for performing LBW, tailored to submarine construction requirements:

Step 1: Preparation of the Workpieces

Before welding, the metal components must be cleaned to remove contaminants such as oil, grease, dirt, and oxide layers. This can be achieved through chemical cleaning or mechanical abrasion. In submarine construction, precise preparation is crucial because any contamination could compromise the strength and corrosion resistance of critical joints, such as pressure hulls or piping systems.

If necessary, workpieces are demagnetized to prevent interference with the laser beam. Parts are also checked for alignment and fit, as LBW relies on tight joints for optimal results.

Step 2: Fixture and Alignment Setup

The workpieces are secured in place using specialized clamps, jigs, or fixtures to prevent movement during welding. Submarine components, such as sections of the pressure hull, often require robust fixtures to maintain proper alignment and ensure precise joints. Automated positioning systems, such as robotic arms or CNC-controlled platforms, are frequently used to guide the laser along the weld path accurately.

Step 3: Selection of Laser Parameters and Shielding Gas

The operator sets the laser parameters, including laser power, pulse duration, beam focus, and welding speed, based on the material type, thickness, and joint configuration. In submarine construction, where materials such as stainless steel, titanium, and nickel alloys are common, fine-tuning these parameters is essential to avoid defects.

A shielding gas, typically argon, helium, or nitrogen, is selected to protect the molten weld pool from oxidation. This step is particularly important for submarine components that will be exposed to seawater, where corrosion resistance is critical.

Step 4: Calibration and Beam Alignment

The laser beam is carefully aligned with the joint to ensure precise targeting. For submarine applications, alignment must be accurate to the micron level, especially when welding delicate components like piping systems or sensor housings. Test welds are often conducted on sample materials to verify the settings and alignment before proceeding with the actual welding.

Step 5: Creation of the Weld Joint

Once the alignment is confirmed, the operator initiates the laser welding process. The high-intensity laser beam melts the metal along the joint line, creating a molten weld pool. In submarine construction, continuous or pulsed laser modes may be used depending on the type of joint and material.

If automated systems are employed, the CNC or robotic controller guides the laser along the joint with precise control of speed and positioning. The laser's energy is concentrated in a small area, minimizing the heat-affected zone and reducing the risk of warping or distortion, which is critical for components like hull sections.

Step 6: Monitoring the Weld Process

During welding, sensors and cameras monitor the weld pool and beam path to ensure consistency. Automated systems may adjust the parameters in real-time to maintain weld quality. In submarine construction, real-time monitoring is essential to detect any irregularities that could compromise the structural integrity of the component.

Step 7: Post-Weld Cooling and Cleaning

After welding, the component is allowed to cool naturally. The shielding gas continues to flow until the weld pool has solidified, preventing oxidation. In submarine construction, proper cooling is crucial to avoid residual stresses and ensure the longevity of the weld.

Once cooled, any residual slag, spatter, or oxide layer is removed using wire brushes or chemical cleaning agents, ensuring the weld surface is smooth and free from defects.

Step 8: Inspection and Non-Destructive Testing (NDT)

Every weld in submarine construction undergoes rigorous inspection to verify its quality. Non-destructive testing (NDT) methods, such as ultrasonic testing (UT) and radiography, are used to detect internal defects like cracks or voids. Fluorescent penetrant testing may also be employed for surface inspections. These inspections are essential for ensuring that critical welds, especially those in pressure hulls and high-pressure systems, meet stringent safety and performance standards.

Step 9: Repairs and Adjustments (if Required)

If defects are found during inspection, corrective measures are taken. Minor surface imperfections may be repaired by grinding and re-welding, while significant defects might require re-welding the entire joint. In submarine construction, these repairs are critical to maintaining the vessel's operational safety and reliability.

Step 10: Documentation and Quality Control

The final step involves documenting the welding process, including the parameters used, inspection results, and any repairs performed. This documentation is essential for quality control and traceability, ensuring that the submarine meets regulatory and safety standards.

Resistance Spot Welding

Resistance spot welding (RSW) is a widely utilized technique for joining thinner metal sheets and internal components in various applications, including submarines. The process involves pressing two metal surfaces together while an electrical current flows through them, generating heat due to electrical resistance, which forms a welded joint. This method is particularly advantageous for assembling non-structural components and electrical enclosures due to its speed and efficiency, making it a preferred choice in high-volume production environments [421, 422].

The effectiveness of resistance spot welding in producing reliable joints is underscored by its application in the automotive and aerospace industries, where it is employed for fabricating sheet metal assemblies [423, 424]. While RSW is not typically used for critical hull sections of submarines, it plays a crucial role in the assembly of internal systems that require strong and dependable joints. The process's ability to create robust connections is enhanced by optimizing welding parameters such as current, time, and electrode pressure, which significantly influence the mechanical properties of the welds [422, 425].

Moreover, the adaptability of resistance spot welding to automation allows for consistent quality in mass production, which is essential for the manufacturing of complex internal components in submarines [426]. The technique's rapid execution and minimal skill requirements further contribute to its widespread adoption in industries that demand high efficiency and reliability [427]. However, challenges such as the need for precise control of welding parameters and the potential for inconsistent weld quality due to the complexity of the process must be addressed to ensure optimal performance [428].

In submarine construction, RSW is particularly useful for joining sheet metal structures, smaller components, and complex assemblies where precision and efficiency are required. It is used in areas that do not demand deep penetration welds, such as non-structural frameworks, interior fittings, and equipment mounts.

1. **Hull Sub-Structures and Internal Panels:** RSW is applied to join secondary structures, such as panels, frames, and bulkheads, to the internal surfaces of the pressure hull. These components provide reinforcement, support, and mounting points for equipment inside the submarine. Since RSW produces minimal distortion, it ensures that these panels and frameworks maintain their precise shape and alignment.

2. **Electrical Equipment and Control Panels:** Submarines contain various electrical systems and control panels that require stable and conductive joints. RSW is used to assemble enclosures, metal casings, and connectors in these systems. It ensures good electrical conductivity while creating strong mechanical bonds, which are essential for ensuring reliable operation in underwater environments.

3. **Hydraulic and Pneumatic System Components:** RSW is employed to fabricate parts of hydraulic and pneumatic systems used within submarines. Thin-walled tubing and fittings are spot-welded to minimize leaks and maintain the integrity of high-pressure systems. This welding method provides secure joints without introducing excess heat that could distort or weaken sensitive parts.

4. **Assembly of Air Handling Systems and Ductwork:** The interior of a submarine requires air handling systems and ventilation ducts for crew safety and equipment cooling. RSW is ideal for welding the sheet metal components of these systems, as it produces airtight joints necessary to maintain proper airflow and prevent leaks, even under the demanding conditions of a submerged vessel.

5. **Mounting Points and Brackets for Equipment:** RSW is often used to secure brackets, mounting points, and clips for holding equipment, cabling, and piping in place. This helps ensure that critical systems remain stable, even during manoeuvres or under high-pressure conditions. The use of spot welding for these components reduces the overall assembly time and improves efficiency during construction.

RSW offers several advantages for submarine construction. The process is fast and efficient, capable of creating multiple welds in quick succession. It produces minimal heat, which reduces the risk of thermal distortion and preserves the mechanical properties of the base materials. Furthermore, RSW does not require filler materials, resulting in clean welds with minimal surface preparation and post-weld cleanup.

In submarine construction, the precision and repeatability of RSW are essential. Automated spot welding systems can be programmed to make consistent welds, ensuring uniformity across assemblies. Additionally, because RSW produces localized heat, it is well-suited for confined spaces, such as those found within submarine compartments and control panels.

Resistance Spot Welding (RSW) relies on specialized tools and equipment to generate precise welds through the application of electrical current and pressure. These components work together to create efficient and reliable welds for thin metal assemblies, making RSW ideal for various applications, including submarine construction. Each piece of equipment plays a crucial role in maintaining the quality and strength of the welds, ensuring that they meet the stringent requirements of underwater environments.

The RSW process requires a robust power source to deliver the necessary electrical current. A transformer steps down the high-voltage input to a low-voltage, high-current output suitable for spot welding. The current used in RSW can range from several hundred to thousands of amperes, depending on the material thickness and joint configuration. Submarine construction often uses programmable power supplies to precisely control the timing and intensity of the weld current, ensuring consistent results across multiple welds.

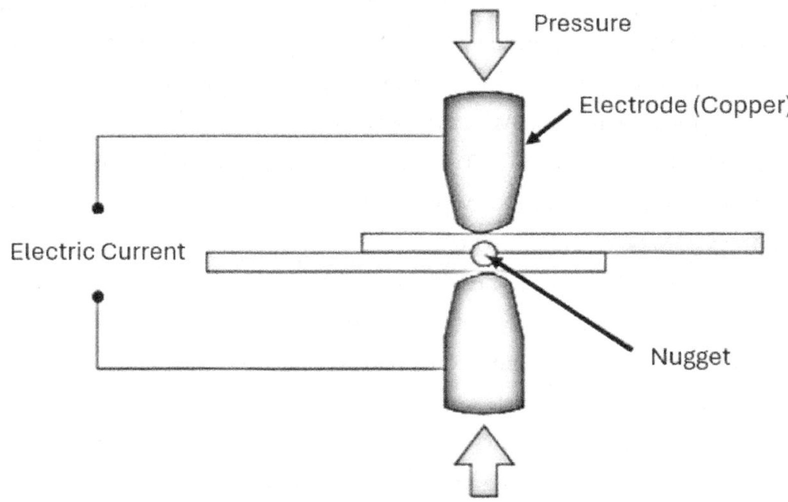

Figure 114: Schematic of Resistance Spot Welding process.

The electrodes used in RSW serve two purposes: they apply mechanical pressure to the workpieces and conduct electrical current to the welding zone. Electrodes are typically made from copper or copper alloys due to their excellent conductivity and wear resistance. In submarine applications, the electrodes may have specific shapes—such as flat, domed, or pointed tips—to accommodate different joint geometries and ensure proper contact with the workpieces. Water-cooled electrodes are often used to prevent overheating, allowing for continuous welding without compromising weld quality.

Electrode holders and arms support the welding electrodes and maintain alignment throughout the welding process. They are adjustable to ensure that the electrodes are positioned correctly over the joint. These components must also be rigid and durable to handle the mechanical pressure required for each weld. In submarine construction, where access to certain parts may be limited, specialized arms or robotic systems are sometimes employed to extend the reach of the welding equipment.

The weld timer and control unit are critical for regulating the duration and sequence of the welding process. They control the application of current and pressure, ensuring that each weld is consistent in strength and appearance. In submarine construction, programmable controllers are often used to store welding parameters for different materials and joint types, facilitating repeatable results across large assemblies. Some systems also integrate real-time monitoring to adjust parameters mid-weld and prevent defects.

RSW generates significant heat due to the high electrical current passing through the workpieces. A cooling system is essential to prevent the electrodes, holders, and transformer from overheating during extended welding sessions. Water-cooled electrodes and transformers are commonly used to maintain the operating temperature of the equipment. In submarine construction, the cooling system is particularly important, as welding is often performed continuously on large assemblies that require multiple spot welds.

Proper alignment and pressure are essential for successful RSW. Clamping mechanisms, such as pneumatic or hydraulic clamps, hold the workpieces firmly together during welding to ensure that the molten metal forms a strong bond. Fixtures may also be used to stabilize and align components, especially in submarine construction, where precision is critical. These fixtures prevent movement during the welding process, ensuring that each weld is properly placed.

In large-scale operations, including submarine construction, robotic arms and automated welding systems are often employed to perform RSW. These systems can execute multiple welds with high precision and consistency, reducing the chance of human error. Automated systems are particularly useful for welding repetitive or intricate assemblies, such as sheet metal panels and internal supports, improving efficiency while maintaining weld quality.

Although RSW does not typically require filler materials, it is essential to keep the workpieces free of contaminants to ensure weld quality. Cleaning tools, such as wire brushes and degreasers, are used to prepare the metal surfaces before welding. In submarine construction, surface preparation is particularly critical because any contamination could compromise the strength and corrosion resistance of the weld, leading to potential failure in service.

The following provides a step-by-step process for performing LBW, tailored to submarine construction requirements:

Step 1: Preparing the Workpieces

Proper surface preparation is essential for achieving strong spot welds. In submarine construction, workpieces are often made of steel or high-strength alloys, and any contamination can compromise weld quality.

- **Clean the surfaces**: Use wire brushes, degreasers, or grinding tools to remove dirt, rust, paint, or other contaminants.

- **Ensure proper alignment**: Align the metal sheets or components to be welded. Fixtures and clamps may be required to maintain precise alignment, especially in large hull sections.

This meticulous preparation ensures better electrical conductivity, reducing the chance of weld defects.

Step 2: Setting Up the Welding Equipment

RSW requires a well-calibrated power source, precise control systems, and appropriate electrodes to ensure high-quality welds.

- **Select the correct power settings**: Adjust the current and voltage on the welding machine based on the material type and thickness. Submarine components often require higher currents to weld thick steel sections effectively.

- **Choose the right electrode**: For submarine construction, copper-alloy electrodes are commonly used to maintain electrical conductivity and minimize wear.

- **Install cooling systems:** In continuous welding of large assemblies, use water-cooled electrodes and transformers to prevent overheating.

Step 3: Positioning the Components

Proper positioning is crucial, especially for welding submarine structures that must withstand extreme underwater pressure.

- **Use clamping systems**: Pneumatic or hydraulic clamps are employed to hold the metal components tightly together, ensuring minimal gaps.
- **Apply fixtures or robotic arms**: In complex assemblies or confined spaces within submarines, automated systems position the welding electrodes accurately to ensure weld consistency.

Step 4: Initiating the Welding Process

Once the components are securely positioned, the welding cycle can begin. This step involves the precise control of current, pressure, and timing.

- **Lower the electrodes**: The upper and lower electrodes are brought into contact with the workpieces, applying mechanical pressure to ensure a secure connection.
- **Activate the current**: A high electrical current is passed through the electrodes, generating heat due to resistance at the contact point between the sheets. This melts the metal at the interface, forming a weld nugget.
- **Control the weld time**: In submarine construction, programmable weld timers are used to ensure consistent heat input and prevent over- or under-welding, which could affect the weld's strength.

Step 5: Cooling and Electrode Release

After the welding current is switched off, the molten metal cools rapidly under pressure, forming a solid bond.

- **Maintain pressure during cooling**: Keeping the pressure on the weld during cooling ensures that the weld nugget solidifies properly, resulting in a strong, uniform joint.
- **Release the electrodes**: Once the weld has solidified, the electrodes are lifted, and the workpiece is ready for inspection.

Step 6: Inspecting the Welds

Inspection is critical in submarine construction to ensure that each weld can withstand underwater pressures and extreme conditions.

- **Visual inspection**: Check for any visible defects such as cracks, uneven surfaces, or incomplete fusion.
- **Non-destructive testing (NDT)**: Use ultrasonic testing (UT) or radiographic methods to detect internal flaws that could compromise the weld's integrity.

Step 7: Post-Weld Treatment and Cleaning

Any remaining slag or residue must be removed to maintain the submarine's corrosion resistance and structural integrity.

- **Clean the weld area**: Use wire brushes or grinding tools to smooth the weld surface.
- **Apply protective coatings**: In submarine construction, protective coatings are often applied to welded joints to

prevent corrosion from seawater exposure.

Step 8: Conducting Repairs (If Needed)

If any defects are identified during inspection, repairs must be carried out to meet strict safety standards.

- **Re-weld defective areas**: If necessary, the defective welds are ground out and re-welded using RSW or other appropriate methods.
- **Re-inspect repaired welds**: Ensure that the repaired areas meet the required quality standards before proceeding.

Special Considerations for Submarine Construction

In submarine construction, RSW is particularly useful for joining thin sheets of steel, internal frameworks, and components that require high precision. The confined spaces and complex geometries involved demand meticulous attention to detail.

- **Use of robotic arms**: Automated systems help access tight spaces within the submarine, such as ballast tanks or compartments.
- **Cooling systems**: Continuous water-cooling ensures that the equipment does not overheat during long welding sessions.
- **Strict quality control**: Given the critical role of submarines, every weld must undergo rigorous testing to ensure reliability and compliance with safety standards.

Non-Destructive Testing and Quality Control

Given the extreme environments submarines operate in, every welded joint undergoes rigorous non-destructive testing (NDT) to confirm the absence of defects that could compromise performance. The submerged nature of submarines and the high pressures they endure require flawless welding, as even minor imperfections in the joints could result in catastrophic failures under deep-sea conditions. As a result, comprehensive inspection techniques are essential to identify any internal voids, cracks, or misalignments within the welds.

X-ray inspection is a key NDT method used to detect internal flaws that are not visible to the naked eye. By passing X-rays through the weld joint, technicians can create a detailed image that reveals imperfections such as porosity, inclusions, or incomplete fusion. This method is particularly effective for inspecting thick sections of the pressure hull, ensuring that the welds meet the stringent requirements for withstanding deep-sea pressures. Radiographic testing allows inspectors to assess internal quality without compromising the weld or requiring disassembly, making it an invaluable tool for maintaining efficiency during submarine construction.

Ultrasonic testing (UT) is another vital technique used in submarine welding inspections. This method relies on high-frequency sound waves that penetrate the welded material, identifying internal defects through changes in wave reflections. Ultrasonic testing offers exceptional precision, making it ideal for locating very small cracks or discontinuities that could develop into more significant issues under stress. Given the high-performance demands on submarines, particularly for pressure hulls and internal frameworks, ultrasonic testing ensures that every weld achieves the necessary strength and reliability.

Magnetic particle inspection (MPI) is commonly used to identify surface or near-surface defects in ferromagnetic materials, such as the steel alloys typically employed in submarines. During MPI, the weld area is magnetized, and iron particles are applied to reveal disruptions in the magnetic field caused by defects. This method is particularly useful for detecting surface cracks, ensuring that external components are free of flaws that could compromise structural integrity or create corrosion points.

In addition to these testing methods, stringent qualifications are required for the welders involved in submarine construction. Welding in submarines often takes place in confined spaces or under challenging conditions, demanding a high level of skill, precision, and experience. Welders must meet industry standards, such as ISO 9606-1 or ASME Section IX, which assess their competency through practical testing. These qualifications ensure that only highly trained professionals perform critical welds, such as those on the submarine's pressure hull or high-pressure piping systems. Their expertise minimizes the risk of defects and ensures the safety and durability of the vessel.

Welding Challenges in Submarine Construction

Welding submarine components presents unique challenges due to the extreme operational conditions submarines face. The welded joints must endure immense hydrostatic pressures as submarines dive to great depths, where external pressure significantly increases, exerting intense compressive forces on the hull and welded joints. This necessitates that these joints maintain structural integrity to prevent catastrophic failures, ensuring the safety of the vessel and its crew [429]. The structural integrity of welded joints is critical, as any failure could lead to severe consequences, including loss of life and equipment.

In addition to static pressures, submarines are subjected to dynamic forces that complicate the welding process. Underwater explosions, such as depth charges, generate shockwaves that propagate through the hull, subjecting welded joints to rapid deformation [429, 430]. Furthermore, vibrations from engines and propellers contribute to the stress on these joints, which can lead to fatigue cracking over time if not properly managed [431]. The accumulation of residual stresses during the welding process can exacerbate these issues, making it essential for welders to focus on stress distribution and management techniques to enhance the durability of the joints [431].

Minimizing residual stress is crucial for the longevity of submarine welds. Techniques such as post-weld heat treatment (PWHT) are employed to relieve residual stresses and improve the toughness of welded joints [432]. Additionally, specialized welding sequences and preheating methods are utilized to prevent uneven thermal expansion, which can lead to distortion or cracking [431]. Peening, a process that mechanically compresses the weld surface, is also employed to reduce tensile stresses and enhance fatigue resistance [431]. These strategies are vital for ensuring that welded components can withstand the repeated cycles of stress encountered throughout the submarine's operational lifespan.

Corrosion resistance is another critical factor in submarine welding, as the constant exposure to seawater accelerates the risk of corrosion, particularly at joints where dissimilar metals are used [433]. Selecting materials with high corrosion resistance, such as stainless steel and nickel alloys, is essential to prevent deterioration of critical components [432]. The welding process must also be meticulously controlled to minimize contamination and ensure a smooth, defect-free weld area. Rough or porous welds can trap seawater, leading to localized corrosion and compromising joint strength [433]. Furthermore, corrosion-resistant coatings and cathodic protection systems are integral to the submarine's maintenance strategy, safeguarding against long-term degradation [433].

The welding of submarine components demands not only technical expertise but also careful planning to address environmental challenges and operational requirements. Welders must adhere to stringent qualification standards to perform complex welding operations under confined and challenging conditions. Each weld undergoes rigorous non-destructive testing to identify potential defects and ensure compliance with safety and performance standards [429, 430]. This comprehensive approach is essential for maintaining the structural integrity and reliability of submarines, enabling them to execute critical missions in deep-sea environments.

Fabrication Methods

Submarine construction is a complex process that integrates advanced fabrication methods to meet the stringent structural, operational, and environmental demands of underwater vessels. The pressure hull, outer structures, and internal systems of submarines must be engineered to withstand extreme conditions, including high hydrostatic pressures, temperature fluctuations, and corrosive marine environments. This necessitates the use of specialized techniques in the fabrication process, particularly in the rolling and forming of steel plates, which are critical for the integrity of the submarine's hull.

One of the primary steps in submarine construction involves the fabrication of the pressure hull, where steel plates, typically several inches thick, are rolled and formed into cylindrical sections. This process is crucial as the circular cross-section of the hull is the most effective shape for resisting external pressures encountered at depth. The use of large industrial rolling machines is essential, as they apply significant force to bend the plates into precise curves, ensuring that any deviations in curvature are minimized. Such deviations could compromise the vessel's ability to withstand hydrostatic pressure, leading to potential structural failures [433].

The materials used in submarine construction also play a significant role in ensuring durability and resistance to environmental factors. Research indicates that the corrosion resistance of welded steels used in underwater marine systems is a critical parameter for maintaining the integrity of submarines over time [433]. The selection of appropriate materials, along with advanced welding techniques, is necessary to enhance the longevity and safety of the submarine's structure under harsh conditions. Furthermore, the mechanical properties of materials, including fibre-reinforced polymers, are being explored to improve the overall performance of submarine components [434].

In addition to the mechanical and material considerations, the operational aspects of submarine design must also be taken into account. The hydrodynamic response of submarines during manoeuvres is a vital area of study, as it influences the design of the hull and the overall stability of the vessel [435]. Computational fluid dynamics (CFD) simulations are increasingly being utilized to analyse these dynamics, allowing for better predictions of how submarines will behave under various operational conditions [436]. This integration of advanced simulation techniques into the design process underscores the importance of a multidisciplinary approach in submarine construction.

Figure 115: The Virginia-class attack submarine Pre-Commissioning Unit (PCU) North Dakota (SSN 784) is rolled out of an indoor shipyard facility. U.S. NAVY, Public Domain, Public Domain Media Search Engine.

Modern submarines are increasingly constructed using a modular approach, which significantly enhances the efficiency and effectiveness of the shipbuilding process. This modular construction involves the fabrication of individual sections or modules that contain specific subsystems, such as propulsion units, crew quarters, and weapons systems. The primary advantage of this technique is that it allows multiple teams to work on different modules simultaneously, thereby reducing overall construction time and improving productivity in shipyards [437].

The modular design philosophy has been notably implemented in advanced submarine projects, such as the U.S. Virginia-class submarines, which fully embrace this approach by developing functional modules tailored to specific missions. This concurrent design of various functional modules facilitates a more streamlined assembly process, ensuring that each module is fabricated with precision to guarantee seamless integration during the final assembly phase [437]. Furthermore, the use of modular construction techniques not only shortens production cycles but also enhances the adaptability of the submarine design, allowing for easier upgrades and maintenance over the vessel's operational life [437].

In addition to improving construction efficiency, modular construction also contributes to the overall performance and reliability of submarines. The modular approach allows for the incorporation of advanced technologies and systems within each module, which can be tested independently before final assembly. This independent testing ensures that each subsystem meets stringent performance standards, thereby enhancing the reliability of the entire vessel [437]. Moreover, the flexibility inherent in modular design enables shipbuilders to respond more effectively to changing operational requirements and technological advancements, making it a preferred method in modern submarine construction [437].

Figure 116: High-angle view of the US Navy (USN) New Attack Submarine (NAS) Pre-Commissioning Unit (PCU) VIRGINIA (SSN 774), under construction inside the Electric Boat Corporation of Connecticut facility, located at Groton Shipyard, Connecticut (CT). The U.S. National Archives, Public Domain, via Picryl.

The construction of submarine components necessitates high-precision machining to meet stringent specifications, particularly for complex parts such as periscopes, propellers, valves, and other mechanical systems. The utilization of Computer Numerical Control (CNC) machines is paramount in this context, as they are designed to ensure tight tolerances essential for components that must function reliably under high-pressure underwater conditions. CNC machining allows for the fabrication of intricate geometries and the maintenance of high accuracy, which is critical in the production of submarine components that operate in challenging environments [438, 439].

Recent advancements in machining technologies, including the application of linear motor feed systems in high-speed CNC machines, have significantly enhanced machining precision and efficiency. These systems offer advantages such as increased speed, acceleration, and thrust, which are vital for meeting the high demands of precision manufacturing in the aerospace and marine industries [440]. Furthermore, non-traditional machining processes, such as wire-electric discharge machining (WEDM), have gained popularity for their ability to fabricate complex shapes from hard-to-machine materials, ensuring high accuracy and precision under tight tolerances [438, 439]. This is particularly relevant for materials like Ti6Al4V, which are commonly used in submarine applications due to their strength and corrosion resistance [439].

The importance of precision in underwater applications extends beyond machining to include the design and fabrication of components that must withstand the unique challenges of underwater environments. For instance, the sensitivity analysis of waveguide structures highlights the necessity for high precision in CNC milling to achieve the desired performance specifications [441] This precision is not only crucial for the mechanical integrity of the components but also for the overall functionality of systems such as sonar and navigation technologies that rely on accurately machined parts [442, 443].

Casting and forging are two pivotal manufacturing processes employed in the production of large metal components, such as propellers, engine parts, and structural supports. Each method has distinct advantages that cater to specific material requirements and performance criteria.

Casting is particularly advantageous for creating complex shapes and large components. Specialized alloys are often utilized in casting processes to enhance properties such as corrosion resistance and fatigue strength, which are critical for components like propellers [444]. The casting process allows for the production of intricate geometries that would be challenging to achieve through forging alone. Advanced casting techniques, such as squeeze casting, combine the benefits of both casting and forging, producing high-quality components with minimal defects [445]. This method not only reduces material waste but also improves the mechanical properties of the final product, making it suitable for demanding applications in the automotive and aerospace industries [445, 446].

On the other hand, forging is recognized for its ability to produce components with superior mechanical properties, such as strength and toughness. Forged parts, such as drive shafts and structural reinforcements, exhibit enhanced performance characteristics due to the refined microstructure resulting from the forging process [447]. The elimination of casting defects and the optimization of the metal flow during forging contribute to the integrity and reliability of the components [447, 448]. Recent advancements in forging technologies, including the use of cast preforms, have further improved the efficiency and quality of forged products, allowing for the manipulation of less ductile alloys with better tensile properties [449, 450].

Moreover, the integration of casting and forging processes, such as the Integrated Casting and Forging Process (ICFP), has been explored to leverage the strengths of both methods. This innovative approach enables the production of components that possess the complex shapes achievable through casting while maintaining the superior mechanical properties associated with forging [451]. Such advancements are crucial for industries that demand high-performance materials capable of withstanding significant loads and stresses, particularly in aerospace and automotive applications [452].

Surface treatment and coatings are critical in enhancing the durability and performance of submarines, particularly in combating corrosion and biofouling resulting from prolonged exposure to seawater. The application of specialized coatings, such as anti-corrosion paints and anti-fouling layers, is essential for maintaining the structural integrity of submarine hulls. These coatings serve as barriers against corrosive agents and biological growth, which can significantly impair operational efficiency.

One effective approach to corrosion protection is the use of cold-sprayed Zn-Al composite coatings, which have demonstrated long-term corrosion resistance in marine environments due to their ability to form a protective layer that retards corrosive medium permeation [453]. Additionally, the incorporation of advanced materials like graphene into coatings has been shown to enhance corrosion resistance by providing a passivation mechanism applicable to various metallic surfaces [454]. The performance of coatings can be further improved through the use of multilayer systems, such as those combining zinc-aluminium alloys with epoxy and polyurethane layers, which exhibit superior anti-corrosion and impact resistance properties [455].

Moreover, the application of cathodic protection systems is a widely adopted method to prevent electrochemical corrosion. These systems work by using sacrificial anodes that corrode preferentially, thus protecting the underlying metal [456]. The effectiveness of these coatings is often influenced by their thickness and uniformity, as inadequate coating can lead to premature failure and increased susceptibility to corrosion [457]. In marine environments, organic coatings are particularly favoured due to their high efficiency and ease of application, making them suitable for various naval applications [458].

Biofouling, which refers to the accumulation of microorganisms, plants, algae, and animals on submerged surfaces, poses another significant challenge for submarines. The presence of biofouling can increase surface roughness, leading to higher drag and reduced hydrodynamic efficiency during underwater travel [459]. Recent advancements in anti-fouling coatings aim to mitigate this issue by employing nature-inspired strategies that enhance the resistance of surfaces to biological attachment [460]. For instance, superhydrophobic coatings have been developed to create surfaces that repel water and contaminants, thereby reducing the likelihood of biofouling [461].

The use of advanced composites in submarine construction has gained significant attention due to their ability to reduce weight and enhance stealth capabilities. Composite materials, particularly those reinforced with fibres such as carbon or glass, are increasingly utilized in non-load-bearing structures, including the sail (conning tower), propeller blades, and sonar domes of submarines. The fabrication process of these composites involves layering fibres within a resin matrix, followed by curing, which results in lightweight and durable components that are essential for modern submarine design [18, 462].

The mechanical properties of composite materials are heavily influenced by the arrangement and type of fibres used. For instance, the tensile strength of layered carbon fibre composites can vary significantly based on the number of layers and the orientation of the fibres [462]. This is critical in submarine applications where structural integrity and performance are paramount. The use of various fibre types, including glass and carbon fibres, allows for tailored mechanical properties that meet specific design requirements [18, 463]. Moreover, the integration of advanced composite materials in submarine construction not only contributes to weight reduction but also minimizes acoustic signatures, making submarines more difficult to detect [18].

The fabrication of these composites often employs techniques such as vacuum infusion, which enhances the quality and consistency of the final product [462]. This method allows for better control over the resin distribution within the fibre layers, leading to improved mechanical properties and durability. Additionally, the use of hybrid composites, which combine different types of fibres, can further optimize performance by balancing stiffness and damage tolerance [464, 465]. The application of these materials extends beyond structural components; they are also crucial in enhancing the stealth capabilities of submarines by reducing their radar and acoustic signatures [18].

The assembly and integration of submarine modules is a critical phase in submarine construction, involving the precise alignment and connection of various systems within the hull. This process typically begins after individual modules have been fabricated and pre-tested, ensuring that each component meets the necessary specifications before integration. Specialized cranes and lifting systems are employed to position these large sections accurately, which is crucial for the structural integrity and operational efficiency of the submarine [31, 34].

During the assembly phase, various mechanical systems, electronics, and propulsion units are integrated into the submarine's hull. This integration is not merely a physical assembly but also involves complex engineering processes to ensure that all systems function cohesively. For instance, the pressure hull, which houses critical components such as personnel, propulsion machinery, and weapon systems, must be designed to withstand significant hydrostatic pressures encountered during operation [32, 34, 38]. The materials used in constructing the hull, such as advanced composites or high-strength steels, are selected based on their ability to endure these pressures while minimizing weight [31].

Furthermore, the installation of electrical wiring, piping systems, and communication networks is a vital part of the integration process. These systems must be meticulously planned and executed to avoid interference and ensure reliability under operational conditions. The integration phase also includes thorough testing to verify the operational readiness of all systems. This testing phase is essential to identify any potential issues that could affect the submarine's performance or safety [466, 467].

In addition to the mechanical and electrical integration, the hydrodynamic aspects of the submarine's design are also optimized during assembly. Computational Fluid Dynamics (CFD) tools are often utilized to simulate and analyse the submarine's performance under various conditions, ensuring that the hull design minimizes drag and maximizes stability [24, 468]. This optimization is crucial for enhancing the submarine's stealth capabilities and overall operational effectiveness [60].

Figure 117: Starboard bow view of the US Navy (USN) New Attack Submarine (NAS) Pre-Commissioning Unit (PCU) VIRGINIA (SSN 774), as it is moved out doors for the first time. The U.S. National Archives, Public Domain, via NARA & DVIDS Public Domain Archive Public Domain Search.

Throughout the fabrication process of submarines, rigorous testing and quality assurance protocols are implemented to ensure compliance with stringent military and engineering standards. These measures are essential for guaranteeing the safety, reliability, and performance of the submarine, as even minor defects can compromise the vessel's ability to operate under extreme underwater conditions.

Non-destructive testing (NDT) methods play a critical role in verifying the integrity of welds and ensuring material quality. Ultrasonic inspection is used to detect internal flaws within welded joints, such as cracks or voids, by transmitting high-frequency sound waves into the material. Radiography, or X-ray testing, provides detailed images of the interior of welds and components, making it possible to identify defects that might not be visible to the naked eye. Dye penetrant testing is also performed on surfaces to reveal hairline cracks and porosities that could weaken structural components.

In addition to NDT, pressure testing is conducted on critical systems to simulate the conditions that submarines encounter at great depths. This process involves exposing the vessel or its components to high pressures in controlled environments, confirming that they can withstand the forces experienced during deep-sea operations. Such tests are vital for verifying the hull's ability to maintain structural integrity and prevent water ingress under extreme pressures.

Both static and dynamic tests are performed on various systems during the quality assurance phase. Static testing involves examining components in fixed positions to confirm their mechanical properties and ensure they meet design specifications. Dynamic testing, on the other hand, assesses the functionality of systems under simulated operating conditions. For example, propulsion systems, ballast tanks, and control surfaces are tested to ensure they respond correctly and reliably during submersion, surfacing, and manoeuvring.

These quality assurance procedures are carried out meticulously throughout the entire fabrication and assembly process. Every stage of production undergoes inspection to ensure that all components meet the highest standards. This thorough approach to testing and verification ensures that the final product—the submarine—can operate safely and efficiently, maintaining its structural and operational integrity even in the harsh and unpredictable environment of the deep sea.

After the assembly of a submarine, the vessel undergoes a critical phase of hydrostatic testing to verify its structural integrity and identify any potential leaks. This process involves immersing the submarine in water or placing it inside a pressure chamber to simulate the intense hydrostatic pressures it will encounter at operational depths. During hydrostatic testing, all compartments, seals, and joints are carefully inspected to ensure they can withstand external pressure without deformation or leakage. Engineers monitor the hull for any signs of stress, as even minor weaknesses could lead to catastrophic failures during deep-sea operations. Passing this phase is essential, as it certifies the vessel's ability to maintain its structural integrity under extreme underwater conditions.

Once the submarine successfully completes hydrostatic testing, it progresses to sea trials, where its systems are evaluated in real-world maritime conditions. Sea trials represent one of the most important stages in the commissioning process, as they allow engineers, operators, and inspectors to verify the performance of the submarine's key components. During these trials, the propulsion system is thoroughly tested to assess the vessel's speed, acceleration, and endurance under various operational scenarios. Engineers ensure that the submarine's engines and propulsion systems function reliably over extended periods without overheating or breakdowns.

Manoeuvrability is another critical aspect evaluated during sea trials. The submarine's ability to perform precise manoeuvres, such as dives, ascents, and course changes, is tested to confirm the responsiveness of its control surfaces, including the rudder and diving planes. This phase ensures that the submarine can navigate effectively, even in tight underwater spaces or under challenging conditions.

Sonar performance is also a focal point of sea trials, as it plays a crucial role in the submarine's ability to detect other vessels and underwater threats. Both passive and active sonar systems are tested to verify their range, accuracy, and sensitivity. Operators assess the submarine's ability to remain stealthy while using sonar, ensuring that it can gather critical information without compromising its location. Additionally, weapons systems are rigorously tested to confirm their operational readiness. Torpedoes, missile launch systems, and other onboard armaments are evaluated to ensure that they can be deployed safely and effectively when needed.

Sea trials provide the final opportunity to identify and resolve any issues before the submarine is officially commissioned for service. Every system is monitored and scrutinized under real-world conditions to ensure that the vessel meets the rigorous standards required for naval operations. Only after successfully completing these trials is the submarine deemed ready for deployment, ensuring that it can operate reliably and safely in the demanding and unpredictable marine environment.

Corrosion Resistance and Maintenance

Steel and its alloys are the primary structural materials used in constructing ships, submarines, and offshore platforms due to their strength, availability, and versatility. However, these structures are constantly exposed to harsh marine environments, which subject them to various chemical and biological factors, such as corrosion and biofouling. These threats can significantly degrade the performance and lifespan of naval vessels, leading to substantial financial losses. In fact, billions of dollars are spent globally each year to combat the damaging effects of corrosion and fouling in marine environments, underscoring the importance of effective protective measures [469].

To address these challenges, organizations like the Naval Materials Research Laboratory (NMRL) have developed advanced technologies to protect naval platforms, ensuring their durability and operational readiness. One of the primary strategies for corrosion prevention is the use of protective coatings. Protective coatings, often composed of organic polymers with various functional additives, serve as a physical barrier that isolates the metal substrate from the corrosive elements in the external environment. These coatings are applied to the surface of steel and its alloys to prevent direct contact with seawater, oxygen, and other corrosive agents that can trigger electrochemical reactions [469].

The coating system typically consists of an organic binder, such as epoxy or polyurethane, which forms the primary matrix. Embedded within this matrix are inorganic additives that enhance the protective properties of the coating. These additives may include corrosion inhibitors, pigments, and fillers that provide improved resistance to the harsh conditions of the marine environment. The carefully formulated coating system ensures long-term protection by reducing the risk of rust, pitting, and other forms of corrosion that can compromise the structural integrity of the vessel.

In addition to corrosion resistance, protective coatings also play a crucial role in mitigating biofouling—the accumulation of marine organisms like barnacles, algae, and molluscs on submerged surfaces. Biofouling increases drag, reducing the vessel's efficiency and increasing fuel consumption. Advanced coatings with antifouling properties are designed to prevent or minimize the adhesion of these organisms, ensuring smooth hull surfaces and enhancing the vessel's hydrodynamic performance.

The development and application of protective coatings require rigorous testing to meet the specific operational demands of naval platforms. Coatings must withstand not only prolonged exposure to seawater but also the mechanical stresses associated with underwater operations, including varying temperatures and pressure changes. Specialized formulations are developed for different sections of the vessel, such as the hull, ballast tanks, and propulsion systems, to optimize performance across all areas [469].

Heavy-duty non-skid paint plays a crucial role in ensuring the safety and functionality of naval vessels by providing anti-slip surfaces on critical decks and areas. Applied on flight decks, helicopter decks, and weather-exposed platforms, this specialized paint helps prevent personnel and aircraft from skidding during operations. The formulation of this paint allows it to withstand extreme thermal stress, enduring intermittent heating up to 250°C. Beyond external decks, the non-skid paint is also applied on engine room floors and other areas where anti-slip properties are necessary, enhancing the safety of personnel working in high-risk environments aboard naval vessels.

High-performance exterior paint is essential for protecting naval ships from the hostile conditions of the marine environment. This paint is applied to exposed bulkheads, the superstructure, and other parts of the ship that are constantly exposed to harsh elements such as saltwater, ultraviolet rays, and extreme weather. Its superior formulation offers excellent gloss retention, ensuring that the ship maintains its appearance and structural integrity over time. The paint also exhibits strong resistance to chalking, a phenomenon where the surface deteriorates and becomes powdery under exposure to the elements. The use of high-performance paint minimizes maintenance and enhances the longevity of naval vessels.

In environments where fire hazards pose significant risks, intumescent fire retardant paint provides essential protection. This specially designed paint forms a protective char layer when exposed to extreme heat or flames, insulating the underlying substrate and preventing the spread of fire. This thermal insulation is critical for safeguarding interior compartments, electrical installations, and essential equipment aboard naval ships. The technology is not limited to naval applications but also extends to offshore structures, high-rise buildings, and railway compartments, ensuring fire safety across a range of sectors. Its induction into naval services reflects the importance of fire mitigation strategies for personnel safety and structural integrity [469].

Anti-corrosive and anti-fouling paints are pivotal in maintaining the performance of submerged naval structures and hulls. These paints address two critical challenges: corrosion caused by continuous exposure to seawater and biofouling from marine organisms such as algae and barnacles. The anti-corrosive properties protect the metal surfaces from rust and degradation, while the anti-fouling component prevents marine organisms from adhering to the hull, thereby improving hydrodynamic efficiency. A key innovation of this technology is the in-house-developed application device, which allows for the direct application of the paint on submerged structures. This enables on-site maintenance without the need to bring the ship to dock, reducing downtime and operational costs.

Collectively, these specialized paint technologies developed for naval applications enhance the safety, durability, and performance of vessels. They protect against environmental hazards, prevent accidents, mitigate fire risks, and extend the service life of ships operating in harsh marine environments.

Traditional paint systems used for marine applications typically consist of multiple layers, including a primer, tie coat, and several topcoats, each applied sequentially. This conventional approach requires strict adherence to intercoat intervals, which can range from minutes to hours, making the process time-consuming and labour-intensive. Furthermore, the use of volatile organic compounds (VOCs) in these coatings has raised environmental concerns. To address these challenges, self-stratifying coating technology has been developed as a cutting-edge solution [469].

Self-stratifying coatings allow for the formation of two distinct, interconnected layers with complementary properties through a single application. This eliminates the need for multiple coating steps, reducing time, labour, and environmental impact. The innovation lies in how the coating segregates into separate functional layers as it cures, creating a synergistic system that maintains both protective and functional properties. This breakthrough simplifies the application process while enhancing performance.

Building on the self-stratification concept, a self-cleaning coating system has been developed and successfully transferred to industries. This system integrates hydrophobic polydimethylsiloxane (PDMS) resin with epoxy resin into a single-layer application. The hydrophobic top surface prevents the accumulation of dirt and contaminants by enabling self-cleaning, while the underlying epoxy layer provides strong anti-corrosive properties and excellent adhesion to metallic substrates. This dual-functionality makes the coating ideal for application on the superstructures and interior compartments of ships, offering both protection and cleanliness in one step. Additionally, the technology can be applied in various civilian settings, further demonstrating its versatility.

An additional advancement from self-stratifying technology is the development of a foul-release and anti-corrosion coating. Similar to the self-cleaning coating, this system features two layers that form automatically upon application. The top layer possesses foul-release characteristics, ensuring that marine organisms, such as algae and barnacles, can be easily removed with a simple water wash. Beneath this, the anti-corrosive layer ensures excellent adhesion to metal surfaces, providing long-lasting protection against corrosion. This dual-purpose coating not only protects the hulls of marine vessels but also contributes to improved performance by reducing maintenance needs.

Beyond protection, the technology has been adapted for functional applications, particularly for drag reduction in high-speed underwater vessels. The drag-reducing variant of the coating is designed to release a high molecular weight, water-soluble polymer upon contact with water, which modifies the turbulent boundary layer around the vessel. Experimental testing in a high-speed towing tank confirmed that the coating could reduce skin friction drag by 10 percent at a towing speed of 4 meters per second. Building on this success, a second-generation Mk-II version of the coating has been developed, optimized for speeds up to 20 meters per second, aligning with real-time operational conditions of high-speed marine vessels [469].

The advancements in self-stratifying coatings mark a significant leap in marine coating technology by combining protection, functionality, and environmental sustainability. These coatings reduce labour and environmental impact while offering enhanced performance and durability for naval vessels and other high-performance applications.

In addition to protective coatings, ship hulls are equipped with advanced corrosion control systems known as Impressed Current Cathodic Protection (ICCP). The ICCP system ensures long-term corrosion resistance by making the ship's hull act as the cathode in an electrochemical circuit. This form of cathodic protection involves either sacrificial anodes or the use of an ICCP system, with the latter offering more precise and controllable protection.

The ICCP system developed by the Naval Materials Research Laboratory (NMRL) is composed of several key components, including an Auto Control Unit (ACU), Platinized Titanium (Pt-Ti) anodes, and Silver-Silver Chloride (Ag/AgCl) reference electrodes. The system works by impressing a controlled current from the Pt-Ti anode to the ship's hull, maintaining it within an optimal protective potential range of -800 to -850 mV relative to the Ag/AgCl reference electrode. This process ensures that the hull remains in a cathodic state, preventing electrochemical corrosion [469].

NMRL has taken the ICCP system a step further by developing a modular, microcontroller-based ACU. This advanced ACU is designed to ensure uniform hull potential, essential for consistent protection. The modular structure integrates multiple components within a single cabinet, with each module containing its own anode, reference electrode, power supply, and controller. This design enhances the system's flexibility and reliability, allowing it to deliver precise current control across different sections of the hull.

Beyond corrosion prevention, the ICCP system also plays a critical role in minimizing the Extremely Low Frequency Electromagnetic (ELFE) signatures emitted by ships. ELFE signatures can be detected by enemy vessels, posing a threat to naval operations. By managing the electrical currents on the hull, the ICCP system not only protects against corrosion but also reduces these electromagnetic signatures, enhancing stealth capabilities.

The NMRL-developed ICCP system is now deployed across the entire Indian Naval fleet, underscoring its importance in naval operations. This technology ensures that vessels remain structurally sound and operational over extended periods while also contributing to the fleet's stealth and defensive capabilities. The system exemplifies a combination of advanced engineering and innovative materials science, providing both corrosion resistance and strategic advantages in maritime operations.

Applying Protective Coatings

During submarine construction, protective coatings are applied in a meticulous, multi-step process to ensure that the vessel is protected from corrosion, biofouling, and environmental degradation. These coatings are essential for maintaining the submarine's structural integrity, performance, and longevity, particularly in the harsh marine environment. The process of applying these coatings involves surface preparation, layering of specialized coatings, and rigorous quality inspections at each stage to ensure adherence to engineering standards.

The first step in the application of protective coatings is surface preparation. All metal surfaces, especially the steel hull, must be cleaned thoroughly to remove contaminants such as dirt, oil, rust, and mill scale. This is typically achieved through abrasive blasting or water jetting, which creates a clean and roughened surface that promotes better adhesion of the coating layers. In submarine construction, achieving proper surface roughness is crucial, as it ensures a strong bond between the metal substrate and the protective coating. Following surface preparation, the surfaces are carefully inspected to verify cleanliness and roughness, and any imperfections are repaired before proceeding.

The next phase involves the application of primer coatings, which form the first layer of protection. Primers contain anti-corrosive pigments that inhibit rust and protect the substrate from moisture. In submarine construction, specialized primers such as zinc-rich coatings are often used to provide cathodic protection. The primers are applied using spray equipment to ensure even coverage and are allowed to cure before the subsequent layers are applied.

After the primer layer, intermediate or tie coats are applied to provide additional corrosion protection and serve as a bonding layer between the primer and the topcoat. These tie coats are essential to ensure compatibility between layers and enhance the adhesion of the topcoat. The coatings are carefully inspected between applications to identify any defects, such as pinholes or blisters, which are repaired promptly to maintain coating integrity.

The final stage of the process involves applying topcoats that provide the submarine with its primary protection against environmental factors, such as seawater, UV radiation, and biofouling. Depending on the area of the submarine, different types of topcoats are used. For external surfaces, high-performance epoxy or polyurethane coatings are applied to resist abrasion and chemical attack. On areas prone to fouling, such as the underwater hull, specialized anti-fouling coatings are applied to prevent the attachment of marine organisms. These coatings contain biocides that gradually leach into the water to inhibit the growth of barnacles, algae, and other organisms. Additionally, some submarines are coated with stealth materials, such as acoustic rubber tiles, which reduce sonar reflections and enhance stealth capabilities.

In some cases, self-stratifying or self-cleaning coatings are applied to minimize maintenance. These advanced coatings combine multiple protective properties, such as anti-corrosive and hydrophobic qualities, in a single layer. The coatings are applied using specialized spray systems that ensure precision and uniformity, especially in confined spaces and hard-to-reach areas of the submarine.

Throughout the coating process, strict quality control measures are followed. Inspectors conduct visual inspections, adhesion tests, and thickness measurements at each stage to ensure that the coatings meet the required standards. Non-destructive testing methods, such as holiday testing, are also employed to detect any breaks or voids in the coating layers. These inspections are critical in submarine construction, as any defects in the coatings could compromise the vessel's performance and lifespan.

Once all layers of coatings are applied and fully cured, the submarine undergoes final inspections and touch-ups to address any minor imperfections. The coated surfaces are inspected again before launching the vessel to ensure the protective systems are intact. In some cases, areas that experience high levels of stress or abrasion may receive additional layers of protective coatings for enhanced durability. This rigorous process ensures that the submarine is ready to withstand the harsh conditions of the marine environment, providing long-term protection, efficiency, and operational reliability.

Figure 118: Sonar Technician 2nd Class Jacob Stich touches up a fresh coat of paint on the sail of the Virginia class fast-attack submarine USS North Carolina (SSN 777). U.S. NAVY, Public Domain, via Picryl.

Chapter 11

Designing for Stealth and Acoustic Signatures

Acoustic Quieting and Sound Management

Acoustic quieting is a fundamental aspect of submarine operations, as it directly impacts a vessel's ability to avoid detection by sonar systems. Submarines face the constant challenge of remaining undetectable while operating in environments monitored by increasingly advanced passive and active sonar. To reduce their acoustic signatures, modern submarines employ a combination of structural design, machinery isolation, and innovative propulsion technologies.

A key strategy in acoustic quieting is optimizing the submarine's hull design. The shape of the hull plays a crucial role in minimizing hydrodynamic drag and turbulence as the submarine moves through water. Smoother hull surfaces and carefully contoured designs ensure that water flows efficiently around the vessel, minimizing noise. Additionally, specialized coatings are often applied to reduce friction and dampen vibrations, further lowering the acoustic footprint.

Another vital technique involves isolating internal machinery from the submarine's hull. Vibrations generated by engines, pumps, and other mechanical systems can propagate through the hull and radiate into the water, making the submarine detectable. To address this, submarines use vibration-isolating mounts and sound-damping materials that absorb or reduce mechanical vibrations. These mounts physically separate noisy equipment from the hull, preventing sound transmission to the external environment.

Propulsion systems also receive significant attention in noise management efforts. Propeller blades, for example, are designed with advanced shapes to prevent cavitation—a phenomenon where vapor bubbles form and collapse, generating noise. Some submarines employ pump-jet propulsion systems that operate more quietly than traditional propellers. Additionally, electric propulsion systems are increasingly used to eliminate the noise generated by conventional engines.

Acoustic quieting also extends to sound-absorbing tiles, often made from rubber composites, which are affixed to the submarine's outer hull. These tiles help absorb and dissipate sound waves, both from the submarine's internal operations and from external sonar systems. This reduces the vessel's overall acoustic signature and makes it more difficult to detect with passive sonar.

The use of advanced computer systems for real-time monitoring and adjustment of onboard machinery further enhances acoustic quieting. These systems allow crews to manage equipment settings dynamically, minimizing noise output when operating near enemy sonar arrays. Submarines may also employ passive sonar systems to detect incoming sonar pings, giving them the opportunity to adjust their positioning or operational parameters to avoid detection.

Through these sophisticated techniques, submarines achieve the necessary stealth to conduct missions covertly. Modern acoustic quieting systems represent the culmination of decades of research, reflecting the ongoing arms race between submarine technologies and sonar detection systems.

A moving submarine generates various types of noise, which can compromise its stealth and be used by adversaries to detect, track, and even identify the vessel. Noise generated by a submarine is classified based on the mechanisms responsible for sound emission, such as hull vibration, mechanical equipment, propulsion systems (especially propellers), and hydrodynamic flow noise [470].

The concept of source level (SL) is used to describe the intensity of noise emitted by a submarine, measured in decibels (dB) relative to a reference intensity—either 1 pascal (Pa) for U.S. standards or 20 μPa for Russian standards. This difference in measurement standards can result in discrepancies of around 27 dB between the two systems. The SL of a submarine provides critical insights into its detectability across various frequency bands [470].

Submarines emit both continuous and discrete noise across their frequency spectrum. Continuous noise, which dominates at lower frequencies, typically peaks between 50-100 Hz, a frequency range often affected by vibrations from the hull or power plant equipment. Beyond 200 Hz, SL tends to drop by 6 dB each time the frequency doubles, meaning that higher frequencies contribute less to the acoustic footprint [470].

The discrete components of a submarine's noise spectrum, such as those generated by rotating machinery like propellers and generators, are crucial for identification. These discrete elements form the "acoustic signature" or "acoustic portrait" of a submarine, enabling enemy sonar systems to classify and track individual vessels. Propeller noise, in particular, emits at very low frequencies (0.1-10 Hz), making it detectable over long distances, as the ocean absorbs low-frequency sound minimally [470].

The rotational speed of the propeller creates predictable, discrete noise components, while mechanical vibrations from generators and turbines contribute harmonics in the frequency range of 50-60 Hz, corresponding to the electrical systems used on board [470].

The submarine's speed plays a critical role in determining its noise emissions. In "ultraquiet" mode, submarines reduce their operational speed to around 4 knots, minimizing noise emissions by restricting mechanical activities. However, in this mode, noise from essential systems like the propeller and power plant is still present. As the submarine's speed increases to approximately 8 knots—referred to as the "maximum low-noise speed"—the SL increases by 5-10 dB [470].

During transit or long-distance patrols, submarines operate at higher speeds, typically around 15 knots. At these speeds, hydrodynamic flow noise becomes the primary source of noise, as turbulent water movement around the hull generates sound. The relationship between speed and SL is exponential; a doubling of speed results in an 18 dB increase in SL, as noise scales with the sixth power of speed [470].

A submarine's noise emissions are not static over its operational lifespan. Wear and tear on mechanical components, misalignments, and material fatigue can gradually increase noise levels by up to 5 dB. Ensuring low noise levels over time requires consistent maintenance and the replacement of worn-out components [470].

Diesel-electric submarines operating on battery power are quieter than nuclear-powered submarines because nuclear reactors generate additional mechanical noise, even when running at minimal speeds. Submarines with double-hull designs—such as Russian SSBNs—tend to be noisier due to the added mechanical complexity and larger displacement, which necessitates more powerful engines and propulsion systems [470].

Over time, submarine builders have achieved significant noise reductions by improving precision manufacturing techniques. For example, the accuracy of gear manufacturing for turbine assemblies can drastically reduce mechanical noise. Russian submarines, such as the Delta IV-class (667 BDRM) and Typhoon-class, have achieved lower SLs through enhanced manufacturing processes, active noise suppression technologies, and improved design features [470].

Acoustic quieting is essential for modern submarines to remain undetected during operations. Submarine designers focus on minimizing noise by optimizing the hull shape, employing advanced materials, and refining propulsion systems. However, maintaining low noise levels over time requires careful maintenance and operational strategies tailored to specific conditions, including adjusting speeds to manage hydrodynamic and mechanical noise [470].

Hull Design

Optimizing the hull design of submarines for acoustic quieting is a multifaceted challenge that integrates advanced engineering principles aimed at minimizing hydrodynamic drag, turbulence, and noise transmission. The shape and surface treatments of the hull are critical factors that influence the submarine's operational stealth and efficiency. A streamlined, smooth hull shape is essential for reducing turbulent flow, which is known to generate vortices that produce noise detectable by passive sonar systems. Research indicates that modern submarines often employ teardrop-shaped hulls, which are designed to promote laminar flow—characterized by smooth water movement along the hull—thereby minimizing drag and enhancing speed and fuel efficiency [61, 63].

The exterior surface treatment of the submarine plays a pivotal role in noise reduction. Advanced materials, such as elastomeric polymers and sound-absorbing rubber tiles, are utilized to create a smoother surface that reduces friction and dampens mechanical vibrations. These coatings not only facilitate more efficient water flow but also serve as a barrier against vibrations from internal machinery, preventing these vibrations from propagating through the hull and into the surrounding water [471, 472]. Anechoic tiles, which are specifically designed to absorb and scatter sound waves, are a notable example of such coatings. These tiles help to reduce sonar echoes and self-noise, effectively masking internal sounds that could compromise the submarine's stealth [471].

To further refine the hull design and surface features, engineers employ advanced computational fluid dynamics (CFD) models. These simulations allow for detailed analysis of water flow patterns and turbulence around the submarine at various speeds, facilitating the optimization of the design to achieve minimal acoustic disturbance [63, 468]. The integration of CFD with multi-objective optimization techniques has proven effective in enhancing the hydrodynamic performance of submarine hulls, ensuring that the design not only meets stealth requirements but also operational efficiency [63].

The combination of a smooth hull contour, low-friction coatings, and vibration-damping materials is vital for maintaining stealth in submarine operations. These design features collectively ensure that submarines can navigate through water with minimal noise, making them less detectable by enemy sonar systems and thereby enhancing their operational effectiveness [473].

Isolating Internal Machinery from the Submarine's Hull

In submarine construction, isolating internal machinery from the hull is a critical strategy for noise reduction. The mechanical vibrations generated by engines, pumps, and other onboard equipment can easily transmit through the hull into the surrounding water, thereby increasing the vessel's acoustic signature and making it more detectable to enemy sonar systems. To mitigate this issue, submarines employ a combination of vibration-isolating mounts, sound-damping materials, and flexible couplings to effectively decouple mechanical noise from the hull [474, 475].

Submarines utilize various vibration-isolation techniques to enhance stealth capabilities by minimizing noise generated by internal machinery, such as engines and pumps. This is crucial because any noise transmitted through the hull can significantly increase the submarine's acoustic signature, making it more detectable by adversaries. A notable example of this is the Virginia-class submarines, which incorporate urethane tiles and advanced vibration-dampening materials strategically placed between machinery and the hull [476]. These materials are designed to absorb mechanical vibrations, preventing them from radiating into the surrounding water, thereby reducing the submarine's noise profile [477].

Vibration-isolating mounts are specifically designed to absorb mechanical energy and reduce the transmission of vibrations from noisy equipment to the submarine's hull. These mounts function as buffers between machinery and the hull by suspending equipment on flexible supports made from materials such as rubber, elastomers, or specialized polymers. This design allows the mounts to act as shock absorbers, dissipating vibrational energy across multiple axes, which prevents the hull from acting as a sound conductor [474, 475]. Advanced mounts often incorporate multi-layered damping systems that combine rubber and metal to balance load-bearing capacity with vibration absorption, thus ensuring controlled movement of machinery while minimizing mechanical energy transmission ([474, 475].

Moreover, the engineering of vibration isolators in modern submarines is tailored to target specific frequency ranges associated with the machinery they support, such as diesel generators or pumps. This targeted approach enhances the effectiveness of vibration isolation, ensuring that the submarine maintains a low acoustic signature during operations [474, 475].

In addition to vibration-isolating mounts, submarines utilize sound-damping materials within machinery compartments and between equipment and the hull. These materials absorb vibrational energy and convert it into heat, thereby reducing noise emissions. Commonly used materials include viscoelastic polymers, rubber-based compounds, and specialized foams that are designed to function effectively under marine conditions [475].

Flexible couplings are also critical in rotary systems, such as pumps and drive shafts, where rigid connections could transmit vibrations. These couplings, made from elastomeric elements or multi-layered composites, allow for movement between connected parts without generating noise, which is particularly important in systems where misalignment or thermal expansion could introduce vibrational noise [474, 475].

In nuclear submarines, isolating the reactor plant and propulsion system presents unique challenges due to the size and complexity of these components. Power plants are typically mounted on floating decks or isolated platforms within the hull, providing an additional layer of separation. Advanced designs utilize mass-loaded mounts to further reduce low-frequency vibrations, which are particularly difficult to suppress but critical for maintaining stealth [474, 475].

By effectively isolating noisy equipment from the hull, submarines minimize the likelihood of detection by enemy sonar systems. This design ensures that the submarine's acoustic signature remains within acceptable levels across various operating speeds. However, maintaining precise alignment and the integrity of these vibration-damping systems is essential for sustaining performance over the submarine's operational life, as the deterioration of isolators can gradually increase noise levels [474, 475].

Propulsion System Quietening

Propulsion systems in submarines are critical for managing noise, which significantly influences the vessel's acoustic signature. The design of propeller blades is a key factor in minimizing noise levels, particularly through advanced shapes that aim to reduce cavitation. Cavitation occurs when vapor bubbles form around the blades and subsequently collapse, generating sound. This phenomenon is typically exacerbated at high speeds or under specific pressure conditions, necessitating the use of skewed blades and optimized designs in modern submarines to facilitate smoother water flow and minimize bubble formation [478].

Designing submarine propeller blades to minimize noise and reduce cavitation involves advanced engineering techniques that focus on controlling water flow and pressure around the blades. Cavitation occurs when rapid changes in pressure create vapor bubbles, which collapse, generating noise detectable by sonar systems. To avoid this, blade geometry is carefully optimized.

One effective approach is using skewed blades, which curve backward along their length. This design reduces sudden pressure changes as the blade rotates, distributing load more evenly and preventing cavitation. Additionally, high-aspect ratio blades, which are long and narrow, enhance hydrodynamic efficiency by reducing drag and improving the smoothness of water flow over the surface.

The leading and trailing edges of the blades are refined and polished, reducing turbulence and friction. Designers also incorporate variable pitch blades, which adjust their angle based on the speed and load, further minimizing noise at different operational conditions. Hydrodynamic simulation tools and computational fluid dynamics (CFD) software are often used to predict water flow and cavitation patterns during the design phase, ensuring the propeller operates quietly under various speeds and depths.

In some modern submarines, pump-jet propulsion systems replace traditional propellers, using enclosed impellers that produce minimal noise. This technology is especially effective at preventing cavitation in high-speed operations, offering stealth advantages crucial in military applications.

Pump-jet propulsion systems are used in several modern submarines due to their stealth and efficiency. These systems offer reduced cavitation noise, making them ideal for high-speed, covert operations.

Examples include:

- **Astute-class (UK)**: These submarines employ pump-jet propulsion to ensure quiet operations, enhancing their stealth capabilities.

- **Virginia-class (USA)**: The use of pump-jets allows these nuclear-powered submarines to remain undetected, especially during covert missions.

- **Suffren-class (France)**: This class integrates pump-jet technology to improve acoustic stealth and manoeuvrability.

These systems make modern submarines significantly quieter and more effective in naval operations.

Designing submarine propellers differs significantly from designing those for surface vessels because the primary focus is on minimizing the acoustic signature to enhance stealth. While efficiency is still important, it is secondary to noise reduction, as even minimal noise emissions can compromise a submarine's ability to remain undetected. Submarine propellers must also balance durability, propulsion power, and manoeuvrability, ensuring reliable performance across a variety of operating conditions, including deep diving and surface cruising. This need for versatility introduces additional complexities, requiring specialized designs to meet stealth and performance goals [479].

Noise management is a critical factor in submarine propeller design, as the propeller is a significant source of underwater noise. Thrust noise typically dominates at lower frequencies, below 100 Hz, and is caused by thrust fluctuations due to uneven wake fields around the blades. At higher frequencies, trailing-edge noise becomes more prominent. This noise results from turbulence and vortex shedding at the blade's trailing edge and can extend to frequencies up to 1000 Hz. Mitigating this noise requires optimizing the geometry of the trailing edges to control vortex behaviour. Another challenge is broadband noise, which arises from turbulent inflow and creates acoustic disturbances that can interfere with a submarine's sonar system, diminishing its detection capabilities [479].

Cavitation control plays a pivotal role in propeller design. Cavitation occurs when low-pressure zones on the blade surface cause water to vaporize into bubbles, which subsequently collapse, generating noise and potentially damaging the blades. To prevent cavitation, designers focus on adjusting the pitch, camber, and surface geometry of the propeller blades. Suction-side cavitation, tip vortex cavitation, and hub vortex cavitation are some of the forms addressed through these optimizations. The use of non-planar lifting surfaces, like the Kappel propeller, demonstrates how advanced designs can improve load distribution, reducing the likelihood of cavitation and enhancing hydrodynamic efficiency [479].

As an example of the Kappel propeller, an 8-bladed design developed to minimize acoustic emissions and stabilize thrust performance. Compared to a 7-bladed variant, the 8-bladed version demonstrated superior noise reduction while maintaining efficient propulsion. These advancements align with the fundamental goal of minimizing a submarine's detectability while ensuring that propulsion performance is not compromised. The Kappel propeller exemplifies how cutting-edge design techniques can deliver both stealth and operational efficiency [479].

The propeller design process is iterative, with continual refinement based on hydrodynamic and acoustic testing. Each iteration examines factors such as the diameter and number of blades, which influence thrust fluctuations and noise levels. Skew and rake adjustments enhance cavitation resistance by controlling how water flows over the blades, while carefully selected materials and anti-singing edges minimize tonal noise generated by vortex shedding. This detailed approach ensures that every aspect of the propeller contributes to the submarine's stealth and performance, meeting the demanding requirements of modern naval operations [479].

In addition to traditional propeller systems, pump-jet propulsion systems represent a significant advancement in noise reduction technology. Unlike conventional propellers, pump-jets utilize a ducted impeller surrounded by a stator, which effectively eliminates tip vortices and reduces cavitation. This design allows for quieter operation, making pump-jets particularly suitable for stealth-oriented submarines such as the UK's Astute-class and France's Barracuda-class [106]. The integration of such systems underscores the importance of innovative design in enhancing submarine stealth capabilities.

Electric propulsion systems are increasingly prevalent in modern submarines due to their minimal noise output. These systems replace traditional mechanical drives powered by diesel or steam turbines with electric motors, which provide smoother and quieter operation. For instance, diesel-electric submarines like the German Type 212 can achieve near-silent running when operating on battery power, significantly complicating detection efforts. Similarly, nuclear-powered submarines utilize electric drives, where the reactor generates electricity to power propulsion motors, thereby reducing mechanical noise associated with traditional transmission systems [109, 480]. The shift towards electric propulsion not only enhances stealth but also aligns with broader trends in naval engineering towards more efficient and environmentally friendly technologies.

The advancements in submarine propulsion systems reflect a concerted effort to address the challenges of noise management in underwater operations. By leveraging innovative designs in propeller blade geometry, adopting pump-jet systems, and transitioning to electric propulsion, modern submarines can significantly mitigate their acoustic signatures, thereby enhancing their operational effectiveness in stealth missions [339, 481].

Sound-Absorbing Tiles

Acoustic quieting in submarines is a critical aspect of naval stealth technology, primarily achieved through the application of sound-absorbing tiles affixed to the vessel's outer hull. These tiles, often composed of specialized rubber composites, are engineered to absorb and dissipate sound waves effectively. By minimizing both the internal noise generated by the submarine and the external sound waves reflecting off its hull, these tiles significantly reduce the submarine's acoustic signature, making it more challenging for enemy forces to detect the vessel using passive sonar systems, which rely on listening for acoustic signatures rather than actively emitting sound [482, 483].

The mechanism by which these tiles operate involves converting sound energy into small amounts of heat, thereby reducing the sound that escapes from the hull into the surrounding water. This function is crucial for stealth operations, as it not only absorbs internal noise but also mitigates the reflections of incoming sonar waves. This dual-purpose capability is essential for maintaining operational stealth, as it complicates the detection process for passive sonar systems, which are designed to pick up the acoustic signatures of vessels [482, 483]. Furthermore, the effectiveness of these tiles is contingent upon their design and application; even minor imperfections, such as gaps between tiles or improper bonding, can compromise the vessel's acoustic stealth [482, 483].

The rubber composites used in these sound-absorbing tiles are meticulously engineered to withstand the extreme conditions submarines encounter, including high pressure at depth, temperature fluctuations, and prolonged exposure to seawater. Continuous research and development efforts are focused on optimizing these materials for durability and effectiveness, ensuring that they retain their sound-dampening properties under challenging conditions [482, 483]. Additionally, the tiles must be resistant to biofouling, as the accumulation of marine organisms can alter the acoustic properties of the hull, further complicating stealth operations [482, 483].

Advanced naval vessels, particularly nuclear submarines, rely heavily on these sound-absorbing tiles to maintain their operational stealth. The evolution of tile composition and installation techniques has significantly contributed to the reduction of acoustic signatures over time, allowing submarines to navigate quietly through hostile waters and minimizing the risk of detection [482, 483]. The ongoing advancements in acoustic stealth technologies underscore the importance of these materials in modern naval warfare, as they enhance the effectiveness of submarines in evading detection by enemy sonar systems [482, 483].

Real-Time Monitoring and Adjustment of Onboard Machinery

Advanced computer systems are integral to enhancing acoustic quieting in submarines, primarily through real-time monitoring and adjustment of onboard machinery. These systems facilitate dynamic management of equipment settings to minimize noise output, especially in proximity to hostile sonar arrays. The continuous flow of real-time data from sensors embedded throughout the submarine's machinery—including machinery vibrations, power plant operations, and propulsion systems—enables the crew to respond swiftly to any noise-generating anomalies. For instance, if a component begins to produce excessive noise, the system can alert the crew, allowing them to make adjustments such as reducing power, altering rotation speeds, or temporarily shutting down non-essential equipment to maintain stealth [351, 484].

The automation provided by these advanced systems significantly enhances operational efficiency while ensuring acoustic stealth. They are capable of detecting abnormal vibrations in real-time and can adjust the operating conditions of engines or pumps to suppress these vibrations before they propagate through the hull. This proactive approach is crucial in reducing the likelihood of noise transmission that could compromise the submarine's stealth capabilities [24, 485]. Moreover, the integration of vibration-isolating technologies within these systems ensures that noise is not transferred to the external environment, even when operational adjustments are made [200, 486].

Another essential feature of these advanced computer systems is their integration with the submarine's passive sonar capabilities. Passive sonar systems are designed to continuously monitor for incoming sonar pings or other acoustic signatures without emitting sound, thus preserving the submarine's stealth. When a hostile sonar ping is detected, the crew can quickly modify the vessel's operational parameters or adjust its positioning. This may involve altering speed, shifting the propeller's rotational rate, or transitioning to a quieter operational mode, such as the "ultraquiet" mode, where only essential equipment operates at minimal noise levels [25, 487]. Such adaptability is critical for maintaining a strategic advantage in stealth operations, allowing submarines to remain undetected while effectively responding to evolving tactical conditions [55, 488].

Through continuous monitoring and adaptive control, these advanced computer systems enable submarines to operate safely and quietly, even in high-risk environments. This capability minimizes the chances of detection by hostile sonar systems, thereby enhancing the overall effectiveness of submarine missions [489, 490]. The integration of these technologies not only supports stealth operations but also contributes to the strategic flexibility required in modern naval warfare [491, 492].

Electromagnetic Signature Reduction

A submarine's electromagnetic (EM) signature refers to the electromagnetic emissions and disturbances generated by the vessel's equipment and systems. These signatures are detectable by sensors and can compromise the stealth of a submarine. The EM signature encompasses several key components, including electric and magnetic fields resulting from onboard machinery, electric currents, and communication systems. In military applications, minimizing a submarine's EM signature is crucial to avoiding detection by adversaries using electromagnetic sensors.

Ships and submarines are increasingly vulnerable to detection by enemy defence systems due to their magnetic signatures, which are generated by the disturbance they create in the Earth's magnetic field. This phenomenon occurs primarily because the steel hulls of these vessels have different magnetic properties compared to the surrounding seawater, leading to localized disturbances in the magnetic field as the submarine moves through the water [493-495]. The detection of these magnetic signatures is facilitated by advanced magnetic sensors and tracking equipment, which can identify the subtle variations caused by the presence of ferromagnetic materials in the hulls of submarines [494, 496].

The design, construction, and operation of submarines necessitate a thorough numerical analysis of these magnetic signatures. Engineers utilize sophisticated simulation techniques to model how a submarine's magnetic properties interact with the environment, which is crucial for developing effective countermeasures to reduce detectability [495, 497]. The challenges in this analysis stem from the thin metallic structure of submarine hulls, which complicates the accurate capture of magnetic interactions. Specialized methods, such as the integral equation method combined with fast multipole algorithms, have been employed to compute the magnetic anomalies associated with complex geometries of submarines [497].

Submarine electromagnetic signatures refer to the unique electromagnetic emissions that submarines produce, which can expose them to detection by enemy systems. Understanding these emissions and minimizing them is crucial for maintaining stealth. These electromagnetic signatures comprise several key components:

The magnetic signature, or magnetostatic signature, results from the interaction between the submarine's ferromagnetic hull and the Earth's magnetic field. As the submarine moves, it distorts the ambient magnetic field, creating a detectable anomaly. Anti-submarine warfare (ASW) aircraft often deploy Magnetic Anomaly Detection (MAD) systems to detect such disturbances, making it vital for submarines to minimize their magnetic footprint through techniques like degaussing, which cancels out the magnetic fields produced by the hull.

Electric field signatures arise from the electrical currents used in the submarine's propulsion system and onboard electrical power distribution. These currents create localized electric fields in the surrounding water, which can be detected by specialized ASW equipment. The electric field component of the electromagnetic signature is particularly relevant when the submarine operates electrical systems at high loads, making precise management of power distribution essential to stealth.

Electromagnetic radiation occurs when the submarine uses communication systems, radar, or antennas. These devices emit electromagnetic waves, and while submarines typically limit emissions to remain covert, any surface activity involving communications or periscope operations increases the risk of detection by long-range electromagnetic sensors.

Corrosion-related electrical signals are generated by electrochemical processes on the submarine's hull, particularly in saltwater environments. Cathodic protection systems, such as Impressed Current Cathodic Protection (ICCP), are installed to prevent hull corrosion, but these systems can also produce electrical emissions. Advanced ICCP systems aim to manage these emissions effectively to prevent compromising the submarine's stealth capabilities.

To mitigate the magnetic footprint of submarines, techniques such as degaussing are employed. Degaussing involves generating counteracting magnetic fields to compensate for the disturbances caused by the vessel's hull [496]. This process can be achieved through the use of cables carrying electric currents or coils integrated into the submarine's structure. The effectiveness of these countermeasures relies heavily on precise numerical analyses that account for the submarine's magnetic properties and their interactions with the surrounding seawater [495, 497]. As adversaries develop more sophisticated sensor systems, the need for effective signature reduction strategies becomes increasingly critical for maintaining the stealth of naval vessels [493, 498].

Figure 119: The US Navy (USN) Sea Wolf Class Attack Submarine USS JIMMY CARTER (SSN 23), sits moored in the magnetic silencing facility at Naval Base (NB) Kitsap, Bangor, Washington (WA), for a Magnetic Treatment (DEPERM) procedure which involves charging, or flashing, the hull with electricity to minimise the electronic signature as it travels through the water thereby reducing it's vulnerability to underwater magnetic mines. The U.S. National Archives, Public Domain, via NARA & DVIDS Public Domain Archive.

Mitigation is accomplished through several strategic methods, addressing both the submarine's magnetic and electromagnetic emissions.

Deperming is a critical technique for reducing a submarine's magnetic signature. Over time, the submarine's ferromagnetic hull can accumulate a magnetic field by interacting with the Earth's natural magnetism. This accumulated magnetization makes it easier for magnetic anomaly detectors (MAD) to detect the vessel. Deperming involves systematically demagnetizing the hull, ensuring that the vessel generates minimal magnetic disturbance during operations. Specialized facilities apply alternating magnetic fields along the hull, effectively neutralizing the magnetization and reducing the magnetic anomaly signature.

Deperming is a critical process employed to mitigate the magnetic signature of submarines, thereby enhancing their stealth capabilities against detection by magnetic anomaly detectors (MAD). This specialized procedure involves wrapping coils around the submarine and applying a controlled series of alternating magnetic fields. The primary aim is to neutralize the submarine's magnetic field, effectively removing residual magnetism from the hull as well as environmental magnetization effects caused by the Earth's magnetic field [496]. The significance of deperming is underscored by its necessity for submarines that do not possess degaussing systems, which are typically used to continuously manage magnetic signatures during operations [499].

The deperming process is conducted at dedicated facilities designed for magnetic treatment. It involves a systematic reduction of magnetic field strength over several cycles, which serves to erase the magnetic memory embedded within the metal structure of the submarine [496]. This gradual approach is essential for ensuring that the submarine's magnetic signature is minimized, thereby enhancing its operational stealth. For submarines lacking advanced degaussing capabilities, the importance of deperming becomes even more pronounced [499]. Protocols like the Flash-D method have been developed specifically for these submarines, allowing for effective demagnetization without the need for continuous degaussing systems during missions [496].

Moreover, the efficiency of deperming protocols has been a subject of research, with studies demonstrating that applying an alternating magnetic field with a linearly decreasing amplitude can effectively achieve the desired demagnetization [496]. This method not only enhances the stealth of submarines but also ensures that minimal magnetic signals remain detectable in sensitive environments, which is crucial for maintaining operational security in hostile settings. The implications of these techniques extend beyond mere stealth; they also play a vital role in the overall effectiveness and survivability of submarine operations in modern naval warfare.

The use of advanced materials plays a pivotal role in minimizing both magnetic and electromagnetic signatures. Submarines increasingly rely on non-magnetic materials, such as titanium alloys and composites, to construct parts of the hull and key internal components. These materials reduce magnetic anomalies and offer corrosion resistance, which is crucial for long-term stealth performance. Employing non-ferrous materials ensures that submarines avoid generating detectable magnetic disturbances during interactions with the Earth's magnetic field.

Electromagnetic (EM) shielding is another essential strategy. Submarines feature specialized enclosures around electrical equipment and wiring to prevent electromagnetic leakage. Without adequate shielding, onboard electrical systems—such as propulsion systems and power distribution networks—could emit EM fields detectable by surveillance systems. Careful system design ensures that these emissions are contained, and shielded compartments isolate sensitive equipment, minimizing the risk of detection.

Low Probability of Intercept (LPI) communications are employed to reduce the submarine's electromagnetic footprint when transmitting data or communicating. Traditional radio signals can be easily detected by enemy sensors, so submarines use highly directional antennas, ultra-low frequency (ULF) transmissions, and other covert communication methods to remain undetected. ULF transmissions, for example, travel over long distances through water, but with minimal energy loss, making them ideal for stealth communications. Additionally, directional antennas focus signal energy in narrow beams, reducing the risk of detection by nearby enemy surveillance systems.

Together, these mitigation strategies ensure that submarines remain covert even when engaging in complex operations or near hostile forces. The combination of deperming, advanced materials, EM shielding, and LPI communications enhances the vessel's stealth capabilities, making it difficult for surveillance systems to detect, track, or intercept the submarine. This technological sophistication is critical in modern naval warfare, where stealth and the ability to evade detection can provide a decisive strategic advantage.

Minimizing the magnetic signature is a continuous process. In some instances, submarines may employ real-time systems that adjust the magnetic compensation as the vessel's operational conditions change. The strategic importance of magnetic signature reduction extends beyond stealth; it ensures the submarine can evade detection from increasingly sensitive enemy defence systems. Using advanced simulations during the design phase helps engineers achieve optimal magnetic performance, providing submarines with the stealth necessary for modern naval operations.

Submarine Stealth Technologies

In the realm of naval warfare, stealth is a critical component for submarines, enabling them to move undetected and carry out strategic missions, such as launching surprise attacks and gathering intelligence. These underwater vessels, often referred to as the 'silent killers' of the deep, rely on sophisticated technologies and meticulous design to maintain stealth. Their ability to remain concealed in the vast ocean is a result of engineering that minimizes sound emissions and enhances hydrodynamic efficiency, countering the inherent challenges posed by underwater acoustics [500].

The science of sound in water plays a pivotal role in submarine stealth. Sound waves travel much faster and farther in water than in air, with a speed of approximately 1,500 meters per second in water compared to 343 meters per second in air. This property of water makes noise management essential for submarines, as even minor sounds can quickly propagate through the ocean, revealing their location to enemy sonar systems. Submarines must therefore be designed to minimize all forms of acoustic emissions, reducing the likelihood of detection by active or passive sonar arrays used by surface ships, other submarines, and aircraft [500].

Design plays a crucial role in achieving the stealth of submarines. The hull of a stealth submarine features a streamlined, teardrop shape to reduce drag and turbulence as it moves through water. This design minimizes noise generated by water resistance, allowing the vessel to glide silently beneath the surface. A bulbous bow, often integrated at the front, further enhances efficiency by reducing wave resistance, making it easier for the submarine to maintain low-speed operations with minimal noise. Additionally, some submarines are equipped with X-shaped rudders, which provide superior manoeuvrability while limiting cavitation, a phenomenon where bubbles form and collapse, generating noise that can betray the submarine's presence [500].

Another essential aspect of submarine stealth is noise reduction. One of the most effective methods to limit acoustic emissions is the application of anechoic tiles on the exterior hull. These rubber-based tiles absorb sound waves, reducing the reflection of sonar signals and making it more difficult for adversaries to detect the submarine through active sonar. Inside the submarine, sound-dampening materials such as rubber linings and acoustic foam absorb mechanical noise and reduce vibrations, preventing sounds generated by machinery and crew activities from leaking into the water. Advanced submarines also employ active noise cancellation technology, where the submarine emits sound waves that neutralize the noise produced by onboard systems, effectively masking its acoustic signature.

Propulsion systems are another critical component of submarine stealth. Electric motors, favoured for their quiet operation, are often used when submarines cruise at slow speeds or remain submerged. These motors generate significantly less noise than traditional diesel engines, making them ideal for stealth missions. Furthermore, many modern submarines employ pump-jet propulsion systems, which encase the propeller within a shroud, reducing cavitation and enhancing efficiency. These pump-jets not only minimize noise but also enable smoother operation, further contributing to the vessel's stealth capabilities.

The evolution of submarine stealth technology has been driven by advancements in materials science, propulsion, and computational modelling. Innovations in hull coatings, noise-cancellation systems, and engine designs have resulted in quieter and more efficient submarines, far surpassing the capabilities of earlier generations. However, as stealth technology advances, so do the detection methods employed by rival navies. New and sophisticated sonar technologies continuously push the boundaries of anti-submarine warfare, necessitating a constant refinement of stealth techniques to ensure submarines remain undetectable [500].

In this ongoing arms race between stealth and detection, modern submarines are at the forefront of technological innovation. They are not only integral to strategic military operations but also symbolize the pinnacle of stealth engineering. The continuous evolution of submarine technology reflects the dynamic nature of naval warfare, where maintaining an edge in stealth ensures dominance beneath the ocean's surface [500].

Naval submarines were developed to capitalize on the obscurity provided by the underwater environment, which offers a significant tactical advantage over more powerful surface vessels. The primary strength of submarines lies in their ability to operate undetected. While submarines possess various capabilities, their effectiveness ultimately hinges on stealth. The reason for this focus is that submarines, once discovered, become highly vulnerable to attacks from surface ships, other submarines, and aircraft. Consequently, stealth has remained a central feature of submarine design and development. Technologies focused on improving submarine stealth have been heavily researched, often kept under strict confidentiality, as they aim to stay ahead of advancements in anti-submarine warfare [501].

Submarine stealth must be approached as a comprehensive system, integrating multiple technologies and strategies. This involves two principal categories: acoustic and non-acoustic stealth. Acoustic stealth addresses the sounds generated by a submarine's movement through water, as well as how sound waves reflect off its surface when detected by an adversary's sonar. Non-acoustic stealth, on the other hand, encompasses reducing other detectable aspects, such as the submarine's radar profile, infrared visibility, and electromagnetic emissions, as well as enhancing its endurance underwater. Every element of submarine design—its hull shape, machinery, communication equipment, sensors, and weapons—plays a role in maintaining this low visibility [501].

The acoustic stealth of a submarine begins with the design of its hull and propellers. In earlier designs, submarines had less streamlined shapes, making them easier to detect. However, advances in computational fluid dynamics (CFD) and optimization algorithms have allowed modern submarines to reduce water resistance and noise. Double-hull construction techniques also enhance stealth by minimizing noise from machinery through better shape optimization. Propeller design has similarly evolved, with innovations like integrated propulsors and polymer ejection systems, which aim to reduce turbulence and further decrease noise levels during movement [501].

During World War II, submarines were constructed with metal hulls similar to warships, which reflected sonar waves effectively, making them easy targets. Modern submarines use coatings such as anechoic tiles to dampen sound emissions and disrupt sonar reflections. However, advancements in low-frequency sonar have forced further innovation. A more recent breakthrough is the development of active tiles, which use giant magnetostrictive materials to emit altered sonar signals, confusing surface ship operators and enhancing stealth capabilities [501].

Submarine machinery, including main engines and auxiliary systems, constitutes another major source of noise. As such, extensive efforts are directed towards minimizing noise from these components. Early noise reduction efforts involved simple rubber mountings, but these have evolved into advanced systems using piezoelectric materials that actively counteract vibrations. Additionally, isolated structures and raft mountings inside submarines further decrease acoustic signatures. Condition-based monitoring (CBM) systems, which continuously monitor the condition of onboard equipment, are now standard features, helping maintain operational stealth by detecting potential issues early [501].

A significant challenge for conventional submarines lies in their need to surface periodically to recharge batteries, a process known as "snorting." This requirement compromises stealth. The introduction of air-independent propulsion (AIP) systems has addressed this problem by extending underwater endurance and reducing the frequency of surfacing. However, AIP systems still have limitations, such as slower underwater speeds and reduced endurance compared to nuclear-powered submarines. Ongoing research aims to develop more efficient and reliable AIP systems to enhance stealth capabilities further. Fuel cell technology, in particular, offers promising advancements, allowing submarines to remain submerged for longer periods with minimal acoustic signatures. This technological evolution is critical, especially as the demands on modern submarines increase to support diverse and complex missions [501].

Weapons systems and communication technologies also play crucial roles in maintaining stealth. During operations, even minor noises during weapon launches can compromise the submarine's covert position. Therefore, modern submarines employ noise-dampening technologies for torpedo and missile handling systems, reducing transient noises during launches. In communication, satellite antennas and wire-trailing systems are engineered to minimize acoustic and radar signatures, ensuring that essential communication does not compromise stealth. Additionally, the quieter a submarine remains, the more effective its onboard sensors, such as sonar arrays, become. Innovations in sensor design include embedding sonar systems directly onto the hull instead of mounting them on the bow, which not only enhances stealth but also improves hydrodynamics [501].

The introduction of off-board vehicles is another cutting-edge development in submarine operations. These remotely operated or autonomous vehicles extend the submarine's operational reach while keeping the main vessel undetected. These off-board systems can perform reconnaissance, deploy sensors, or even act as decoys, ensuring the submarine maintains a tactical advantage [501].

Investing in stealth technologies is vital for navies around the world, especially as anti-submarine warfare capabilities become more advanced. Nations without adequate submarine stealth capabilities risk being easily neutralized in potential conflicts. Submarines are not just tactical assets but also play strategic roles, such as second-strike capabilities in nuclear deterrence. This makes stealth critical to ensuring survivability and effectiveness in a high-threat environment.

The future of submarine stealth will require ongoing research and development across multiple disciplines, including CFD modelling, non-acoustic technologies, and improved propulsion systems. As anti-submarine detection methods improve, submarines must also evolve to remain a step ahead. For countries like India, where submarines are integral to both tactical operations and strategic deterrence, continuous investment in stealth technologies is not just desirable but essential. Developing expertise in these fields ensures that the submarine fleet remains a formidable tool in both warfighting and deterrence strategies. Maintaining this advantage will require innovation, collaboration, and sustained investment to outpace the rapid technological advancements in surveillance and detection technologies worldwide [501].

As examples, the stealth features on the Chinese Type-039C Yuan-class submarine are optimized to counter medium-frequency sonar systems, which are crucial for identifying and classifying targets. These systems can distinguish between objects based on sound reflections, making it possible to identify the presence and type of a submarine. The angled sail on the Yuan-class works by scattering sonar waves, reducing the "signal strength" returned to the enemy. While it only decreases the sonar return by a few decibels, this reduction can make a critical difference in evading classification and delaying enemy reactions [502].

However, the stealth design is not fully effective against all sonar types. Low-frequency sonar, with its long wavelengths, can still detect the presence of an object, although it struggles to classify it accurately. Conversely, short-wavelength sonar, like those used in torpedoes, is better at identifying targets at close range, posing a distinct threat. The Yuan-class addresses this vulnerability with an additional layer of stealth technology—rubberized anechoic coatings along the hull. These coatings absorb sound waves, preventing them from bouncing back to the sonar source, further reducing the vessel's acoustic signature [502].

Figure 120: U 35, one of the newest submarines of the 212A class in the German Navy. Tomasz Przechlewski, CC BY 2.0, via Store norske leksikon.

The implementation of advanced sail shaping and sound-absorbing materials reflects a strategic shift in submarine design toward evading medium-frequency sonar detection. This trend is driven by the growing importance of active sonar in modern undersea warfare, where passive listening methods are less effective against newer, quieter submarines. While the shaping sacrifices some hydrodynamic efficiency, the trade-off is considered worthwhile, especially in scenarios where stealth is paramount. Similar developments are seen in other submarines, such as Sweden's A26 and Germany's Type-212CD, but the Chinese Yuan-class represents one of the first operational applications of this technology [502].

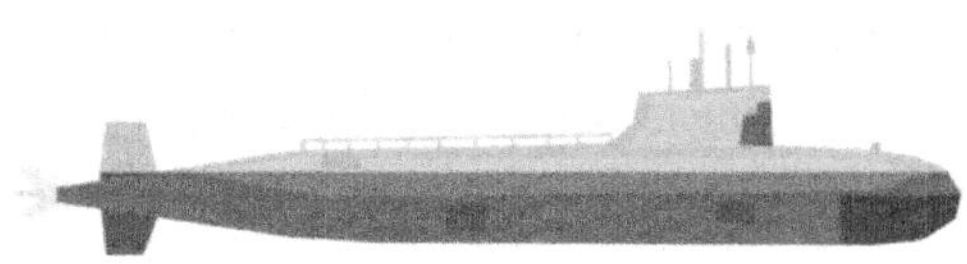

Chapter 12

Autonomous and Unmanned Submarine Systems

Design of Autonomous Underwater Vehicles (AUVs)

Autonomous Underwater Vehicles (AUVs) are sophisticated unmanned systems designed to operate independently in underwater environments, performing a variety of missions such as scientific research, military reconnaissance, and environmental monitoring. Their design and operational capabilities are influenced by several critical factors, including propulsion systems, navigation technologies, and sensor integration. The unique challenges posed by the underwater environment, such as high pressure, salinity variations, and limited communication, necessitate a comprehensive approach to AUV development.

Autonomous Underwater Vehicles (AUVs) have become increasingly popular in underwater search and exploration missions due to their cost-effectiveness compared to manned submersibles. Over the years, these unmanned systems have evolved significantly, enabling them to undertake complex operations in fields such as oceanography, military applications, environmental monitoring, and commercial activities. The focus of recent development has been on creating AUVs capable of long-term data collection to support ocean exploration, extraction programs, and coastal management.

In commercial industries, particularly oil and gas, AUVs are indispensable for mapping the seafloor before the installation of pipelines and other underwater infrastructure. These vehicles allow for highly precise bathymetric surveys that would otherwise be costly or impractical using traditional methods. AUVs also conduct post-lay inspections of pipelines and other structures, ensuring the infrastructure is correctly installed and maintained. Furthermore, the potential of AUVs extends to seabed mining operations, as they are used to explore areas rich in polymetallic nodules—valuable for future mineral extraction.

In the field of scientific research, AUVs are widely deployed for studying aquatic environments, including lakes and oceans, and gathering data on environmental conditions. Researchers equip AUVs with sensors such as CTDs (conductivity-temperature-depth sensors), fluorometers, and pH meters to measure chemical and physical parameters across different depths. Some AUVs act as towed vehicles, delivering specialized sensors to targeted locations. Notably, the Seaglider AUV, developed by the University of Washington, is an example of a platform initially designed for research but now also utilized by industries and militaries due to its low cost and high reliability.

In environmental conservation, AUVs play a unique role. For example, the Crown-of-Thorns Starfish Robot (COTSBot), developed by the Queensland University of Technology (QUT), is an innovative AUV programmed to identify and eliminate crown-of-thorns starfish, which threaten the Great Barrier Reef. Another QUT creation, the RangerBot, was designed for easy deployment and helps monitor reefs globally, providing real-time vision and obstacle detection capabilities for management tasks.

The hobbyist community has also embraced AUV technology, with enthusiasts building their own vehicles to participate in competitions. Although these homemade AUVs are generally less capable than commercial models, they demonstrate the growing accessibility of underwater robotics. Hobbyist designs typically use simple components like microcontrollers and PVC housings, and some leverage open-source software to enhance functionality.

Beyond legitimate uses, AUV technology has found applications in illicit activities, such as drug trafficking. Smugglers have developed GPS-guided autonomous submarines to transport contraband over long distances. In addition, AUVs have proven valuable in air crash investigations, helping locate wreckage in difficult-to-access areas. For instance, the AUV ABYSS assisted in locating debris from Air France Flight 447, and the Bluefin-21 was used in the search for the missing Malaysia Airlines Flight 370.

Figure 121: The Bluefin 21, Artemis autonomous underwater vehicle. Defense Visual Information Distribution Service, Public Domain, via Picryl.

The Bluefin-21, also referred to as the Artemis autonomous underwater vehicle (AUV), is a modular unmanned underwater platform developed by Bluefin Robotics. This AUV is designed to perform a wide variety of underwater tasks, including deep-sea surveys, salvage operations, marine archaeology, oceanography, and military missions like mine countermeasures.

The Bluefin-21 is notable for its involvement in the search for the wreckage of Malaysia Airlines Flight 370 in 2014. During this mission, it was deployed from the Australian vessel *ADV Ocean Shield* to conduct side-scan sonar surveys in the Southern Indian Ocean, reaching depths of up to 4,695 meters. It covered significant areas of the seabed—up to 250 square miles—while accumulating hundreds of operational hours across multiple search missions. This mission exemplified the AUV's endurance and adaptability for high-profile, time-sensitive operations.

The AUV features a torpedo-shaped design that enhances hydrodynamic efficiency. It is equipped with modular payload bays, allowing it to carry various sonar systems, such as the EdgeTech 2200-M side-scan sonar, and other instruments like sub-bottom profilers and multi-beam echo sounders. This modularity makes the Bluefin-21 versatile for both civilian and defence applications. It can operate at speeds of up to 4.5 knots and has an endurance of up to 25 hours, powered by nine lithium-polymer battery modules.

The Bluefin-21's navigation system integrates inertial navigation for precision and uses an ultra-short baseline (USBL) system to maintain accuracy during missions. This configuration ensures reliable data collection and effective navigation, even in complex underwater environments. Its compact design allows for easy deployment from ships, increasing its flexibility across different types of operations.

Given its modularity, endurance, and operational depth capacity, the Bluefin-21 continues to be a highly capable AUV, deployed in both scientific and military contexts to explore challenging underwater environments efficiently. It highlights the growing importance of autonomous systems in addressing complex underwater challenges, both for defence and exploration purposes.

The military has extensively adopted AUV technology, with applications ranging from mine countermeasures and anti-submarine warfare to intelligence gathering and reconnaissance. The U.S. Navy, for example, uses AUVs like the Mk 18 Mod 1 Swordfish for shallow-water reconnaissance. These vehicles fall into distinct categories based on size and capability, including man-portable, lightweight, heavyweight, and large vehicles, such as the Orca UUV, which is the first unmanned submarine designed with combat capability.

The Mk 18 Mod 1 Swordfish is an autonomous underwater vehicle (AUV) specifically designed for shallow-water reconnaissance and mine countermeasure operations. Built on the Hydroid REMUS 100 platform, the Mk 18 Mod 1 is compact, lightweight, and highly portable, making it ideal for rapid deployment in expeditionary missions. Its primary purpose is to operate in very shallow waters (10 to 40 feet), conducting hydrographic surveys, mine hunting, and reconnaissance activities in environments that might otherwise be inaccessible to larger underwater systems.

One of the key strengths of the Mk 18 Mod 1 is its ability to collect and map detailed bathymetric data, which helps create baseline maps of the seafloor. In subsequent missions, it can identify changes in the underwater terrain, such as the appearance of mines or other obstructions, which are critical for ensuring the safety of naval operations. The vehicle supports the identification and localization of underwater hazards, after which divers or remotely operated vehicles (ROVs) can be deployed to neutralize threats.

The Mk 18 Mod 1's small size and man-portability enable it to be launched from small boats or even helicopters, providing flexibility in a variety of operational contexts. This feature allows military forces to rapidly respond to evolving situations, such as clearing mines from harbors or securing beachheads for amphibious landings. The U.S. Navy uses the Swordfish AUV extensively, often working in tandem with the larger Mk 18 Mod 2 Kingfish, which operates in slightly deeper waters, providing a complementary capability across different marine environments.

This AUV has been a crucial tool for the Navy's Explosive Ordnance Disposal units, particularly in operations like the Arctic Expeditionary Capabilities Exercise (AECE), where it demonstrated its ability to function effectively in extreme environments. The Mk 18 Mod 1's deployment reflects the growing importance of unmanned systems in modern naval warfare, enhancing operational safety while reducing risks to personnel.

Design

The propulsion and structural design of AUVs are fundamental to their operational efficiency. AUVs typically utilize thrusters and control surfaces to manoeuvre through water, which must be optimized for the specific missions they undertake [503]. The integration of robust control systems is essential for maintaining stability and navigation accuracy in dynamic underwater conditions. For instance, adaptive robust control techniques have been shown to enhance the performance of AUVs in unpredictable ocean environments [503, 504]. Furthermore, the use of advanced navigation algorithms, such as the Unscented Kalman Filter, allows AUVs to effectively navigate using acoustic signals, which are crucial for underwater positioning where GPS is unavailable [504, 505].

Navigation technology is a critical component of AUV functionality, as it directly impacts their ability to operate autonomously. AUVs employ various navigation methods, including inertial navigation systems (INS), which are often combined with acoustic positioning systems to improve accuracy [289, 506]. The challenges of underwater navigation include the degradation of signals due to environmental factors, which can lead to significant localization errors [507, 508]. Recent advancements in sensor technology and data fusion techniques have improved the reliability of AUV navigation, enabling them to perform complex tasks such as seabed mapping and underwater inspections [509, 510].

In addition to navigation, AUVs are equipped with a range of sensors that facilitate data collection for scientific and industrial applications. These sensors can include sonar systems, cameras, and environmental sensors, which are essential for tasks such as marine biology studies and underwater infrastructure inspections [510, 511]. The integration of machine learning and artificial intelligence in sensor data processing has further enhanced the capabilities of AUVs, allowing for real-time decision-making and adaptive responses to environmental changes [505, 512].

Moreover, the operational scope of AUVs has expanded significantly, driven by advancements in technology and increasing demand for underwater exploration and monitoring. AUVs are now employed in diverse fields, including oil and gas exploration, archaeological surveys, and environmental assessments, demonstrating their versatility and effectiveness in various underwater tasks [512, 513]. The ongoing development of cooperative multi-AUV systems is also noteworthy, as these systems can perform complex missions that require coordinated efforts, such as pipeline inspections and environmental monitoring [514, 515].

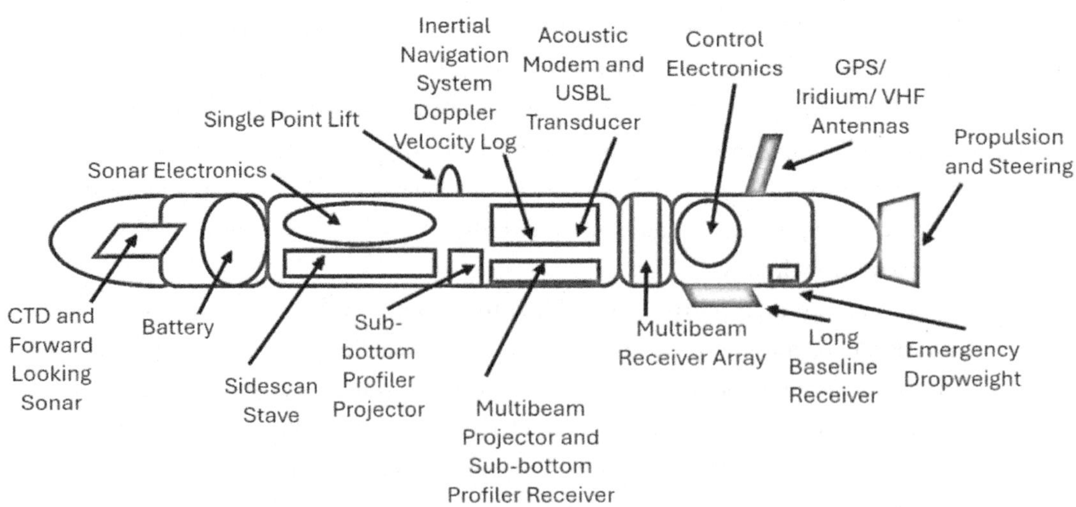

Figure 122: Typical structure of an AUV.

Hull Design and Structure

The design of the hull in Autonomous Underwater Vehicles (AUVs) is pivotal in determining their hydrodynamic efficiency, stability, and manoeuvrability. A well-engineered hull must align with the specific mission requirements of the AUV, balancing factors such as speed, endurance, and operational flexibility. For instance, torpedo-shaped hulls, characterized by their streamlined cylindrical forms, are particularly effective in minimizing water resistance and drag, which is essential for achieving high speeds and operational efficiency. This design is especially beneficial in military applications where stealth, speed, and extended operational range are paramount [516-518]. The reduction in drag not only enhances the AUV's speed but also conserves energy, thereby increasing its endurance and enabling it to cover greater distances without the need for frequent recharging or refuelling [519, 520].

In contrast, box or open-frame AUVs are tailored for missions that prioritize high manoeuvrability over speed. These vehicles are often employed in research and underwater inspections, where stability and precision are more critical than velocity. The open-frame design allows for the attachment of various external instruments, making these AUVs highly adaptable for diverse tasks [521, 522]. However, this adaptability comes at the cost of increased drag, which can limit the speed and operational range of these vehicles [523]. Despite these limitations, the ability to carry multiple payloads enhances the utility of box-framed AUVs for complex missions that require precise control and a variety of sensors [524, 525].

The selection of materials for the hull is another crucial aspect, as AUVs must withstand the extreme pressures of deep-sea environments while remaining lightweight for efficient operation. Titanium is frequently chosen for its superior strength-to-weight ratio and corrosion resistance, making it ideal for harsh marine conditions [526, 527]. Aluminium alloys present a lightweight and cost-effective alternative, although they may not match the durability of titanium [528]. Additionally, advanced composites such as carbon fibre are gaining traction due to their lightweight properties and structural strength, which contribute to reducing the overall weight of the AUV and enhancing energy efficiency without sacrificing durability [529, 530].

To ensure the safety and functionality of onboard electronics and payloads, AUVs often incorporate pressure-resistant designs within their hulls. Internal pressure vessels are commonly utilized to protect sensitive components from the crushing pressures encountered at significant depths, creating a stable environment for electronic systems and ensuring their reliability throughout the mission [531, 532]. The integration of robust materials, hydrodynamic design, and pressure-resistant construction is essential for enabling AUVs to perform complex tasks in some of the most challenging and inaccessible oceanic regions. This thoughtful synthesis of hull design, material science, and engineering principles ensures that AUVs remain resilient, efficient, and adaptable across a wide spectrum of underwater missions [533, 534].

Figure 123: "Vityaz-D" autonomous underwater vehicle during the "Armiya 2021" exhibition. Kirill Borisenko, CC BY-SA 4.0, via Wikimedia Commons.

As a case example of an AUV, the Vityaz-D is a cutting-edge autonomous underwater vehicle (AUV) developed by the Central Design Bureau for Marine Engineering (RUBIN) in collaboration with Russia's Advanced Research Foundation. It is notable for its ability to reach full-ocean depths and operate autonomously, representing a significant leap in underwater exploration and military applications. The Vityaz-D achieved international attention when it reached a depth of 10,028 meters in the Mariana Trench in May 2020, becoming the first fully autonomous AUV to explore this extreme environment without human intervention.

The Vityaz-D measures 5.7 meters in length, with a diameter of 1.3 meters, and weighs around 5.7 tons. It can operate at depths of up to 12,000 meters, making it suitable for deep-sea missions. It incorporates four electric propulsors and ten manoeuvring thrusters, enabling fine-tuned control and propulsion at a speed of approximately one meter per second. The vehicle has an operational endurance of 24 hours, during which it can conduct research and reconnaissance missions.

This AUV is equipped with advanced sensors, including side-scan sonar, echo sounders, video cameras, and environmental sensors, which allow it to collect valuable scientific data. Additionally, it employs artificial intelligence (AI) to autonomously navigate complex underwater environments, avoid obstacles, and discover confined spaces. The AI system supports its ability to perform missions independently, following pre-programmed routes and adapting to changing conditions along its path.

The Vityaz-D also features a sophisticated communication system, utilizing acoustic channels that allow real-time communication with its mother ship during deep dives. This system ensures the transmission of data and commands even at extreme depths, where traditional communication methods are ineffective. During its mission to the Mariana Trench, the Vityaz-D installed a pennant to commemorate the 75th anniversary of the Soviet victory in World War II, demonstrating both technical achievement and symbolic intent.

The Vityaz-D program was officially unveiled at the Army-2020 International Military and Technical Forum. Its design reflects a shift towards greater autonomy in underwater systems, aligning with broader trends in unmanned naval technologies. The Russian Navy plans to incorporate the Vityaz-D into its operations, potentially replacing older deep-sea submersibles with this advanced platform for both scientific exploration and strategic missions.

Propulsion System

Propulsion systems are critical components in Autonomous Underwater Vehicles (AUVs), as they provide the necessary mobility and control for executing complex underwater missions. The choice of propulsion mechanism is influenced by the vehicle's design, operational requirements, and the environmental conditions it will encounter. The most prevalent type of propulsion in AUVs is the propeller system, which generates forward thrust through rotational motion. Propellers are favoured for their simplicity and efficiency, making them well-suited for long-range missions where maintaining steady speeds with minimal energy consumption is essential [535, 536].

Figure 124: OZZ-2 UUV(REMUS 600) propeller in JS Awaji(MSO-304) at Hanshin Base. Hunini, CC BY-SA 4.0, via Wikimedia Commons.

For tasks that demand precise manoeuvrability, AUVs often employ thrusters—small motors that enable multi-directional movement. This capability allows for fine adjustments in positioning, which is particularly beneficial for sensitive operations such as inspections and data collection in confined spaces [537, 538]. The integration of advanced control algorithms further enhances the manoeuvring capabilities of AUVs, allowing them to navigate complex underwater environments effectively [538, 539].

In recent years, there has been a growing interest in bio-inspired propulsion systems that mimic the natural movements of marine animals. These designs, which utilize oscillating fins or tails, can produce quieter and more efficient movement through water, thus minimizing acoustic signatures. This feature is particularly advantageous for stealth operations in military applications or for research missions in sensitive ecosystems [540-542]. The study of marine animals, such as rays and fish, has provided insights into how their unique locomotion strategies can be translated into effective propulsion systems for AUVs [541, 542].

Powering these propulsion systems typically involves lithium-ion batteries, which are favoured for their high energy density and widespread use across AUV platforms [543, 544]. However, the energy demands of extended missions have prompted research into alternative power sources, such as fuel cells and ocean-based energy harvesting systems. These innovations aim to enhance the operational endurance of AUVs, enabling them to conduct longer missions without frequent recharging [543-545].

AUV propulsion systems can also be categorized based on their mechanical configurations. Direct-drive motors are often preferred due to their high efficiency and low noise levels, while some designs incorporate additional components like gearboxes and seals to improve performance [535, 546]. In advanced configurations, AUV thrusters may include redundant sealing systems to prevent water ingress, ensuring continued operation even in the event of a seal failure during a mission [537, 539].

Certain AUVs, known as underwater gliders, utilize a different approach to propulsion. Instead of relying on traditional propulsion methods, these vehicles generate forward motion by adjusting their buoyancy and pitch. By repeatedly ascending and descending, the glider's wings convert vertical motion into horizontal movement. This method allows gliders to achieve exceptional endurance, often lasting several months and completing transoceanic missions while consuming minimal energy [547]. Their energy-efficient design makes them particularly effective for long-term environmental monitoring and oceanographic research [547].

The propulsion systems of AUVs are diverse and tailored to meet specific operational needs. From traditional propeller systems to innovative bio-inspired designs and energy-efficient gliders, these propulsion mechanisms are integral to the effectiveness and versatility of AUVs in various underwater applications.

Navigation and Control Systems

Autonomous Underwater Vehicles (AUVs) are increasingly essential for various underwater missions, necessitating reliable navigation and control systems to operate effectively in environments where traditional GPS signals are unavailable. The primary navigation technology employed in AUVs is the Inertial Navigation System (INS), which utilizes a combination of accelerometers and gyroscopes to estimate the vehicle's position, velocity, and orientation through dead reckoning. This method is particularly crucial in underwater settings, as GPS signals cannot penetrate water, making INS indispensable for maintaining navigational accuracy over extended periods [289, 548].

To enhance the accuracy of navigation, AUVs often integrate acoustic positioning systems. The Long Baseline (LBL) navigation system employs a network of transponders on the seafloor that communicate with the AUV to triangulate its position via acoustic signals. In scenarios where a support vessel is present, Ultra-Short Baseline (USBL) systems can be utilized to track the AUV by measuring the range and bearing to the surface ship, which has access to GPS coordinates. This real-time tracking capability is vital for precise underwater operations, particularly in complex environments [508, 549].

In addition to INS and acoustic systems, many AUVs are equipped with Doppler Velocity Logs (DVLs), which measure the vehicle's velocity relative to the seafloor. This information is critical for correcting the vehicle's trajectory and minimizing drift that can accumulate over time due to dead reckoning. The DVL also aids in depth and altitude measurements, ensuring that the AUV maintains its operational depth effectively [289, 549]. Furthermore, pressure sensors play a significant role in depth control by measuring hydrostatic pressure and converting it into accurate depth readings. The integration of data from DVLs, INS, and pressure sensors through filtering algorithms provides an optimized navigation solution, enabling the AUV to maintain stability and follow its intended path throughout its mission [548, 549].

Control systems are equally important for ensuring the stability and responsiveness of AUVs. Automatic controllers, such as Proportional-Integral-Derivative (PID) controllers, are employed to monitor and adjust the AUV's propulsion and buoyancy systems in real-time, ensuring that the vehicle maintains the correct depth, orientation, and trajectory despite dynamic underwater conditions [550-552]. Advanced AUVs also incorporate artificial intelligence-based path planning algorithms, allowing them to adapt autonomously to environmental changes or mission-specific requirements. This adaptability is crucial for conducting surveys or avoiding obstacles in unpredictable underwater environments [553, 554].

In missions where navigation accuracy is paramount, such as mine countermeasures or precision mapping, AUVs may periodically surface to acquire GPS fixes. This practice allows the vehicle to correct any cumulative errors from dead reckoning, ensuring that its navigation systems remain synchronized and accurate. The combination of these sophisticated navigation and control technologies enables AUVs to perform complex, autonomous missions with minimal human intervention, significantly enhancing their effectiveness across scientific, military, and commercial applications [548, 549, 554].

Communication Systems

Communication systems are indeed a critical component of Autonomous Underwater Vehicles (AUVs), enabling operators to monitor the vehicles and exchange data during missions. The challenges of underwater communication arise primarily from the limitations of radio waves, which do not propagate effectively through water, necessitating the use of alternative communication methods. Acoustic modems are the predominant technology utilized by AUVs during submerged operations. These modems transmit data via sound waves, which can travel considerable distances underwater, albeit at lower data rates compared to terrestrial communication systems. This characteristic makes acoustic modems essential for relaying commands, status updates, and sensor readings between the AUV and its operators, despite their bandwidth limitations that hinder the transmission of large datasets or high-frequency communications [555, 556].

The NATO standard ANEP-87 JANUS exemplifies recent advancements in acoustic communication, facilitating interoperable communication between different systems at data rates of approximately 80 bits per second, which is critical for military and scientific operations [555]. Furthermore, when AUVs surface, they can utilize satellite communication to transmit collected data and update their position using GPS. This hybrid communication approach, combining acoustic systems underwater and satellite networks on the surface, ensures efficient data transmission over long missions, particularly in remote locations [557, 558]. The integration of satellite modems allows for high-bandwidth communication, enabling operators to download mission data or reconfigure the AUV for subsequent phases of operation [557, 558].

In addition to acoustic and satellite communication, optical modems present another method for underwater communication, offering significantly higher data rates than acoustic systems. However, their effectiveness is constrained by water clarity and range, limiting their use to short distances and controlled environments. In clear water conditions, optical systems can transmit large volumes of data in real-time, making them suitable for specific tasks such as close-proximity vehicle coordination or harbor inspections [557, 559, 560]. The development of underwater wireless optical communication (UWOC) technologies has gained traction, as they provide high bandwidth and low latency, addressing some of the limitations associated with acoustic communication [557, 559, 560].

AUVs are designed to operate autonomously, executing missions based on pre-programmed instructions and onboard decision-making algorithms. This autonomy allows AUVs to adapt to environmental changes and navigate obstacles without direct human intervention [558]. Recent research also explores multi-modal communication solutions that combine acoustic, optical, and even radio frequency (RF) techniques, depending on mission requirements [561]. Such advancements aim to utilize existing underwater infrastructure, such as pipelines or fibre-optic cables, to enhance operational flexibility and communication capabilities [561, 562]. Through these diverse communication strategies, AUVs maintain operational effectiveness across various underwater environments, with acoustic and satellite systems ensuring basic command and data exchange, while optical systems enhance capabilities for specific tasks [557, 558].

Sensors and Payloads

Autonomous Underwater Vehicles (AUVs) are increasingly recognized for their versatility and capability in performing complex underwater tasks autonomously. These vehicles are equipped with a diverse array of sensors and instruments tailored to specific mission requirements, enabling them to execute a variety of operations ranging from environmental monitoring to military reconnaissance. Among the most critical components of AUVs are sonar systems, which are essential for seafloor mapping, obstacle detection, and target tracking. Multi-beam sonar, for instance, provides highly detailed three-dimensional images of the ocean floor, proving invaluable for topographic mapping and underwater exploration [563, 564]. Side-scan sonar, on the other hand, is often employed for broader area surveys, delivering high-resolution images that can detect objects or anomalies such as shipwrecks or debris fields [563, 565].

In addition to sonar technologies, AUVs are frequently equipped with optical sensors, including high-definition cameras and laser scanners, which are crucial for capturing visual data necessary for underwater inspections and scientific research. These visual systems are particularly effective in challenging underwater conditions, providing clear imagery that supports various applications such as underwater infrastructure inspection, marine archaeology, and habitat monitoring [566]. The integration of these optical systems allows researchers and engineers to gain critical insights into underwater environments, enhancing the overall utility of AUVs in scientific exploration [567].

Furthermore, AUVs are equipped with environmental sensors that collect vital oceanographic data, including temperature, salinity, and pH levels. Instruments such as Conductivity-Temperature-Depth (CTD) sensors and fluorometers are commonly utilized to monitor marine ecosystems and detect environmental changes [568, 569]. These measurements are essential for climate research, coastal management, and tracking the health of sensitive ecosystems like coral reefs [570]. For military applications, AUVs may also carry magnetometers to detect magnetic anomalies on the seafloor, which is crucial for mine countermeasures and locating submerged wreckage [571].

The modular design of many AUV platforms allows for the flexible swapping of sensors and instruments, enhancing their adaptability for various missions. This flexibility ensures that a single AUV can be reconfigured for different tasks, such as switching from environmental monitoring to military reconnaissance or pipeline inspection [572]. Advanced AUVs can even tow specialized equipment, such as hydrophone arrays, without compromising their operational efficiency [573]. For example, a demonstration in Monterey Bay showcased a 21-inch-diameter AUV successfully towing a 400-foot hydrophone array while maintaining a speed of six knots, illustrating the capability of these systems to integrate complex payloads [573].

Through the integration of diverse sensors and instruments, AUVs serve as powerful tools for scientific exploration, environmental monitoring, and defence operations. Their ability to autonomously gather detailed data, combined with modular payload configurations, ensures that they remain versatile assets across a wide range of underwater applications [568, 574].

Autonomy and Artificial Intelligence (AI)

The autonomy of Autonomous Underwater Vehicles (AUVs) is significantly enhanced by advanced software systems that manage mission planning, navigation, and decision-making processes. These systems are crucial for enabling AUVs to operate independently for extended durations, particularly in challenging underwater environments where direct communication with operators is often limited. For instance, AUVs utilize sophisticated algorithms for path planning, which allow them to navigate safely and efficiently, avoiding obstacles and adapting to dynamic conditions in real-time [575-577]. The integration of machine learning algorithms further enhances this autonomy, enabling AUVs to dynamically adjust their operations based on environmental changes, such as rerouting in response to unexpected obstacles [578, 579].

Moreover, the implementation of cooperative swarming behaviour among multiple AUVs represents a significant advancement in autonomous operations. This approach allows AUVs to function as a coordinated unit, sharing information and optimizing their paths to cover larger areas more efficiently. Such collaborative behaviour is particularly beneficial in complex missions like underwater mapping and environmental monitoring, where multiple AUVs can divide tasks, thereby reducing redundancy and increasing mission effectiveness [580]. The ability to operate in swarms not only enhances efficiency but also improves the precision of operations in expansive or intricate environments [581].

In addition to operational autonomy, artificial intelligence (AI) plays a pivotal role in the post-mission phase of AUV operations. AUVs equipped with onboard data processing capabilities can analyse large datasets, filtering out irrelevant information and identifying critical findings before transmitting results to human operators. This capability is essential for managing data overload, especially in missions that generate substantial amounts of sonar, video, or environmental data [582, 583]. By streamlining data processing, AUVs alleviate the workload for human analysts, allowing them to concentrate on significant insights and accelerating the overall decision-making process [584].

The incorporation of AI-driven autonomy not only enhances the operational flexibility of AUVs but also expands their potential for undertaking complex missions. With advanced systems in place, AUVs can perform real-time environmental assessments or deep-sea explorations with minimal human oversight. This autonomy is particularly advantageous in remote or hazardous locations where human involvement is challenging, thus making AUVs more effective in various applications, including scientific research, commercial endeavours, and military operations [585-587]. As AI technologies continue to evolve, the capabilities of AUVs are expected to improve further, enhancing their performance across diverse applications.

Energy Management and Buoyancy Control

Energy management is indeed a critical aspect of Autonomous Underwater Vehicle (AUV) design, as these vehicles must efficiently power their propulsion systems alongside various onboard sensors, electronics, and communication systems. To enhance energy efficiency, designers focus on optimizing propulsion mechanisms to minimize drag, employing low-power electronics, and utilizing sophisticated energy management algorithms. These algorithms are essential for real-time monitoring of power usage, enabling the AUV to allocate energy effectively across different functions based on mission-specific requirements. For instance, the integration of energy harvesting technologies in advanced AUV models allows them to extend operational endurance by capturing energy from environmental sources, such as thermal gradients in ocean layers or harnessing ocean currents for battery recharging during missions, thereby reducing the need for surface interventions [543, 588].

Buoyancy control is equally vital for AUV operations, as it allows the vehicle to maintain a stable depth without the need for continuous propulsion, which is crucial for conserving energy during long-duration missions. AUVs typically employ variable buoyancy systems that adjust the vehicle's density by taking in or expelling water. This capability enables the AUV to achieve neutral buoyancy, thereby conserving power while maintaining the desired operational depth [543, 588]. The interplay between energy management and buoyancy control is fundamental to the performance and reliability of AUVs, allowing them to operate autonomously over extended distances and durations with minimal human intervention.

Most AUVs utilize rechargeable batteries, particularly lithium-ion or lithium-polymer batteries, due to their high energy density and reliability. Battery management systems are critical for monitoring charge levels and preventing conditions that could impair battery performance, such as overcharging or deep discharge [115, 544, 589]. While some AUVs may use nickel-metal hydride batteries, these typically offer lower energy density compared to lithium-based options. For missions requiring extended endurance, primary batteries may be employed, although they come with higher operational costs and reduced reusability [589]. The exploration of hybrid systems that combine traditional batteries with supercapacitors is also noteworthy, as supercapacitors can quickly store and release energy, making them suitable for tasks requiring sudden power bursts, such as rapid manoeuvring or data transmission [543, 588].

Historically, semi-fuel cells made from aluminium were considered for AUV power systems due to their high energy density; however, they posed challenges such as frequent maintenance and waste management [543, 588]. As battery technologies have advanced, the trend has shifted towards more efficient and sustainable power solutions. Lithium-ion batteries have become the standard due to their high energy density and performance, although ongoing research continues to explore alternatives, including lithium-sulphur batteries, which offer significant potential advantages in terms of energy density and weight [589, 590]. The integration of advanced battery management systems is crucial for optimizing the performance and longevity of these power sources, ensuring that AUVs can meet the demands of complex underwater missions effectively [544, 591].

Challenges in AUV Design

Designing AUVs involves addressing several complex engineering challenges. One of the primary difficulties is ensuring that the vehicle can withstand the immense pressures encountered at great ocean depths. As depth increases, the surrounding water exerts intense pressure on the AUV's hull, requiring the use of high-strength materials such as titanium, aluminium alloys, or advanced composites like carbon fibre. These materials provide both durability and corrosion resistance, which are essential for withstanding the harsh underwater environment. Pressure-resistant designs, such as internal pressure vessels, are also implemented to protect sensitive electronics and sensors from being damaged by deep-sea conditions.

Operating in unpredictable underwater environments introduces additional challenges. AUVs must navigate through complex terrains, such as rugged seabeds, coral reefs, or debris-laden areas, without external assistance. These vehicles rely on advanced navigation systems, including inertial navigation and Doppler velocity logs, to manoeuvre accurately despite limited visibility and currents. Furthermore, AUVs need to perform reliably in varied and dynamic conditions, such as temperature fluctuations, changing currents, and varying salinity levels, all of which could affect their operation and sensor performance.

Communication is another significant challenge, as radio waves cannot penetrate water effectively, limiting bandwidth and range. This limitation necessitates a high degree of autonomy in AUVs. To ensure mission success, AUVs are equipped with robust decision-making algorithms that allow them to operate independently. These algorithms enable real-time obstacle avoidance, path recalibration, and adaptive behaviour in response to unexpected environmental changes. AI-based systems are also increasingly integrated into AUVs, allowing them to carry out complex missions autonomously while reducing the need for constant operator input.

Energy efficiency is a crucial factor in AUV design, as limited power availability directly impacts operational endurance. AUVs must balance propulsion, sensor usage, and communication systems efficiently to extend mission duration. Designers incorporate lightweight materials and hydrodynamic hulls to reduce energy consumption, while battery management systems ensure optimal power use throughout the mission. Emerging energy solutions, such as hybrid power systems and energy harvesting technologies, are being explored to enhance endurance and reduce downtime for recharging. These innovations are particularly important for long-duration missions, where the ability to remain operational for extended periods is critical for mission success.

The combination of these design challenges makes AUV development a highly specialized field that requires expertise across multiple disciplines, including material science, robotics, marine engineering, and computer science. As AUVs continue to evolve, advances in autonomous decision-making, energy management, and pressure-resistant materials will play an essential role in expanding their capabilities and applications.

Applications and Future Directions

Autonomous Underwater Vehicles (AUVs) have emerged as pivotal tools across various sectors, including oceanographic research, underwater archaeology, oil and gas exploration, and military surveillance. Their autonomous capabilities enable them to conduct complex missions with minimal human intervention, which has led to their increasing adoption in diverse applications. For instance, AUVs are utilized for environmental monitoring, resource exploration, and military operations, demonstrating their versatility and effectiveness in challenging underwater environments [592-596].

As technology progresses, AUVs are becoming more sophisticated, capable of executing longer and more intricate missions. This evolution is driven by advancements in materials science, artificial intelligence, and energy management systems. Enhanced AI capabilities are particularly noteworthy, as they allow AUVs to operate with greater autonomy, making real-time decisions based on environmental data [597, 598]. Furthermore, the development of bio-inspired propulsion systems is gaining traction, promising to improve the efficiency and manoeuvrability of AUVs in complex marine environments [599]. These innovations are critical for expanding the operational capabilities of AUVs, enabling them to undertake tasks such as underwater mapping, surveillance, and ecological monitoring more effectively [594, 600].

Future trends in AUV design are expected to focus on improving energy efficiency and management solutions. As AUVs often rely on onboard batteries or tethered power sources, research is ongoing to develop more sustainable energy systems that can support extended missions without frequent recharging [601, 602]. The integration of advanced energy management technologies will be essential for maximizing the operational range and duration of AUVs, thereby enhancing their utility in both scientific and military applications [603, 604].

The continued evolution of AUV technology holds the potential to revolutionize underwater operations, providing unprecedented opportunities for exploration, research, and defence. With the integration of advanced materials, AI, and energy systems, AUVs are poised to play an increasingly critical role in enhancing our understanding of oceanic environments and supporting global maritime operations [593, 596, 600]. As these technologies mature, the capabilities of AUVs will likely expand, allowing for more ambitious missions that can address pressing challenges in marine science and security [595, 605].

Role of Artificial Intelligence in Submarines

Artificial intelligence (AI) is increasingly recognized as a transformative force in the operation of modern submarines, enhancing their capabilities across various domains such as navigation, target detection, data processing, and autonomous operations. The complexity of underwater environments, characterized by limited visibility and dynamic conditions, necessitates advanced technological solutions that AI can provide. For instance, AI algorithms significantly improve navigation by analysing underwater terrain, currents, and environmental factors to chart optimal routes. This capability is crucial for submarines, which must navigate safely through challenging underwater landscapes while avoiding obstacles in real-time [606].

In the realm of target detection and classification, AI plays a pivotal role in processing sonar data. Submarines utilize sophisticated sonar systems to detect nearby objects, but interpreting this data can be labour-intensive and prone to human error. AI algorithms streamline this process by efficiently analysing acoustic signals, distinguishing between various objects, including marine life and potential threats such as enemy vessels or underwater mines. This automation not only accelerates decision-making but also enhances operational precision in high-stress scenarios [315, 607, 608]. The integration of AI in sonar systems has been shown to improve the accuracy of underwater target recognition, which is vital for maintaining operational effectiveness [609, 610].

Moreover, AI contributes to predictive maintenance and system diagnostics within submarines. These vessels are equipped with numerous mechanical and electronic systems that require constant monitoring. AI-powered predictive maintenance tools analyse data from onboard sensors to identify potential issues before they escalate into failures. This proactive approach minimizes downtime and extends the operational lifespan of submarine systems, ensuring that they remain mission-ready during extended deployments [611]. The ability to anticipate maintenance needs is particularly beneficial in the context of long missions where access to repair facilities may be limited.

In combat operations, AI enhances tactical decision-making by providing real-time recommendations based on evolving battlefield conditions. Advanced submarines are increasingly equipped with AI-driven systems that analyse enemy behaviour patterns, predict potential threats, and suggest countermeasures. This capability not only improves the effectiveness of military operations but also paves the way for the development of fully autonomous submarines capable of executing missions with minimal human intervention [612, 613]. The future of undersea warfare may see AI systems coordinating with both unmanned and manned naval assets, further enhancing operational synergy.

AI also plays a crucial role in data processing and communications. Submarines generate vast amounts of data from various sensors, and AI algorithms help streamline this information by identifying relevant insights while filtering out redundant data. This optimization reduces bandwidth usage and ensures that critical information is transmitted efficiently to command centres. In stealth operations, where communication must be minimized, AI can facilitate local decision-making, allowing submarines to maintain a low profile while achieving mission objectives [614].

Overall, the integration of AI into submarine operations is revolutionizing their functionality, making them more autonomous, efficient, and responsive to dynamic underwater environments. As research and development in AI continue to advance, the potential for further reducing crew workload and enhancing strategic decision-making will solidify AI's role as a cornerstone of modern undersea warfare [615].

Artificial Intelligence (AI) is increasingly recognized for its transformative role in enhancing the capabilities of modern submarines, particularly in areas such as autonomous navigation, target detection, predictive maintenance, and decision-making. Submarines equipped with AI-based navigation systems demonstrate significant advancements in autonomous route planning. These systems enable submarines to autonomously chart optimal paths, adapt to environmental changes, and avoid obstacles, which is crucial during prolonged submerged operations. The ability to operate with reduced human input allows for precise navigation through complex underwater environments, thereby enhancing operational effectiveness and safety [616, 617].

Moreover, AI-powered systems significantly improve target detection capabilities by analysing sonar data in real-time. Submarines utilize machine learning algorithms to differentiate between marine life, natural obstacles, and potential threats, such as enemy vessels or mines. This automated detection process minimizes false positives and facilitates quicker, more reliable responses in high-pressure scenarios, which is essential for maintaining tactical advantage during missions [315, 618]. The integration of AI in sonar systems allows for enhanced situational awareness and operational readiness, which are vital in modern naval warfare [619].

Predictive maintenance is another critical application of AI in submarine operations, ensuring the operational readiness of vessels. By analysing sensor data from various onboard systems, AI algorithms can identify patterns indicative of potential equipment failures before they occur. This proactive maintenance approach minimizes downtime and extends the lifespan of critical components, reducing the need for unplanned maintenance and ensuring that submarines remain fully operational during missions [620, 621]. The ability to predict maintenance needs not only enhances reliability but also contributes to cost savings and improved mission planning [622].

In terms of mission planning and decision-making, AI provides real-time recommendations by processing vast datasets, including environmental data, enemy activity, and historical mission outcomes. These insights support submarine commanders in making informed decisions, thereby optimizing mission outcomes. Additionally, AI enhances communication within submarines and with command centres through tools like natural language processing, which streamlines crew interactions and improves operational coordination [623]. As research progresses, AI is poised to further revolutionize submarine operations, potentially leading to fully autonomous underwater vessels that can operate in conjunction with manned units. The capacity of AI to optimize energy management, enhance surveillance, and automate routine tasks ensures that submarines maintain strategic superiority in an increasingly complex maritime environment [624].

Future Trends in Unmanned Submarine Development

The rapid advancement of Unmanned Underwater Vehicles (UUVs) is significantly influenced by technological innovations in autonomy, artificial intelligence (AI), and energy management. These developments are reshaping the operational capabilities of UUVs, with substantial implications for both military and scientific applications.

Enhanced Autonomy and Swarm Operations: A critical trend in UUV development is the enhancement of autonomy, which allows these vehicles to conduct complex missions with minimal human intervention. Advanced AI algorithms play a pivotal role in enabling real-time decision-making, facilitating navigation in dynamic underwater environments where traditional communication methods may falter [625]. The integration of swarm technology further enhances operational capabilities, allowing multiple UUVs to collaborate on reconnaissance, search, and mapping missions. This coordinated operation not only improves mission effectiveness by covering larger areas but also reduces the risk of detection in hostile environments [626]. The ability of UUVs to operate autonomously in challenging conditions is essential for effective marine resource exploration and surveillance [625].

Integration with Networked Naval Operations: UUVs are increasingly being integrated into broader networked naval operations, functioning alongside manned submarines and unmanned surface vehicles. This integration enhances their role as offboard sensors or decoys, particularly in anti-submarine warfare (ASW) scenarios. For example, UUVs equipped with bistatic sonar systems can significantly improve target detection capabilities by illuminating targets with sonar signals while coordinating data processing with other platforms [627]. This synergy not only augments situational awareness but also extends the operational reach of manned submarines, as demonstrated by the U.S. Navy's Virginia-class fleet [627]. The collaborative capabilities of UUVs within a networked framework are crucial for modern naval operations, enhancing both strategic and tactical advantages [627].

Energy Efficiency and Endurance Improvements: A primary focus for future UUV designs is extending mission endurance. Researchers are exploring hybrid energy systems and ocean-based energy harvesting technologies to reduce reliance on conventional batteries. Fuel cells, which generate power through chemical reactions, represent a promising avenue for prolonging operational time [628]. Additionally, energy harvesting from thermal gradients or ocean currents is being investigated as a means to sustain UUV operations for extended periods without the need for recharging [629]. These innovations are essential for enabling long-term surveillance missions, thereby enhancing the strategic utility of UUVs in both military and scientific contexts [629].

Modular Design and Flexible Payloads: The trend towards modular design in UUVs allows for rapid customization based on specific mission requirements. This flexibility enables a single platform to perform various roles, such as underwater inspection, mine detection, or environmental monitoring. For instance, Boeing's Orca XLUUV exemplifies this modularity, capable of carrying diverse payloads, including sensors and torpedoes tailored to operational needs [630].

Such adaptability enhances the value of UUVs across military and civilian applications, from maritime security to scientific exploration [630].

AI-Driven Predictive Maintenance and Data Processing: AI is also instrumental in enhancing UUV maintenance and data management. Predictive maintenance algorithms analyse real-time sensor data to identify potential issues before they escalate into failures, thereby reducing downtime and ensuring continuous operation [631]. Moreover, AI-driven data processing systems enable UUVs to filter and prioritize information, transmitting only the most relevant findings to operators. This capability is particularly beneficial in bandwidth-limited environments, facilitating rapid decision-making during missions [631].

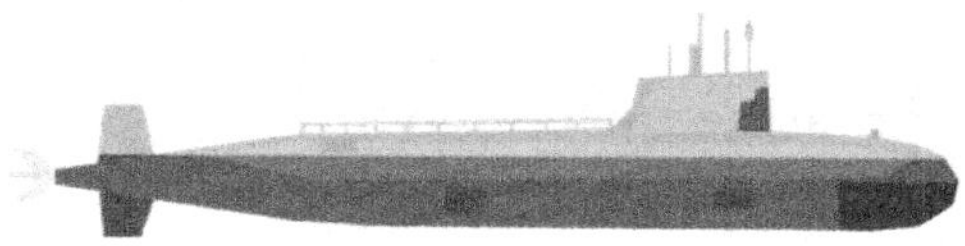

Chapter 13

Environmental and Operational Constraints

Arctic and Deep-Sea Submarine Design

Designing submarines for Arctic and deep-sea environments presents unique challenges that require specialized technologies to operate under extreme conditions. In the Arctic, submarines need to navigate under thick sea ice, withstand freezing temperatures, and surface safely through ice-covered waters. To address these challenges, these submarines are equipped with robust hulls, often reinforced with HY-80 or HY-100 steel, to withstand collisions with ice formations [632]. They also incorporate upward-looking sonar to detect polynyas (areas of thin ice) and leads, which are crucial for safe surfacing. Sails, the vertical structures on submarines, are often specially hardened to punch through ice, though some components like sonar windows remain more vulnerable to damage during such operations [633, 634].

HY-80 and HY-100 are high-yield, low-alloy steels that are specifically designed for use in environments requiring both high strength and resistance to extreme conditions, such as submarine construction. These steels are named for their yield strength, with HY-80 having a yield strength of 80,000 psi and HY-100 at 100,000 psi. The "HY" designation refers to "high-yield" steel, which is known for its ability to maintain structural integrity under pressure, making it ideal for critical applications such as naval vessels.

Both HY-80 and HY-100 are alloyed with elements like nickel, molybdenum, and chromium, giving them excellent toughness and corrosion resistance, even in harsh marine environments. They exhibit high fracture toughness, meaning that they can absorb significant stress without cracking—a key property for submarines that operate under extreme underwater pressure.

These steels also offer enhanced weldability compared to other high-strength steels, making them easier to use in large structural components. This feature is essential for submarine construction, where complex hull shapes and modular designs must be assembled with high precision through welding. Additionally, their resistance to brittle fractures at low temperatures ensures that they perform well in Arctic waters, where submarines are exposed to freezing conditions.

In submarines, HY-80 and HY-100 are primarily used for the pressure hull, which is the structural core that maintains the vessel's integrity under immense underwater pressure. The steel ensures that the hull can withstand both the mechanical stresses from deep-sea environments and potential collisions with ice in Arctic waters. HY-80 was initially adopted in the construction of early nuclear-powered submarines, such as the U.S. Navy's Skipjack-class and Los Angeles-class submarines, due to its superior strength and fracture resistance.

As submarine designs evolved, HY-100 began to replace HY-80 in newer vessels, offering even greater strength. This allowed for deeper operational depths while maintaining hull thicknesses similar to or thinner than those constructed with HY-80. Using HY-100 also contributes to reducing the overall weight of the submarine, which improves efficiency without compromising safety.

Both materials are integral not just to the pressure hull but also to other components that need to endure harsh underwater conditions, such as ballast tanks, bulkheads, and key structural supports. Their role in submarine design is critical, especially for vessels operating in the Arctic, where the pressure hull must be able to withstand impacts from sea ice without fracturing.

The integration of upward-looking sonar systems is another vital technological advancement in Arctic submarine design. These sonar arrays are essential for scanning the ice above the vessel, enabling the crew to identify polynyas—areas of thin ice—or open leads where the submarine can surface safely. This capability is crucial for preventing unplanned surfacing through thick ice, which could lead to significant damage and operational delays [633, 635]. Furthermore, the sonar systems provide critical navigational data, reducing the risks of collision with submerged ice ridges or other underwater obstacles [633, 634].

In addition to structural integrity and sonar capabilities, submarines operating in polar regions must contend with freezing temperatures. Specialized coatings and thermal management systems are employed to prevent the freezing of critical systems and maintain operational readiness [633, 635]. The design must also ensure precise manoeuvrability in confined spaces, such as marginal ice zones, where ice ridges can extend deep below the surface. Auxiliary propulsion systems, including retractable thrusters, are often integrated to enhance agility at low speeds, facilitating movement between ice formations and positioning for surfacing [633, 635].

Deep-sea submarines, on the other hand, focus on withstanding immense water pressure at depths that can reach thousands of meters. For these vehicles, materials such as titanium and advanced composites are used to construct pressure-resistant hulls. Submersibles designed for deep-sea exploration, such as the DSV Alvin or China's new mini-submarines, prioritize modular design and include specialized instruments for oceanographic studies. These vessels often operate from research ships or icebreakers, such as China's Xue Long, which facilitate operations in polar regions and deep-sea environments.

Energy management is also critical for both Arctic and deep-sea operations. In these remote environments, submarines rely on highly efficient power systems, such as lithium-ion batteries or experimental fuel cells, to ensure they can operate for extended periods without needing frequent resupply. For Arctic navigation, submarines must carefully manage their paths to avoid sonar detection when passing through chokepoints like the Bering Strait, where both ice and enemy surveillance present challenges.

These Arctic and deep-sea designs reflect a blend of durability, autonomy, and environmental adaptability. As geopolitical interest grows in the Arctic, along with scientific exploration in deep-sea environments, future submarines will increasingly rely on advanced technologies to perform complex missions safely and efficiently in these harsh conditions.

Figure 125: The Royal Navy hunter killer submarine HMS Trenchant (S 91) surfaces in the Arctic Circle in support of Ice Exercise (ICEX) 2018, March 11. Defense Visual Information Distribution Service, Public Domain, via Picryl.

The most significant challenge for Arctic submarine operations is the ever-present threat of sea ice. In densely packed ice, surfacing can be either limited or entirely prevented, forcing submarines to remain submerged for extended periods. This creates serious operational risks, especially for diesel-electric submarines, which require periodic surfacing to recharge their batteries. The harsh conditions of Arctic waters have made nuclear-powered submarines the preferred choice since the 1950s. Nuclear propulsion allows these vessels to remain submerged indefinitely, making them far more capable of conducting sustained operations under the ice. However, this creates limitations for navies like China's PLAN, where the majority of the fleet consists of diesel-electric submarines, leaving only a small number of nuclear submarines capable of Arctic missions [636].

Recent advancements in air-independent propulsion (AIP) offer a potential solution to extend the submerged endurance of conventional submarines. AIP systems allow vessels to operate for several weeks without surfacing by using chemical reactions that do not rely on atmospheric oxygen. One example is the Stirling engine, first used by Sweden in the 1980s, which generates energy through the combustion of liquid oxygen and diesel. Other emerging technologies, such as hydrogen fuel cells, provide even quieter propulsion with fewer maintenance requirements but at the cost of increased complexity. AIP-equipped submarines, such as the Type 041 Yuan-class, have demonstrated the ability to stay underwater for extended periods, making them more viable for Arctic or near-Arctic missions [636].

Despite these advancements, AIP systems remain limited in their ability to meet the demands of Arctic operations. While AIP extends underwater endurance, these systems do not provide sufficient power for high-speed transits or long-range missions, which are essential in the Arctic environment. Submarines using AIP would still need access to air after exhausting their fuel supply, forcing them to rely on open water areas or polynyas—gaps in the ice where submarines can surface. Navigating between these areas requires constant monitoring using sonar or light sensors, which can reveal the vessel's location to potential adversaries [636].

Non-nuclear submarines also face structural limitations in the Arctic. These submarines tend to be smaller, reducing their ability to push through thick ice when surfacing. The pressure of ice cover requires specialized design features, such as reinforced sails and retractable diving planes, to minimize damage. During the Cold War, adjustments to American submarines like the Sturgeon class demonstrated how such modifications allowed vessels to surface through ice by rotating their diving planes vertically. However, this process was time-consuming and posed risks if the planes became stuck in the ice. Modern designs place diving planes on the forward hull for improved safety and efficiency, a feature seen in later variants of the Los Angeles-class submarines [636].

Surfacing through ice remains a delicate operation requiring reinforced sails to protect critical equipment such as periscopes, communication antennas, and sonar arrays. Some submarines, like the USS Skate, were equipped with hardened steel to penetrate up to 18 inches of ice. However, not all components, particularly sonar windows, can be reinforced, making them vulnerable to damage. Propellers, another critical component, are especially susceptible to damage from ice. Historical examples, such as USS Skate's experience in the Kara Sea, highlight the challenges of maintaining operational capabilities after encountering ice damage [636].

Navigating Arctic waters also involves significant risks, particularly in regions like the Bering Strait, where shallow waters and compressed ice ridges create natural obstacles. Submarines must use upward- and forward-looking sonar to detect hazards, but using active sonar increases the risk of detection by enemy forces. Additionally, the narrow straits require precise manoeuvring, further complicating transit for larger submarines. Smaller submarines, with retractable auxiliary motors for low-speed manoeuvring, offer advantages in these environments, allowing them to position themselves more effectively under ice for both offensive and defensive operations [636].

Ultimately, while advancements in AIP technology narrow the endurance gap between conventional and nuclear submarines, nuclear propulsion remains essential for sustained Arctic operations. The growing number of nuclear submarines in China's fleet will enhance its Arctic capabilities, but achieving a continuous presence in polar waters would strain even a modernized fleet. The operational demands of Arctic missions, including the need to minimize detection and maintain high-speed transit, make nuclear propulsion the only viable option for long-term strategic operations [636].

Accounting for Freezing Temperatures

Designing submarines for polar waters necessitates a comprehensive understanding of the extreme environmental conditions encountered in these regions, particularly the challenges posed by freezing temperatures and ice formation. Ice accumulation on external components and critical systems can severely impair functionality, increase operational risks, and compromise readiness. To mitigate these issues, submarines intended for Arctic operations employ specialized coatings on exposed surfaces, such as hulls and periscopes, which are engineered to resist ice formation even under extreme conditions. These coatings are crucial for ensuring smooth operations during navigation and surfacing, as they significantly reduce the likelihood of ice-related obstructions that could hinder operational capabilities [637].

In addition to protective coatings, submarines utilize advanced thermal management systems to maintain the functionality of onboard electronics and propulsion systems. These systems are designed to strategically distribute heat throughout the vessel, preventing vital components from freezing and ensuring uninterrupted operation during extended missions beneath the ice [106]. The importance of thermal management is underscored by the unique thermal dynamics of polar environments, where the risk of freezing can be exacerbated by the submarine's operational profile and the surrounding icy waters [638].

Navigating under ice introduces further complexities, particularly in marginal ice zones where ice ridges can extend deep beneath the surface. These regions require precise manoeuvrability to avoid collisions and position the vessel for surfacing through polynyas or thin ice. Standard propulsion systems, which are typically optimized for open-water operations, may be inadequate in these scenarios. Consequently, submarines designed for Arctic missions are equipped with auxiliary propulsion systems, such as retractable thrusters, which enhance manoeuvrability at low speeds. These thrusters allow submarines to execute lateral movements and reposition themselves effectively between ice formations, thereby improving operational safety and efficiency [639].

The significance of these auxiliary systems becomes particularly evident during missions that necessitate delicate manoeuvring, such as surveillance near ice-covered shores or surfacing through narrow openings in the ice. Retractable thrusters provide enhanced control, especially in situations requiring sudden changes in direction to avoid underwater hazards. This capability is essential for safely surfacing in unpredictable ice conditions, where the risk of collision with ice formations is heightened [640]. Furthermore, maintaining precise depth control is critical in these environments; submarines must carefully manage their buoyancy to maintain stable positions under the ice without excessive propulsion, thereby conserving energy and minimizing noise emissions [641].

Submarine Operations in Littoral Zones

Submarine operations in littoral zones present distinct challenges and require specialized tactics due to the complex environmental conditions these coastal waters impose. Unlike the open ocean, where water depths are vast, littoral zones are characterized by shallow waters, dynamic currents, varying salinity, and high levels of ambient noise from natural and human activities. These conditions can complicate the use of traditional sonar and navigation systems, making stealth operations, navigation, and target detection much more difficult.

Littoral zones, defined as coastal regions where land meets the sea, are characterized by shallow waters that extend from the high-water mark to the depth where sunlight no longer penetrates to the seabed. These zones encompass a variety of marine environments, including estuaries, lagoons, bays, intertidal areas, and coastal waters up to the edge of the continental shelf. The ecological significance of littoral zones is well-documented; they are crucial habitats for a diverse array of organisms, including fish, invertebrates, and aquatic plants. Studies indicate that littoral zones support a majority of species in lakes, highlighting their importance in maintaining biodiversity [642, 643]. The availability of sunlight in these shallow waters fosters the growth of aquatic plants and algae, contributing to the high productivity of these ecosystems [643].

From a biological perspective, littoral zones serve as vital breeding and feeding grounds for many marine organisms. They are often rich in biodiversity, providing habitats for various species, including coral reefs and mangroves, which are essential for ecological stability [644, 645]. The presence of structured habitats in littoral zones not only supports a diverse range of species but also influences their distribution and abundance. For example, the abundance of certain fish species in the littoral zone is positively correlated with the availability of structured habitats that reduce predation risk [646]. Moreover, the ecological dynamics within these zones are shaped by factors such as water level fluctuations, which can significantly impact the structure and functioning of these ecosystems [643, 647].

In military and naval contexts, littoral zones hold strategic importance due to their proximity to the coast. These areas are where naval operations, including surveillance and reconnaissance, are often conducted. The shallow depths of littoral zones present unique challenges for submarines and other naval vessels, as navigation can be complicated by underwater obstacles and high levels of ambient noise [648]. The operational dynamics in these regions are influenced by the need for manoeuvrability and the risks posed by coastal defence systems, making littoral zones critical for military strategy [648].

One of the key challenges in littoral environments is navigating shallow waters. Submarines must carefully manage their depth to avoid running aground or colliding with underwater obstacles like sandbars, reefs, and wrecks. Additionally, the proximity to shorelines increases the risk of detection by coastal radar, anti-submarine warfare (ASW) patrols, and aircraft. This proximity requires submarines to operate at low speeds and maintain a low acoustic signature to avoid being tracked by active sonar systems deployed by surface vessels or coastal installations.

The high levels of ambient noise in littoral zones, caused by wave action, sediment movement, and marine traffic, present both challenges and opportunities for submarine operations. While the noise can help mask the sounds of a submarine, it also interferes with sonar effectiveness. Submarines must rely on passive sonar systems to detect enemy vessels without revealing their own presence, but the cluttered acoustic environment makes distinguishing between legitimate targets and background noise more difficult. In response, submarines operating in these zones often use advanced signal processing and acoustic filters to enhance target identification while maintaining stealth.

Another important factor in littoral operations is the increased risk of detection due to the proximity to enemy forces and surveillance systems. Coastal waters are often heavily monitored, making it easier for adversaries to track submarine movements. Submarines must employ tactics such as bottoming out—resting quietly on the seabed—to avoid detection. This tactic requires precise navigation to avoid underwater hazards while positioning the vessel effectively for potential ambushes or surveillance missions.

Submarines operating in littoral zones must also account for dynamic water conditions. Tidal currents, river outflows, and thermoclines (sharp changes in water temperature) can all affect a submarine's manoeuvrability and sonar performance. These factors require continuous monitoring and adjustments to navigation systems to maintain control and effective mission execution. Littoral submarines often use Doppler Velocity Logs (DVLs) and advanced inertial navigation systems to counteract these environmental factors and ensure precise movement through the water.

Operational planning for littoral missions also emphasizes coordinated efforts with other naval and air assets. Submarines are often deployed as part of a larger network, working with surface ships, aircraft, and unmanned underwater vehicles (UUVs) to perform tasks such as reconnaissance, mine clearance, and anti-submarine warfare. Coordination with these units ensures that submarines can operate effectively in the confined and complex littoral environment while minimizing risks.

Design

Submarines designed for littoral zone operations incorporate several specialized design features to address the unique challenges of shallow coastal waters, such as manoeuvrability, stealth, and resilience in dynamic environments. These features allow them to navigate confined spaces, avoid detection, and operate effectively near shorelines, where traditional deep-sea submarines might struggle.

Littoral submarines are often smaller than ocean-going counterparts, improving their ability to navigate in confined spaces, such as narrow channels, harbors, or near shorelines. A smaller size also allows these submarines to manoeuvre between underwater obstacles, such as sandbars, reefs, or wreckage, with greater agility.

To enhance manoeuvrability at low speeds, littoral submarines feature auxiliary propulsion systems, such as retractable thrusters. These thrusters provide precise control for operations in confined areas and assist in delicate movements, such as docking or positioning under enemy ships. This feature is crucial in environments where fine adjustments are needed to avoid obstacles or maintain stealth.

Many littoral submarines use AIP systems to reduce the need for surfacing, allowing them to remain submerged for extended periods. This capability is critical in coastal areas where frequent surfacing increases the risk of detection.

AIP-powered submarines can perform stealth operations, including surveillance and reconnaissance, without exposing themselves to enemy forces.

Littoral zones are acoustically complex due to noise from waves, marine life, and human activity. To navigate effectively, submarines are equipped with advanced sonar systems, such as forward- and upward-looking sonar, which help detect obstacles, underwater mines, and potential threats. Doppler Velocity Logs (DVLs) enhance precision by tracking movement relative to the seafloor, ensuring accurate navigation in strong currents and shallow waters.

Littoral submarines rely on stealth to operate undetected in areas with high levels of surveillance. They are equipped with anechoic tiles that absorb sound waves, reducing the acoustic signature. Engines and propulsion systems are designed to run quietly, with vibration isolation technologies minimizing noise generated by onboard machinery. This ensures the submarine remains difficult to detect in noisy coastal environments.

In coastal waters, maintaining communication with command centres without exposing the submarine's position is essential. Littoral submarines often use towed communication buoys or low-frequency antennas to transmit data while remaining submerged. This allows them to relay critical information without the need to surface, preserving operational security.

Littoral submarines are designed to be flexible, with modular compartments that can accommodate mission-specific equipment. These submarines can be configured for a variety of operations, including mine detection, special forces deployment, or intelligence gathering. This versatility allows them to adapt quickly to changing mission requirements in complex coastal environments.

The following are examples of littoral submarines used by different navies, specifically designed to operate in shallow coastal waters where stealth and manoeuvrability are essential:

1. **Gotland Class (Sweden)** – Known for its stealth capabilities, this class uses Air-Independent Propulsion (AIP), allowing it to remain submerged for extended periods. It operates effectively in both littoral and open-sea environments, with minimal noise emissions.

2. **Sōryū Class (Japan)** – These advanced diesel-electric submarines feature large lithium-ion batteries, enhancing underwater endurance. Although designed for various missions, the Sōryū class performs well in littoral waters with its manoeuvrability and quiet propulsion system.

3. **Type 212 (Germany and Italy)** – This class employs AIP technology, making it one of the quietest submarines in operation. With excellent stealth features, the Type 212 is ideal for operations near coastlines, performing reconnaissance and special forces missions.

4. **Scorpène Class (France, India, Brazil, and others)** – Modular and versatile, this class of submarines excels in littoral zones due to its size and stealth technology. It is widely used for surveillance, anti-submarine warfare, and special operations.

5. **Yuan Class (China)** – This modern class, equipped with AIP, represents a significant development in the Chinese Navy's ability to operate in confined coastal environments. Its design focuses on enhanced stealth and extended underwater endurance.

6. **Archer Class (Singapore)** – Adapted from the Swedish Västergötland-class submarines, the Archer class is optimized for operations in the busy and shallow waters around Singapore. These submarines have upgraded systems and can conduct stealthy missions in littoral zones.

Figure 126: The Royal Malaysian Navy's second Scorpène-class submarine, KD Tun Abdul Razak docked at its naval jetty in Awana Porto Malai, Langkawi. Rizuan, CC BY-SA 3.0, via Wikimedia Commons.

These submarines are tailored to the challenges of operating in confined, shallow waters, where larger, deep-sea vessels struggle. Littoral submarines play a critical role in regional defence, reconnaissance, and special operations, offering stealth and flexibility near coastlines.

Environmental Impacts and Sustainability

The environmental impact and sustainability of submarines are critical considerations due to their operation in sensitive marine ecosystems and the substantial energy and resources they consume throughout their lifecycle. The direct and indirect effects of submarines on the environment include noise pollution, chemical discharges, energy consumption, and impacts on marine biodiversity, all of which necessitate a thorough understanding and implementation of sustainable practices in submarine development and operations.

One of the most pressing environmental concerns associated with submarines is the noise they generate, particularly from their propulsion systems and sonar equipment. This noise pollution can significantly disrupt marine ecosystems, especially for marine mammals that rely on sound for communication, navigation, and hunting. Studies have shown that active sonar systems emit high-intensity sound waves that can lead to behavioural changes and even strandings in species such as whales and dolphins [649]. Furthermore, mechanical noise from submarine engines can alter the natural soundscape of the ocean, posing additional risks to marine life in sensitive habitats [650]. The cumulative effects of such disturbances can lead to long-term changes in marine biodiversity and ecosystem health [650].

Submarines, similar to other maritime vessels, can release various chemicals into the marine environment during routine operations. Lubricants, coolants, and fuel residues may leak into the ocean, particularly during maintenance or surfacing operations [651]. Diesel-electric submarines contribute to air and water pollution through emissions from their diesel engines, while nuclear submarines face unique challenges related to the management of radioactive waste generated by their reactors [652]. The protocols for preventing environmental contamination from these discharges are critical, as improper management can lead to significant ecological damage [653].

Submarines are inherently energy-intensive, with nuclear-powered vessels requiring complex reactors and fuel systems. Although nuclear submarines offer extended operational endurance, the processes involved in uranium extraction, reactor operation, and fuel management contribute to their overall carbon footprint [654]. In contrast, diesel-electric submarines rely on fossil fuels, emitting greenhouse gases during surface operations [655]. To mitigate these environmental impacts, modern submarine designs are increasingly incorporating Air-Independent Propulsion (AIP) systems and exploring hybrid energy solutions to enhance efficiency and reduce emissions [656].

The operations of submarines, particularly in shallow and coastal waters, can disrupt marine ecosystems and habitats. Activities such as manoeuvring can disturb seabeds, coral reefs, and breeding grounds for various marine species [657]. The noise and vibrations generated by submarines can alter animal behaviour and lead to habitat fragmentation, further threatening marine biodiversity [658]. The ecological role of submarine canyons, which serve as critical habitats, is particularly vulnerable to such disturbances [659].

Sustainability is becoming increasingly important in submarine design and lifecycle management. Innovations aimed at reducing energy consumption include improved hydrodynamics and quieter propulsion systems, as well as the use of lightweight materials to minimize resource use [660]. The decommissioning of submarines, especially nuclear-powered ones, presents significant environmental challenges, necessitating careful planning and management of radioactive waste to prevent contamination [661]. Effective strategies for dismantling submarines and managing their lifecycle are essential for minimizing their ecological footprint [152].

The submarine industry is beginning to adopt more sustainable practices, such as the development of hybrid-electric submarines that utilize lithium-ion batteries and renewable energy sources [662]. Research into ocean-based energy harvesting is also underway, aiming to extend underwater endurance without reliance on fossil fuels or nuclear power [663]. Enhanced recycling processes during decommissioning and the implementation of stricter environmental policies are vital for reducing the environmental impact of submarine operations over time [664].

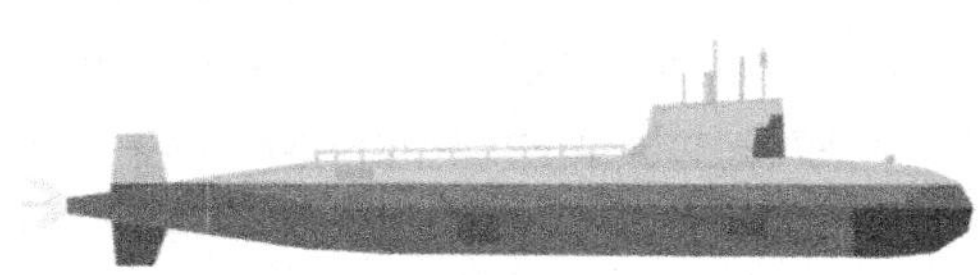

Chapter 14

Project Management in Submarine Construction

Design and Development Process

The design and development of submarines is a highly complex, multi-disciplinary process that involves extensive planning, engineering, testing, and integration. This process can span several years or even decades due to the specialized technologies, materials, and systems involved. Submarine development requires collaboration between naval architects, engineers, defence contractors, governments, and naval forces to ensure the vessel meets operational requirements while maintaining safety, stealth, and sustainability.

Design and development phases:

1. Conceptual Design and Feasibility Studies

The development process begins with defining the mission and operational requirements. Naval forces outline the capabilities needed based on anticipated missions, such as anti-submarine warfare, surveillance, or special operations. These requirements guide initial feasibility studies, where naval architects evaluate whether the desired features can be practically implemented given current technologies and budget constraints. During this phase, designers also explore basic configurations for size, propulsion, endurance, and payloads.

This phase often begins with a systems analysis methodology that evaluates various design alternatives based on operational needs, technological capabilities, and cost considerations [665]. During this phase, hydrodynamic performance is a primary concern, as it significantly influences the submarine's speed, manoeuvrability, and stealth capabilities. Advanced computational tools and numerical simulations are employed to assess hydrodynamic characteristics under various operating conditions, including submerged and at snorkel depth [24, 666].

Moreover, the integration of multi-objective optimization techniques, such as genetic algorithms, allows designers to refine the submarine hull shape to minimize drag while maximizing internal volume [63]. This optimization is essential for enhancing the submarine's operational efficiency and overall performance.

2. Preliminary Design

In the preliminary design phase, engineers create detailed drawings and digital models, focusing on hydrodynamics, hull shape, and general system layouts. Computational Fluid Dynamics (CFD) simulations are used to optimize the hull design for reduced drag and improved underwater performance. Critical elements like ballast tanks, control surfaces (rudders and diving planes), and propulsion systems are outlined. Trade-offs between speed, stealth, and operational endurance are carefully assessed. Submarines designed for Arctic or littoral environments undergo specialized analysis to ensure the vessel can operate in challenging conditions.

3. Detailed Design and Systems Integration

Once the preliminary design is approved, the detailed design process begins. During this phase, every aspect of the submarine is carefully planned, including propulsion systems (such as diesel-electric engines or nuclear reactors), navigation systems, sonar arrays, communication systems, and life-support equipment. Designers create 3D models and blueprints for every component, which must fit precisely within the confined space of the submarine's hull.

The detailed design phase involves the creation of comprehensive specifications and engineering drawings. This stage incorporates advanced modelling techniques, including finite element analysis (FEA) and computational fluid dynamics (CFD), to evaluate structural integrity and fluid interactions [666, 667]. The use of model-based systems engineering (MBSE) tools is also prevalent, as they facilitate the translation of functional requirements into detailed design specifications, ensuring that all aspects of the submarine's systems are cohesively integrated [668].

The detailed design phase in submarine development involves creating precise engineering drawings and technical specifications that ensure all components are built and integrated correctly. This phase utilizes advanced computational tools to model, analyse, and refine every aspect of the submarine. This includes:

Step 1: Developing Comprehensive Engineering Drawings and Specifications

The first task is to translate the high-level design concept into detailed blueprints. These documents include specifications for:

- **Hull structure** (including shape, materials, and dimensions).
- **Propulsion systems** (nuclear reactors or diesel-electric systems).
- **Navigation and control systems** (rudders, diving planes, and sonar arrays).
- **Electrical and life-support systems**.

Every system and component must be drawn in 2D and 3D CAD software to ensure precision. The detailed drawings must adhere to regulatory standards and account for both safety and functionality.

Step 2: Finite Element Analysis (FEA)

Once the engineering drawings are created, Finite Element Analysis (FEA) is used to assess the structural integrity of the submarine under various conditions. This technique involves breaking down complex structures into smaller elements to simulate how they respond to:

- **Pressure at operational depths**.
- **Thermal expansion** from nuclear or diesel propulsion systems.
- **Stress from collisions** or other mechanical forces.

FEA helps identify weak points in the design so that modifications can be made before physical construction begins.

Finite Element Analysis (FEA) is a computational technique used to predict how a structure or component will respond to external forces, pressure, heat, and other physical phenomena. It involves dividing complex structures into smaller, manageable elements and analysing each element to determine the overall behaviour of the system. FEA is widely used across industries such as aerospace, automotive, and marine engineering, particularly in submarine design, to ensure that structures can withstand high stress, pressure, and environmental conditions.

FEA is essential for optimizing structural integrity, as it allows engineers to evaluate potential weak points and make informed design decisions before physical prototypes are built. By simulating various scenarios, FEA can identify stress concentrations, predict material deformation, and assess how the design performs under dynamic loads or extreme conditions, such as those encountered by submarines at great depths.

Performing Finite Element Analysis entails:

1. Model Creation

The first step in FEA is to create a digital model of the structure or component that needs to be analysed. This model is typically built using Computer-Aided Design (CAD) software. For complex systems like submarines, models include components such as hull sections, bulkheads, and control surfaces.

2. Mesh Generation

The model is then divided into smaller discrete elements connected at nodes, a process known as meshing. The mesh can consist of various element types, such as triangles or quadrilaterals in 2D, or tetrahedrons and hexahedrons in 3D. The mesh quality directly impacts the accuracy of the analysis: finer meshes provide more precision but require more computational resources, while coarser meshes run faster but may compromise accuracy.

3. Assigning Material Properties

Next, engineers define the material properties of the components in the model, such as density, Young's modulus (elasticity), Poisson's ratio, and thermal conductivity. These properties reflect the real-world behaviour of the material under different loading conditions. For example, submarine hull materials like HY-80 steel must be accurately represented to predict stress under pressure.

4. Applying Loads and Boundary Conditions

In this step, external forces and constraints are applied to the model to simulate real-world conditions. Loads might include water pressure at different depths, temperature gradients, or mechanical forces such as vibration from the engine. Boundary conditions—such as fixed points or restricted movement along certain axes—are also applied to represent how the structure interacts with its environment (e.g., attachments to other components).

5. Solving the Model

Once the setup is complete, the analysis is run using specialized FEA software (e.g., ANSYS, Abaqus, or COMSOL). The software solves a set of mathematical equations for each element, based on the applied loads and boundary conditions. This process calculates displacements, stresses, and strains across the entire structure. Large or complex models may require substantial computational power and time to solve.

6. Post-Processing and Results Interpretation

After the analysis, the results are visualized using the software's post-processing tools. Engineers can inspect graphical outputs, such as color-coded stress maps, deformation animations, and load distribution plots. These insights help identify areas where the structure may fail or require reinforcement. For example, if stress concentrations appear around the submarine's dive planes, engineers might redesign those components to distribute forces more evenly.

7. Validation and Refinement

The results from FEA are often validated through comparison with experimental data or analytical calculations. If discrepancies are found, the model may need to be refined by adjusting the mesh density, boundary conditions, or material properties. Multiple iterations may be required to achieve a design that meets performance and safety criteria.

<u>Step 3: Computational Fluid Dynamics (CFD) for Hydrodynamics</u>

Submarines must operate efficiently underwater, requiring streamlined shapes to reduce drag. Computational Fluid Dynamics (CFD) simulations model how water flows around the hull and other external components like rudders and propellers. CFD analysis helps:

- Optimize the hull shape to reduce resistance.
- Minimize cavitation, a phenomenon where bubbles form around propellers, which can increase noise and compromise stealth.
- Ensure stability and manoeuvrability in complex environments like littoral zones.

The results of these simulations guide design adjustments to improve hydrodynamics and propulsion efficiency.

Computational Fluid Dynamics (CFD) is a branch of fluid mechanics that uses numerical analysis and algorithms to solve and analyse problems involving fluid flows. Engineers and scientists use CFD to simulate the behaviour of fluids (such as air or water) when they interact with surfaces or objects, helping them understand complex phenomena like drag, turbulence, and pressure distribution. In submarine design, CFD is crucial for optimizing hull shapes, reducing drag, minimizing noise, and improving the efficiency of propulsion systems, which are essential for stealth and performance.

CFD replaces the need for expensive physical models by enabling engineers to perform virtual experiments, allowing them to predict how fluids behave under various conditions. This approach is widely used in industries such as aerospace, automotive, and marine engineering.

To perform CFD:

1. Geometry Creation

The first step is to create a 3D digital model of the object or system being analysed. This model could represent the submarine hull, propeller, or other components interacting with the fluid. CAD (Computer-Aided Design) software like SolidWorks or AutoCAD is often used to generate these models.

2. Meshing

Meshing is the process of dividing the geometry into a network of small elements, called a mesh. Each element represents a part of the fluid domain and the surface with which the fluid interacts. A finer mesh with more elements results in higher accuracy but requires more computational resources, while a coarser mesh reduces computation time but may compromise precision. Engineers must strike a balance between mesh resolution and computational cost.

3. Defining the Physics of the Problem

Next, the fluid properties and environmental conditions are specified. For example:

- **Fluid type** (e.g., water or air).
- **Density and viscosity** values, which describe the fluid's resistance to flow.
- **Boundary conditions**, such as flow velocity, temperature, or pressure at the inlets and outlets of the system.
- **Turbulence models**, which simulate chaotic fluid behaviour often encountered in real-world scenarios.

CFD can simulate different types of flows, including laminar flow (smooth and orderly) or turbulent flow (chaotic and irregular). The choice of turbulence model, such as the k-epsilon or Reynolds-Averaged Navier-Stokes (RANS) models, is critical for capturing the fluid behaviour accurately.

4. Solver Setup

The CFD solver software (such as ANSYS Fluent, OpenFOAM, or COMSOL) is configured to perform the calculations. This software uses mathematical models—specifically, the Navier-Stokes equations—to describe the motion of fluids. These equations are discretized and solved iteratively to calculate parameters like pressure, velocity, and temperature at each point in the mesh.

The solver must be configured carefully, including:

- Setting time steps for transient analysis (if studying time-dependent flows).
- Specifying the number of iterations the solver will perform to achieve convergence.
- Ensuring stability by monitoring for any errors or non-physical results during computation.

5. Running the Simulation

The simulation is executed, with the software computing fluid behaviour over time. Depending on the complexity of the model and the mesh size, simulations can take anywhere from a few hours to several days. Engineers typically use high-performance computers or cloud-based computing platforms to speed up the process.

6. Post-Processing and Visualization

Once the simulation completes, the results are visualized through post-processing tools. Engineers use software like ANSYS or ParaView to generate color-coded contour plots, vector fields, and 3D animations. These visualizations help in understanding:

- How fluid flows over the submarine's surface.
- Areas of high drag or pressure.
- Potential turbulence zones that could cause noise or instability.

Animations and plots allow engineers to see how changes in design affect flow behaviour and performance, helping them make data-driven design improvements.

7. Validation and Iteration

The final step involves validating the CFD results by comparing them with experimental data or analytical calculations. If discrepancies arise, engineers refine the model, adjust mesh density, or modify boundary conditions. This iterative process continues until the simulation closely matches expected real-world behaviour.

Step 4: Model-Based Systems Engineering (MBSE)

MBSE tools play a crucial role in the detailed design phase by enabling the integration of functional requirements into the design process. Using MBSE:

- Engineers create virtual models that represent the submarine's systems and subsystems, ensuring all parts work together seamlessly.
- The tool ensures traceability of requirements, linking design elements to the operational needs defined during earlier stages.
- It facilitates collaboration across teams by providing a unified framework for testing and validating system interactions.

MBSE tools also allow for iterative design, meaning that engineers can simulate the performance of individual systems and modify them before physical construction begins.

Model-Based Systems Engineering (MBSE) is an approach to systems engineering that emphasizes the use of models rather than traditional document-based methods to support the design, development, and management of complex systems. MBSE uses visual models to capture the structure, behaviour, and interactions of system components, which helps engineers understand, analyse, and manage the complexity inherent in modern projects. These models ensure that all stakeholders—engineers, designers, managers, and customers—have a consistent view of the system throughout its lifecycle.

MBSE is used across industries such as aerospace, automotive, and defence to design systems that require integration of hardware, software, and human components. By focusing on models, MBSE provides better traceability of requirements, improves system validation, and enhances collaboration across disciplines.

Performing MBSE includes:

1. Define System Requirements and Objectives

The MBSE process starts with identifying the requirements and objectives of the system. Engineers gather inputs from all stakeholders, such as customers, designers, and operators, to define what the system must achieve. These requirements form the foundation of the models, ensuring that every part of the system aligns with the desired functionality.

For example, in the case of a submarine, the requirements might include stealth capabilities, endurance, propulsion specifications, and safety features. These objectives are translated into system-level goals that guide the modelling process.

2. Create System Architecture Models

Next, system architecture models are developed to represent the overall structure and behaviour of the system. Tools such as SysML (Systems Modelling Language) are often used to build these models, which provide diagrams representing:

- **System components** and their interactions.
- **Behavioural models**, such as flow diagrams and activity charts.
- **Functional models** that describe how the system achieves its objectives.

The architecture models ensure that engineers have a clear understanding of how different components work together and help identify potential design conflicts early in the development process.

3. Use Simulation and Analysis for Validation

MBSE allows for simulations and analyses at various stages of the design. Engineers can use the models to:

- Simulate system behaviour under different scenarios.
- Perform trade-off analyses to evaluate multiple design options.
- Identify potential failures through fault analysis and mitigate risks.

For instance, in submarine development, engineers might simulate how the propulsion system interacts with control surfaces during manoeuvres. This ensures that design choices align with operational requirements and performance standards.

4. Ensure Traceability and Documentation

One of the key benefits of MBSE is that it provides traceability across the entire development lifecycle. Each component in the model is linked to the requirements it fulfills, ensuring that design decisions are traceable back to stakeholder needs. This makes it easier to manage changes, track progress, and ensure that all requirements are met.

In addition to traceability, the models serve as living documentation, which evolves as the system develops. This eliminates the need for large, static documents that quickly become outdated.

5. Collaborate Across Disciplines Using Shared Models

MBSE fosters collaboration by providing a common language and platform for different teams. Design, electrical, mechanical, and software engineers all work with the same models, ensuring consistency and reducing misunderstandings. Cloud-based platforms or collaborative MBSE tools allow for seamless sharing of models in real-time, promoting effective teamwork.

6. Perform Verification and Validation (V&V)

The MBSE process involves continuous verification and validation (V&V). Verification ensures that the models are built correctly, while validation confirms that the models represent the intended system accurately. Engineers use model-based simulations to validate that the system meets requirements under various conditions and identify areas where improvements are needed.

For example, the submarine's noise-reduction systems can be validated by simulating operational scenarios to ensure that they perform as expected in real environments.

7. Manage the Lifecycle with Continuous Updates

MBSE supports the entire system lifecycle, from design and development to deployment, maintenance, and decommissioning. As changes occur—whether due to new requirements, component upgrades, or operational adjustments—the models are continuously updated. This ensures that the system documentation remains accurate and relevant throughout its operational life.

<u>Step 5: Iterative Testing and Refinement</u>

Throughout the detailed design phase, virtual prototypes are created and tested. The design is refined based on the insights gained from FEA, CFD, and MBSE simulations. Engineers also review potential risks, like failure modes, and adjust specifications to mitigate them.

Regular design reviews are conducted, bringing together experts from multiple disciplines (e.g., structural, electrical, and mechanical engineering) to ensure every aspect is aligned with the operational goals and safety standards.

This stage also includes systems integration, where engineers ensure that all subsystems—such as propulsion, control, weapon systems, and sensors—work harmoniously. Testing individual components in isolation is essential to confirm compatibility and reliability. For nuclear submarines, additional focus is placed on reactor shielding, cooling systems, and nuclear safety protocols.

The manoeuvrability of submarines, particularly under intact and degraded conditions, is analysed using six-degree-of-freedom (6-DoF) simulations, which provide insights into the submarine's handling characteristics [669].

4. Material Selection and Hull Construction

The materials used in submarine construction are carefully chosen for strength, corrosion resistance, and low magnetic signatures to avoid detection. HY-80 and HY-100 steel are commonly used for pressure hulls due to their ability to withstand the immense pressures at depth. The construction of the hull involves welding the steel sections together with precision to ensure no weak points compromise the vessel's integrity.

Hull construction takes place in specialized shipyards, often using modular techniques. Submarines are built in sections, which are later assembled into larger modules. These modules contain internal components, such as propulsion units and living quarters, which are pre-fabricated to simplify the final assembly process.

5. Prototyping and Testing

The prototyping phase is essential for validating design choices and assessing performance in real-world conditions. This phase typically involves the construction of scale models or full-scale prototypes, which undergo rigorous testing in controlled environments, such as towing tanks or specialized test facilities. The testing phase evaluates various performance metrics, including hydrodynamic efficiency, stability, and manoeuvrability [24, 666].

Furthermore, the integration of advanced technologies, such as autonomous systems and advanced materials, is tested to ensure reliability and effectiveness. For instance, the use of composite materials in pressure hull design is explored to achieve optimal buoyancy and structural integrity under extreme conditions [667].

6. Production

Once the design has been validated through testing, the production phase commences. This phase involves the actual manufacturing of the submarine, where precision engineering and quality control are paramount. The production process must adhere to stringent safety and performance standards, particularly for nuclear-powered submarines, where safety protocols are critical due to the complexities involved in nuclear systems [670].

Lifecycle management practices are also implemented to ensure that the submarine remains operationally viable throughout its service life. This includes regular maintenance, upgrades, and modifications based on evolving technological advancements and operational requirements [671].

Figure 127: USS Arkansas (SSN-800) Under Construction. Huntington Ingalls Industries, CC BY-SA 4.0, via Wikimedia Commons.

5. Testing and Sea Trials

After the submarine is assembled, it undergoes extensive testing to validate its design and performance. These tests begin with static tests in dry docks, where engineers check the integrity of the hull and systems. Pressure tests are performed to ensure the submarine can withstand the forces it will encounter at operational depths.

The submarine then undergoes sea trials, where it is tested in real-world conditions. Trials assess propulsion, buoyancy control, noise levels, sonar performance, and navigational systems. Submarines designed for stealth are subjected to acoustic trials to measure their noise signature. Any issues encountered during trials are addressed, with necessary modifications made to ensure safety and performance.

6. Crew Training and Certification

Submarines are highly complex vessels, requiring specialized training for crew members. Before the submarine is fully operational, its crew must be trained to handle every system on board, including propulsion, navigation, and emergency procedures. This training is conducted through a combination of classroom instruction, simulators, and hands-on experience during sea trials. Nuclear-powered submarines also require additional certification to ensure compliance with nuclear safety standards.

Figure 128: A damaged Advanced SEAL Delivery System is converted to a training mini-sub at the Pearl Harbor Naval Shipyard and Intermediate Maintenance Facility. The submarine is used for hands-on training of multiple trades that work on submarines. Defense Visual Information Distribution Service, Public Domain, via GetArchive.

7. Commissioning and Deployment

Once the submarine has passed all tests and the crew is fully trained, the vessel is officially commissioned into service. During commissioning, the submarine is formally accepted by the navy and integrated into the operational fleet. At this point, the submarine becomes available for deployment, either for training exercises or real-world missions, such as patrolling territorial waters or conducting surveillance.

8. Maintenance and Lifecycle Management

Submarines require regular maintenance throughout their operational lives to ensure safety and reliability. Maintenance involves scheduled dry-docking, system upgrades, and repairs. Nuclear-powered submarines must also undergo reactor refuelling and overhauls at specific intervals. The submarine's lifecycle is carefully managed, with upgrades to electronics, weapons, and propulsion systems to keep the vessel operational for several decades.

9. Decommissioning and Disposal

When a submarine reaches the end of its operational life, it is decommissioned. For nuclear submarines, decommissioning involves safely dismantling the reactor and disposing of radioactive waste. This process is complex and requires specialized facilities to ensure environmental safety. Conventional submarines are either scrapped, converted for civilian use, or preserved as museum pieces.

A prime example of the submarine design and development process in action is the AUKUS SSN-AUKUS project, a collaborative effort between the United Kingdom, Australia, and the United States. This ambitious initiative aims to deliver a new class of conventionally-armed, nuclear-powered attack submarines to replace the UK's Astute-class submarines and introduce similar capabilities to Australia's Royal Navy.

The SSN-AUKUS program is currently progressing through the detailed design phase. This stage follows extensive conceptual work to align the requirements of the three nations involved, focusing on enhancing stealth, endurance, and operational effectiveness in contested maritime environments. The program emphasizes modular design, incorporating both American and British technologies, and aims to address evolving geopolitical challenges.

The design will use advanced computational tools to optimize hydrodynamics and ensure operational readiness across a variety of environments, including deep-sea operations and littoral zones. One of the project's goals is to create a future-proof platform capable of operating autonomously with minimal crew, integrating cutting-edge sensor suites and propulsion systems, and leveraging AI for navigation and data analysis.

In terms of construction, the project will rely heavily on BAE Systems, Babcock, and Rolls-Royce for critical components and infrastructure, such as nuclear reactors. Initial construction is expected to occur in Barrow-in-Furness, UK, with Australia simultaneously building up its industrial capacity to support future domestic submarine manufacturing. This effort includes a focus on workforce development, with Australian personnel training on UK and US vessels to gain operational experience with nuclear-powered systems.

As part of the testing and trials phase, the submarines will undergo extensive sea trials, where every component—from propulsion to stealth—will be evaluated under real-world conditions. Following these trials and necessary refinements, the first SSN-AUKUS submarines are scheduled to enter service with the Royal Navy by the late 2030s, with Australia receiving its variants in the early 2040s.

The AUKUS program also includes lifecycle management plans to ensure the submarines remain operationally relevant for decades. This approach includes continuous upgrades to electronics, weapon systems, and reactor technologies, along with maintenance schedules to optimize vessel availability.

This submarine program demonstrates the intricate and collaborative nature of modern submarine development, combining expertise from multiple nations to address both technical challenges and strategic objectives in the evolving global security landscape. The SSN-AUKUS project not only advances submarine technology but also strengthens military alliances between Australia, the UK, and the US, ensuring they remain capable of countering future maritime threats effectively.

Development Timeframes and Human Resource Requirements

The production timeline for submarines, whether conventional diesel-electric or nuclear-powered, is influenced by various factors including design complexity, technological requirements, and the specific capabilities desired. Generally, the entire process from initial design to commissioning can take anywhere from 6 to 15 years. This timeframe encompasses several phases, including conceptual design, detailed engineering, construction, testing, and final delivery [240].

For conventional diesel-electric submarines, the production timeline tends to be on the shorter end of this spectrum, often around 6 to 10 years. This is primarily due to their simpler design and fewer technological complexities compared to nuclear-powered submarines. In contrast, nuclear submarines typically require a more extended development period, often reaching up to 15 years or more. This is attributed to the advanced technology involved in nuclear propulsion systems, stringent safety protocols, and the need for extensive testing and certification processes [240, 672].

The human resources required for each phase of submarine production also vary significantly. The design phase necessitates a multidisciplinary team of engineers, including naval architects, mechanical engineers, and systems engineers, often comprising hundreds of personnel. During the construction phase, this number can increase dramatically, as shipyards employ thousands of workers, including skilled tradespeople such as welders, electricians, and quality assurance inspectors. The testing phase, which is critical for ensuring the submarine meets all operational specifications, requires additional specialized personnel, including those with expertise in systems integration and underwater acoustics [240, 672, 673].

Moreover, the complexity of the submarine's mission profile can further influence the required human resources. For instance, submarines designed for advanced stealth and combat capabilities may require additional personnel for specialized systems such as sonar and weapons systems, thus increasing the overall workforce needed throughout the production cycle [240, 672].

The following provides a breakdown of each phase's estimated duration and the human resources typically involved.

1. Design and Planning

- **Duration:** 1 to 3 years
- **Key Personnel:**
    - Naval architects
    - Systems engineers
    - Computational analysts (CFD and FEA experts)
    - Project managers
    - Government and military stakeholders

This phase involves defining operational requirements, planning the system's architecture, and running simulations to optimize the hull and propulsion systems. Collaboration between military forces and contractors begins here, setting the groundwork for the remaining phases.

2. Modular Construction and Hull Fabrication

- **Duration:** 3 to 7 years
- **Key Personnel:**
    - Structural engineers
    - Welders and steelworkers

    - Electricians
    - Shipyard workers
    - Quality assurance specialists

This is the most labour-intensive phase. Workers fabricate individual hull sections and assemble internal systems like propulsion units, ballast tanks, and life-support systems. Modular construction allows simultaneous work on different sections of the submarine, speeding up the process.

3. Systems Integration

- **Duration:** 2 to 4 years (often overlaps with construction)
- **Key Personnel:**
    - Integration engineers
    - Control systems specialists
    - Electrical engineers
    - Software developers
    - Technicians for sonar and weapon systems

Engineers ensure that all the submarine's subsystems—propulsion, navigation, sonar, and weapon systems—are integrated and operate as intended. Extensive testing at this stage ensures that all components function harmoniously.

4. Testing and Sea Trials

- **Duration:** 1 to 2 years
- **Key Personnel:**
    - Marine engineers
    - Test pilots and operators
    - Naval officers
    - Technicians for acoustic analysis and buoyancy tests

The submarine undergoes dockside and sea trials to test performance under real-world conditions. Stealth systems are assessed during acoustic trials, and engineers fine-tune the propulsion and control systems. Any malfunctions are corrected before moving to commissioning.

5. Crew Training and Certification

- **Duration:** 6 months to 1 year (sometimes overlaps with testing)
- **Key Personnel:**
    - Naval instructors

- Simulation specialists
- Reactor engineers (for nuclear submarines)
- Command officers

Submarine crews are trained using simulators and real equipment to ensure they are familiar with every system on-board. Nuclear submarines require additional training for handling reactors, safety protocols, and emergency procedures.

6. Commissioning and Deployment

- **Duration:** 6 months
- **Key Personnel:**
  - Naval officers
  - Logistics teams
  - Government representatives

Commissioning involves finalizing the submarine for active service, conducting ceremonies, and integrating the vessel into the fleet. Logistic teams coordinate supplies and deployment strategies to ensure the submarine is mission-ready.

7. Maintenance and Lifecycle Management

- **Ongoing (up to 30 years or more)**
- **Key Personnel:**
  - Maintenance engineers
  - Technicians for periodic upgrades
  - Nuclear engineers (for refuelling reactors)
  - Logistics support teams

Throughout its operational life, the submarine undergoes routine maintenance and system upgrades to ensure it remains effective. Nuclear submarines also require reactor refuelling every **10-15 years**.

8. Decommissioning and Disposal

- **Duration:** 1 to 2 years
- **Key Personnel:**
  - Nuclear disposal specialists (for reactors)
  - Environmental engineers
  - Shipyard workers

At the end of its service life, the submarine is dismantled or recycled. Nuclear submarines require special handling to safely dispose of radioactive components. Conventional submarines may be scrapped, repurposed, or turned into museum exhibits.

The production process for submarines requires thousands of specialized workers across multiple disciplines, including structural engineering, electronics, software, and nuclear science. The entire process can take over a decade, with ongoing upgrades and maintenance extending the submarine's lifespan to several decades. Modular construction and integration allow for some phases to overlap, ensuring smoother workflows and reducing delays.

Virginia-Class Submarine Development

The Virginia-class submarine development offers a detailed case study on the intricate, multi-stage production process involved in building modern nuclear-powered submarines. The construction and assembly of these submarines are handled by two major shipyards: General Dynamics Electric Boat and Newport News Shipbuilding. Each yard contributes specific modules of the submarine, with one focusing on sections such as the stern and torpedo room and the other on the engine room and control room.

The Virginia-class submarine program represents a significant advancement in the U.S. Navy's undersea capabilities, initiated to replace the aging Los Angeles-class submarines. Officially launched in 1998, the program aimed to enhance the Navy's operational effectiveness in various mission profiles, including anti-submarine warfare, intelligence gathering, and special operations support. The first submarine of this class, USS Virginia (SSN-774), was commissioned in 2004, marking a pivotal shift in submarine technology and design [674].

The production of each Virginia-class submarine typically takes around five to seven years, including construction, sea trials, and testing phases.

Figure 129: The Virginia-class attack submarine Minnesota (SSN 783) is under construction at Huntington Ingalls Newport News Shipbuilding. U.S. NAVY, Public Domain, via Picryl.

The Virginia-class submarines are constructed using a modular design approach, which facilitates an iterative block-building strategy. This method allows for the introduction of technological advancements with each new block, enhancing the submarines' capabilities over time. For instance, Block III introduced new payload tubes, significantly increasing the versatility of the submarine's armament, while Block V incorporated the Virginia Payload Module (VPM), which expands missile-carrying capacity to offset the retirement of Ohio-class submarines [674, 675]. The modular construction not only streamlines the building process but also allows for concurrent production of different submarine sections, thereby optimizing the overall timeline from construction to commissioning, which typically spans five to seven years [676].

The U.S. Navy's strategy for the Virginia-class program includes plans for continuous production into the 2030s, with an expected output of approximately two submarines per year to meet operational demands. As of 2024, more than 26 Virginia-class submarines have been commissioned, demonstrating the program's success in enhancing the Navy's undersea fleet [671, 674]. Furthermore, the Navy is already planning for the next generation of attack submarines, designated SSN(X), with procurement anticipated to begin around 2040, ensuring that the U.S. remains at the forefront of naval capabilities [674].

The development process of the Virginia-class submarines emphasizes integrated product and process development (IPPD), which ensures that design considerations are aligned with construction realities. This approach incorporates inputs from various stakeholders, including engineers and naval operators, early in the design phase, thereby enhancing the overall efficiency and effectiveness of the program [675]. The use of commercial off-the-shelf (COTS) technologies has also played a crucial role in reducing costs and facilitating system upgrades throughout the submarines' operational life [677].

Budgeting, Scheduling, and Risk Management

Project Budget

Developing a budget for a submarine construction project is indeed a highly complex process that necessitates meticulous planning, accurate forecasting, and effective coordination among multiple stakeholders. The intricacies involved in this process are underscored by the need to balance operational, safety, and technological requirements against the constraints imposed by national defence budgets.

Stakeholder engagement plays a pivotal role in this budgeting process. Early involvement of stakeholders during the project definition phase is essential to ensure that all parties have a clear understanding of the project's scope and objectives, which can significantly influence budget accuracy and project success [678]. The complexity of submarine construction projects often requires contractual agreements that delineate responsibilities and expectations among stakeholders, thereby facilitating smoother collaboration and reducing the likelihood of cost overruns [678]. Furthermore, effective stakeholder management can lead to improved project performance by fostering better relationships and reducing uncertainties during the pre-construction phase [679].

Accurate cost estimation is another critical aspect of budgeting for submarine construction. Various methodologies exist for estimating project costs, and the choice of method can significantly impact the reliability of budget forecasts. For instance, probabilistic techniques have been shown to enhance the accuracy of cost assessments by accounting for uncertainties inherent in construction projects [680]. Moreover, the integration of advanced technologies, such as Building Information Modelling (BIM), can streamline the cost estimating process by providing reliable data that aligns with the project scope and level of development [681]. This is particularly relevant in the context of submarine construction, where the complexity and uniqueness of each project necessitate a robust estimation framework to avoid budget discrepancies and ensure financial viability [682].

The design phase also significantly influences budget outcomes. Research indicates that dedicating adequate time and resources to the early design stages can lead to more accurate budget estimates, as it allows for a thorough understanding of the project's complexity and scope [683]. This is particularly important in submarine projects, where technological advancements and operational requirements can introduce additional layers of complexity that must be accounted for in the budget [684]. Furthermore, the application of advanced cost control methods, such as those based on artificial intelligence, can enhance the precision of budget forecasts by analysing various influencing factors and providing insights into potential cost overruns [685].

1. Defining Project Requirements and Scope

The process begins with defining the scope and mission requirements of the submarine. This is done in collaboration between the naval force, defence contractors, and government agencies. Requirements include specifications for stealth, speed, payload, endurance, and operating environments (e.g., Arctic or littoral zones).

- These requirements are translated into technical specifications, such as materials, propulsion systems, and onboard weaponry.
- The project scope helps to outline the complexity and expected timeline, forming the basis for the budget.

This phase involves close collaboration between multiple stakeholders, including naval forces, defence contractors, and government agencies. Their combined input ensures that the vessel aligns with the strategic needs of the military while accounting for constraints such as cost and technological feasibility.

The mission requirements typically outline the roles the submarine will perform, such as anti-submarine warfare, intelligence gathering, or missile deployment. Other considerations include operating environments—whether the submarine will function primarily in littoral zones (shallow coastal waters) or open oceans, or whether it requires special capabilities for Arctic operations. The mission profile informs the technical aspects needed, such as stealth levels, speed, endurance, and payload capacity.

These requirements are translated into technical specifications. For example, endurance requirements influence whether the submarine will use nuclear propulsion for long missions or diesel-electric propulsion for shorter durations. Stealth considerations will dictate the materials used, such as HY-80 steel or advanced composites with low magnetic signatures, and shape decisions to minimize acoustic signatures. The payload capacity and weapons systems are also defined here, determining the type and number of missiles, torpedoes, or unmanned vehicles the submarine will carry.

The project scope lays the foundation for the entire lifecycle of the project, detailing the scale and complexity of the submarine's construction. It outlines the major milestones, such as design phases, system integration, sea trials, and delivery timelines. This helps in formulating an initial timeline and budget estimate, ensuring that resources are allocated efficiently and timelines are realistic.

In essence, this phase provides the blueprint for success by ensuring that the technical features of the submarine align with the strategic goals of the navy. A clear and well-defined scope minimizes risks, prevents scope creep during later stages, and ensures that all stakeholders are aligned before significant investments are made.

2. Estimating Costs by Component

A submarine project is divided into major cost categories, such as:

1. **Hull construction** (e.g., materials like HY-80 steel)
2. **Propulsion systems** (nuclear reactors, diesel-electric engines)
3. **Weapons systems and payload** (torpedoes, missile systems)
4. **Electronics and sonar systems** (navigation, communications)
5. **Crew accommodations** and life-support systems

Each component requires detailed cost estimation, based on market prices, vendor contracts, labour costs, and past project data. These estimates are aggregated to form an initial cost projection.

The **cost estimation process for submarine construction** is a detailed and methodical effort, dividing the project into major components. Each of these components represents a significant portion of the overall budget, and their costs must be calculated accurately to ensure the project remains on schedule and within financial limits. Below is an in-depth breakdown of how costs are estimated for each of these key components.

- **Hull Construction:** Hull construction involves the fabrication of the pressure hull from specialized materials, such as HY-80 or HY-100 steel, which provide the necessary strength to withstand high underwater pressures. The cost includes not only the raw materials but also the welding, shaping, and assembly of hull sections, which require skilled labour. Submarine hulls are often fabricated in modular sections, and each module must meet strict tolerances to ensure structural integrity. Historical data from similar projects and labour costs at shipyards are used to estimate the expense for this critical component.

- **Propulsion Systems:** The propulsion system represents one of the largest cost drivers. For nuclear-powered submarines, this includes the nuclear reactor, cooling systems, and steam turbines, along with safety mechanisms. In diesel-electric submarines, the costs are associated with diesel engines, batteries, and electric motors. Each propulsion system has its own set of challenges. For example, nuclear reactors require extensive regulatory oversight, specialized components, and long-term fuel storage, all of which add to the cost. These costs are estimated based on vendor contracts, current market prices, and historical data from previous submarines with similar propulsion systems.

- **Weapons Systems and Payload:** Submarines carry a variety of weapons systems and payloads, such as torpedoes, cruise missiles, and ballistic missiles. The cost of weapons is influenced by the number of weapons bays, torpedo tubes, and specialized missile silos required for the design. In the case of advanced submarines, the integration of launch control systems and automated weapon handling adds to the cost. This category also considers unmanned underwater vehicles (UUVs) if they are included as part of the payload, reflecting the growing trend toward autonomous systems.

- **Electronics and Sonar Systems:** The electronics onboard a submarine play a crucial role in navigation, communication, and combat systems. These include advanced sonar arrays, radar, satellite communication links, and fire control systems. Estimating the cost of these systems requires consultation with specialized vendors and accounting for custom software development and integration. Costs may also reflect maintenance agreements with suppliers to ensure the systems remain operational over the submarine's lifecycle.

- **Crew Accommodations and Life-Support Systems:** Designing a submarine to accommodate crew members for long missions involves significant expenses. Life-support systems include air purification systems, oxygen generators, fresh water supplies, waste management, and temperature control. In addition, the layout of living quarters, dining areas, medical facilities, and recreation spaces impacts the final cost. These elements must be optimized to ensure comfort and operational efficiency, especially on nuclear submarines where missions may last several months without surfacing.

Aggregating Component Costs for the Initial Projection

Each component's estimated cost is aggregated to form the initial budget projection. This involves not only direct expenses but also labour costs, vendor pricing, and long-term maintenance agreements. Project managers leverage historical data from past submarine programs (such as Virginia-class or Columbia-class submarines) to ensure accurate estimates. They also consider inflation and potential supply chain risks to avoid future cost overruns.

The aggregated cost estimates are presented as a comprehensive budget proposal, which forms the basis for securing government funding. This initial budget is also used to set contingency funds to cover unexpected expenses during construction. Regular monitoring and financial tracking throughout the project ensure that the project remains on target, with adjustments made as necessary.

3. Use of Cost Models and Historical Data

Budget planners rely on cost models and historical data from previous submarine programs (e.g., Virginia-class). These models help predict expenses for materials, labour, testing, and maintenance. For example:

- Nuclear submarines typically involve higher initial capital costs due to reactors and specialized components.

- Programs like the Virginia-class or Columbia-class provide useful benchmarks to estimate procurement and maintenance costs for new submarine designs.

Budget planners rely heavily on cost models and historical data from previous submarine programs to predict expenses for new submarine designs. These models are essential tools for estimating not only construction costs but also long-term operational and maintenance expenses. Programs such as the Virginia-class and Columbia-class submarines provide valuable insights into how budgets are formed and adjusted over time to meet both financial constraints and operational needs.

Capital Costs of Nuclear Submarines

Nuclear submarines generally require higher upfront capital investments compared to diesel-electric models. The primary reason for these elevated costs lies in the inclusion of nuclear reactors and their associated safety and regulatory systems. Nuclear reactors also demand specialized shielding, cooling systems, and skilled labour for construction, all of which increase the project's complexity and cost. Planners must account for both the reactor's lifecycle costs (including future refuelling) and the expense of maintaining strict nuclear safety protocols. These elements are incorporated into the cost models to create accurate predictions for future nuclear-powered submarines.

Benchmarking with Previous Submarine Programs

Historical data from Virginia-class and Columbia-class submarines serve as benchmarks for estimating procurement and maintenance costs for new submarine projects. For instance, the Virginia-class program, which relies on modular construction, helps planners understand the cost-saving benefits of building multiple submarines in parallel. Data from these programs provide insights into the price of modular hull sections, sonar systems, and propulsion components, which can be applied when planning the budget for new submarines with similar configurations.

Additionally, submarine programs are typically designed to last for several decades, requiring predictions not just for construction costs but also for long-term maintenance, upgrades, and overhauls. These historical models help planners estimate future operational expenses, such as reactor refuelling for nuclear-powered submarines, as well as periodic technological upgrades needed to maintain combat effectiveness.

Cost Modelling Techniques

Planners use parametric cost models that incorporate data from earlier submarine programs. These models take into account variables such as labour hours, material costs, and system complexity to forecast the financial requirements of new designs. For example, if the historical data from the Virginia-class submarines show an average construction cost of $4.5 billion per unit, planners use this figure as a baseline, adjusting for inflation, technology improvements, and scope changes for the new project.

These predictive models are also essential for identifying potential cost overruns early in the planning process, allowing for contingency funds to be set aside. Historical data enables planners to recognize areas prone to delays or increased expenses, such as new technologies or unforeseen supply chain disruptions.

4. Factoring in Contingencies and Risk Management

Given the high degree of uncertainty in complex defence projects, budgets include contingency funds for unexpected costs, such as delays, material shortages, or design changes. Planners use risk analysis tools to identify potential risks and allocate funds accordingly.

Factoring in Contingencies and Risk Management in Submarine Budgeting

In complex defence projects like submarine construction, contingency funds and risk management practices are essential components of the budgeting process. These large-scale projects are inherently uncertain, with potential risks arising from technological challenges, supply chain issues, delays, or evolving mission requirements. Factoring in contingencies ensures that the project can accommodate unforeseen events without exceeding the allocated budget or compromising quality.

Contingency Funds for Unforeseen Costs

Given the high degree of uncertainty in submarine construction, budgets typically include contingency funds to cover unexpected expenses. These funds serve as financial buffers, addressing costs that were not identified in the initial estimates, such as:

- Delays in delivery due to shipyard bottlenecks or labour shortages.
- Material shortages or price fluctuations in critical components, such as specialized steels or advanced electronics.
- Design changes introduced during construction, either due to technical issues or updated mission requirements.

The amount allocated to contingencies varies based on the project's complexity and timeline, but it is usually expressed as a percentage of the total budget, typically ranging between 10-20%. Planners may increase the contingency percentage for projects with innovative technologies or novel components, as these introduce additional uncertainties.

Risk Analysis and Identification Tools

Risk management plays a critical role in determining how much contingency funding is necessary. Project managers rely on risk analysis tools to systematically identify potential risks and their likelihood. Some common tools used include:

- Risk matrices to categorize risks by their probability and impact on cost or schedule.
- Monte Carlo simulations, which generate multiple scenarios to predict potential cost variations over time.
- Expert panels that assess the technical, operational, and financial risks associated with new technologies or complex components.

These tools help planners prioritize high-risk areas, enabling them to allocate contingency funds effectively to areas most likely to experience overruns.

Dynamic Risk Management During Construction

The dynamic nature of submarine construction requires continuous risk monitoring throughout the project's lifecycle. As the project progresses through various phases—design, construction, testing, and commissioning—new risks may emerge, requiring adjustments

Risk Management Throughout the Project Lifecycle

Managing risks dynamically is essential throughout the submarine's development and construction. As the project progresses, new risks can emerge, necessitating ongoing adjustments to the contingency fund. For example, issues identified during early testing phases may require changes to designs, systems, or materials. Project managers conduct risk reassessments periodically to reflect current conditions and reallocate resources if needed. This ensures that potential issues are addressed proactively, minimizing disruptions during construction.

Additionally, real-time monitoring tools are used to track critical milestones and detect deviations from the project plan. In cases where multiple contractors and shipyards are involved—such as with the Virginia-class submarine program—communication between stakeholders becomes vital. Early warning systems provide alerts if labour shortages, supply chain disruptions, or technical challenges arise, allowing time for risk mitigation actions.

Allocating Contingency Funds Strategically

Contingency funds are not applied uniformly across the budget; instead, they are allocated strategically based on identified risk areas. For example:

- Advanced technologies or untested components might receive a larger share of contingency funds due to higher uncertainty.
- Critical-path activities—those that directly affect the project's timeline—may be allocated additional funds to avoid cascading delays if issues occur.
- Regulatory and compliance risks, particularly in nuclear-powered submarines, may require additional budget buffers to accommodate safety inspections or design adjustments based on evolving safety standards.

The goal is to minimize the financial impact of risks while keeping the project on schedule. If risks do not materialize, the unused contingency funds can be reallocated to other parts of the project or reduce the overall project cost.

5. Procurement Planning and Supplier Contracts

The procurement phase involves engaging with contractors and suppliers to acquire components. Naval shipyards, such as General Dynamics Electric Boat and Huntington Ingalls Industries, provide estimates for labour and material costs. Long-term contracts may offer bulk discounts, but they also require careful cash flow management.

Some costs include:

- Sole-source suppliers for specialized components (e.g., sonar arrays, reactor parts).
- Labor costs for skilled workers, such as welders, engineers, and electricians.
- Shipyard facilities and infrastructure costs for modular construction.

The procurement phase of submarine construction is a critical part of the development process, requiring coordination between shipyards, contractors, and suppliers to acquire the specialized components and systems needed. This phase ensures that the necessary materials, labor, and infrastructure are available to meet production timelines. Naval shipyards, such as General Dynamics Electric Boat (GDEB) and Huntington Ingalls Industries (HII), take the lead in this phase, working closely with government agencies and vendors to secure contracts, manage costs, and ensure efficient cash flow throughout the construction process.

Engagement with Contractors and Suppliers

During procurement, shipyards engage with a variety of suppliers, many of whom provide highly specialized components that are essential for submarine operations. These may include:

- Sonar arrays, which are crucial for navigation and threat detection.
- Nuclear reactor parts used in nuclear-powered submarines, which require precise engineering and compliance with safety standards.

Some of these components can only be obtained from sole-source suppliers—vendors that are the exclusive providers of a specific technology or part. Engaging with sole-source suppliers necessitates careful negotiation and planning, as delays or disruptions in these partnerships can severely impact the entire project timeline.

Labor Costs and Skilled Workforce

The procurement phase also involves estimating and managing labour costs. Submarine construction requires a skilled workforce, including welders, engineers, electricians, and machinists. These workers are essential not only for fabricating the submarine's hull but also for assembling propulsion systems, weapons platforms, and onboard electronics.

Labor-intensive tasks, such as welding HY-80 steel for the pressure hull, demand expertise and precision. Therefore, labour costs must account for the availability of skilled personnel, potential overtime, and union agreements. Shipyards often employ thousands of workers for submarine projects, with employment spanning several years from construction to sea trials.

Shipyard Facilities and Modular Construction Costs

Modern submarines, such as those from the Virginia-class program, are built using modular construction techniques. In this method, different sections of the submarine, or modules, are constructed independently and then assembled at the shipyard. This approach speeds up the construction process and allows multiple teams to work simultaneously on different parts of the submarine.

Shipyards such as GDEB and HII invest heavily in infrastructure to support this process. This includes specialized dry docks, fabrication facilities, and transport systems to move large modules. The costs associated with maintaining and upgrading these facilities are significant and are factored into the procurement budget.

Long-Term Contracts and Cash Flow Management

During the procurement phase, shipyards often negotiate long-term contracts with suppliers to ensure a steady flow of materials and components. These contracts offer several advantages, such as bulk discounts on raw materials, including steel and composite materials. However, managing these contracts requires careful oversight to ensure the project maintains cash flow throughout the construction timeline.

Cash flow management is particularly important for multi-year projects, as delays in payment from government agencies or suppliers could slow production. Shipyards use financial planning tools to schedule payments and ensure that funds are available when needed to avoid bottlenecks in the supply chain.

6. Phased Budget Development

The budget is typically divided into phases corresponding to key milestones:

1. Research and Development (R&D): Covers early design and feasibility studies.

2. Construction: Includes material procurement, labour, and assembly.

3. Testing and Sea Trials: Allocates funds for trials, repairs, and adjustments.

4. Commissioning: Final checks and crew training.

5. Operations and Maintenance: Ensures lifecycle support, upgrades, and refuelling for nuclear submarines.

In a submarine construction project, the budget is divided into phases, corresponding to the major milestones throughout the vessel's lifecycle. This phased budget development approach ensures that funds are allocated appropriately at each stage, reducing the risk of overspending and ensuring that the project stays on track. Each phase has distinct financial needs, from research and development through to commissioning, operations, and long-term maintenance. This structure provides better financial oversight and allows for adjustments as the project progresses.

1. Research and Development (R&D): The R&D phase focuses on early design work, feasibility studies, and proof-of-concept testing. During this phase, funds are allocated for conceptual design, mission analysis, and preliminary technical studies. Naval engineers collaborate with defence contractors to explore new technologies and assess the feasibility of meeting mission requirements within budget constraints. Computational modelling and simulations—such as Computational Fluid Dynamics (CFD) and Finite Element Analysis (FEA)—are often used during this phase.

This phase is essential for identifying potential challenges early, reducing risks in later stages. R&D can represent a significant percentage of the total budget, especially for nuclear submarines, as they require research into advanced materials, propulsion technologies, and weapons systems.

2. Construction: The construction phase is the most capital-intensive part of the project. It involves procuring materials, such as HY-80 steel for the pressure hull, and securing components like propulsion systems and sonar equipment. Labor costs account for a large portion of the budget, as shipyards employ skilled workers (e.g., welders, engineers, and electricians) to assemble the submarine's modules.

Construction uses a modular approach, with different sections built separately and then integrated into the final vessel. Effective budget management during this phase involves long-term contracts with suppliers to ensure material availability and prevent cost fluctuations. The construction budget also includes funds for infrastructure costs, such as dry docks and specialized shipyard facilities.

3. Testing and Sea Trials: Once assembled, the submarine undergoes extensive testing and sea trials to validate performance. This phase requires budget allocations for repairs and adjustments, as issues are often identified during trials. Propulsion, buoyancy, sonar systems, and onboard electronics are all tested in real-world conditions, with funds allocated for troubleshooting and redesign if necessary.

This phase also involves acoustic signature testing to ensure that the submarine meets stealth requirements. Submarines designed for Arctic or deep-sea operations may require specialized tests, adding to the budget. The goal is to identify and resolve issues before commissioning, ensuring that the vessel is safe and combat-ready.

4. Commissioning: The commissioning phase marks the official entry of the submarine into service. Budget allocations during this phase focus on final checks, certifications, and crew training. Specialized training programs ensure that the crew can operate the submarine's systems safely and effectively. For nuclear submarines, additional certification is required to comply with nuclear safety protocols.

Funds are also allocated for equipping the submarine with initial payloads, such as torpedoes or missiles, and conducting final onboard inspections. At the end of this phase, the submarine becomes part of the operational fleet, ready for deployment.

5. Operations and Maintenance: Once the submarine is operational, it enters the operations and maintenance (O&M) phase, which ensures its continued performance over the years. This phase requires ongoing funding for routine maintenance, system upgrades, and repairs. Nuclear submarines, in particular, need reactor refuelling and overhauls at set intervals, which represent a significant portion of the O&M budget.

The O&M budget also accounts for crew training and mission-specific upgrades to keep the submarine technologically up to date. For example, new sonar systems or upgraded communication equipment may be integrated as needed. The lifecycle of a submarine can span several decades, and regular maintenance ensures the vessel remains reliable and combat-ready throughout its service life.

7. Approval and Government Oversight

Once the budget is developed, it is submitted to the appropriate government bodies for approval. In the U.S., this involves multiple stages of review by the Department of Defense (DoD), Congress, and the Navy. Budget planners must ensure that the proposed submarine fits within broader defence spending priorities.

- Multi-year procurement strategies may be employed to spread costs over several fiscal years.
- Budgets must align with long-term shipbuilding plans to maintain fleet readiness and capability.

After the budget for a submarine construction project is developed, it undergoes a rigorous approval process involving multiple layers of government oversight. In the U.S., this process is essential to ensure that the project aligns with the nation's defence priorities and financial constraints. The Department of Defense (DoD), Congress, and the U.S. Navy each play critical roles in reviewing, approving, and managing the budget over the project's lifecycle. This oversight ensures that taxpayer money is spent effectively and that the submarine program supports the strategic goals of the armed forces.

The approval process begins within the Department of Defense, where the submarine budget is reviewed to ensure it aligns with broader military objectives. The DoD evaluates whether the proposed submarine project supports key missions, such as deterrence, sea control, and undersea warfare capabilities. Once approved by the DoD, the budget proposal is submitted to Congress as part of the annual defence budget request.

Congress then holds hearings and deliberations, where lawmakers review the cost, scope, and necessity of the submarine program. They assess whether the project fits within the overall defence spending priorities and whether it aligns with national security strategies. Budget planners may need to make revisions based on Congressional feedback before the budget is approved and allocated for spending.

The U.S. Navy plays an integral role throughout this process, ensuring that the submarine's design and operational capabilities meet the Navy's long-term goals. The Navy also communicates the operational needs to Congress, justifying the importance of the program.

Given the high cost and long timelines associated with submarine construction, the project budget is often divided across multiple fiscal years through multi-year procurement strategies. This approach allows the Navy to spread costs over time, easing the financial burden on annual defence budgets. For example, the Virginia-class submarine program utilizes multi-year procurement contracts, which help stabilize production schedules and lower costs by providing suppliers with guaranteed demand over several years.

Multi-year strategies also allow shipyards to optimize labour and resources, making the construction process more efficient. However, these contracts require Congressional approval, as they involve long-term financial commitments that must be justified within future defence budgets.

Submarine budgets must align with the Navy's long-term shipbuilding plans to ensure the fleet remains operational and capable. The Navy's 30-year shipbuilding plan outlines the number and types of submarines needed to maintain fleet readiness and strategic deterrence. Any new submarine project, such as the Columbia-class ballistic missile submarine, must fit within this broader framework to maintain the desired number of vessels over the long term.

The budget approval process also considers maintenance and modernization plans for existing submarines. As new vessels are commissioned, older submarines are decommissioned or upgraded, ensuring the Navy maintains an optimal mix of capabilities. Planners must carefully balance the funding for new construction with the costs of maintaining and upgrading the current fleet.

8. Monitoring and Adjustments During Construction

During construction, budget tracking and monitoring are essential. Project managers use cost-tracking software and performance metrics to ensure the project stays on schedule and within budget. If costs exceed expectations, contractors may negotiate additional funds or adjust project timelines.

During the construction phase of a submarine project, budget tracking and continuous monitoring are essential to ensure the project stays on schedule and within financial constraints. Submarine construction is highly complex, involving multiple components, contractors, and phases, which makes precise monitoring crucial. Project managers use specialized tools and processes to track costs, monitor performance metrics, and ensure compliance with agreed timelines. These systems allow for early identification of potential risks, enabling timely adjustments to avoid delays and cost overruns.

<u>Cost-Tracking Software and Performance Metrics</u>

Effective cost management during construction requires the use of cost-tracking software. These tools enable project managers to monitor real-time expenses against the approved budget and identify variances before they escalate. The software integrates data from various sources—such as labour costs, material purchases, and contractor invoices—and provides comprehensive dashboards that display key financial metrics.

Performance metrics such as earned value management (EVM) are often employed to assess progress. EVM compares the value of work completed to the planned budget and timeline, helping managers determine whether the project is on track. Metrics like cost variance (CV) and schedule variance (SV) highlight areas where spending or progress deviates from expectations, prompting immediate corrective actions.

Addressing Cost Overruns

Despite careful planning, cost overruns can occur due to various factors, such as material shortages, labour inefficiencies, or unexpected technical challenges. In such cases, contractors may request additional funds through change orders or renegotiate certain terms of the contract. If the budget cannot accommodate additional expenses, the project team may need to adjust timelines, prioritize certain components, or delay non-essential phases of the construction.

Project managers also reallocate contingency funds to address unforeseen issues. Regular financial reviews help ensure that the remaining contingency budget is managed carefully. In some cases, changes to procurement schedules or labour shifts are required to address bottlenecks without inflating costs further.

Collaboration Between Stakeholders

Monitoring during construction requires close collaboration between contractors, suppliers, and naval oversight teams. Weekly or monthly meetings allow stakeholders to review progress, address challenges, and make data-driven decisions. Any deviations from the original plan must be documented, with justifications provided to the project's oversight committee. Contractors must also submit progress reports detailing how funds are being utilized, ensuring accountability throughout the project lifecycle.

Flexibility for Continuous Adjustments

Given the long timelines associated with submarine construction, flexibility is built into the budget and project schedule to accommodate adjustments. For example, if a new technology becomes available or certain components face delays, the project team may need to reorganize priorities to stay on track. Some parts of the construction process may run in parallel to compensate for lost time, requiring revised labour allocations and updated timelines.

Modular construction, used in programs like the Virginia-class submarine, helps with such adjustments. Modular sections can be built independently and assembled later, allowing for efficient integration even if certain modules face delays. The ability to shift resources between modules ensures smoother adjustments during construction.

9. Reporting and Auditing

Governments conduct audits to ensure transparency and accountability. Submarine programs like the Virginia-class undergo periodic reviews to assess spending efficiency and identify areas for improvement. These reviews help avoid cost overruns and ensure funds are spent responsibly.

The reporting and auditing phase of submarine construction is essential for maintaining accountability, transparency, and efficiency in the use of public funds. Given the substantial financial commitments involved in building advanced submarines, such as those within the Virginia-class program, governments ensure that all stages of the project undergo regular audits and reviews. This process allows for the detection of inefficiencies, the prevention of cost overruns, and ensures that funds are allocated and spent responsibly.

Government audits serve as a financial oversight mechanism to ensure that public funds are managed in compliance with relevant laws, policies, and regulations. For submarine programs, this is especially important as these projects often span several years and involve multiple stakeholders, including defence contractors, shipyards, and government agencies.

Audits verify that each party meets its financial obligations, ensuring that contract terms are honoured and that no funds are misused.

Additionally, audits help ensure that funding aligns with intended outcomes, confirming that the allocated budget contributes directly to the submarine's operational capabilities, safety, and delivery timeline. Submarine programs like the Virginia-class undergo multiple audit cycles to assess whether spending follows pre-approved financial plans and whether the project remains on budget.

Submarine programs are subject to periodic financial and performance reviews during construction, testing, and operational phases. These reviews involve detailed reporting by contractors to the responsible government agencies, such as the U.S. Navy and the Department of Defense (DoD). The reports track project progress, budget status, and any deviations from the original plan. They also provide insight into potential challenges, such as delays or rising material costs, which can trigger further audits or corrective actions.

These periodic reviews help project managers and government officials make informed decisions about whether additional funding or timeline extensions are needed. For example, in the Virginia-class program, regular audits have highlighted areas where cost-saving measures could be implemented, such as improved modular construction techniques and streamlined procurement processes.

Auditing plays a key role in preventing cost overruns by ensuring that expenses are tracked meticulously. If discrepancies between the approved budget and actual spending are identified, auditors work with project managers to investigate and address the causes. Early detection of budget mismanagement helps prevent delays and ensures the submarine can be delivered on time.

Audits also ensure contract compliance, confirming that vendors and contractors are adhering to agreed-upon schedules and quality standards. By holding all parties accountable, audits help ensure the submarine meets its performance requirements without requiring excessive additional funding.

The insights gained from audits and reviews contribute to continuous improvement in submarine construction. Lessons learned from each program—such as cost-saving measures or the impact of specific risks—inform future projects, ensuring that budget forecasting becomes more accurate over time. In large, ongoing programs like the Virginia-class or Columbia-class, feedback from audits ensures that resources are reallocated efficiently for future builds.

Government agencies also use these reports to communicate with Congress and other oversight bodies, providing transparency about how defence budgets are being spent. This accountability strengthens public trust in military procurement programs and ensures that submarine construction projects remain aligned with national defence priorities.

As such, developing a budget for a submarine construction project is a multi-faceted process that integrates technical requirements, risk assessments, and procurement strategies. It involves careful planning, coordination between shipyards and contractors, and oversight from government bodies to ensure efficient use of resources. As submarine programs often span decades, budgets must be flexible enough to accommodate changes while maintaining financial discipline throughout the vessel's lifecycle.

As a case example, the Virginia-class nuclear-powered attack submarines (SSNs) have been a cornerstone of the U.S. Navy's submarine fleet since their procurement began in FY1998. These submarines are designed to perform multiple roles, including surveillance, special operations support, anti-submarine warfare, and sea control. However, the Navy is now preparing for the next-generation attack submarine, known as SSN(X), slated to succeed the Virginia class. Initially scheduled for procurement in FY2035, the Navy's FY2025 budget submission has deferred the acquisition of the first SSN(X) to FY2040, citing broader budget constraints [686].

The U.S. Navy's submarine force is composed of ballistic missile submarines (SSBNs), cruise missile and special operations submarines (SSGNs), and attack submarines (SSNs). The Virginia-class SSNs serve as general-purpose submarines capable of conducting various missions, from peacetime surveillance to wartime operations. A Virginia-class submarine, when equipped with the Virginia Payload Module (VPM), currently costs over $4.5 billion per boat when procured at a rate of two per year [686].

Construction of these submarines is handled by General Dynamics Electric Boat (GD/EB) and Huntington Ingalls Industries (HII/NNS). These two shipyards are the only facilities in the U.S. capable of building nuclear-powered submarines, forming a crucial part of the nation's submarine construction industrial base [686].

The SSN(X) program aims to address emerging threats from near-peer competitors and enhance the Navy's undersea capabilities. The Navy envisions the SSN(X) to feature [686]:

- Greater speed and enhanced payload capacity.
- Improved acoustic stealth and non-acoustic signature reduction.
- Higher operational availability for extended missions.
- Coordination capabilities with off-hull vehicles, sensors, and unmanned systems.

Navy officials have indicated that the SSN(X) will likely incorporate elements from past submarine designs [686]:

- Speed and payload capacity from the Seawolf-class (SSN-21).
- Acoustic stealth and sensors from the Virginia class.
- Extended operational availability modelled on the Columbia-class ballistic missile submarines.

The result will likely be a larger submarine than the original Virginia-class, with a projected submerged displacement exceeding 10,100 tons, based on the Congressional Budget Office's (CBO) projections [686].

The deferment of SSN(X) procurement to FY2040 introduces significant challenges for the submarine design industrial base. With the Columbia-class SSBN program scheduled to complete before SSN(X) construction begins, the Navy will need to manage an extended design gap. This delay risks creating workforce disruptions, reducing expertise retention, and straining the supply chain of specialized components [686].

The SSN(X) is projected to have higher procurement costs than its predecessors. The Navy estimates each SSN(X) will cost $6.7 billion to $7 billion per boat (in constant FY2023 dollars), while the CBO's estimate is $7.7 billion to $8 billion, reflecting a 14-15% increase over the Navy's projections. This price difference stems from assumptions about the increased displacement, enhanced capabilities, and evolving technologies that will be incorporated into the SSN(X) [686].

The Navy's FY2025 budget has requested $586.9 million in research and development (R&D) funding to support early SSN(X) development. However, this is $208 million less than originally proposed, highlighting ongoing budgetary constraints. Congressional committees have reviewed and adjusted the Navy's funding requests, reflecting concerns about the project's scope and feasibility [686].

Several issues have emerged that will shape the future of the SSN(X) program [686]:

1. Capability Requirements vs. Cost: Congress must assess whether the Navy has accurately identified the SSN(X)'s required capabilities and if those capabilities justify the projected costs.
2. Impact on Broader Defence Spending: There are concerns about the impact of SSN(X) on other Navy programs, especially if the CBO's higher cost estimates prove more accurate.
3. Maintaining Submarine Superiority: Delaying SSN(X) procurement could compromise the Navy's ability to

maintain undersea dominance in the future.

4. Industrial Base Management: Managing the gap between Columbia-class construction and SSN(X) development will require strategic planning to maintain the submarine design workforce and supply chain.

5. Reactor Technology and Nuclear Policy: Discussions continue about whether the SSN(X) should adopt low-enriched uranium (LEU) reactors instead of highly enriched uranium (HEU). However, the Navy has expressed concerns that LEU reactors would negatively impact performance, endurance, and overall force structure.

The SSN(X) program represents the next evolution in the U.S. Navy's submarine fleet, with ambitious goals to meet the challenges of modern undersea warfare. While the program aims to incorporate the best features from previous classes, the deferral of procurement to FY2040 introduces significant risks for maintaining operational readiness and industrial base stability. With estimated costs ranging between $6.7 billion and $8 billion per boat, careful oversight will be required to balance capability development with budgetary realities. Congress will need to weigh these factors carefully to ensure the SSN(X) program meets the Navy's needs while maintaining fiscal responsibility [686].

Below is a detailed sample budget breakdown for a nuclear submarine construction project. This example is based on typical cost structures seen in modern submarine programs such as the Virginia-class SSNs and Columbia-class SSBNs, adjusted to reflect major components, labour costs, R&D needs, and project phases.

Sample Nuclear Submarine Construction Project Budget

Category	Estimated Cost (USD)	Details
1. Research & Development (R&D)	$2.0 billion	Includes conceptual design, feasibility studies, simulations (CFD and FEA), prototyping, and testing.
2. Hull Construction	$1.5 billion	HY-80 or HY-100 steel procurement, modular assembly, and welding at shipyard facilities.
3. Propulsion System	$3.5 billion	Nuclear reactor (e.g., HEU core), reactor shielding, steam turbines, and auxiliary generators.
4. Weapons Systems	$1.2 billion	Torpedo launch tubes, VLS (vertical launch system), and missile payload modules (e.g., Tomahawk or Trident).
5. Electronics & Sonar Systems	$900 million	Sonar arrays, communications equipment, radar, and combat control systems.
6. Life-Support & Crew Systems	$600 million	Environmental control, oxygen generation, crew quarters, and emergency systems.
7. Labor Costs & Shipyard Operations	$2.8 billion	Skilled labour (welders, engineers, electricians) and operational expenses at GD Electric Boat or Huntington Ingalls.
8. Testing and Trials	$800 million	Sea trials, acoustic tests, performance verification, and adjustments.
9. Project Management & Oversight	$500 million	Contractor oversight, reporting, and compliance with naval standards.
10. Contingency Fund	$1.0 billion	Allocated for unexpected delays, material shortages, and design changes.
11. Commissioning & Crew Training	$300 million	Final system checks, crew simulations, and on-board training.
12. Operations & Maintenance Planning	$600 million	Establishing lifecycle support, spare parts inventory, and maintenance protocols.

Figure 130: Budget Summary.

Budget Breakdown by Phase

1. Research and Development Phase:

- **Estimated Cost:** $2 billion
- Includes the conceptual design, computational modelling (CFD and FEA), and initial prototypes. This phase ensures that all systems will function cohesively and meet the Navy's operational requirements.

2. Construction Phase:

- **Estimated Cost:** $7.8 billion

- The construction phase involves **hull assembly**, **reactor installation**, and integration of propulsion, electronics, and weapon systems. This phase leverages modular construction techniques to improve efficiency.

3. Testing and Sea Trials:

- **Estimated Cost:** $800 million
- This phase validates the submarine's performance and ensures compliance with safety and operational standards. Testing includes deep-sea trials, noise signature tests, and combat system evaluations.

4. Commissioning and Crew Training:

- **Estimated Cost:** $300 million
- Commissioning prepares the submarine for operational service, involving system finalization and crew readiness through simulations and hands-on experience.

5. Operations & Maintenance Planning:

- **Estimated Cost:** $600 million
- Covers the establishment of **lifecycle support** to maintain the submarine over its 30-40 year service life, including reactor refuelling and routine maintenance scheduling.

6. Contingency Fund:

- **Allocated Amount:** $1 billion
- This reserve ensures that **unexpected challenges**, such as supply chain disruptions or technical redesigns, do not derail the project timeline or budget.

Key Considerations

- **Nuclear Propulsion Costs:** The reactor and propulsion systems make up a significant portion of the budget due to the complex engineering and regulatory requirements involved in nuclear technology.
- **Labor and Shipyard Costs:** Shipyard labour is highly specialized, and the construction process requires a large workforce of skilled tradespeople working over several years. The collaboration between General Dynamics Electric Boat and Huntington Ingalls drives these costs.
- **Weapons and Payload Costs:** With increasing demand for advanced missile systems, the weapons budget includes costs for torpedo launch systems, vertical launch tubes, and specialized payloads such as Tomahawk cruise missiles.

Final Projected Cost: $11.3 - $12.0 Billion Per Submarine

Given the complexity and scale of nuclear submarine projects, budgets can fluctuate based on technological advancements, regulatory requirements, and supply chain challenges. Regular auditing and reporting ensure that spending remains within approved limits, while contingency funds cover unforeseen expenses.

For further context, the AUKUS submarine program, a trilateral security pact involving Australia, the United States, and the United Kingdom, has garnered significant attention due to its ambitious budget and strategic implications. The program is projected to cost up to $368 billion AUD (approximately $235 billion USD) over several decades, encompassing the development, procurement, and construction of nuclear-powered submarines in Australia. This substantial investment also includes enhancements to local shipbuilding capabilities and workforce development, reflecting a commitment to bolster Australia's defence infrastructure and technological capabilities [687-689].

Australia's initial commitment of $3 billion USD to support the U.S. and U.K. submarine industrial bases is a critical aspect of the AUKUS agreement. This funding aims to enhance production capabilities and facilitate collaborative efforts among the three nations, thereby ensuring a more robust and efficient submarine manufacturing process [690, 691]. The strategic rationale behind AUKUS is not merely to enhance military capabilities but also to counterbalance China's growing influence in the Indo-Pacific region. The partnership is seen as a pivotal move to maintain regional security and stability, as well as to reinforce the existing alliances among these nations [692-694].

Furthermore, the AUKUS pact signifies a shift in the geopolitical landscape, with implications for international security dynamics. The arrangement is designed to foster technological sharing and maritime cooperation, which are essential for addressing contemporary security challenges in the region [689, 694, 695]. Critics, however, caution that while AUKUS aims to deter aggression, it could inadvertently escalate tensions and contribute to an arms race in the Indo-Pacific [694, 696]. The legal and operational frameworks established by AUKUS will also play a crucial role in determining its long-term effectiveness and impact on regional stability [697, 698].

Project Scheduling

The scheduling of a submarine construction project is a complex endeavour that encompasses numerous phases, interdependencies, and a stringent timeline. This process typically spans several years, beginning from the concept design phase and extending through to commissioning. The intricacies involved necessitate meticulous coordination among various stakeholders, including shipyards, contractors, and naval authorities.

One of the primary challenges in scheduling submarine construction projects is the interdependence of tasks. The Dependency Structure Matrix (DSM) has been recognized as an effective tool for managing such complex projects, as it accurately represents the interdependencies between different components of the project [699]. This matrix facilitates the identification of critical paths and potential bottlenecks, thereby enhancing the overall scheduling process. Furthermore, the literature indicates that the interaction between cooperative and competitive group processes can significantly influence scheduling outcomes, particularly in environments characterized by high task interdependency [700]. This highlights the importance of collaborative efforts among all parties involved in the construction process.

A Dependency Structure Matrix (DSM) is an advanced project management tool used to map and analyse dependencies among various tasks or components in a complex project, such as submarine construction. It ensures that interdependent activities are managed efficiently, preventing delays and cost overruns.

How DSM works in submarine construction:

1. **Identify Key Tasks and Modules:** In a submarine construction project, these tasks might include hull fabrication, propulsion system integration, weapons installation, and sonar system testing. Each task or component is represented as a row and column in the matrix.

2. **Map Dependencies:** A DSM highlights which tasks depend on others. For example, reactor integration must occur after the hull section is complete, while propulsion system testing might rely on earlier installation phases. Dependencies are marked within the matrix where tasks intersect.

3. **Analyse Task Flow:** The matrix shows sequential, parallel, and iterative dependencies. If multiple tasks can occur in parallel (e.g., fabricating different modules simultaneously), they are mapped accordingly, ensuring efficiency. Iterative loops indicate areas where rework may be needed if dependencies are tightly linked, such as

the iterative testing and adjustment of sonar systems.

4. **Optimize the Sequence of Work:** The DSM helps identify critical paths and areas of potential bottlenecks. If two dependent tasks have conflicting schedules, the matrix flags these conflicts early, allowing managers to re-sequence tasks or allocate more resources.

5. **Manage Change and Uncertainty:** Submarine construction often involves unforeseen issues, such as supply chain delays or technical redesigns. With a DSM, managers can assess the ripple effects of changes by adjusting the matrix. For example, a delay in sonar installation will automatically highlight subsequent delays in testing and integration phases.

Below is a simplified Dependency Structure Matrix (DSM) example for a submarine construction project. In this matrix, tasks are organized along both the rows and columns to reflect interdependencies. Each cell indicates whether a task is dependent on another. The values inside the cells can be either binary (1 for dependency, 0 for no dependency) or can represent more complex relations like dependency strength.

Dependency Structure Matrix (DSM) Example for Submarine Construction

Tasks	Concept Design	Hull Fabrication	Reactor Installation	Propulsion Integration	Sonar System Testing	Sea Trials
Concept Design	0	1	0	0	0	0
Hull Fabrication	0	0	1	1	0	0
Reactor Installation	0	0	0	1	0	0
Propulsion Integration	0	0	0	0	1	0
Sonar System Testing	0	0	0	0	0	1
Sea Trials	0	0	0	0	0	0

Figure 131: Dependency Structure Matrix (DSM) Example for Submarine Construction.

Explanation of the matrix:

- **Concept Design**: This task initiates the project and is a prerequisite for all subsequent tasks. Therefore, Hull Fabrication depends on Concept Design (marked as 1).

- **Hull Fabrication**: Must be completed before the Reactor Installation and Propulsion Integration.

- **Reactor Installation**: Cannot proceed until the hull is completed and assembled. It influences propulsion system integration.

- **Propulsion Integration**: After propulsion is integrated, the sonar system can be tested.
- **Sonar System Testing**: Must be finished before conducting Sea Trials.
- **Sea Trials**: Final task without any further dependencies.

This DSM helps project managers identify the critical path and dependencies among different tasks. It shows that any delay in Hull Fabrication will directly impact Reactor Installation and Propulsion Integration, and ultimately the entire project. The matrix also ensures efficient resource allocation by highlighting tasks that can run concurrently (e.g., some parts of propulsion and sonar systems can be integrated simultaneously).

By analysing this DSM, adjustments can be made in scheduling to avoid bottlenecks and ensure the project is completed on time.

Moreover, the timeline for submarine construction is often dictated by stringent regulatory and operational requirements. For instance, the Indian Navy's submarine development program illustrates the necessity for a long-term planning approach, as it encompasses a 30-year timeline for the construction of multiple submarines [701]. Such extensive planning is essential to ensure that all phases of construction are aligned with national defence strategies and technological advancements.

Key Phases and Milestones in Submarine Construction

Submarine construction is a long, multi-stage process that spans years. Each phase builds upon the previous, ensuring that the vessel is developed systematically, meeting safety, performance, and operational goals. The main phases include:

- **Research and Development (R&D) Phase:** The R&D phase typically lasts 2 to 4 years. During this time, the focus is on conducting feasibility studies, defining technical and operational specifications, and creating early prototypes. This phase includes intensive simulations, computer modelling, and initial testing to assess if the design can meet performance requirements. The goal is to reduce technical risks before construction begins and ensure that the design aligns with strategic naval needs.
- **Construction Phase:** The construction phase spans 5 to 7 years, involving the fabrication of the hull, assembly of modular components, and the integration of propulsion, electronics, weapons systems, and life-support equipment. Modern modular construction techniques allow different parts of the submarine to be built simultaneously at separate locations and later assembled at the shipyard. Shipyards coordinate closely to ensure precise installation of sensitive components, such as nuclear reactors or sonar arrays, which require specialized infrastructure.
- **Testing and Trials Phase:** The testing and trials phase lasts about 1 to 2 years. Once construction is complete, the submarine undergoes rigorous trials to ensure every system functions as intended. This includes sea trials, where the vessel's propulsion, buoyancy, sensors, weapons, and stealth capabilities are tested under real-world conditions. Acoustic tests are conducted to measure noise signatures, ensuring the submarine can operate covertly. Any performance issues identified are addressed through repairs or modifications during this phase.
- **Commissioning Phase:** The commissioning phase spans several months and marks the final step before the submarine enters operational service. It involves thorough inspections, system validations, and crew training to ensure that the vessel is ready for deployment. Training focuses on operating the submarine's systems, practicing emergency procedures, and familiarizing the crew with the unique demands of underwater missions. After commissioning, the submarine is formally integrated into the naval fleet, ready for both training missions and real-world operations.

These phases together ensure that submarines meet the highest standards of safety, performance, and strategic readiness. Each phase requires coordination between shipyards, contractors, and the navy to adhere to strict schedules and avoid delays, with the overall process often spanning 8 to 15 years from initial concept to final deployment.

Critical Path Management in Submarine Construction Projects

Critical Path Management (CPM) plays a pivotal role in scheduling submarine construction, ensuring the timely completion of interconnected tasks. A Critical Path Schedule (CPS) identifies the longest sequence of tasks that must be completed on time to prevent project delays. This sequence outlines the essential activities without which the submarine cannot be delivered within the set timeline.

The CPM methodology allows project managers to identify the longest sequence of dependent tasks that determine the minimum project duration. By focusing on these critical tasks, project managers can allocate resources more effectively and prioritize activities that directly impact the project's timeline [702, 703]. In the context of the Arihant-class project, the use of CPM would facilitate the identification of critical milestones, such as hull construction, reactor installation, and systems integration, which are essential for the submarine's operational readiness.

Moreover, the integration of risk management practices within the CPM framework is crucial for addressing uncertainties that could affect project outcomes. Studies indicate that effective risk management can lead to significant improvements in project performance, particularly in high-stakes environments like defence projects [704, 705].

The most current submarine class in the Indian Navy is the Arihant-class nuclear-powered ballistic missile submarine (SSBN). The commissioning of INS Arighaat, the second submarine of this class, is expected in 2024, representing a significant enhancement of India's strategic nuclear capabilities. This induction strengthens India's nuclear triad, providing a credible second-strike potential that is crucial for maintaining deterrence in a region characterized by rising geopolitical tensions and adversarial posturing, particularly from neighbouring countries like China and Pakistan [706].

The Arihant-class SSBNs are indigenously developed under the Advanced Technology Vessel (ATV) Project, showcasing India's growing self-reliance in defence technology. INS Arighaat, similar to its predecessor INS Arihant, has a displacement of approximately 6,000 tons and is equipped to launch both K-4 ballistic missiles, which have a range of 3,500 km, and K-15 missiles with a shorter range of 750 km. These capabilities are vital for India's strategic deterrence posture, allowing it to respond effectively to potential threats and enhancing its overall military readiness [706].

The construction and commissioning of the Arihant-class SSBNs involved intricate planning and efficient Critical Path Management (CPM). The complexity of the project necessitated meticulous scheduling to ensure that interdependent systems were aligned and integrated without delays. Each phase, from conceptual design to final commissioning, was carefully orchestrated to prevent cascading delays that could jeopardize the operational readiness of these submarines. This level of project management reflects India's commitment to advancing its naval capabilities and ensuring that its strategic assets are maintained at the highest standards [707].

For the Arihant-class project, identifying potential risks—such as technical challenges, supply chain disruptions, or regulatory hurdles—would enable project managers to develop mitigation strategies that ensure adherence to timelines and budget constraints [708].

The application of Critical Chain Project Management (CCPM), an extension of CPM that incorporates resource constraints and buffers, could further enhance the effectiveness of project management in the Arihant-class project. CCPM emphasizes the importance of managing uncertainties and optimizing the flow of work, which is particularly relevant in a project characterized by complex interdependencies and limited resources [709, 710]. By implementing strategic buffering techniques, project managers can better absorb delays in non-critical tasks without impacting the overall project schedule.

Furthermore, the collaborative nature of the Arihant-class project, involving multiple stakeholders from various sectors, necessitates robust communication and coordination mechanisms. The Project Management Institute (PMI) methodology, which emphasizes stakeholder engagement and proactive risk management, aligns well with the requirements of such complex projects [711, 712]. Effective communication among team members and stakeholders is essential for ensuring that all parties are aligned with project goals and timelines, thereby reducing the likelihood of misunderstandings that could lead to delays.

In submarine construction, modular assembly is the core method, with different sections—like the reactor core, propulsion systems, and sonar units—being developed simultaneously at different facilities. Each component has dependencies. For instance, reactor installation cannot begin until hull fabrication is complete, and propulsion testing must wait for the final reactor configuration. If a critical task such as hull assembly faces delays, it affects dependent tasks like system integration and sea trials, slowing down the entire project.

Effective CPM helps identify such dependencies and potential bottlenecks early. This allows project managers to reallocate resources, adjust timelines, or shift tasks that are not on the critical path to ensure the project stays on schedule. Additionally, with the complexity involved in testing and trials, the CPS ensures these processes are accounted for with sufficient time to address any issues without affecting final commissioning. By managing the critical path carefully, submarine builders reduce the risk of cascading delays and improve the likelihood of meeting tight operational timelines.

Critical Path Management Example for Submarine Construction

Consider a nuclear submarine construction project with major tasks such as hull fabrication, reactor installation, propulsion system integration, sonar installation, and sea trials. Below is an example of how Critical Path Management (CPM) is applied:

1. **Hull Fabrication (6 months)**: This is the starting point. All internal systems depend on the completion of the hull.

2. **Reactor Installation (4 months)**: Can only begin once the hull is complete. Any delay in hull fabrication delays the reactor installation.

3. **Propulsion System Integration (5 months)**: Starts after the reactor is installed, as it connects directly to the reactor for testing.

4. **Sonar Installation and Testing (3 months)**: Runs in parallel with the final stages of propulsion integration but must be completed before sea trials.

5. **Sea Trials (2 months)**: The final critical task, dependent on all prior phases being completed on time.

In this scenario, if hull fabrication is delayed by a month, it cascades down the critical path, pushing back reactor installation and subsequent tasks, including sea trials and commissioning. To manage the timeline, project managers may decide to overlap non-critical tasks, such as starting early sonar work or pre-assembling propulsion modules, to reduce bottlenecks.

This modular construction with overlapping schedules helps streamline the timeline. However, CPM ensures the key dependencies—like reactor integration post-hull fabrication—remain prioritized, avoiding cascading delays. Through this structured approach, submarine builders maintain control over the project schedule, ensuring timely delivery.

Modular construction is an innovative approach used in modern submarine projects like the Virginia-class submarines to streamline the building process. This method involves dividing the submarine into various subassemblies or modules, each of which is fabricated separately. Once completed, these modules are then integrated to form the final structure. Modular construction has revolutionized submarine manufacturing by enhancing efficiency, reducing time, and allowing for better resource allocation.

One of the key advantages of modular construction is that it allows multiple shipyards to work on different sections of the submarine simultaneously. For example, the hull can be built in one shipyard, while systems like reactors, sonar systems, propulsion units, or crew accommodations are fabricated in another facility. These separate modules are then transported to a central assembly location, where they are integrated. This method minimizes bottlenecks because various workstreams progress in parallel, rather than waiting for one task to finish before starting another.

In the Virginia-class submarine program, shipyards like General Dynamics Electric Boat and Huntington Ingalls Industries divide responsibilities for different modules. This distribution of tasks reduces the overall construction timeline by enabling each facility to focus on its specific section of the submarine, making the assembly process more streamlined and manageable.

Modular construction drastically reduces the construction time of complex systems like submarines. Instead of a linear process where each step must follow the previous one, modularity introduces flexibility by allowing the overlapping of tasks. This flexibility also applies to repairs and upgrades. Modules can be replaced or upgraded without disassembling the entire submarine, which is particularly useful for submarines designed for long service lives.

For example, once the hull and propulsion sections are ready, they are brought together with weapons systems, electronics, and command structures, significantly reducing assembly time. Modular construction also simplifies testing and quality control, as each module can be tested individually before integration, reducing the chances of errors that would otherwise be harder to identify in a fully assembled submarine.

The integration of the modules is a complex but efficient process. Once the modules are brought together, engineers fit them with precision to ensure that the submarine's hull is watertight and its systems are seamlessly connected. This integration requires advanced technologies and skilled labour, but it also benefits from pre-fabricated interfaces between the modules that are designed to fit together with minimal rework.

For submarines like the Virginia-class, which incorporates advanced stealth and propulsion technologies, modular construction ensures that the submarine remains on the cutting edge of technology. New systems can be integrated into existing designs without requiring a complete overhaul of the vessel, allowing for technological upgrades and enhancements over time.

Workforce and Resource Planning in Submarine Construction

Effective workforce and resource planning is critical for ensuring the success of submarine construction projects. The complexity of building a submarine—whether nuclear-powered or diesel-electric—demands a diverse range of skilled labour, precise resource management, and careful scheduling. Failure to manage these aspects can lead to costly delays and inefficiencies, especially in large-scale projects where every task is interconnected.

A submarine construction project requires a highly skilled workforce, including welders, electricians, engineers, pipefitters, and mechanics, to perform specialized tasks. For example, welders are crucial for assembling the pressure hull, while electricians focus on wiring critical systems like sonar and communication arrays. Each of these roles requires advanced training, especially in the context of a nuclear submarine, where precision and safety are paramount.

In planning, project managers must ensure that the right number of skilled workers are available at different phases of the project. This means identifying periods of peak demand—such as during hull fabrication or reactor installation—and ensuring enough workers are scheduled to meet those demands without overwhelming the workforce during less intensive phases, like testing or final assembly.

Managing labour shifts also plays a significant role in preventing fatigue and ensuring consistent quality. Submarine construction often runs 24/7 in shipyards, requiring careful scheduling of shifts and rotations. This helps maintain productivity while ensuring that the complex, detail-oriented work is completed to the highest standard. When these shifts are not planned adequately, it can lead to worker burnout, mistakes in critical tasks, or slowdowns in the overall construction timeline.

Resource planning involves the timely procurement of materials, such as HY-80 or HY-100 steel, required for building the hull, electronic systems, and components for propulsion and weapons systems. Submarines are often built using cutting-edge materials and technologies, which may involve long lead times for procurement.

Planning must include timely ordering of materials to avoid shortages that can halt or slow down construction. For example, delays in acquiring specific sonar or reactor components can cascade into delays in system integration and testing phases. Ensuring a reliable supply chain for these specialized materials is crucial for keeping the project on track.

In submarine construction projects, many critical components, such as reactors or ballistic missile launch systems, come from sole-source suppliers due to the highly specialized nature of the materials. Delays or disruptions with these suppliers can have a significant impact, so building strong relationships and having contingency plans is essential.

The shipyard infrastructure itself also plays a key role in ensuring project success. Facilities must be maintained to support the construction of submarines, especially those requiring dry docks, cranes, and assembly halls capable of handling large, heavy components. Regular maintenance of these facilities ensures that they are fully operational when needed, and failure to do so can lead to significant downtime.

For example, delays in dry dock availability can prevent modular sections from being assembled, or issues with overhead cranes can slow down the process of integrating large sections like the reactor or propulsion unit into the submarine hull. Ensuring that shipyard facilities are properly maintained and capable of supporting the construction schedule is a crucial element of resource planning.

All these elements—workforce management, material procurement, and facility maintenance—need to be synchronized to ensure the project remains on track. This requires integrated project scheduling tools, often supported by software that allows project managers to visualize task dependencies and resource allocation. For example, if a delay is detected in the procurement of a key material, labour resources can be reallocated to other tasks to avoid downtime and ensure progress continues in other areas of the project.

Effective planning also incorporates buffers for unexpected delays in material delivery or labour shortages, ensuring that the project can absorb these challenges without derailing the entire schedule. By balancing the demand for labour with available resources and ensuring the timely procurement of materials, project managers can maintain steady progress toward project completion.

Risk Mitigation and Flexibility in Submarine Construction Projects

Risk mitigation and flexibility are integral components of any submarine construction project, given the complexity and high stakes involved. The construction process spans several years, involves numerous interdependent systems, and is highly susceptible to risks such as material shortages, technical failures, and testing issues. Effective risk mitigation strategies help ensure that the project stays on schedule and within budget, while built-in flexibility allows teams to respond to unforeseen challenges without derailing the entire project.

One of the primary ways to manage risk in submarine construction is by incorporating buffers in the schedule. Buffers are periods of extra time added to accommodate potential delays without affecting the overall timeline. For instance, during critical phases such as reactor integration or sea trials, buffers ensure that if something goes wrong, there is time to address the issue without pushing back the entire project.

For example, in a scenario where material shortages delay the procurement of critical components like HY-80 steel for the pressure hull, having an extra few weeks in the schedule allows the project to absorb the delay without disrupting subsequent phases. These buffers are not uniform across the project; they are strategically placed where risks are highest, such as during system integration, where technical compatibility issues between different modules can arise, or during the final testing phase, when failures in systems like propulsion or weaponry may require redesigns or repairs.

Certain phases of submarine construction are especially sensitive to delays, such as sea trials and reactor integration. These are highly technical and safety-critical phases that often involve cutting-edge technologies. During sea trials, the submarine's performance in real-world conditions is tested, including acoustic quieting, propulsion efficiency, and combat systems. Failures during sea trials can require a return to dry dock, where engineers must troubleshoot and fix any issues that arise. Without contingency plans or buffers, such failures can result in costly delays.

Similarly, reactor integration for nuclear-powered submarines is a highly specialized phase where any missteps can jeopardize the entire project. The integration and testing of nuclear reactors require stringent safety protocols, and any issues found during reactor operation tests could mean revisiting earlier phases of construction. Contingency planning for these high-risk phases typically includes identifying potential failure points early, ensuring access to critical spare parts, and having backup teams on standby to manage unexpected complications.

Material shortages are another significant risk factor, particularly given the specialized components used in submarine construction, such as sonar arrays, propulsion systems, and advanced materials for the hull. Many of these components are sourced from sole-source suppliers, meaning that any disruption in the supply chain can cause delays.

To mitigate these risks, submarine construction projects often establish long-term contracts with suppliers to guarantee the availability of critical materials. In addition, some projects adopt a multi-supplier strategy where feasible, ensuring that alternative sources for materials are available if the primary supplier experiences disruptions. In cases where material shortages still arise, procurement buffers ensure that alternative plans can be put into action, either by finding substitute materials or by rescheduling the integration of other components to keep the project moving forward.

The use of modular construction in submarine projects also adds flexibility to the schedule. By dividing the submarine into distinct modules—such as the hull, propulsion, weapons systems, and electronic systems—each module can be built and tested separately. This means that delays in one module do not necessarily halt progress on others. For example, if an issue arises with the reactor module, work on the sonar or missile systems can continue, allowing the overall project to stay on track. Once the delayed module is completed, it can be integrated without causing significant delays to the overall timeline.

This modular approach also helps in mitigating risks during integration, as individual modules are tested and validated before being brought together. If a problem arises during testing, it can be fixed within the module, avoiding the need to disassemble a fully built submarine.

Risk monitoring is a continuous process throughout the submarine's construction. Project managers use real-time data from various sources, including material inventories, workforce availability, and task progress, to track the health of the project. Software tools can help visualize potential risks and assess their impact on the project timeline. If delays occur or risks materialize, adjustments can be made to reallocate resources, shift labour to high-priority areas, or modify the sequence of tasks to prevent further delays.

For instance, if testing reveals flaws in the sonar system, engineers might focus on resolving those issues while other teams move ahead with crew training or weapon system installation. This flexible approach minimizes the impact of localized issues on the broader project schedule.

Tools and Technologies Used in Submarine Construction

Submarine construction projects are complex endeavours that require meticulous planning, resource management, and precision in execution. Various tools and technologies are employed to ensure the project stays on track and maintains the highest standards of quality. Among these tools are project management software, Earned Value Management (EVM), and digital twins, which collectively help streamline operations, improve efficiency, and reduce risks.

Project Management Software

One of the essential tools used in submarine construction projects is project management software. This technology helps project managers track timelines, resources, and budgets in real-time, providing a centralized platform for planning and monitoring every phase of the project. Given the interdependent nature of submarine construction, where delays in one phase can affect the overall timeline, real-time tracking is crucial for maintaining visibility into the project's progress.

These software platforms allow project managers to create detailed Gantt charts or Critical Path Schedules (CPS) to visualize key milestones and deadlines. By monitoring labour availability, materials procurement, and the progress of modular construction, project managers can quickly address any potential bottlenecks or delays. Additionally, the software's resource allocation features ensure that skilled workers, such as welders or electricians, are efficiently assigned to the tasks they are needed for at each stage of the construction process. By integrating data from various sources, such as shipyard operations and procurement systems, the software also helps manage cash flow, monitor supplier contracts, and maintain financial oversight.

Earned Value Management (EVM)

Earned Value Management (EVM) is another critical tool used in submarine construction to measure performance and ensure that the project stays on schedule and within budget. EVM combines data on project scope, cost, and schedule to provide a comprehensive assessment of how the project is progressing compared to the original plan. This method offers insight into whether the project is ahead or behind schedule and over or under budget, allowing for proactive adjustments.

In submarine construction, EVM helps identify potential cost overruns or delays early in the process, ensuring that corrective actions can be taken before problems escalate. For example, if the installation of a propulsion system is running behind schedule, EVM will flag this delay, prompting managers to either reallocate resources to accelerate progress or adjust downstream tasks to accommodate the delay. This level of insight is particularly valuable in large-scale projects where individual components—such as nuclear reactors, sonar arrays, and weapon systems—are being developed and integrated simultaneously.

Digital Twins

The use of digital twins is a cutting-edge technology that enhances submarine construction by creating a virtual replica of the submarine. Digital twins allow engineers and designers to simulate assembly processes and predict how different components will perform before they are physically constructed. This reduces the need for trial and error during the actual construction phase, which can be time-consuming and costly.

In the case of submarine construction, digital twins are used to model critical systems like propulsion units, reactors, and weapons systems. By simulating these systems in a virtual environment, engineers can test various configurations and identify potential issues before physical assembly begins. This ensures that when the actual components are built and integrated, they function as intended without the need for extensive rework.

Digital twins also play a role during testing and sea trials. By comparing the real-world performance of the submarine against the virtual model, engineers can identify discrepancies and refine the design. This approach is particularly valuable for ensuring stealth capabilities and minimizing the submarine's acoustic signature, as the digital twin allows for fine-tuning of hull shapes, material properties, and noise-reduction technologies.

These tools—project management software, Earned Value Management, and digital twins—are essential for navigating the complexities of submarine construction. They enable real-time tracking, performance assessment, and predictive modelling, ensuring that each phase of the project is completed efficiently and to the highest standards. By leveraging these technologies, submarine construction projects can meet stringent operational requirements while mitigating risks and avoiding costly delays.

Government Oversight and Reporting in Submarine Construction Projects

Government oversight and reporting are critical components of submarine construction projects, given the immense public investment and national security interests involved. These projects require close alignment between naval authorities, government bodies, defence contractors, and shipyards. To maintain transparency, ensure accountability, and keep projects on track, regular progress reports must be submitted at every significant milestone. These reports serve as a formal way to verify that the project aligns with the budget allocations and schedules initially approved, while also highlighting potential challenges such as delays or cost overruns.

During the construction of submarines, the project is divided into multiple phases and milestones. These milestones might include completing key design stages, initiating hull construction, integrating propulsion systems, or beginning sea trials. Each of these milestones is typically tied to specific budget allocations, meaning that once a milestone is achieved, additional funding is unlocked for the next phase of the project. Therefore, progress reports must provide detailed insights into whether the project is on track, covering aspects such as timeline adherence, cost analysis, and resource utilization.

If a project hits a milestone ahead of schedule or under budget, it serves as an indicator of efficiency. However, if delays occur—due to technical challenges, procurement issues, or resource shortages—those delays must be documented in the progress reports, along with the associated budgetary implications. Timely reporting allows both the navy and government authorities to anticipate potential problems and make informed decisions on how to proceed.

Submarine construction is a complex endeavour, and delays or cost overruns are not uncommon. When issues arise, schedule adjustments and, in some cases, the renegotiation of contracts are necessary to prevent the project from stalling. For instance, delays in the procurement of critical components like nuclear reactors or sonar systems may lead to timeline shifts for other dependent tasks. Such delays must be communicated transparently, so naval authorities can decide whether to reallocate resources or adjust the project's scope.

In cases of cost overruns, progress reports must outline the specific reasons for the additional expenses—whether due to increases in labour costs, changes in material prices, or unforeseen technical difficulties. Contractors may be required to justify these overruns and present plans for mitigating further financial impacts. Renegotiating contracts or adjusting the scope of work may be necessary to keep the project aligned with available budgets. Earned Value Management (EVM) is often used to assess whether the project is within acceptable cost and time parameters, helping decision-makers adjust as necessary.

Transparency in tracking and reporting is not only essential for project management but also for maintaining public and governmental trust. Given that submarine construction often involves billions of dollars in public funds, government bodies like Congress or the Department of Defense (DoD) expect clear, consistent updates on progress and spending. Regular audits and independent reviews may also be conducted to ensure compliance with the approved budget and to check that the project is adhering to all safety and regulatory standards.

In large projects such as the Virginia-class or Columbia-class submarine programs, reporting and oversight are ongoing throughout the project's lifecycle. Contractors must ensure that there are no hidden costs or unapproved modifications to the design or construction process. Government audits help to ensure that funds are spent effectively, that safety protocols are strictly adhered to, and that the final product meets all operational requirements without compromising on quality.

In addition to internal oversight, government audits provide another layer of scrutiny. These audits are designed to ensure that spending is efficient and that every dollar allocated is used responsibly. Regular reviews and assessments also help identify areas where efficiencies can be improved or where costs may be reduced without compromising the final product. These reviews may also provide recommendations for contract modifications, particularly when unexpected challenges arise.

For instance, during the construction of a new submarine class, issues with advanced materials or propulsion systems could lead to design changes, which, in turn, could necessitate renegotiation of the contract's terms. These renegotiations are informed by the findings from independent reviews and government oversight bodies, ensuring that decisions are based on thorough analysis rather than reactive measures.

Risk Management

Risk management in submarine construction projects is of paramount importance due to the intricate nature of these endeavours, which encompass advanced technologies, substantial financial investments, extended timelines, and significant implications for national security. The complexity of submarine construction projects necessitates a comprehensive approach to risk management, which involves identifying potential risks, assessing their likelihood and impact, and implementing effective mitigation strategies. This multifaceted process is essential to ensure that projects are completed on time, within budget, and in accordance with specified requirements.

The construction of submarines is inherently risky, as it involves advanced engineering and technology that must operate reliably under extreme conditions. For instance, the construction and operation of submarine cables, which are critical for communication and power transmission, require rigorous monitoring and risk assessment to prevent failures that could have severe consequences for national security and infrastructure [713]. The vulnerability of submarine cable networks, particularly in the context of geopolitical tensions, underscores the need for effective risk management strategies [714]. Moreover, the operational risks associated with submarine cables, such as damage from external forces, highlight the necessity of implementing robust protective measures [715].

Effective risk management in submarine construction also draws from lessons learned in other engineering fields. For example, the methodologies developed for assessing risks in submarine tunnel construction can be adapted to submarine projects, emphasizing the importance of cumulative risk control and the application of analytical hierarchy processes (AHP) for risk assessment [716]. Furthermore, the insights gained from the study of submarine geological hazards can inform the development of risk zonation strategies that enhance the safety and reliability of submarine construction projects [717].

Identifying Risks

In the construction of submarines, risk identification is crucial to ensuring that projects proceed smoothly, within budget, and on schedule. Given the complexity and the scale of these undertakings, identifying and addressing potential risks early in the process is critical to the success of the project. These risks can be categorized into several major types: technical risks, schedule risks, financial risks, safety risks, and supply chain risks. Below is a detailed exploration of each type.

<u>Technical Risks</u>

Technical risks are among the most significant challenges in submarine construction due to the sophisticated technologies involved. Submarines, especially modern nuclear-powered vessels, integrate advanced systems like nuclear propulsion, sonar arrays, stealth technologies, and weaponry. Each of these components relies on the latest engineering and innovation, which inherently carries the risk of technical failure or delays in integration.

For example, a cutting-edge nuclear propulsion system might encounter difficulties during initial testing phases, potentially related to the complex design or unanticipated performance issues under operational conditions. Similarly, advanced sonar systems, which are critical for submarine stealth and threat detection, might fail to integrate smoothly with other systems or require more time for refinement than initially expected. These technical delays not only affect timelines but also have a ripple effect on the coordination of related systems, potentially delaying the overall completion of the submarine.

Addressing technical risks requires extensive testing during the early stages of development and constant monitoring during integration. The use of digital twins—virtual models of submarine components—can help simulate performance and detect potential issues before physical assembly, thus mitigating some of the technical risks early on.

Schedule Risks

Submarine construction projects can span over a decade, and during this time, numerous variables can influence the timeline. Schedule risks occur when delays in one phase affect subsequent phases, leading to cascading timeline shifts. These delays can stem from technical difficulties, material shortages, or unforeseen changes in the design specifications. The challenge with schedule risks is their ability to escalate quickly, often affecting the entire project.

For instance, a delay in procuring materials—such as specialized metals for the submarine's hull—can push back the assembly phase, which, in turn, delays the integration of critical systems like propulsion or weaponry. In some cases, design changes or testing failures can force the re-engineering of key components, further compounding delays.

Effective schedule risk management involves building buffers into the project timeline and employing Critical Path Management (CPM) to identify key milestones that must be achieved to keep the project on track. The use of project management software is critical for tracking these milestones and addressing issues before they lead to significant delays.

Financial Risks

Submarine construction is a multi-billion-dollar endeavour, with individual submarines often costing upwards of $4 billion to $5 billion. Financial risks are common, especially in long-term projects where material prices fluctuate and labour costs increase over time. Additionally, delays in construction can lead to significant cost overruns, forcing contractors and governments to renegotiate contracts or adjust budgets.

For instance, a sharp rise in the cost of specialized materials such as HY-80 or HY-100 steel—used for constructing submarine pressure hulls—can push the project budget beyond its initial estimates. Labor shortages or unexpected needs for skilled workers can also increase overall costs. Furthermore, procurement delays or design changes necessitated by unforeseen technical challenges can lead to additional expenses, as systems need to be reworked or alternative components sourced.

Financial risks are managed through detailed cost forecasting, contingency budgets, and regular financial audits. Governments and contractors must keep financial planning flexible enough to accommodate unexpected expenditures while keeping the project within the approved budget.

Safety Risks

Building and operating submarines—especially nuclear-powered submarines—involves a range of significant safety risks. Strict adherence to safety protocols is required to ensure the safety of both the crew and the environment. Any failure in the construction or operation of safety-critical systems—such as the nuclear reactor, ballast systems, or emergency life-support systems—can lead to catastrophic outcomes. This includes equipment failures, reactor accidents, or collisions, which can result in both human casualties and severe legal or financial consequences.

Safety risks are mitigated by employing redundant safety systems, conducting thorough quality control checks during construction, and adhering to international safety standards. Crew training on safety procedures and regular emergency drills are also essential to ensure that submarines are operated safely.

Supply Chain Risks

Submarine construction projects rely on a vast and intricate **supply chain** for specialized materials and components. Supply chain risks arise when delays or failures in sourcing these materials disrupt the construction timeline. Many of these materials, such as high-strength steel, nuclear reactor components, and advanced electronics, are sourced from sole-source suppliers. If these suppliers experience production issues, financial troubles, or shipping delays, the entire project can come to a halt.

For example, if the supplier of reactor components faces production delays or issues in delivering parts to the shipyard, the integration of the nuclear propulsion system could be delayed, significantly impacting the project's timeline. Material shortages or difficulties in acquiring high-strength materials used for hull construction can similarly disrupt production schedules.

To mitigate supply chain risks, submarine projects often establish long-term contracts with suppliers and, where possible, diversify their supplier base. Contingency plans, such as identifying alternative sources for critical components, are also essential for minimizing supply chain disruptions.

Assessing Risks in Submarine Construction Projects

After identifying potential risks in a submarine construction project, the next crucial step is to assess these risks in terms of their likelihood and impact. This process is essential for determining which risks require immediate attention and resources and which can be monitored more passively. The most common tool used in this phase is the risk assessment matrix, which categorizes risks into levels of low, medium, or high for both their probability of occurring and the severity of their impact on the project. The goal is to prioritize risks so that resources can be allocated effectively to prevent major disruptions.

Risk Assessment Matrix: Probability and Impact

The risk assessment matrix is a simple yet effective tool for visualizing and categorizing risks. The matrix typically plots likelihood (how probable it is that a risk will occur) on one axis and impact (the severity of the consequences if the risk occurs) on the other axis. This creates a grid that enables project managers to assess risks at a glance and decide which ones need proactive mitigation and which ones can be monitored over time.

For instance, a risk with high impact and high likelihood will fall into a critical category requiring immediate action, while a risk with low likelihood but high impact might be monitored closely but may not need immediate intervention unless it starts to materialize. The same applies to low-impact, low-likelihood risks, which are generally of lower priority and may require fewer resources for mitigation.

High-Likelihood, High-Impact Risks

An example of a high-impact, high-likelihood risk in submarine construction could be delays in the procurement of essential materials such as specialized steel for the hull or reactor components. The likelihood of procurement delays is high because of the limited availability of such materials and the reliance on sole-source suppliers. The impact of these delays can be severe because they affect the entire construction timeline, potentially halting progress on multiple fronts, such as modular assembly and the integration of key systems like propulsion units.

In such cases, risk mitigation would likely involve setting up contingency plans with backup suppliers, establishing long-term procurement contracts, or building schedule buffers to accommodate potential delays. These proactive steps ensure that a high-likelihood, high-impact risk does not derail the project.

Low-Likelihood, High-Impact Risks

A low-likelihood but high-impact risk might be the catastrophic failure of a newly developed propulsion system during testing. While the likelihood of such a failure may be low—especially with rigorous testing and simulation protocols in place—the impact could be enormous. A propulsion failure could result in massive project delays, design rework, and significant financial loss, as well as reputational damage if it affects the safety of the vessel.

For such risks, mitigation may include extensive testing and redundancy in system design. Backup systems may be developed in parallel, and digital twin simulations may be used to identify potential issues before physical testing begins. This approach helps to manage risks that are unlikely but could be disastrous if they materialize.

Medium-Risk Categories

Medium-risk categories, often characterized by medium probability and medium impact, can include issues such as labour shortages or minor technical challenges in the integration of systems. While these risks are important, they may not need immediate attention if they do not threaten the overall project timeline or budget.

For example, temporary labour shortages may delay certain sections of assembly, but these delays may be absorbed by other activities that continue in parallel. Monitoring these risks closely through project management software and schedule tracking tools ensures that they are managed without overwhelming resources dedicated to more critical risks.

Prioritization for Risk Mitigation

The main advantage of using a risk assessment matrix is that it helps in prioritizing risks. High-likelihood, high-impact risks are addressed immediately, with detailed mitigation strategies and resource allocation. Medium risks may require monitoring and periodic review to ensure they do not escalate, while low-impact, low-likelihood risks may not need direct intervention but should remain on the radar for future assessments.

Prioritization is key to effective risk management because it ensures that resources—whether financial, technical, or labour-related—are applied efficiently. Focusing on the most pressing risks first ensures that the project remains on track and avoids major disruptions while allowing for flexibility in handling lower-priority risks.

Following is a sample Risk Assessment Matrix that can be used in the context of a submarine construction project. This matrix helps assess risks based on their likelihood (probability of occurrence) and impact (consequences if the risk materializes).

Risk Description	Likelihood	Impact	Risk Category	Mitigation Strategy
Delay in Hull Material Procurement	High	High	Critical	Establish backup suppliers, contract negotiation
Failure in Nuclear Propulsion Testing	Low	High	Major	Implement redundant testing and backup systems
Labor Shortage	Medium	Medium	Moderate	Workforce planning, flexible labour shifts
Supply Chain Disruption for Sonar Tech	Medium	High	Critical	Build inventory buffer, alternate supplier agreements
Design Changes in Control Systems	High	Medium	Major	Build design flexibility, fast-track testing phases
Unexpected Technical Challenges	Low	Medium	Minor	Regular system testing, contingency funds
Regulatory Delays in Testing Phase	Medium	High	Critical	Early regulatory engagement, buffer timeline

Figure 132: Sample Risk Assessment Matrix.

Risk Categories:

1. **Critical**: These risks require immediate attention due to their high likelihood and severe impact on the project.

2. **Major**: Medium likelihood with a significant impact, requiring close monitoring and planned mitigation.

3. **Moderate**: Medium risks that may cause moderate disruption and can be managed through regular monitoring.
4. **Minor**: Low likelihood and impact, usually monitored but without needing immediate resource allocation.

Legend for Likelihood and Impact:

- **Likelihood:**
  - **High**: Likely to occur (e.g., greater than 50% probability).
  - **Medium**: Possible occurrence (e.g., 20%-50% probability).
  - **Low**: Unlikely to occur (e.g., less than 20% probability).
- **Impact:**
  - **High**: Severe consequences on budget, timeline, or safety.
  - **Medium**: Noticeable impact that may cause moderate disruption.
  - **Low**: Minor impact with little to no disruption to the overall project.

By using this **Risk Assessment Matrix**, project managers can better identify which risks require immediate action and which can be monitored for potential escalation.

Mitigating Risks in Submarine Construction Projects

After identifying and assessing risks in submarine construction, the next critical step is developing mitigation strategies. These strategies are designed to reduce either the likelihood of the risk occurring or its potential impact if it does occur. Each significant risk requires a mitigation plan tailored to its specific characteristics. Effective risk mitigation ensures the project continues with minimal disruptions, delays, or cost overruns. Below are detailed explanations of common mitigation strategies used in large-scale submarine construction projects.

Technical Mitigations

For technical risks, mitigation often involves creating parallel development paths or backup systems to ensure that technological challenges do not halt the overall progress. For example, a major technical risk in submarine construction could be the integration of a newly developed sonar system. If integration issues arise, they can delay the construction process and affect downstream tasks. In such cases, a mitigation strategy would be to have a previously validated sonar system ready for temporary installation. This allows the construction process to proceed while the technical team resolves the integration issues with the new system. By implementing these types of backup solutions, potential bottlenecks are avoided, ensuring that construction and system development progress simultaneously.

Another aspect of technical mitigation involves rigorous testing during development phases. This includes using digital tools like computational fluid dynamics (CFD) or finite element analysis (FEA) to simulate performance before physical testing begins. Such tools help identify potential technical failures early on, reducing the risk of delays during the actual construction phase.

Schedule Buffers

Delays are a common risk in large, complex projects like submarine construction, and managing schedule risks is crucial for keeping the project on track. One of the most effective strategies for mitigating schedule risks is to incorporate buffers into the project timeline. These buffers act as a cushion, providing extra time for critical phases that are likely to encounter challenges.

For example, the integration of a nuclear reactor or the testing of sonar systems are phases where unforeseen technical difficulties could arise, causing delays. By building in time buffers, the project schedule remains flexible, allowing for adjustments without pushing the overall timeline too far behind.

Buffers are especially useful during sea trials, where the submarine undergoes real-world testing. In these trials, unexpected technical or performance issues may necessitate adjustments or even minor redesigns, and having additional time planned for these contingencies ensures that they don't lead to extended delays.

Financial Contingencies

Financial risks are inevitable in high-cost projects like submarine construction. Fluctuations in material prices, labour shortages, or changes in the scope of the project can lead to cost overruns. To manage these risks, project managers often set aside contingency budgets. These are funds specifically reserved for unforeseen expenses, ensuring that the project can continue even if certain costs exceed initial estimates.

For instance, a sudden rise in the price of high-strength steel, used for submarine hulls, could significantly increase material costs. A contingency fund would cover this additional expense, preventing the need for renegotiating contracts or halting production. Similarly, if labour costs increase due to shortages of skilled workers, the contingency budget can absorb this without affecting the overall project's financial health.

By planning for such financial fluctuations, the project can proceed without requiring constant budget revisions, which could otherwise cause delays or diminish trust between contractors and the government.

Supplier Agreements

The supply chain is one of the most critical and vulnerable aspects of submarine construction. The specialized materials and components required often come from a small pool of suppliers, and delays from a single supplier can affect the entire project. To mitigate supply chain risks, long-term agreements with suppliers are established. These agreements often include provisions for priority production, fixed pricing, and bulk discounts, which help secure the necessary materials within the project's timeline and budget.

Additionally, maintaining a network of alternative suppliers is a key mitigation strategy. For example, if a supplier of nuclear reactor components is unable to meet deadlines due to manufacturing issues, an alternative supplier can step in, minimizing the delay. This network of suppliers ensures that if one source fails, another is available to continue the project, reducing the risk of halting production.

Maintaining inventory buffers of critical components also helps in cases where sudden supply disruptions occur, ensuring that the project has a stock of necessary materials to continue work in the short term while new supplies are sourced.

Continuous Monitoring and Risk Adjustments

In submarine construction, risk management is an ongoing, dynamic process that requires continuous monitoring throughout the lifecycle of the project. The complexity and scale of submarine construction mean that new risks can emerge, while previously identified risks can evolve in their nature and impact as the project progresses. It is essential to have mechanisms in place to track these risks in real time and adjust the project plan accordingly to mitigate their effects.

Real-time tracking tools are commonly employed to ensure that all aspects of the project are progressing according to schedule and within the budget. These tools enable project managers to maintain an overview of various milestones and components being developed simultaneously. For example, if there is a delay in the fabrication of the submarine's hull due to material shortages or supplier issues, the real-time tracking system can immediately flag this issue. This allows the management team to address the problem promptly, whether by reallocating resources, adjusting the timeline for other related components, or securing alternative material sources.

Earned Value Management (EVM) is another critical tool used in continuous monitoring. EVM is an integrated system that measures project performance in terms of cost and schedule. It allows project managers to compare the actual work completed against the planned work and budgeted costs, providing an early indication of whether the project is on track or if there are deviations that need to be corrected. By using key performance indicators (KPIs) such as cost variance and schedule variance, project managers can quickly assess whether risks are materializing and take steps to manage them before they cause significant disruption.

For example, if EVM data indicates that the schedule variance is growing due to delays in one area of construction, project managers can investigate the root cause—be it material delays, labour shortages, or technical difficulties—and take corrective actions. These might include extending work hours, increasing workforce allocations, or modifying project sequences to ensure other parts of the project continue while the issue is addressed.

When risks are identified or begin to materialize, the project team may need to make rapid adjustments in scheduling, budgeting, or resource allocation. This flexibility is essential in submarine construction, where risks can cascade and impact multiple phases of the project if left unaddressed. A material shortage affecting hull construction, for example, might impact the installation of propulsion systems or internal components if not managed promptly.

Risk adjustments are made based on the severity and likelihood of the identified risks. For instance, a minor delay in the delivery of non-essential components may be manageable without significant schedule shifts, whereas a delay in delivering critical components like the nuclear reactor core could necessitate major adjustments, including rescheduling or reallocating resources to keep other sections of the project moving forward.

Contingency Plans and Flexibility in Submarine Construction Projects

In large-scale submarine construction, contingency planning and flexibility are essential components of effective risk management. Despite the best mitigation efforts, some risks will inevitably materialize, causing delays, cost overruns, or technical setbacks. To prepare for these eventualities, contingency plans are developed during the planning stages, ensuring that the project can continue smoothly even when unexpected disruptions occur. These plans focus on alternative solutions, backup resources, and flexible schedules to reduce the impact of issues on the overall construction timeline and budget.

Alternative Design Solutions

One of the key components of a contingency plan in submarine construction is having alternative design solutions ready to deploy if new technologies fail to perform as expected. For example, submarines often rely on cutting-edge technologies, such as advanced propulsion systems or stealth features. However, these systems can sometimes encounter technical issues during testing or integration. If a newly developed propulsion system aimed at reducing noise fails to meet performance criteria, a previously tested and validated system can be used as a temporary or permanent solution to prevent further delays.

This flexibility is particularly important in a project where the introduction of new technologies is a high-risk area. Having a set of pre-approved alternative designs or fallback options allows the construction process to continue without halting progress. While the technical team works on resolving the issues with the new system, the alternative can ensure that deadlines are met, and the submarine remains on schedule for delivery. This method of integrating backup technologies is often discussed and tested during the early design phases, ensuring that the team is prepared for any technological failure that could arise.

Alternative Suppliers

Another critical aspect of contingency planning is having alternative suppliers to minimize supply chain disruptions. In submarine construction, specific components—such as reactor parts, specialized steel, or electronic systems—are often sourced from specialized suppliers. However, relying on a single supplier for these critical materials can introduce significant risk. If a supplier experiences delays, such as production issues or transportation problems, the entire project can be affected.

To mitigate this risk, submarine projects establish relationships with multiple suppliers for key components. Having a diverse supply network means that if one supplier fails to meet their delivery deadlines, another supplier can step in, ensuring that production continues with minimal disruption. For instance, if a supplier for the submarine's sonar system components cannot deliver due to unforeseen circumstances, an alternative supplier can provide the parts within the required timeframe.

This strategy not only ensures timely material procurement but also provides the flexibility needed to handle other issues such as price fluctuations or quality concerns. By diversifying the supply chain, the project becomes more resilient to external disruptions, and the risk of major delays is significantly reduced.

<u>Flexibility in Scheduling and Resource Allocation</u>

Contingency plans also incorporate flexibility in scheduling and resource allocation to accommodate potential disruptions. By building buffer time into the project timeline, particularly for critical phases like sea trials or system integration, project managers ensure that there is room to address issues without jeopardizing the entire schedule. These buffers are planned during the project scheduling phase and are essential for absorbing delays caused by unexpected problems like material shortages or technical failures.

In addition to time buffers, having resource flexibility is another way to ensure smooth operations. For example, if one phase of the construction is delayed due to a material shortage, workers and engineers can be reassigned to other sections of the project that are progressing as planned. This adaptive resource management ensures that productivity remains high, even when certain components or sections of the project are delayed.

<u>Financial Flexibility</u>

Along with technical and scheduling flexibility, financial flexibility is also built into contingency plans. This often takes the form of contingency budgets, which are set aside to cover unforeseen costs that arise during the construction process. These could include labour cost increases, material shortages, or additional testing required due to design changes or technical failures. Having a financial buffer ensures that the project can continue without requiring major budget renegotiations or additional government approval, which could otherwise delay progress.

Risk Communication in Submarine Construction Projects

In complex projects such as submarine construction, effective risk communication is critical for ensuring that all stakeholders—government bodies, naval authorities, contractors, and suppliers—are informed about potential risks, current mitigation efforts, and any changes to the project's timeline or budget. Transparent and continuous communication not only keeps all parties aligned but also enhances the capacity to make timely decisions when risks escalate or evolve.

Throughout the construction lifecycle, regular risk reports are generated and shared during project review meetings. These reports provide up-to-date information on potential risks, their likelihood, and impact. The reports also outline the mitigation strategies in place and whether adjustments need to be made based on emerging risks.

For example, if delays are anticipated due to material shortages, stakeholders can assess how these shortages will affect the timeline, the budget, or critical project milestones such as system integration or sea trials. By keeping everyone informed, risk communication facilitates proactive responses rather than reactive ones, preventing smaller risks from growing into major issues that could derail the project.

These review meetings serve as a formal platform where all stakeholders can share insights, ask questions, and agree on necessary adjustments. The regular nature of these reports and reviews ensures that risk management remains a priority throughout the project, and no risks go unnoticed or unaddressed.

When risks escalate, having a strong communication framework in place enables rapid decision-making. If a risk increases in likelihood or impact—such as the failure of a key supplier to deliver critical components—this information must be communicated immediately to enable quick resource reallocation or timeline shifts. Transparent communication among contractors, naval authorities, and suppliers ensures that the project can pivot in real-time, minimizing the overall disruption.

For example, if the integration of a new propulsion system encounters technical difficulties, contractors can communicate these issues to the naval authorities and government bodies early. This early communication allows them to decide whether to allocate additional resources to address the technical issues or revert to an alternative system to keep the project moving forward. Clear and timely communication is essential to preventing cascading delays, as it allows for the quick implementation of contingency plans.

Submarine construction projects are often subject to strict government oversight, especially given their strategic and financial importance to national defence. This oversight ensures that risks are communicated clearly and that all parties remain accountable for managing and mitigating risks according to the agreed-upon project plans.

Government agencies play a critical role in overseeing how budgets are allocated and ensuring that risk mitigation strategies align with national defence priorities. If a project experiences cost overruns or significant delays, government bodies must be informed promptly, and the necessary approvals for additional funding or schedule adjustments must be sought. This keeps the project aligned with broader defence goals and ensures that financial accountability is maintained throughout the construction lifecycle.

Collaboration with Military and Civil Authorities

In the design and construction of submarines, the collaboration between military authorities, civil agencies, and defence contractors is essential for the successful execution of these complex and resource-intensive projects. This partnership is primarily driven by the need for alignment among the operational requirements of the navy, the technical expertise of defence contractors, and the regulatory and environmental considerations imposed by civil authorities. Such collaboration is crucial for optimizing the design process, managing project timelines, and ensuring compliance with both national and international laws.

The integration of civil and military sectors has been recognized as a vital component in various defence-related projects, including submarine construction. For instance, the concept of civil-military integration emphasizes the importance of collaboration between military enterprises and civil agencies to enhance operational efficiency and innovation in defence projects [718]. This integration allows for the pooling of resources and expertise, which is particularly beneficial in the context of submarine design, where advanced technologies and regulatory compliance are paramount [719]. Furthermore, the establishment of a collaborative innovation system can facilitate the sharing of knowledge and best practices, ultimately leading to improved project outcomes [718].

Moreover, environmental considerations play a significant role in submarine projects, necessitating close cooperation between military and civil authorities. Research has shown that risk assessment methodologies, which incorporate environmental legislative frameworks, are essential in submarine outfall projects [720]. Such methodologies ensure that environmental impacts are systematically evaluated and mitigated, thus aligning with the regulatory requirements set forth by civil agencies. This alignment not only aids in compliance but also enhances the sustainability of submarine operations [721].

Additionally, the strategic outsourcing of certain functions to private contractors has been a topic of discussion in the context of civil-military integration. This approach allows for greater flexibility and responsiveness in project execution, as defence contractors can leverage their specialized knowledge and capabilities to meet the unique demands of submarine construction [722]. The adaptation of defence firms to changes in funding and research priorities further underscores the necessity of collaborative partnerships in navigating the complexities of modern defence projects [723].

The establishment of military requirements and operational input for submarines is a critical aspect of naval defence strategy. This process is heavily influenced by the strategic objectives of the military, which dictate the operational capabilities necessary for submarines to fulfill their roles in surveillance, nuclear deterrence, anti-submarine warfare, and special operations. The military's input is essential in defining specifications related to stealth, speed, payload capacity, endurance, and adaptability to various environments, including Arctic and littoral zones [724].

Naval authorities collaborate closely with defence contractors to ensure that submarine designs align with these operational requirements. This collaboration is vital for integrating advanced technologies and adapting to evolving geopolitical circumstances that may alter capability needs. As technology matures, operational requirements may shift, necessitating design changes to accommodate new functionalities or operational profiles that emerge during actual service [724]. The continuous feedback loop between military needs and design processes ensures that submarines are equipped to meet both tactical and strategic objectives effectively.

Moreover, the military's operational input extends throughout the design and construction phases, ensuring that submarines are built to withstand the rigors of warfare. This includes considerations for survivability and the ability to maintain operational effectiveness under various threat conditions. For example, the design of electrical power systems on submarines is directly linked to operational needs, emphasizing the importance of survivability metrics and the continuity of service [725]. Additionally, advancements in simulation technologies allow for the evaluation of operational effectiveness in naval tasks, further enhancing the military's ability to define and refine operational requirements [726].

The involvement of civil authorities in the development of submarines, particularly nuclear-powered ones, is crucial for ensuring compliance with safety regulations and environmental standards. Agencies such as the Nuclear Regulatory Commission (NRC) in the United States play a pivotal role in overseeing the safety of nuclear reactors on submarines during both their construction and operational phases. The NRC establishes stringent safety requirements that must be met to protect both personnel and the environment from potential hazards associated with nuclear energy.

In addition to nuclear oversight, environmental agencies are actively engaged in the submarine construction process to mitigate potential ecological impacts. These agencies collaborate with military and civilian contractors to ensure that submarine operations do not adversely affect marine ecosystems. For instance, environmental impact assessments are required to evaluate the potential effects of submarine construction activities on sensitive coastal regions, particularly in light of international treaties aimed at pollution control and underwater noise reduction. Such assessments are essential for balancing military needs with environmental protection, as they help identify and implement measures to minimize harm to marine life during construction and operation.

Moreover, the construction of submarine power cables and other underwater infrastructure is also subject to civil authority regulations. These regulations ensure that such projects adhere to environmental standards and do not disrupt marine habitats. The placement of submarine cables is influenced by government policies that aim to protect marine environments while facilitating technological advancements. The integration of environmental considerations into the planning and execution of submarine projects reflects a growing recognition of the need for sustainable practices in maritime operations.

In the context of defence contracting, particularly for submarine construction, the engagement between contractors and suppliers is critical for translating operational requirements into functional designs. Major defence contractors like General Dynamics Electric Boat and Huntington Ingalls Industries must navigate complex relationships with both military and civil authorities to ensure compliance with technical specifications and regulatory standards. This involves adhering to military specifications for weapon systems and sonar arrays, as well as engaging with civil agencies to meet environmental regulations, which is a significant aspect of modern defence contracting [727, 728].

The collaboration between contractors and suppliers is essential for the successful integration of specialized components, such as nuclear reactors and propulsion systems. Research indicates that effective contractor-supplier relationships can enhance project performance, particularly when suppliers are involved early in the design process. This early engagement allows suppliers to align their capabilities with the contractor's needs, thereby fostering innovation and improving overall project outcomes [729, 730]. Furthermore, the emphasis on environmental capabilities among suppliers has become increasingly relevant, as contractors seek to implement green supply chain management practices in their projects [728, 731].

Moreover, the dynamics of contractor-supplier engagement are influenced by the need for quality and safety standards, which are often overseen by military and civil authorities. These authorities play a crucial role in certifying materials and ensuring that all parties adhere to established regulations. The complexity of these relationships necessitates a strategic approach to supplier selection and management, emphasizing the importance of collaboration and communication throughout the supply chain [732, 733]. Studies have shown that a more integrated approach to supply chain management, which includes fostering partnerships rather than merely transactional relationships, can lead to improved performance and reduced costs in construction projects [730, 734].

Government oversight and accountability play a crucial role in the design and construction of submarines, particularly in ensuring that these projects adhere to budgetary constraints and national defence priorities. The U.S. Department of Defense (DoD) is primarily responsible for overseeing defence acquisitions, which includes substantial investments in submarine programs. It is reported that over one-third of the DoD's annual budget, averaging more than $150 billion since 2005, is allocated to acquiring materiel, systems, and services, which underscores the complexity and political nature of defence procurement [735]. Effective program management within the DoD is essential for maintaining accountability and ensuring that taxpayer funds are utilized efficiently [736].

Moreover, legislative bodies, including Congress, play a significant role in monitoring defence spending. They ensure that funds are allocated appropriately and that projects, such as submarine construction, are completed on time and within budget. This oversight is critical not only for financial accountability but also for ensuring that the submarines meet the necessary operational requirements and comply with international treaties and defence regulations. The collaboration between military and civil authorities is vital in this context, as military authorities define the tactical capabilities required, while civil authorities enforce the legal frameworks governing these capabilities [238].

In addition to financial oversight, government accountability extends to compliance with international treaties, particularly those related to nuclear weapons and arms control. The deployment of nuclear submarines raises significant legal and ethical considerations, necessitating close cooperation between military and civilian entities to ensure adherence to international norms and regulations [238]. The integration of Complex System Governance (CSG) principles can further enhance the management of these complex interdependent systems, thereby reducing risks associated with emergent challenges in submarine development [737].

Collaboration during the testing and commissioning phases of submarines is crucial for ensuring operational performance and safety compliance. As submarines transition to testing, military and civil authorities play pivotal roles. Military authorities oversee sea trials to validate that submarines meet operational performance standards, while civil authorities, particularly those responsible for nuclear safety, ensure that the submarine's nuclear propulsion system adheres to regulatory standards before the vessel is cleared for active duty. This dual oversight is essential for maintaining both military readiness and public safety, particularly in the context of nuclear-powered submarines.

Effective collaboration is facilitated through transparent communication among all stakeholders, including military, civil, and contractor entities. Regular project reviews, risk assessments, and updates are vital for maintaining alignment among these groups. Such practices foster effective decision-making and enable timely resolution of issues that may arise during the testing phase. For instance, the integration of hardware-in-the-loop (HIL) simulations allows for real-time testing and validation of control systems, which can significantly reduce risks associated with submarine commissioning [738]. This approach exemplifies how collaborative efforts can enhance the reliability and safety of submarine operations.

Moreover, the importance of quality control and risk management during the early stages of design and planning cannot be overstated. Effective management of risks associated with offshore installations has been shown to prevent incidents that could compromise safety and operational integrity [739]. The lessons learned from offshore oil and gas practices can be applied to submarine commissioning processes, emphasizing the need for a comprehensive approach to quality assurance that involves all stakeholders from the outset [739].

References

1.Bila, T.V.S., W. Widodo, and B.A. Yulianto, *Policy Review of the Effectiveness of the Submarine Technology Transfer Roadmap Between Indonesia and South Korea.* Jurnal Pertahanan Media Informasi TTG Kajian & Strategi Pertahanan Yang Mengedepankan Identity Nasionalism & Integrity, 2023. **9**(3): p. 467-478.

2.Joiner, K.F., et al., *Cybersecurity for Allied Future Submarines.* World Journal of Engineering and Technology, 2018. **06**(04): p. 696-712.

3.Leorocha, F., et al., *Comparative Study on Maritime Security Theory of Mahan Alfred Thayer and Geoffrey Till on the Strategic and Practical Implications of Constructing a Sea Defense.* International Journal of Progressive Sciences and Technologies, 2023. **38**(1): p. 456.

4.Cunningham, F.S. and M.T. Fravel, *Assuring Assured Retaliation: China's Nuclear Posture and U.S.-China Strategic Stability.* International Security, 2015. **40**(2): p. 7-50.

5.O'Donnell, F. and Y. Joshi, *Lost at Sea: The<i>Arihant</I>in India's Quest for a Grand Strategy.* Comparative Strategy, 2014. **33**(5): p. 466-481.

6.Mehar, A.K. and P. Muralidhar, *Analysis of Submarine With the Study of Mechanical Investigations Using Borei - Class Submarine Model.* International Journal of Engineering Research And, 2020. **V9**(08).

7.Zhang, X., et al., *Modelling Research on Breaking Through the Blockade Area.* Destech Transactions on Engineering and Technology Research, 2018(pmsms).

8.Kang, K. and J. Kugler, *Assessment of Deterrence and Missile Defense in East Asia: A Power Transition Perspective.* International Area Studies Review, 2015. **18**(3): p. 280-296.

9.Herz, M., L. Dawood, and V.C. Lage, *A Nuclear Submarine in the South Atlantic: The Framing of Threats and Deterrence.* Contexto Internacional, 2017. **39**(2): p. 329-350.

10.Masood, M. and M.A. Baig, *Potential Impact of Lethal Autonomous Weapon Systems on Strategic Stability and Nuclear Deterrence in South Asia.* Margalla Papers, 2023. **27**(2): p. 27-43.

11.Raju, N. and T. Erästö, *The Role of Space Systems in Nuclear Deterrence.* 2023.

12.Chakraborty, S., *Introduction to Submarine Design.* 2019, Marine Insight.

13.Zhang, J., *Lead–bismuth eutectic (lbe): a coolant candidate for gen. iv advanced nuclear reactor concepts.* Advanced Engineering Materials, 2013. **16**(4): p. 349-356.

14.Li, B., R. Chiong, and L. Gong, *Search-Evasion Path Planning for Submarines Using the Artificial Bee Colony Algorithm.* 2014.

15.O'Donnell, F. and Y. Joshi, *Lost at Sea: The Arihant in India's Quest for a Grand Strategy.* Comparative Strategy, 2014. **33**(5): p. 466-481.

16.Cao, L., et al., *Review on the Hydro- And Thermo-Dynamic Wakes of Underwater Vehicles in Linearly Stratified Fluid.* Journal of Marine Science and Engineering, 2024. **12**(3): p. 490.

17.Heidemann, J., M. Stojanovic, and M. Zorzi, *Underwater Sensor Networks: Applications, Advances and Challenges.* Philosophical Transactions of the Royal Society a Mathematical Physical and Engineering Sciences, 2012. **370**(1958): p. 158-175.

18.Rubino, F., et al., *Marine Application of Fiber Reinforced Composites: A Review.* Journal of Marine Science and Engineering, 2020. **8**(1): p. 26.

19.Qu, Z. and M. Lai, *A Review on Electromagnetic, Acoustic, and New Emerging Technologies for Submarine Communication.* Ieee Access, 2024. **12**: p. 12110-12125.

20.Mahmutoğlu, Y., K. Türk, and C. Albayrak, *Investigation of Underwater Wireless Optical Communication Channel Capacity for Different Environment and System Parameters.* Hittite Journal of Science & Engineering, 2020. **7**(4): p. 279-285.

21.Polmar, N.C. and N. Friedman, *submarine*, in *Britannica*. 2024.

22.Rayner, J., *'The Deadliest Thing That Keeps the Seas': The Technology, Tactics and Terror of the Submarine in <i>The War Illustrated</I> Magazine.* Journal for Maritime Research, 2017. **19**(1): p. 1-22.

23.Winokur, R.S., *Naval Oceanography Contributions to Underwater Acoustics—The Cold War Era.* The Journal of the Acoustical Society of America, 2015. **137**(4_Supplement): p. 2306-2307.

24.Toxopeus, S., et al., *Submarine Hydrodynamics for Off-Design Conditions.* Journal of Ocean Engineering and Marine Energy, 2022. **8**(4): p. 499-511.

25.Fay, D., N.A. Stanton, and A.P. Roberts, *All at Sea With User Interfaces: From Evolutionary to Ecological Design for Submarine Combat Systems.* Theoretical Issues in Ergonomics Science, 2019. **20**(5): p. 632-658.

26.Kong, D., et al., *An Improved Wake Vortex-Based Inversion Method for Submarine Maneuvering State.* Computational Intelligence and Neuroscience, 2023. **2023**(1).

27.Empyrea Consulting, *Under Pressure: The Science of Submarine Hulls.* 2023.

28.Davis, B., *How thick is a submarine's hull? What material is it made out of?* 2022, Medium.

29.Szturomski, B. and R. Kiciński, *Material Properties of HY 80 Steel After 55 Years of Operation for FEM Applications.* Materials, 2021. **14**(15): p. 4213.

30.Lee, J., et al., *Application of Macro-Instrumented Indentation Test for Superficial Residual Stress and Mechanical Properties Measurement for HY Steel Welded T-Joints.* Materials, 2021. **14**(8): p. 2061.

31.Fathallah, E., et al., *Optimal Design Analysis of Composite Submersible Pressure Hull.* Applied Mechanics and Materials, 2014. **578-579**: p. 89-96.

32.Helal, M., et al., *Numerical Analysis of Sandwich Composite Deep Submarine Pressure Hull Considering Failure Criteria.* Journal of Marine Science and Engineering, 2019. **7**(10): p. 377.

33.Jeong, H.Y. and P. Henry, *Optimal Design of Deep-Sea Pressure Hulls Using CAE Tools.* Journal of the Computational Structural Engineering Institute of Korea, 2012. **25**(6): p. 477-485.

34.Fathallah, E. and M. Helal, *Optimum Structural Design of Deep Submarine Pressure Hull to Achieve Minimum Weight.* The International Conference on Civil and Architecture Engineering, 2016. **11**(11): p. 1-22.

35.Fathallah, E., et al., *Optimal Structure Design of Elliptical Deep-Submersible Pressure Hull.* Materials Science Forum, 2015. **813**: p. 85-93.

36.Balaji, V., S.S.A. Ravi, and P. Chandran, *FEM Method Structural Analysis of Pressure Hull by Using Hyper Mesh.* International Journal of Engineering & Technology, 2017. **7**(1.5): p. 258.

37.MacKay, J.R. and F.v. Keulen, *Partial Safety Factor Approach to the Design of Submarine Pressure Hulls Using Nonlinear Finite Element Analysis.* Finite Elements in Analysis and Design, 2013. **65**: p. 1-16.

38.Helal, M., et al., *Minimizing Buoyancy Factor of Metallic Pressure-Hull Subjected to Hydrostatic Pressure.* Intelligent Automation & Soft Computing, 2023. **35**(1): p. 769-793.

39.Sohn, J.M., et al., *Development of Numerical Modelling Techniques for Composite Cylindrical Structures Under External Pressure.* Journal of Marine Science and Engineering, 2022. **10**(4): p. 466.

40.Maalawi, K.Y., *Optimal Buckling Design of Anisotropic Rings/Long Cylinders Under External Pressure.* Journal of Mechanics of Materials and Structures, 2008. **3**(4): p. 775-793.

41.Molavizadeh, A. and A.M. Rezaei, *Progressive Damage Analysis and Optimization of Winding Angle and Geometry for a Composite Pressure Hull Wound Using Geodesic and Planar Patterns.* Applied Composite Materials, 2019. **26**(3): p. 1021-1040.

42.Messager, T., et al., *Optimal Laminations of Thin Underwater Composite Cylindrical Vessels.* Composite Structures, 2002. **58**(4): p. 529-537.

43.Paik, J.K. and P.A. Frieze, *Ship Structural Safety and Reliability.* Progress in Structural Engineering and Materials, 2001. **3**(2): p. 198-210.

44.Chakraborty, S., *Understanding Structure Design of a Submarine*. 2021, Marine Insight.

45.Hsu, C.Y., et al., *The Effect of Roundness on the Buckling Strength for the Submerged Pressure Hull.* Applied Mechanics and Materials, 2014. **644-650**: p. 5133-5137.

46.Pei, Z., T. Xu, and W. Wu, *Progressive Collapse Test of Ship Structures in Waves.* Polish Maritime Research, 2018. **25**(s3): p. 91-98.

47.Xing, J.T., Z. Tian, and X. Yan, *The Dynamics of Ship Propulsion Unit-Large Hull–water Interactions.* Ocean Engineering, 2016. **124**: p. 349-362.

48.Plotkina, V.A., *Reliability Performance of Shipboard Power Complexes With Regard to Ecosystem Effects.* E3s Web of Conferences, 2021. **320**: p. 01005.

49.MacKay, J.R., et al., *Experimental Investigation of the Strength and Stability of Submarine Pressure Hulls With and Without Artificial Corrosion Damage.* Marine Structures, 2010. **23**(3): p. 339-359.

50.Sinaga, L.T.P., *Experimental Analysis on Sinking Time of Littoral Submarine in Various Trim Angle.* Applied Mechanics and Materials, 2018. **874**: p. 128-133.

51.Ibrahim, A.E., M.N. Karsiti, and I. Elamvazuthi, *Experimental Depth Positioning Control for a Spherical Underwater Robot Vehicle (URV).* Applied Mechanics and Materials, 2015. **785**: p. 729-733.

52.Pan, K. and I.M. Chao, *Analysis of Operating Principles and Flow Field Characteristics for a Diving Ballast Tank.* Defence Science Journal, 2020. **70**(5): p. 564-570.

53.Alsubal, S., M.S. Liew, and L.E. Shawn, *Preliminary Design and Dynamic Response of Multi-Purpose Floating Offshore Wind Turbine Platform: Part 1.* Journal of Marine Science and Engineering, 2022. **10**(3): p. 336.

54.Xiang, G., et al., *A Study on the Influence of Unsteady Forces on the Roll Characteristics of a Submarine During Free Ascent From Great Depth.* Journal of Marine Science and Engineering, 2024. **12**(5): p. 757.

55.Zhang, S., et al., *Numerical Simulation Study on the Effects of Course Keeping on the Roll Stability of Submarine Emergency Rising.* Applied Sciences, 2019. **9**(16): p. 3285.

56.Zhang, S., et al., *Experimental Investigation on Roll Stability of Blunt-Nose Submarine in Buoyantly Rising Maneuvers.* Applied Ocean Research, 2018. **81**: p. 34-46.

57.Chakraborty, S., *Understanding Stability of Submarine*. 2021, Marine Insight.

58.Globe Composite, *Buoyancy in Submersibles and Submarines*. 2024, Globe Composite Solutions.

59.Hien Le Tat, N., N. Nguyen Duy Anh, and N. Nguyen Thi Ngoc Hoa, *Numerical Investigate the Effect of Turbulence Models on the CFD Computation of Submarine Resistance.* CFD Letters, 2024. **16**(10): p. 126-139.

60.Lv, B. and Y. Chen, *The Influence of Different Diving Angles on the Hydrodynamic Noise of Submarine Enclosures.* Journal of Physics Conference Series, 2023. **2565**(1): p. 012006.

61.Moonesun, M., et al., *Optimization on Submarine Stern Design.* Proceedings of the Institution of Mechanical Engineers Part M Journal of Engineering for the Maritime Environment, 2016. **231**(1): p. 109-119.

62.Parunov, J., et al., *Hydrodynamic Tests of Innovative Tourist Submarine.* Journal of Marine Science and Engineering, 2023. **11**(6): p. 1199.

63.Vasudev, K.L., R. Sharma, and S.K. Bhattacharyya, *Multi-Objective Shape Optimization of Submarine Hull Using Genetic Algorithm Integrated With Computational Fluid Dynamics.* Proceedings of the Institution of Mechanical Engineers Part M Journal of Engineering for the Maritime Environment, 2017. **233**(1): p. 55-66.

64.Wu, X., et al., *An Effective CFD Approach for Marine-Vehicle Maneuvering Simulation Based on the Hybrid Reference Frames Method.* Ocean Engineering, 2015. **109**: p. 83-92.

65.White, F.M. and H. Xue, *Fluid mechanics.* Vol. 3. 2003: McGraw-hill New York.

66.Ke, L., J. Ye, and Q. Liang, *Experimental Study on the Flow Field, Force, and Moment Measurements of Submarines With Different Stern Control Surfaces.* Journal of Marine Science and Engineering, 2023. **11**(11): p. 2091.

67.Zhao, G., et al., *Submarine Maneuvering Performance Analysis: A Study of the Effect of Different Rudder Angle Settings on the Turning Dynamics of SUBOFF Submarine.* 2024. **1**(2).

68.Wang, W., et al., *A Fault-Tolerant Steering Prototype for X-Rudder Underwater Vehicles.* Sensors, 2020. **20**(7): p. 1816.

69.Wang, C., et al., *Influence of Stern Rudder Type on Flow Noise of Underwater Vehicles.* Journal of Marine Science and Engineering, 2022. **10**(12): p. 1866.

70.Anderson, D.M., et al., *Stability of Floating Objects at a Two-Fluid Interface.* European Journal of Physics, 2024. **45**(5): p. 055001.

71.Sakagami, N., et al., *An Attitude Control System for Underwater Vehicle-Manipulator Systems.* 2010.

72.Li, Q., S. Abdullah, and M.R.M. Rasani, *A Review of Progress and Hydrodynamic Design of Integrated Motor Pump-Jet Propulsion.* Applied Sciences, 2022. **12**(8): p. 3824.

73.Kim, J.-H., et al., *Performance Prediction of Composite Marine Propeller in Non-Cavitating and Cavitating Flow.* Applied Sciences, 2022. **12**(10): p. 5170.

74.Ku, G., et al., *Numerical Investigation of Tip Vortex Cavitation Inception and Noise of Underwater Propellers of Submarine Using Sequential Eulerian–Lagrangian Approaches.* Applied Sciences, 2020. **10**(23): p. 8721.

75.Grządziela, A., et al., *Experimental Validation of an FEM Model Based on Lifting Theory Applied to Propeller Design Software.* Polish Maritime Research, 2024. **31**(2): p. 67-76.

76.Indiaryanto, M., et al., *Design and Analysis of Cavitation on 22 M Submarine Propellers.* Majalah Ilmiah Pengkajian Industri, 2023. **14**(2): p. 163-171.

77.Liu, L., et al., *Full-Scale Simulation of Self-Propulsion for a Free-Running Submarine.* Physics of Fluids, 2021. **33**(4).

78.Özgen, C., *İsrail'in Savunma Stratejisi'nde Denizaltı Filosu'nun Rolü.* Güvenlik Stratejileri Dergisi, 2019. **15**(31): p. 497-545.

79.Duarte, A., et al., *GRASP With Path Relinking Heuristics for the Antibandwidth Problem.* Networks, 2010. **58**(3): p. 171-189.

80.Indiaryanto, M., et al., *Design and Hydrodynamic Model Test of Mini Submarine Propeller With High Efficiency and Low Cavitation.* Epi International Journal of Engineering, 2018. **1**(2): p. 59-64.

81.Putra, Y.D., A.F. Zakki, and A. Trimulyono, *Design of Kaplan-Series Propeller for Commercial Submarine by Varying Rake Angle and Number of the Blade to Obtain the Highest Thrust and Efficiency.* Iop Conference Series Materials Science and Engineering, 2021. **1096**(1): p. 012032.

82.Paredes, R., et al., *Numerical Flow Characterization Around a Type 209 Submarine Using OpenFOAM.* Fluids, 2021. **6**(2): p. 66.

83.Javadi, M., et al., *Experimental Investigation of the Effect of Bow Profiles on Resistance of an Underwater Vehicle in Free Surface Motion.* Journal of Marine Science and Application, 2015. **14**(1): p. 53-60.

84.Moonesun, M., et al., *Computational Fluid Dynamics Analysis on the Added Resistance of Submarine Due to Deck Wetness at Surface Condition.* Proceedings of the Institution of Mechanical Engineers Part M Journal of Engineering for the Maritime Environment, 2016. **231**(1): p. 128-136.

85.Chen, J., et al., *Study on Resistance Characteristics of Submarine Near Water Surface.* Matec Web of Conferences, 2022. **355**: p. 01002.

86.Gourlay, T. and E. Dawson, *A Havelock Source Panel Method for Near-Surface Submarines.* Journal of Marine Science and Application, 2015. **14**(3): p. 215-224.

87.Mitari, I.G., W. Sulistyawati, and P.J. Suranto, *Hydrodynamic Analysis in Redesigning a Monohull Passenger Ship Into a Catamaran.* E3s Web of Conferences, 2021. **328**: p. 07007.

88.Utama, I.K.A.P., et al., *CFD Analysis of Biofouling Effect on Submarine Resistance and Wake.* Journal of Marine Science and Engineering, 2023. **11**(7): p. 1312.

89.Gangemi, P.J., et al., *Active Control of the Accordion Modes of a Submerged Hull.* Journal of Low Frequency Noise Vibration and Active Control, 2011. **30**(3): p. 169-184.

90.Trost, L.C.S., *Moving Towards the Next Milestone of Submarine Design.* Naval Engineers Journal, 2000. **112**(2): p. 53-60.

91.Chen, F., Y. Chen, and H. Hu, *Coupled Vibration Characteristics of a Submarine Propeller-Shaft-Hull System at Low Frequency.* Journal of Low Frequency Noise Vibration and Active Control, 2019. **39**(2): p. 258-279.

92.Merz, S., R. Kinns, and N. Kessissoglou, *Influence of a Resonance Changer on the Sound Radiation of a Submarine.* The Journal of the Acoustical Society of America, 2008. **123**(5_Supplement): p. 3953-3953.

93.The Editors of Encyclopaedia, *naval architecture.* 2018, Britannica.

94.Overpelt, B. and B. Nienhuis, *Bow Shape Design for Increased Surface Performance of an SSK Submarine.* 2014: p. 117-126.

95.Sarraf, S., et al., *Experimental and Numerical Investigation of Squat Submarines Hydrodynamic Performances.* Ocean Engineering, 2022. **266**: p. 112849.

96.Guang-hui, Z. and J. Zhu, *Study on Key Techniques of Submarine Maneuvering Hydrodynamics Prediction Using Numerical Method.* 2010.

97.Pan, Y.-c., H. Zhang, and Q. Zhou, *Numerical Prediction of Submarine Hydrodynamic Coefficients Using CFD Simulation.* Journal of Hydrodynamics, 2012. **24**(6): p. 840-847.

98.Zarnetske, M.R. and J.B. Blottman, *Carbon Nanotube Thermoacoustic Projectors for Undersea Vehicles.* The Journal of the Acoustical Society of America, 2013. **134**(5_Supplement): p. 4091-4091.

99.Gad, A.M., *Antimicrobial and Antifouling Activities of the Cellulase Produced by Marine Fungal Strain; Geotrichum Candidum MN638741.1.* Egyptian Journal of Aquatic Biology and Fisheries, 2021. **25**(6): p. 49-60.

100.Slayter, J., N. McNaughton, and P. Bremner, *How do you design a submarine propulsion system that is efficient, reliable, and stealthy?* 2024, LinkedIn.

101.Defencyclopedia, *Anti-Submarine Warfare (Part-2) : Diesel-Electric Submarines.* 2014, Defencyclopedia.

102.Pekelney, R., *Submarine Main Propulsion Diesels: NavPers 16161.* 2007: San Francisco Maritime National Park Association.

103.Rothmund, M., *Underwater*. 2014, MTU.

104.Friedman, N., *nuclear submarine*, in *Encyclopedia Britannica*. 2024, Encyclopedia Britannica.

105.Nassersharif, B. and D. Thomas, *Nuclear Propulsion*. 2023.

106.Ismaeel, S.M.E., *New Applications for Linear Induction Drives Used for Silent Propulsion Systems in Nuclear Submarines and Supercarriers*. Arab Journal of Nuclear Sciences and Applications, 2020. **0**(0): p. 1-11.

107.World Nuclear Association, *Nuclear-Powered Ships*. 2023, World Nuclear Association.

108.J.M.K.C. Donev et al., *Pressurized water reactor*. 2024, Energy Education.

109.Jeon, B. and M. Khorsand, *Energy Management System in Naval Submarines*. 2020.

110.Phong, C.X., N.Q. Q, and C.X.C.X. P, *Early-Stage Analysis of Air Independent Propulsion Based on Fuel Cells for Small Submarines*. Advances in Military Technology, 2022. **17**(2): p. 457-469.

111.Wien, T., N. Gartner, and R. Geertsma, *Risk Mitigation for Application of Li-Ion Batteries on Submarines by Modelling of Heat and Combustible Gasses Development During a Thermal Runaway*. 2022.

112.Piłat, T., et al., *Implementation of the Assessment Method of the Lead–acid Battery Electrical Capacity in Submarines*. Journal of Marine Engineering & Technology, 2017. **16**(4): p. 326-330.

113.Yamashita, Y., H.N. Giang, and T. Oyama, *Investigating the Performance of Japan's Competitive Grant Grants-in-Aid for Scientific Research System*. International Journal of Higher Education, 2018. **7**(5): p. 167.

114.Depetro, A., G. Gamble, and K. Moinuddin, *Fire Safety Risk Analysis of Conventional Submarines*. Applied Sciences, 2021. **11**(6): p. 2631.

115.Madani, S.S., *Experimental Study of the Heat Generation of a Lithium-Ion Battery*. Ecs Transactions, 2020. **99**(1): p. 419-428.

116.Lai, X., et al., *Mechanism, Modeling, Detection, and Prevention of the Internal Short Circuit in Lithium-Ion Batteries: Recent Advances and Perspectives*. Energy Storage Materials, 2021. **35**: p. 470-499.

117.Sidhu, A., A. Izadian, and S. Anwar, *Adaptive Nonlinear Model-Based Fault Diagnosis of Li-Ion Batteries*. Ieee Transactions on Industrial Electronics, 2015. **62**(2): p. 1002-1011.

118.Ke, K. and S. Pang, *Working Principle and Application of Hydrogen Fuel Cells*. Highlights in Science Engineering and Technology, 2024. **90**: p. 97-104.

119.Wang, F., et al., *Boosting Electrochemical Performance of a Nanocomposite Ni-GDC Anode via Oxygen Vacancy and Ni Dispersion Modulation for Solid Oxide Fuel Cells*. Acs Applied Energy Materials, 2023. **6**(18): p. 9409-9416.

120.Ni, M., Z. Shao, and K.Y. Chan, *Modeling of Proton-Conducting Solid Oxide Fuel Cells Fueled With Syngas*. Energies, 2014. **7**(7): p. 4381-4396.

121.Gao, Z., et al., *A Perspective on Low-Temperature Solid Oxide Fuel Cells*. Energy & Environmental Science, 2016. **9**(5): p. 1602-1644.

122.Huerta, G.V., et al., *Impact of Multi-Causal Transport Mechanisms in an Electrolyte Supported Planar SOFC With (ZrO2)x−1(Y2O3)x Electrolyte*. Entropy, 2018. **20**(6): p. 469.

123.Lust, E., et al., *Development of Medium-Temperature Solid Oxide Fuel Cells and CO₂ and H₂O Co-Electrolysis Cells in Estonia*. Ecs Transactions, 2015. **68**(1): p. 3407-3415.

124.Almutairi, G., F. Alenazey, and Y. Alyousef, *Impact of Changing Mode on the Execution of 100 W Solid Oxide Fuel Cells (SOFCs)*. Journal of New Materials for Electrochemical Systems, 2019. **22**(4): p. 179-184.

125.Li, C.Y. and Z. Lv, *Preparation and Performance of Ni-Cu/SDC Anodes*. Advanced Materials Research, 2010. **113-116**: p. 1951-1954.

126.Bozorgmehri, S. and M. Hamedi, *Modeling and Optimization of Anode-Supported Solid Oxide Fuel Cells on Cell Parameters via Artificial Neural Network and Genetic Algorithm*. Fuel Cells, 2012. **12**(1): p. 11-23.

127.Zhou, F., et al., *Direct Ammonia-Fed Liquid Metal Anode Solid Oxide Fuel Cell for Co-Generation of Hydrogen and Electricity.* Ecs Transactions, 2023. **111**(6): p. 1517-1523.

128.Gadsbøll, R.Ø., et al., *Solid Oxide Fuel Cells Powered by Biomass Gasification for High Efficiency Power Generation.* Energy, 2017. **131**: p. 198-206.

129.Rahmanta, M.A., et al., *Nuclear Power Plant to Support Indonesia's Net Zero Emissions: A Case Study of Small Modular Reactor Technology Selection Using Technology Readiness Level and Levelized Cost of Electricity Comparing Method.* Energies, 2023. **16**(9): p. 3752.

130.Jiang, W., et al., *Numerical Analysis of the Hydrodynamic Performance of RDT With Varying Blade Numbers.* Journal of Physics Conference Series, 2022. **2271**(1): p. 012010.

131.Yan, X., et al., *A Review of Progress and Applications of Ship Shaft-Less Rim-Driven Thrusters.* Ocean Engineering, 2017. **144**: p. 142-156.

132.Dubas, A., N.W. Bressloff, and S.M. Sharkh, *Numerical Modelling of Rotor–stator Interaction in Rim Driven Thrusters.* Ocean Engineering, 2015. **106**: p. 281-288.

133.Zhang, L., et al., *Numerical Prediction of Cavitation Performance for Rim Driven Thruster.* 2018: p. 777-782.

134.Kim, J.-W., et al., *Parametric Study of the Hydrodynamic Characteristics of the Pumpjet Propulsor for the SUBOFF Submarine.* Journal of Marine Science and Engineering, 2023. **11**(10): p. 1926.

135.Kao, J.-H. and Y. Liao, *Discussion of the Fluid Acceleration Quality of a Ducted Propulsion System on the Propulsive Performance.* Computer Modeling in Engineering & Sciences, 2022. **130**(3): p. 1325-1348.

136.Chen, W., T. Liang, and V. Dinavahi, *Comprehensive Real-Time Hardware-in-the-Loop Transient Emulation of MVDC Power Distribution System on Nuclear Submarine.* 2022.

137.Ding, H., et al., *Effect of Frequency–Amplitude Parameter and Aspect Ratio on Propulsion Performance of Underwater Flapping-Foil.* Biomimetics, 2024. **9**(6): p. 324.

138.Han, L., Q. Huang, and G. Pan, *Investigation on the Propulsion of a Pump-Jet Propulsor in an Effective Wake.* Journal of Fluids Engineering, 2022. **144**(5).

139.Chakraborty, S., *Different Systems on a Naval Submarine.* 2021, Marine Insight.

140.Zaccone, R., U. Campora, and M. Martelli, *Optimisation of a Diesel-Electric Ship Propulsion and Power Generation System Using a Genetic Algorithm.* Journal of Marine Science and Engineering, 2021. **9**(6): p. 587.

141.Jafarboland, M. and M. Zadehbagheri, *Modeling of Belt-Pulley and Flexible Coupling Effects on Submarine Driven System Electrical Motors.* Journal of Power Electronics, 2011. **11**(3): p. 319-326.

142.Giovanni Benvenuto and Ugo Campora, *Economic and Environmental Comparison Between Diesel-Electric and Mechanical Propulsion Plants for a Small Cruise Ship.* International Journal of Frontiers in Engineering and Technology Research, 2023. **5**(2): p. 012-029.

143.Altosole, M., et al., *Progress in Marine Hybrid Propulsion Drive Systems.* 2023.

144.Pekelney, R., *Submarine Electrical Installations: NavPers 16112.* 2007: San Francisco Maritime National Park Association.

145.Sterling Thermal Technology, *The challenges of cooling submarine generators and motors.* 2023, Sterling Thermal Technology.

146.Le, H.N. and E. Tedeschi, *Comparative Evaluation of AC and DC Power Distribution Systems for Underwater Vehicles Based on Multiobjective Optimization Techniques.* Ieee Transactions on Power Delivery, 2021. **36**(6): p. 3456-3465.

147.Toshinsky, G.I. and V.V. Petrochenko, *Modular Lead-Bismuth Fast Reactors in Nuclear Power.* Sustainability, 2012. **4**(9): p. 2293-2316.

148.Wang, Y.-F., et al., *Study on ± 500 kV VSC-HVDC Submarine Cable Transmission Bottlenecks Under Different Laying Methods.* 2022.

149.Zhang, H., et al., *Overload Capacity Analysis of Extra High Voltage AC XLPE Submarine Cable.* Frontiers in Energy Research, 2023. **11**.

150.Jang, E.-H., et al., *Experimental Study on the Toxicity Characteristics of Non-Class 1E Cables According to Accelerated Deterioration.* Fire Science and Engineering, 2019. **33**(6): p. 105-113.

151.Youn, D.J., et al., *Controlling Factors of Degassing in Crosslinked Polyethylene Insulated Cables.* Polymers, 2019. **11**(9): p. 1439.

152.Li, Q., et al., *Numerical Study of the Local Scouring Process and Influencing Factors of Semi-Exposed Submarine Cables.* Journal of Marine Science and Engineering, 2023. **11**(7): p. 1349.

153.Wang, W., et al., *Failure of Submarine Cables Used in High-voltage Power Transmission: Characteristics, Mechanisms, Key Issues and Prospects.* Iet Generation Transmission & Distribution, 2021. **15**(9): p. 1387-1402.

154.Zhu, W., et al., *Thermal Effect of Different Laying Modes on Cross-Linked Polyethylene (XLPE) Insulation and a New Estimation on Cable Ampacity.* Energies, 2019. **12**(15): p. 2994.

155.Zhang, Z., et al., *The Study of Burial Depth and Risk Assessment of Submarine Power Cable.* Destech Transactions on Environment Energy and Earth Science, 2019(gmee).

156.Dong, W., et al., *Acetylated SEBS Enhanced DC Insulation Performances of Polyethylene.* Polymers, 2019. **11**(6): p. 1033.

157.Liu, Z., et al., *Morphological, Structural, and Dielectric Properties of Thermally Aged AC 500 kV XLPE Submarine Cable Insulation Material and Its Deterioration Condition Assessment.* Ieee Access, 2019. **7**: p. 165065-165075.

158.AbdelGawad, A.F., M. Elshaher, and H.E. Mostafa, *Stability of Double Base Propellants, a Comparison Between Classical Stability Tests and Modern Analytical Techniques.* The International Conference on Chemical and Environmental Engineering, 2014. **7**(7): p. 1-8.

159.Dai, X., et al., *Multi-Dimensional Analysis and Correlation Mechanism of Thermal Degradation Characteristics of XLPE Insulation for Extra High Voltage Submarine Cable.* Ieee Transactions on Dielectrics and Electrical Insulation, 2021. **28**(5): p. 1488-1496.

160.Wang, L., et al., *Research on Electromagnetic Loss Characteristics of Submarine Cables.* Energy Engineering, 2023. **120**(11): p. 2651-2666.

161.Wang, X., et al., *Physicochemical and Electrical Property of a New AC 500kV XLPE Submarine Cable Insulation With Different Thermal Aging Condition.* 2018.

162.Lee, J.H., H.S. Kim, and I.-S. Lee, *State of Health Monitoring of a Battery Module Using Multilayer Neural Network and Internal Resistance.* International Journal of Engineering Research and Technology, 2020. **13**(11): p. 3240.

163.Gao, Z., et al., *Design and Implementation of a Smart Lithium-Ion Battery System With Real-Time Fault Diagnosis Capability for Electric Vehicles.* Energies, 2017. **10**(10): p. 1503.

164.Skiba, R., *Battery Powered: The Social, Economical, and Environmental Impacts of the Lithium Ion Battery.* 2024: After Midnight Publishing.

165.Tang, Z., et al., *Numerical Analysis of Heat Transfer Mechanism of Thermal Runaway Propagation for Cylindrical Lithium-Ion Cells in Battery Module.* Energies, 2020. **13**(4): p. 1010.

166.Larsson, F. and B.-E. Mellander, *Abuse by External Heating, Overcharge and Short Circuiting of Commercial Lithium-Ion Battery Cells.* Journal of the Electrochemical Society, 2014. **161**(10): p. A1611-A1617.

167.Gao, A., F. Xu, and W. Dong, *The Concept of Early Monitoring and Warning of Thermal Runaway of Lithium-Ion Power Battery Using Parameter Analysis.* Journal of Physics Conference Series, 2022. **2181**(1): p. 012020.

168.Liang, K., Q. Zhu, and X.U.N. Zhou, *Simulation and Characteristic Analysis of High-Temperature Thermal Runaway Process in Ternary Lithium- Ion Batteries.* 2023.

169.Ma, X., et al., *Advances in Simulation Research for Thermal Runaway of Lithium-Ion Batteries.* Destech Transactions on Materials Science and Engineering, 2021(ameme).

170.Kimura, S., et al., *Influence of Aerogel Felt With Different Thickness on Thermal Runaway Propagation of 18650 Lithium-Ion Battery.* Electrochemistry, 2022. **90**(8): p. 087003-087003.

171.Yang, T., et al., *Thermal Performance Analysis of a Prismatic Lithium-Ion Battery Module Under Overheating Conditions.* Batteries, 2024. **10**(3): p. 86.

172.Ling, Z., et al., *A Hybrid Thermal Management System for Lithium Ion Batteries Combining Phase Change Materials With Forced-Air Cooling.* Applied Energy, 2015. **148**: p. 403-409.

173.Wen, L., et al., *Gas Characterization-Based Detection of Thermal Runaway Fusion in Lithium-Ion Batteries.* Electrochemistry, 2023. **91**(5): p. 057006-057006.

174.Li, B., et al., *A Thermoelectric Material Suitable for Early Warning of Thermal Runaway for Energy Storage Batteries: Monolayer GeP₃.* Journal of Physics Conference Series, 2024. **2788**(1): p. 012010.

175.Koch, S., K.P. Birke, and R. Kuhn, *Fast Thermal Runaway Detection for Lithium-Ion Cells in Large Scale Traction Batteries.* Batteries, 2018. **4**(2): p. 16.

176.Snyder, M. and A. Theis, *Understanding and Managing Hazards of L<scp>ithium-ion</Scp> Battery Systems.* Process Safety Progress, 2022. **41**(3): p. 440-448.

177.Ouyang, D., et al., *A Review on the Thermal Hazards of the Lithium-Ion Battery and the Corresponding Countermeasures.* Applied Sciences, 2019. **9**(12): p. 2483.

178.Liu, Y., et al., *Experimental Study on the Efficiency of Dodecafluoro-2-Methylpentan-3-One on Suppressing Lithium-Ion Battery Fires.* RSC Advances, 2018. **8**(73): p. 42223-42232.

179.Ghiji, M., et al., *Lithium-Ion Battery Fire Suppression Using Water Mist Systems.* Frontiers in Heat and Mass Transfer, 2021. **17**.

180.Jung, S., et al., *The Early Detection of Faults for Lithium-Ion Batteries in Energy Storage Systems Using Independent Component Analysis With Mahalanobis Distance.* Energies, 2024. **17**(2): p. 535.

181.Essl, C., A.W. Golubkov, and A. Fuchs, *Influence of Aging on the Failing Behavior of Automotive Lithium-Ion Batteries.* Batteries, 2021. **7**(2): p. 23.

182.Kaliaperumal, M., et al., *Cause and Mitigation of Lithium-Ion Battery Failure—A Review.* Materials, 2021. **14**(19): p. 5676.

183.Feng, X., et al., *Mitigating Thermal Runaway of Lithium-Ion Batteries.* Joule, 2020. **4**(4): p. 743-770.

184.Shen, J. and A. Khaligh, *A Supervisory Energy Management Control Strategy in a Battery/Ultracapacitor Hybrid Energy Storage System.* Ieee Transactions on Transportation Electrification, 2015. **1**(3): p. 223-231.

185.Tshiani, C.T. and P. Umenne, *The Impact of the Electric Double-Layer Capacitor (EDLC) in Reducing Stress and Improving Battery Lifespan in a Hybrid Energy Storage System (HESS) System.* Energies, 2022. **15**(22): p. 8680.

186.Xue, X.D., et al., *Loss Analysis of Hybrid Battery-Supercapacitor Energy Storage System in EVs.* 2017.

187.Park, E.Y. and J. Choi, *The Performance of Low-Pressure Seawater as a CO2 Solvent in Underwater Air-Independent Propulsion Systems.* Journal of Marine Science and Engineering, 2020. **8**(1): p. 22.

188.Banaei, M., et al., *Energy Management of Hybrid Diesel/Battery Ships in Multidisciplinary Emission Policy Areas.* Energies, 2020. **13**(16): p. 4179.

189.Wang, Y., et al., *A Comparative Study of Power Distribution Strategies for Fuel Cell and Supercapacitor Hybrid Power Source System.* Destech Transactions on Environment Energy and Earth Science, 2019(iceee).

190.Karimi, S., M.K. Zadeh, and J.A. Suul, *Evaluation of Energy Transfer Efficiency for Shore-to-Ship Fast Charging Systems.* 2020.

191.Shen, C. and L. Wan, *A Design Methodology for Lithium-Ion Battery Management System and Its Application to an Autonomous Underwater Vehicle.* Advanced Materials Research, 2011. **383-390**: p. 7175-7182.

192.Zhao, Q., et al., *Designing Solid-State Electrolytes for Safe, Energy-Dense Batteries.* Nature Reviews Materials, 2020. **5**(3): p. 229-252.

193.Banerjee, A., et al., *Interfaces and Interphases in All-Solid-State Batteries With Inorganic Solid Electrolytes.* Chemical Reviews, 2020. **120**(14): p. 6878-6933.

194.Fang, Y., et al., *High-Performance Dual-Ion Battery Based on a Layered Tin Disulfide Anode.* Acs Omega, 2022. **7**(9): p. 7616-7624.

195.Makaremi, M., B. Mortazavi, and C.V. Singh, *2D Hydrogenated Graphene-Like Borophene as a High Capacity Anode Material for Improved Li/Na Ion Batteries: A First Principles Study.* Materials Today Energy, 2018. **8**: p. 22-28.

196.Feng, G., *Transformation of Lithium Battery Material Design and Optimization Based on Artificial Intelligence.* Academic Journal of Materials & Chemistry, 2024. **5**(2): p. 18-22.

197.Samin, N.K., et al., *Synthesis and Battery Studies of Sodium Cobalt Oxides, NaCoO₂ Cathodes.* Advanced Materials Research, 2012. **545**: p. 185-189.

198.Sun, X., *Application and Performance Evaluation of Solid State Batteries in Renewable Energy Storage Systems.* 2023. **1**: p. 22-26.

199.Trinca, D., *Near Future Submarine: Development of a Combined Air Independent and Lithium Battery Propulsion System (AI-LiB Propulsion System).* 2022.

200.Lus, T., *Waiting for Breakthrough in Conventional Submarine's Prime Movers.* Transactions on Maritime Science, 2019. **8**(1): p. 37-45.

201.Moody, M. and D. Mitchell, *Battery Technology – Where It Came From and Where Its Going.* 2021.

202.Min, F., et al., *Research on the Heading Control of Underwater Vehicle Under Hover Condition.* Ieee Access, 2020. **8**: p. 220908-220920.

203.Yu, J., et al., *Redundancy Optimization of Standby Phased-Mission Systems.* 2010.

204.Ottaviano, A., et al., *ControlPULP: A RISC-V on-Chip Parallel Power Controller for Many-Core HPC Processors With FPGA-Based Hardware-in-the-Loop Power and Thermal Emulation.* 2023.

205.Zhao, Y., Q. Wang, and D. Ha, *A Sliding-Mode Duty-Ratio Controller for DC/DC Buck Converters With Constant Power Loads.* Ieee Transactions on Industry Applications, 2014. **50**(2): p. 1448-1458.

206.Qun-ying, S., et al., *Power and Energy Management in Integrated Power System.* 2011.

207.Pekelney, R., *Submarine Air Systems: NavPers 16164.* 2007: San Francisco Maritime National Park Association.

208.Wu, Q., et al., *Advances and Status of Anode Catalysts for Proton Exchange Membrane Water Electrolysis Technology.* Materials Chemistry Frontiers, 2023. **7**(6): p. 1025-1045.

209.d'Amore-Domenech, R., et al., *Alkaline Electrolysis for Hydrogen Production at Sea: Perspectives on Economic Performance.* Energies, 2023. **16**(10): p. 4033.

210.Wang, W., et al., *The Oxygen Generation Performance of Hollow-Structured Oxygen Candle for Refuge Space.* Journal of Chemistry, 2018. **2018**: p. 1-9.

211.Goodall, A., *The Development of a Non-Powered Oxygen Generator for Royal Navy Submarines.* 2012.

212.Shu, W., et al., *Study on Chemical Oxygen Source in Underground Emergency Refuge System.* 2014: p. 571-576.

213.Jin, L., et al., *Development of a Low Oxygen Generation Rate Chemical Oxygen Generator for Emergency Refuge Spaces in Underground Mines.* Combustion Science and Technology, 2015. **187**(8): p. 1229-1239.

214.Goeppert, A., et al., *Easily Regenerable Solid Adsorbents Based on Polyamines for Carbon Dioxide Capture From the Air.* Chemsuschem, 2014. **7**(5): p. 1386-1397.

215.Zoannou, K.-S., D.J. Sapsford, and A.J. Griffiths, *Thermal Degradation of Monoethanolamine and Its Effect on CO2 Capture Capacity.* International Journal of Greenhouse Gas Control, 2013. **17**: p. 423-430.

216.Dani, C., *Automated Control of Inspired Oxygen (FiO₂) in Preterm Infants: Literature Review.* Pediatric Pulmonology, 2019. **54**(3): p. 358-363.

217.Lellouche, F. and E. L'Her, *Automated Oxygen Flow Titration to Maintain Constant Oxygenation.* Respiratory Care, 2012. **57**(8): p. 1254-1262.

218.Malli, F., et al., *Automated Oxygen Delivery in Hospitalized Patients With Acute Respiratory Failure: A Pilot Study.* Canadian Respiratory Journal, 2019. **2019**: p. 1-7.

219.Hallenberger, A., et al., *Closed-Loop Automatic Oxygen Control (CLAC) in Preterm Infants: A Randomized Controlled Trial.* Pediatrics, 2014. **133**(2): p. e379-e385.

220.Roca, O., et al., *Closed-Loop Oxygen Control Improves Oxygen Therapy in Acute Hypoxemic Respiratory Failure Patients Under High Flow Nasal Oxygen: A Randomized Cross-Over Study (The HILOOP Study).* Critical Care, 2022. **26**(1).

221.Hollier, C.A., et al., *Moderate Concentrations of Supplemental Oxygen Worsen Hypercapnia in Obesity Hypoventilation Syndrome: A Randomised Crossover Study.* Thorax, 2013. **69**(4): p. 346-353.

222.Vivodtzev, I., et al., *Automated O₂ Titration Improves Exercise Capacity in Patients With Hypercapnic Chronic Obstructive Pulmonary Disease: A Randomised Controlled Cross-Over Trial.* Thorax, 2018. **74**(3): p. 298-301.

223.Lellouche, F., et al., *Automated Oxygen Titration and Weaning With FreeO₂ In Patients With Acute Exacerbation of COPD: A Pilot Randomized Trial.* International Journal of Chronic Obstructive Pulmonary Disease, 2016. **Volume 11**: p. 1983-1990.

224.Kaltsogianni, O., et al., *Does Closed-Loop Automated Oxygen Control Reduce the Duration of Supplementary Oxygen Treatment and the Amount of Time Spent in Hyperoxia? A Randomised Controlled Trial in Ventilated Infants Born at or Near Term.* Trials, 2023. **24**(1).

225.Sturrock, S., et al., *A Randomised Crossover Trial of Closed Loop Automated Oxygen Control in Preterm, Ventilated Infants.* Acta Paediatrica, 2020. **110**(3): p. 833-837.

226.Hansen, E.F., et al., *Automated Oxygen Control With O2matic® During Admission With Exacerbation of COPD.* International Journal of Chronic Obstructive Pulmonary Disease, 2018. **Volume 13**: p. 3997-4003.

227.Cormoş, C.-C., et al., *Assessment of Hybrid Solvent—Membrane Configurations for Post-Combustion CO2 Capture for Super-Critical Power Plants.* Energies, 2021. **14**(16): p. 5017.

228.Capocelli, M. and M.D. Falco, *Generalized Penalties and Standard Efficiencies of Carbon Capture and Storage Processes.* International Journal of Energy Research, 2021. **46**(4): p. 4808-4824.

229.Johnson, N., I. Ikoko, and A.J. Chukwuma, *Driving Sustainability in Power Generation: Amine Scrubbing Integration as a Cost-Effective Measure for Carbon Dioxide Mitigation.* 2024.

230.García-Mariaca, A. and E. Llera-Sastresa, *Energy and Economic Analysis Feasibility of CO₂capture on a Natural Gas Internal Combustion Engine.* Greenhouse Gases Science and Technology, 2022. **13**(2): p. 144-159.

231.Gervasi, J., L. Dubois, and D. Thomas, *Screening Tests of New Hybrid Solvents for the Post-Combustion CO2 Capture Processby Chemical Absorption.* Energy Procedia, 2014. **63**: p. 1854-1862.

232.Al-Saedi, R., *A Review on Modified MOFs as CO2 Adsorbents Using Mixed Metals and Functionalized Linkers.* Samarra Journal of Pure and Applied Science, 2023. **5**(1): p. 1-18.

233.Katayev, N., et al., *An Intelligent Fuzzy-Pid Controller for Supporting Comfort Microclimate in Smart Homes.* International Journal of Advanced Computer Science and Applications, 2024. **15**(2).

234.Tunyagi, A., et al., *Automatic System for Continuous Monitoring of Indoor Air Quality and Remote Data Transmission Under SMART_RAD_EN Project.* Studia Universitatis Babeș-Bolyai Ambientum, 2017. **62**(2): p. 71-80.

235.Chen, Y., *Research on Energy-Saving Conventional Submarine Air-Conditioning System Based on Heat and Humidity Load Calculation.* Highlights in Science Engineering and Technology, 2023. **56**: p. 407-414.

236.Lawson, S., et al., *Amine-Functionalized MIL-101 Monoliths for CO₂ Removal From Enclosed Environments.* Energy & Fuels, 2019. **33**(3): p. 2399-2407.

237.Pekelney, R., *Submarine Refrigerating and Air-Conditioning Systems: NavPers 16163.* 2007: San Francisco Maritime National Park Association.

238.Миан, З., M.V. Ramana, and A.H. Nayyar, *Nuclear Submarines in South Asia: New Risks and Dangers.* Journal for Peace and Nuclear Disarmament, 2019. **2**(1): p. 184-202.

239.Merk, B., et al., *On a Long Term Strategy for the Success of Nuclear Power.* 2017.

240.Hippel, F.v., *Mitigating the Threat of Nuclear-Weapon Proliferation via Nuclear-Submarine Programs.* Journal for Peace and Nuclear Disarmament, 2019. **2**(1): p. 133-150.

241.Hao, X., et al., *Analysis of Heat Stress and the Indoor Climate Control Requirements for Movable Refuge Chambers.* International Journal of Environmental Research and Public Health, 2016. **13**(5): p. 518.

242.Trezza, B.M., et al., *Environmental Heat Exposure and Cognitive Performance in Older Adults: A Controlled Trial.* Age, 2015. **37**(3).

243.Jwo, C.S., et al., *Energy Saving Analysis on Application of Condensing Heat Recovery to Constant Temperature/Humidity System.* Applied Mechanics and Materials, 2013. **291-294**: p. 1805-1811.

244.Li, C.L., X. Zhang, and S.L. Chung, *Temperature and Humidity Control Inside an Automobile During Heating Period.* Journal of the Chinese Institute of Engineers, 2012. **35**(6): p. 641-654.

245.Roberts, A.P., N.A. Stanton, and D. Fay, *Land Ahoy! Understanding Submarine Command and Control During the Completion of Inshore Operations.* Human Factors the Journal of the Human Factors and Ergonomics Society, 2017. **59**(8): p. 1263-1288.

246.Pope, K.A., A.P. Roberts, and N.A. Stanton, *Investigating Temporal Implications of Information Transition in Submarine Command Teams.* 2018: p. 243-253.

247.Loft, S., et al., *The Chronic Detrimental Impact of Interruptions in a Simulated Submarine Track Management Task.* Human Factors the Journal of the Human Factors and Ergonomics Society, 2015. **57**(8): p. 1417-1426.

248.Mitchell, S.W., *Transitioning the <scp>SWFTS</Scp> Program Combat System Product Family From Traditional Document-Centric to Model-Based Systems Engineering.* Systems Engineering, 2013. **17**(3): p. 313-329.

249.Seo, K.-M., et al., *Measurement of Effectiveness for an Anti-Torpedo Combat System Using a Discrete Event Systems Specification-Based Underwater Warfare Simulator.* The Journal of Defense Modeling and Simulation Applications Methodology Technology, 2011. **8**(3): p. 157-171.

250.Liu, X., et al., *Research on the Collaborative Search Problem of Unmanned Surface Vehicles.* 2023.

251.Tatasciore, M., et al., *The Benefits and Costs of Low and High Degree of Automation.* Human Factors the Journal of the Human Factors and Ergonomics Society, 2019. **62**(6): p. 874-896.

252.Asundi, R.V., et al., *Sonar Interface With FCS and Target Detection.* International Journal of Engineering & Technology, 2018. **7**(3.12): p. 541.

253.Kim, J.Y., et al., *Design, Implementation and Navigation Test of Manta-Type Unmanned Underwater Vehicle.* International Journal of Ocean System Engineering, 2011. **1**(4): p. 192-197.

254.Stanton, N.A. and K. Bessell, *How a Submarine Returns to Periscope Depth: Analysing Complex Socio-Technical Systems Using Cognitive Work Analysis.* Applied Ergonomics, 2014. **45**(1): p. 110-125.

255.Costa, E.P.L.D. and D.P. Júnior, *Comparing the Effectiveness of Nuclear and Air-Independent Propulsion Submarine Fleets.* Estudos Internacionais Revista De Relações Internacionais Da Puc Minas, 2023. **10**(2): p. 42-58.

256.Zhao, K., B. Yu, and J. Wang, *Simulations of the Anti-Torpedo Tactic of the Conventional Submarine Using Decoys and Jammers.* Applied Mechanics and Materials, 2011. **65**: p. 165-168.

257.Silva, M.V.M.d., *Safeguards and the Nuclear-Powered Submarines of the NNWS: There Is No Gap; There Is a First Time / Submarinos De Propulsão Nuclear Dos NNWS E as Salvaguardas Da AIEA: Não Há Lacuna; Há Uma Primeira Vez.* Mural Internacional, 2023. **14**: p. e75437.

258.Hause, M. and J. Hallett, *Model-Based Product Line Engineering to Plan and Track Submarine Configuration.* Insight, 2019. **22**(2): p. 57-66.

259.Zheng, Y., et al., *Research on the Interception Capability of Distributed UUV Swarm.* 2024.

260.Zheng, Y., W. Liu, and H.F. Li, *Research on the Effect of Load on the Fatigue Tearing of the Elastic Element.* Advanced Materials Research, 2013. **753-755**: p. 1836-1841.

261.Park, K.-m., et al., *Modeling and Simulation for Anti-Submarine HVU Escort Mission.* Journal of the Korea Society for Simulation, 2014. **23**(4): p. 75-83.

262.Naval History and Heritage Command, *Navy's Use of Torpedoes.* 2024, Naval History and Heritage Command,.

263.Pekelney, R., *The Fleet Type Submarine Online 21-Inch Submerged Torpedo Tubes.* 2007: San Francisco Maritime National Park Association.

264.Zhang, L., et al., *Synthetic Evaluation of Warfare Capability of Submarine Launched Torpedoes Based on Projection Pursuit Model.* Advanced Materials Research, 2012. **490-495**: p. 2956-2960.

265.Korbut, M. and D. Szpica, *A Review of Compressed Air Engine in the Vehicle Propulsion System.* Acta Mechanica Et Automatica, 2021. **15**(4): p. 215-226.

266.Mikalsen, R. and A.P. Roskilly, *The Design and Simulation of a Two-Stroke Free-Piston Compression Ignition Engine for Electrical Power Generation.* Applied Thermal Engineering, 2008. **28**(5-6): p. 589-600.

267.Wang, X.P., J.J. Dang, and Z.C. Zhao, *Research on Closed-Loop Control of the Step Response.* Applied Mechanics and Materials, 2011. **127**: p. 71-76.

268.Liu, Y., et al., *Study of Mechanism of Counter-Rotating Turbine Increasing Two-Stage Turbine System Efficiency.* International Journal of Fluid Machinery and Systems, 2013. **6**(3): p. 160-169.

269.Liu, F., et al., *Active Fault Localization of Actuators on Torpedo-Shaped Autonomous Underwater Vehicles.* Sensors, 2021. **21**(2): p. 476.

270.Huang, C.-N., et al., *Experimental Investigation on the Performance of a Compressed-Air Driven Piston Engine.* Energies, 2013. **6**(3): p. 1731-1745.

271.Budge, K.G., *Torpedoes*, in *The Pacific War Online Encyclopedia.* 2016, The Pacific War Online Encyclopedia.

272.Krzyscik, M.A., Ł. Opaliński, and J. Otlewski, *Novel Method for Preparation of Site-Specific, Stoichiometric-Controlled Dual Warhead Conjugate of FGF2 via Dimerization Employing Sortase a-Mediated Ligation.* Molecular Pharmaceutics, 2019. **16**(8): p. 3588-3599.

273.Yu, W., D.J. Weber, and A.D. MacKerell, *Integrated Covalent Drug Design Workflow Using Site Identification by Ligand Competitive Saturation.* Journal of Chemical Theory and Computation, 2023. **19**(10): p. 3007-3021.

274.Wang, C., T. Ma, and J. Ning, *Experimental Investigation of Penetration Performance of Shaped Charge Into Concrete Targets.* Acta Mechanica Sinica, 2008. **24**(3): p. 345-349.

275.Hong, Z., et al., *Fluid-Structure Interaction Simulation and Accurate Dynamic Modeling of Parachute Warhead System Based on Impact Point Prediction.* Ieee Access, 2021. **9**: p. 104418-104428.

276.Wang, J., et al., *Selective Covalent Targeting of Pyruvate Kinase M2 Using Arsenous Warheads.* Journal of Medicinal Chemistry, 2023. **66**(4): p. 2608-2621.

277.Gehringer, M. and S. Laufer, *Emerging and Re-Emerging Warheads for Targeted Covalent Inhibitors: Applications in Medicinal Chemistry and Chemical Biology.* Journal of Medicinal Chemistry, 2018. **62**(12): p. 5673-5724.

278.Newman, A.M., et al., *Optimizing Assignment of Tomahawk Cruise Missile Missions to Firing Units.* Naval Research Logistics (Nrl), 2011. **58**(3): p. 281-294.

279.Li, R. and Y. Shi, *A Time-Fuel Optimal Control Problem of a Cruise Missile Based on an Improved Sliding Mode Variable Structure Model.* The Anziam Journal, 2009. **51**(2): p. 261-276.

280.None, N. and M. Banasik, *Multi-Domain Concept of Using A2/Ad Capabilities in the Military Strategy of the Russian Federation.* Polish Political Science Yearbook, 2023. **52**(3): p. 119-131.

281.Köklücan, S. and M.K. Leblebicioğlu, *Energy-Optimal Control of a Submarine-Launched Cruise Missile.* 2020.

282.Dwivedi, P.N., A. Bhattacharyya, and R. Padhi, *Improved Capturability of Terminal Angles With Modified Suboptimal Mid-Course Guidance.* Ifac Proceedings Volumes, 2011. **44**(1): p. 3891-3896.

283.Wu, Z., et al., *Three-Dimensional Cooperative Mid-Course Guidance Law Against the Maneuvering Target.* Ieee Access, 2020. **8**: p. 18841-18851.

284.Guo, C., H. Cai, and G.H.M.v.d. Heijden, *Guidance and Control of a Cruise Missile Flying Along a Geomagnetic Isoline.* Proceedings of the Institution of Mechanical Engineers Part G Journal of Aerospace Engineering, 2013. **228**(7): p. 1215-1224.

285.Wu, L., et al., *Dynamic Sequential Radar Cross Section Properties of Airborne Corner Reflector in Array.* Iet Radar Sonar & Navigation, 2023. **17**(9): p. 1405-1419.

286.Sun, K. and W. Yin, *Research on Optimization of Ship Radar RCS Based on PO-MOM Algorithm.* 2024: p. 83.

287.Luo, Y., et al., *Study on the Jamming-Position Maneuver Algorithm of Off-Board Active Electronic Countermeasure Unmanned Surface Vehicles.* Ieee Access, 2021. **9**: p. 61184-61192.

288.Pekelney, R., *Submarine Sonar Operator's Manual: NavPers 16167.* 2007: San Francisco Maritime National Park Association.

289.Eliav, R. and I. Klein, *INS/Partial DVL Measurements Fusion With Correlated Process and Measurement Noise.* 2018. **39**: p. 34.

290.Xia, X. and Q. Sun, *Initial Alignment Algorithm Based on the DMCS Method in Single-Axis RSINS With Large Azimuth Misalignment Angles for Submarines.* Sensors, 2018. **18**(7): p. 2123.

291.Zhang, L., et al., *A Novel Monitoring Navigation Method for Cold Atom Interference Gyroscope.* Sensors, 2019. **19**(2): p. 222.

292.Xu, H., et al., *Application and Development of Fiber Optic Gyroscope Inertial Navigation System in Underground Space.* Sensors, 2023. **23**(12): p. 5627.

293.Guo, R., et al., *Metrology and Measurement Systems.* 2019.

294.Zhang, Q., et al., *An Accurate Calibration Method Based on Velocity in a Rotational Inertial Navigation System.* Sensors, 2015. **15**(8): p. 18443-18458.

295.Klein, I. and R. Diamant, *Observability Analysis of DVL/PS Aided INS for a Maneuvering AUV.* Sensors, 2015. **15**(10): p. 26818-26837.

296.Cha, J., et al., *Integration of Inertial Navigation System With EM-log Using H-Infinity Filter.* E3s Web of Conferences, 2019. **94**: p. 01013.

297.Zhang, K., et al., *Underwater Navigation Based on Topographic Contour Image Matching.* 2010.

298.Ye, K., et al., *A Mobile Prototype-Based Localization Approach Using Inertial Navigation and Acoustic Tracking for Underwater.* Frontiers in Marine Science, 2024. **11**.

299.Akbarian, H. and M.h. Sedaaghi, *Underwater Acoustic Target Recognition Using Spectrogram ROI Approximation With Mobilenet One-Dimensional and Two-Dimensional Networks.* 2023.

300.Peyvandi, H., et al., *SONAR Systems and Underwater Signal Processing: Classic and Modern Approaches.* 2011.

301.Hu, Q., *Sonar Technology.* 2019: p. 1-9.

302.Fernandes, J.d.C.V., N.M.J.N.D. Moura, and J.M. Seixas, *Deep Learning Models for Passive Sonar Signal Classification of Military Data.* Remote Sensing, 2022. **14**(11): p. 2648.

303.Wei, G., X. Yu, and X. Long, *Novel Approach for Identifying Z-Axis Drift of RLG Based on GA-SVR Model.* Journal of Systems Engineering and Electronics, 2014. **25**(1): p. 115-121.

304.Qiao, W. and S. Li, *Submarine Target Recognition Based on GPS Positioning System.* Iop Conference Series Materials Science and Engineering, 2020. **740**(1): p. 012207.

305.Adams, G.W. and M.P. Gokhale, *Fiber Optic Gyro Based Precision Navigation for Submarines.* 2000.

306.Li, K., et al., *Federated Ultra-tightly Coupled GPS/INS Integrated Navigation System Based on Vector Tracking for Severe Jamming Environment.* Iet Radar Sonar & Navigation, 2016. **10**(6): p. 1030-1037.

307.Niu, M., et al., *A New Self-Calibration and Compensation Method for Installation Errors of Uniaxial Rotation Module Inertial Navigation System.* Sensors, 2022. **22**(10): p. 3812.

308.El-Shafie, A., et al., *Performance Enhancement of Underwater Target Tracking by Fusing Data of Array of Global Positioning System Sonobuoys.* Journal of Computer Science, 2009. **5**(3): p. 199-206.

309.El-Shafie, A., et al., *Performance Evaluation of a Non-Linear Error Model for Underwater Range Computation Utilizing GPS Sonobuoys.* Neural Computing and Applications, 2010. **19**(7): p. 1057-1067.

310.Kissai, A. and M.G. Smith, *UAV Dead Reckoning With and Without Using INS/GPS Integrated System in GPS Denied Polar Regions.* International Journal of Aeronautics and Aerospace Engineering, 2019. **1**(2): p. 58-67.

311.Zhang, H., et al., *A Real Time Localization System for Vehicles Using Terrain-Based Time Series Subsequence Matching.* Remote Sensing, 2020. **12**(16): p. 2607.

312.Hide, C., T. Moore, and M.J. Smith, *Adaptive Kalman Filtering for Low-Cost INS/GPS.* Journal of Navigation, 2003. **56**(1): p. 143-152.

313.Wölfl, A.-C., et al., *Seafloor Mapping – The Challenge of a Truly Global Ocean Bathymetry.* Frontiers in Marine Science, 2019. **6**.

314.Heyuan, S., et al., *Bathymetric Prediction Using Multisource Gravity Data Derived From a Parallel Linked BP Neural Network.* Journal of Geophysical Research Solid Earth, 2022. **127**(11).

315.Neupane, D. and J. Seok, *A Review on Deep Learning-Based Approaches for Automatic Sonar Target Recognition.* Electronics, 2020. **9**(11): p. 1972.

316.Song, D., et al., *Construction of Secure Adaptive Frequency Hopping Sequence Sets Based on AES Algorithm.* Iet Communications, 2024. **18**(8): p. 490-502.

317.Ning, B., et al., *Probabilistic Frequency-hopping Sequence With Low Probability of Detection Based on Spectrum Sensing.* Iet Communications, 2017. **11**(14): p. 2147-2153.

318.Yuan, Z., et al., *Intelligent Reception of Frequency Hopping Signals Based on CVDP.* Applied Sciences, 2023. **13**(13): p. 7604.

319.Niu, X., et al., *A Construction of Optimal Frequency Hopping Sequence Set via Combination of Multiplicative and Additive Groups of Finite Fields.* Ieee Transactions on Information Theory, 2020. **66**(8): p. 5310-5315.

320.Peng, Q., et al., *Performance Improvement of Underwater Continuous-Variable Quantum Key Distribution via Photon Subtraction.* Entropy, 2019. **21**(10): p. 1011.

321.Ruan, J.X., Y. Xu, and M. Cui, *Frequency-Hopping Communication Technology Based on Self-Adaptive MIMO-OFDM System.* Advanced Materials Research, 2014. **945-949**: p. 2230-2236.

322.Roberts, A.P. and N.A. Stanton, *Macrocognition in Submarine Command and Control: A Comparison of Three Simulated Operational Scenarios.* Journal of Applied Research in Memory and Cognition, 2018. **7**(1): p. 92-105.

323.Pekelney, R., *Submarine Hydraulic Systems: NavPers 16169.* 2007: San Francisco Maritime National Park Association.

324.Tkáč, Z., et al., *Experimental Hydraulic Device for the Testing of Hydraulic Pumps and Liquids.* Tribology in Industry, 2018. **40**(1): p. 149-155.

325.Hujo, Ľ., et al., *Measurement of Flow Characteristics of a Gear Hydraulic Pump by Simulating the Operating Load of the Tractor's Hydraulic System.* Matec Web of Conferences, 2021. **338**: p. 01010.

326.Quilumba, F.L., et al., *Improving Hydraulic System Energy Efficiency With High-Performance Hydraulic Fluids.* Ieee Transactions on Industry Applications, 2014. **50**(2): p. 1313-1321.

327.Shi, P.C. and Y. Sun, *Hydraulic Fluid Mathematical Modeling.* Applied Mechanics and Materials, 2013. **432**: p. 127-132.

328.Deuster, S. and K. Schmitz, *Holistic Efficiency Measurements of a Mobile Working Machine: Comparison of Conventional Mineral Oils and a Sustainable Water-Based Fluid.* 2024: p. 102-114.

329.Wan, H., J. Fang, and H. Huang, *Numerical Simulation on a Throttle Governing System With Hydraulic Butterfly Valves in a Marine Environment.* Journal of Marine Science and Application, 2010. **9**(4): p. 403-409.

330.Chanbua, W. and U. Pinsopon, *Influence of RBD Palm Olein on Hydraulic Pump Performance.* Advanced Materials Research, 2014. **931-932**: p. 403-407.

331.Majdan, R., et al., *Effect of Ecological Oils on the Quality of Materials of Hydraulic Pump Components.* Advanced Materials Research, 2013. **801**: p. 1-6.

332.Salloom, M.Y. and Z. Samad, *Experimental Test of Magneto-Rheological Directional Control Valve.* Advanced Materials Research, 2011. **383-390**: p. 5409-5413.

333.Wu, S., J. Zhang, and H. Zhu, *An Integrated Simulation and Evolution Platform of Multiple Control Valve Based on AMESim.* 2011.

334.Liu, D.D., C. Tang, and C. Zhao, *The Electro-Hydraulic Control Directional Valve Based on Magneto-Rheological Fluid.* Key Engineering Materials, 2013. **567**: p. 139-142.

335.Hamilton, P.B., K. Strom, and D.C.J.D. Hoyal, *Hydraulic and Sediment Transport Properties of Autogenic Avulsion Cycles on Submarine Fans With Supercritical Distributaries.* Journal of Geophysical Research Earth Surface, 2015. **120**(7): p. 1369-1389.

336.Zhang, F., *Design of Hydraulic Control System for Press Machine and Analysis on Its Fluid Transmission Features.* International Journal of Heat and Technology, 2021. **39**(1): p. 161-169.

337.García-Valdovinos, L.G., et al., *Modelling, Design and Robust Control of a Remotely Operated Underwater Vehicle.* International Journal of Advanced Robotic Systems, 2014. **11**(1).

338.Brandt, H., *Submarine Technologies? A Source for Solutions!* 2012.

339.Piwowarski, M., *The Analysis of Turbine Propulsion Systems in Nuclear Submarines.* Key Engineering Materials, 2013. **597**: p. 99-105.

340.Satish, U., et al., *Is CO ₂ an Indoor Pollutant? Direct Effects of Low-to-Moderate CO ₂ Concentrations on Human Decision-Making Performance.* Environmental Health Perspectives, 2012. **120**(12): p. 1671-1677.

341.Lowe, R., G.M. Huebner, and T. Oreszczyn, *Possible Future Impacts of Elevated Levels of Atmospheric CO₂ on Human Cognitive Performance and on the Design and Operation of Ventilation Systems in Buildings.* Building Services Engineering Research and Technology, 2018. **39**(6): p. 698-711.

342.Perrone, C., *SDO-SuRS Deployable Assets Program.* 2022.

343.Kotsky, M.A., et al., *Opportunity of Liquid Ventilation Method Implementation in Advanced Technologies of Accidental Submarine Crews Rescue.* Russian Journal of Occupational Health and Industrial Ecology, 2022. **62**(9): p. 566-578.

344.Ponton, K., D. Parera, and J. Irons, *The Submarine Habitability Assessment Questionnaire: A Survey of RAN Submariners.* Journal of Marine Science and Engineering, 2021. **9**(1): p. 54.

345.Yang, J. and W. Li, *Reliability Optimization Design of Submarine Free-Running Model Systems.* International Journal of Engineering and Technology, 2016. **8**(5): p. 323-328.

346.Yu, J., S.H. Chen, and G.Q. Huang, *Method Based on Knowledge Modules of Hydraulic Fault and It's Application.* Advanced Materials Research, 2011. **186**: p. 66-70.

347.Tile, P., *Materials and Manufacturing Process used in Submarines.* 2022, Medium.

348.Uppal, R., *Advanced Materials and Technologies Transforming Submarine Hulls for Enhanced Warfighting Capabilities.* 2023, International Defense, Security & technology (CA, USA).

349.Nurulloh, M.I., L. Simbolon, and G.R. Deksino, *Barium Ferrite Magnet as Anti-Radar Material.* Techno (Jurnal Fakultas Teknik Universitas Muhammadiyah Purwokerto), 2022. **23**(1).

350.Airimiţoaie, T.-B., et al., *Algorithms for Adaptive Feedforward Noise Attenuation—A Unified Approach and Experimental Evaluation.* Ieee Transactions on Control Systems Technology, 2021. **29**(5): p. 1850-1862.

351.Li, B., et al., *Research on Characteristics of Flow Noise and Flow-Induced Noise.* Applied Sciences, 2023. **13**(19): p. 11095.

352.He, S., et al., *Study on the Relationship Between Reflectivity and Thickness of Radar-Absorbing Material.* Journal of Physics Conference Series, 2024. **2808**(1): p. 012084.

353.Zhang, H., et al., *Application of the General Matching Law on the Study of Multi-Coated Radar Absorbing Materials.* E3s Web of Conferences, 2020. **165**: p. 05027.

354.Khashaba, U.A., *Toughness, Flexural, Damping and Interfacial Properties of Hybridized GFRE Composites With MWCNTs.* Composites Part a Applied Science and Manufacturing, 2015. **68**: p. 164-176.

355.Jerome, S., et al., *Studies on Mechanical Properties and Wear Behaviour of ≪i>In Situ ≪/I>Al-TiC Composites.* Advanced Materials Research, 2011. **328-330**: p. 1654-1658.

356.Łabuński, P. and L. Witek, *Experimental Analysis of Damping Properties of Viscoelastic Materials.* Acta Metallurgica Slovaca, 2021. **27**(2): p. 63-67.

357.Rafiee, M., F. Nitzsche, and M.R. Labrosse, *Processing, Manufacturing, and Characterization of Vibration Damping in Epoxy Composites Modified With Graphene Nanoplatelets.* Polymer Composites, 2019. **40**(10): p. 3914-3922.

358.Curle, U.A., J.D. Wilkins, and G. Govender, *Industrial Semi-Solid Rheocasting of Aluminum A356 Brake Calipers.* Advances in Materials Science and Engineering, 2011. **2011**: p. 1-5.

359.Duc, D.M. and N.H. Hai, *Study on Rheo-Continuous Casting of Al-Si A356 (EN AC4200) Alloys.* Key Engineering Materials, 2016. **682**: p. 220-225.

360.Möller, H., et al., *Comparison of Heat Treatment Response of Semisolid Metal Processed Alloys A356 and F357.* International Journal of Cast Metals Research, 2010. **23**(1): p. 37-43.

361.Guan, T.Y., et al., *Effects of Annular Electromagnetic Stirring Melt Treatment on Microstructure and Mechanical Properties of 7050 Rheo-Casting.* Solid State Phenomena, 2019. **285**: p. 219-223.

362.Pramanik, R. and A. Arockiarajan, *Effective Properties and Nonlinearities in 1-3 Piezocomposites: A Comprehensive Review.* Smart Materials and Structures, 2019. **28**(10): p. 103001.

363.Sun, R., et al., *Characterization of 1-3 Piezoelectric Composite With a 3-Tier Polymer Structure.* Materials, 2020. **13**(2): p. 397.

364.Mirza, M.S., et al., *Dice-and-Fill Processing and Characterization of Microscale and High-Aspect-Ratio (K, Na)NbO3-Based 1–3 Lead-Free Piezoelectric Composites.* Ceramics International, 2016. **42**(9): p. 10745-10750.

365.Woodward, D.I., et al., *Additively-<scp>m</Scp>anufactured Piezoelectric Devices.* Physica Status Solidi (A), 2015. **212**(10): p. 2107-2113.

366.Yang, H.C., et al., *Crosstalk Reduction for High-Frequency Linear-Array Ultrasound Transducers Using 1-3 Piezocomposites With Pseudo-Random Pillars.* Ieee Transactions on Ultrasonics Ferroelectrics and Frequency Control, 2012. **59**(10): p. 2312-2321.

367.Jiang, L., et al., *Fabrication of a (K,Na)NbO3-Based Lead-Free 1-3 Piezocomposite for High-Sensitivity Ultrasonic Transducers Application.* Journal of Applied Physics, 2019. **125**(21).

368.Park, H. and Y. Roh, *Design of Ultrasonic Fingerprint Sensor Made of 1–3 Piezocomposites by Finite Element Method.* Japanese Journal of Applied Physics, 2017. **56**(7S1): p. 07JD06.

369.Pyo, S. and Y. Roh, *Optimization of the Structure of 1–3 Piezocomposite Materials to Maximize the Performance of an Underwater Acoustic Transducer Using Equivalent Circuit Models and Finite Element Method.* Japanese Journal of Applied Physics, 2015. **54**(7S1): p. 07HB03.

370.Jeronimo, K., et al., *PDMS-ZnO Piezoelectric Nanocomposites for Pressure Sensors.* Sensors, 2021. **21**(17): p. 5873.

371.Zhou, M., et al., *Fabrication and Properties of 1-3-2 Multi-Element Piezoelectric Composite.* Journal of Electroceramics, 2012. **28**(2-3): p. 139-143.

372.Golbabaei, F. and M. Khadem, *Air Pollution in Welding Processes — Assessment and Control Methods.* 2015.

373.Ilić, A., et al., *Analysis of Influence of the Welding Procedure on Impact Toughness of Welded Joints of the High-Strength Low-Alloyed Steels.* Applied Sciences, 2020. **10**(7): p. 2205.

374.Layus, P., et al., *Study of the Sensitivity of High-Strength Cold-Resistant Shipbuilding Steels to Thermal Cycle of Arc Welding.* International Journal of Mechanical and Materials Engineering, 2018. **13**(1).

375.Mraz, L., et al., *Identification of Weld Residual Stresses Using Diffraction Methods and Their Effect on Fatigue Strength of High Strength Steels Welds.* Materials Science Forum, 2013. **768-769**: p. 668-674.

376.Li, Z.L., et al., *Research on Welding Process of Low Alloy High Strength Steel 20MnTiB.* Key Engineering Materials, 2019. **815**: p. 114-119.

377.Fydrych, D. and J. Łabanowski, *Weldability of High Strength Steels in Wet Welding Conditions.* Polish Maritime Research, 2013. **20**(2): p. 67-73.

378.Kornookar, K., et al., *Influence of Heat Input on Microstructure and Mechanical Properties of Gas Tungsten Arc Welded HSLA S500MC Steel Joints.* Metals, 2022. **12**(4): p. 565.

379.Tiwadi, D., et al., *Interrelationship Modeling Among Weld Strength Improvement by Parametric Approach in TIG Welding Using DEMATEL Software.* Evergreen, 2023. **10**(4): p. 2564-2569.

380.Karganroudi, S.S., et al., *Experimental and Numerical Analysis on TIG Arc Welding of Stainless Steel Using RSM Approach.* Metals, 2021. **11**(10): p. 1659.

381.Devakumar, D., et al., *Characterization of Duplex Stainless Steel/Cold Reduced Low Carbon Steel Dissimilar Weld Joints by GTAW.* Applied Mechanics and Materials, 2015. **766-767**: p. 780-788.

382.Vasudevan, M., et al., *Genetic Algorithm for Optimisation of a-Tig Welding Process for Modified 9Cr–1Mo Steel.* Science and Technology of Welding & Joining, 2010. **15**(2): p. 117-123.

383.Mohammed, R., G.M. Reddy, and K.S. Rao, *Effect of Filler Wire Composition on Microstructure and Pitting Corrosion of Nickel Free High Nitrogen Stainless Steel GTA Welds.* Transactions of the Indian Institute of Metals, 2016. **69**(10): p. 1919-1927.

384.Kumar, R., S.C. Vettivel, and H. Kumar Kansal, *Bulletin of the Polish Academy of Sciences: Technical Sciences.* 2021.

385.Peasura, P., *Experiment Design With Full Factorial in Gas Tungsten Arc Welding Parameters on Aluminium Alloy 5083.* Advanced Materials Research, 2013. **711**: p. 183-187.

386.Ridzuan, M.N.H.B., *Mechanical Properties and Microstructural Analysis of 304 Stainless Steel by TIG Welding.* 2017.

387.Tseng, K.-H. and C.-J. Hsu, *Performance of Activated TIG Process in Austenitic Stainless Steel Welds.* Journal of Materials Processing Technology, 2011. **211**(3): p. 503-512.

388.Yuan, X., et al., *Profiling Inclusion Characteristics in Submerged Arc Welded Metals of EH36 Shipbuilding Steel Treated by CaF₂–TiO₂ Fluxes.* Science and Technology of Welding & Joining, 2022. **27**(8): p. 683-690.

389.Vinay Yadav, G.S., Manish Bharti, *Research of Arc Stability in Submerged Arc Welding Based on SiO2 and TiO2 Flux System.* International Journal of Innovative Technology and Exploring Engineering, 2019. **8**(12S): p. 282-286.

390.Gook, S., et al., *Joining 30 Mm Thick Shipbuilding Steel Plates EH36 Using a Process Combination of Hybrid Laser Arc Welding and Submerged Arc Welding.* Journal of Manufacturing and Materials Processing, 2022. **6**(4): p. 84.

391.Yamada, S., et al., *Experimental Study of the Ductility of a Submerged Arc Welded Corner Joint in a High-Performance Steel Built-Up Box Column.* International Journal of Steel Structures, 2020. **20**(5): p. 1454-1464.

392.Kanjilal, P., T.K. Pal, and S. Majumdar, *Combined Effect of Flux and Welding Parameters on Chemical Composition and Mechanical Properties of Submerged Arc Weld Metal.* Journal of Materials Processing Technology, 2006. **171**(2): p. 223-231.

393.Patnaik, A., S. Biswas, and S.S. Mahapatra, *An Evolutionary Approach to Parameter Optimisation of Submerged Arc Welding in the Hardfacing Process.* International Journal of Manufacturing Research, 2007. **2**(4): p. 462.

394.Satheesh, M., et al., *Multi Objective Optimization of Weld Parameters of Boiler Steel Using Fuzzy Based Desirability Function.* Journal of Engineering Science and Technology Review, 2014. **7**(1): p. 29-36.

395.Towsyfyan, H., et al., *Comparing the Regression Analysis and Artificial Neural Network in Modeling the Submerged Arc Welding (SAW) Process.* Research Journal of Applied Sciences Engineering and Technology, 2013. **5**(9): p. 2701-2706.

396.Nowacki, J. and P. Rybicki, *The Influence of Welding Heat Input on Submerged Arc Welded Duplex Steel Joints Imperfections.* Journal of Materials Processing Technology, 2005. **164-165**: p. 1082-1088.

397.Karaoğlu, S. and A. Seçgin, *Sensitivity Analysis of Submerged Arc Welding Process Parameters.* Journal of Materials Processing Technology, 2008. **202**(1-3): p. 500-507.

398.Siddharth, P.N. and C.S. Narayanan, *A Review on Electron Beam Welding Process.* Journal of Physics Conference Series, 2020. **1706**(1): p. 012208.

399.Zheng, Q., et al., *Effect on LY12 Aluminum Alloy Welding Joint Microstructure and Properties With Electron Beam Welding Technical Parameters.* Applied Mechanics and Materials, 2013. **475-476**: p. 1275-1279.

400.Górka, J., S. Błacha, and D. Zagrobelny, *Electron Beam Welding of TMCP Steel S700MC.* Biuletyn Instytutu Spawalnictwa, 2020(4): p. 17-23.

401.Wang, F., et al., *Effect of Electron Beam Welding on Microstructure and Mechanical Properties of Spray-Deposited Al-Zn-Mg-Cu Alloy.* Applied Mechanics and Materials, 2013. **302**: p. 230-235.

402.Singh, J.K., G.G. Roy, and J.D. Majumdar, *Studies on the Effect of Process Parameter on Corrosion Behaviour of Electron Beam Welded Ti-Based Alloy (Ti6Al4V).* Welding in the World, 2023. **67**(12): p. 2731-2747.

403.Guo, R., et al., *Preparation and Welding Performance of Ti–6Al–4V Powder Compact Fabricated by Hot Isostatic Pressing.* Materials Science Forum, 2016. **849**: p. 760-765.

404.Kotlarski, G., et al., *Electron-Beam Welding of Titanium and Ti6Al4V Alloy.* Metals, 2023. **13**(6): p. 1065.

405.Xiao, G., et al., *Bionic Structure on Complex Surface With Belt Grinding for Electron Beam Welding Seam of Titanium Alloy.* Applied Sciences, 2020. **10**(7): p. 2370.

406.Ol'shanskaya, T.V., et al., *Application of Dynamic Beam Positioning for Creating Specified Structures and Properties of Welded Joints in Electron-Beam Welding.* Materials, 2020. **13**(10): p. 2233.

407.Fan, J., et al., *Influence of Multi-Beam Electron Beam Welding Technique on the Deformation of Ti6Al4V Alloy Sheet.* Rare Metal Materials and Engineering, 2017. **46**(9): p. 2417-2422.

408.Karhu, M. and V. Kujanpää, *Controlling Root Penetration in Electron Beam Welding by a Through-Current Feedback.* Welding in the World, 2022. **67**(3): p. 777-791.

409.Nayak, L.J. and G.G. Roy, *Role of Beam Oscillation on Electron Beam Welded Zircaloy-4 Butt Joints.* Science and Technology of Welding & Joining, 2021. **26**(6): p. 478-486.

410.Ltd., E.B.P., *The Electron Beam Welding Process Explained.* 2024, Electron Beam Processes Ltd.

411.Oyyaravelu, R., P. Kuppan, and A. Natarajan, *Metallurgical and Mechanical Properties of Laser Welded High Strength Low Alloy Steel.* Journal of Advanced Research, 2016. **7**(3): p. 463-472.

412.Sisodia, R.P.S., M. Gáspár, and L. Draskóczi, *Effect of Post-Weld Heat Treatment on Microstructure and Mechanical Properties of DP800 and DP1200 High-Strength Steel Butt-Welded Joints Using Diode Laser Beam Welding.* Welding in the World, 2020. **64**(4): p. 671-681.

413.Vemanaboina, H., et al., *Thermal Analysis Simulation for Laser Butt Welding of Inconel625 Using FEA.* International Journal of Engineering & Technology, 2018. **7**(4.10): p. 85.

414.Fahlström, K., et al., *Correlation Between Laser Welding Sequence and Distortions for Thin Sheet Structures.* Science and Technology of Welding & Joining, 2017. **22**(2): p. 150-156.

415.Andersson, O., et al., *Experimental Measurements and Numerical Simulations of Distortions of Overlap Laser-Welded Thin Sheet Steel Beam Structures.* Welding in the World, 2017. **61**(5): p. 927-934.

416.Pala, Z., et al., *Study of Residual Stress Surface Distribution on Laser Welded Steel Sheets.* Applied Mechanics and Materials, 2013. **486**: p. 3-8.

417.Schimek, M., et al., *Laser Bead-on-Plate Welding and Overlap Seams for Increasing the Strength and Rigidity of High Strength Steel.* Advanced Materials Research, 2010. **137**: p. 161-190.

418.Subbaiah, K., et al., *Comparative Evaluation of Tungsten Inert Gas and Laser Beam Welding of AA5083-H321.* Sadhana, 2012. **37**(5): p. 587-593.

419.Landowski, M., *Influence of Parameters of Laser Beam Welding on Structure of 2205 Duplex Stainless Steel.* Advances in Materials Science, 2019. **19**(1): p. 21-31.

420.Buschenhenke, F., T. Seefeld, and F. Vollertsen, *Strategies for Reduced Distortion During Laser Beam Welding of Shaft-hub Joints.* Materialwissenschaft Und Werkstofftechnik, 2012. **43**(1-2): p. 105-111.

421.Hong, J., K.-H. Lee, and C. Lee, *The Frame Optimization and Validation of Resistance Spot Welding Gun.* International Journal for Simulation and Multidisciplinary Design Optimization, 2020. **11**: p. 22.

422.Farrahi, G.H., et al., *Analysis of Resistance Spot Welding Process Parameters Effect on the Weld Quality of Three-Steel Sheets Used in Automotive Industry: Experimental and Finite Element Simulation.* International Journal of Engineering, 2020. **33**(1).

423.Xu, C., et al., *Study on Microstructure and Fracture Morphology of 2205 Duplex Stainless Steel Resistance Spot Welds.* Materials Science Forum, 2014. **804**: p. 289-292.

424.Arabi, S.H., M. Pouranvari, and M. Movahedi, *Pathways to Improve the Austenite–ferrite Phase Balance During Resistance Spot Welding of Duplex Stainless Steels.* Science and Technology of Welding & Joining, 2019. **24**(1): p. 8-15.

425.Chandio, A.D., et al., *Failure Study of Two Dissimilar Steels Joined by Spot Welding Technique.* Key Engineering Materials, 2018. **778**: p. 262-267.

426.Yang, Y., R.J. Zhou, and F.X. Wang, *Numerical Simulation of the Temperature Field During DP1000 Dual-Phase Steel Resistance Spot Welding.* Advanced Materials Research, 2011. **391-392**: p. 666-671.

427.Wang, N.N., et al., *Resistance Spot Welding Between Mild Steel and Stainless Steel.* Applied Mechanics and Materials, 2014. **675-677**: p. 23-26.

428.Viňáš, J. and Ľ. Kaščák, *Analysis of Welds Made by Delta Spot Method.* Materials Science Forum, 2015. **818**: p. 229-232.

429.Gandomkar, M. and M. Moshref-Javadi, *The Crack Resolution Required to Ensure Fatigue Life of a Rescue Submarine's Transfer Skirt.* The International Journal of Maritime Engineering, 2022. **163**(A4).

430.Zhu, Y., J. Yang, and H. Pan, *Three-Dimension Crack Propagation Behavior of Conical-Cylindrical Shell.* Metals, 2023. **13**(4): p. 698.

431.Hu, Y., et al., *Research on Development and Test Analysis of Full-Scale Fatigue Test System of X65 Submarine Pipeline.* E3s Web of Conferences, 2021. **253**: p. 01055.

432.Garcia, J.H.N., et al., *Corrosion Behavior of 316L and Alloy 182 Dissimilar Weld Joint With Post-Weld Heat Treatment.* Matéria (Rio De Janeiro), 2019. **24**(3).

433.Oikonomou, A.G. and G.A. Aggidis, *Determination of the Corrosion Resistance of the Welded Steels Used in Underwater Marine Systems (Including the Submerged Parts of Wave Energy Converters).* Materials Today Proceedings, 2021. **44**: p. 5048-5053.

434.Agarwal, K., et al., *Mechanical Properties of Fiber Reinforced Polymer Composites: A Comparative Study of Conventional and Additive Manufacturing Methods.* Journal of Composite Materials, 2018. **52**(23): p. 3173-3181.

435.Han, K., et al., *Six-Dof CFD Simulations of Underwater Vehicle Operating Underwater Turning Maneuvers.* Journal of Marine Science and Engineering, 2021. **9**(12): p. 1451.

436.Zhang, N. and S. Zhang, *Numerical Simulation of Hull/Propeller Interaction of Submarine in Submergence and Near Surface Conditions.* Journal of Hydrodynamics, 2014. **26**(1): p. 50-56.

437.Yu, P., et al., *Simulation Analysis of Modular Assembly Accuracy of Marine Power Shafting Based on 3DCS.* 2022.

438.Farooq, M.U., et al., *Process Parameters Optimization and Performance Analysis of Micro-Complex Geometry Machining on Ti6Al4V.* International Journal on Interactive Design and Manufacturing (Ijidem), 2024.

439.Farooq, M.U. and S. Anwar, *Investigations on the Surface Integrity of Ti6Al4V Under Modified Dielectric(s)-Based Electric Discharge Machining Using Cryogenically Treated Electrodes.* Processes, 2023. **11**(3): p. 877.

440.Yang, X., et al., *A Novel Evaluation Method on the Precision of Linear Motor Feed System in High-Speed Machine Tools.* Materials Science Forum, 2016. **836-837**: p. 220-227.

441.Rao, J.M., et al., *Sensitivity Analysis of a Double Corrugated Waveguide Slow Wave Structure for a 151 – 161.5 GHz TWT.* Asian Journal of Physics, 2023. **32**(9-12): p. 559-571.

442.Zhou, Y., et al., *Fast Anti-Turbidity Underwater Topography Measurement Based on Rotating Structured Light.* 2024.

443.Miller, A.B., B.W. Miller, and G. Miller, *Navigation of Underwater Drones and Integration of Acoustic Sensing With Onboard Inertial Navigation System.* Drones, 2021. **5**(3): p. 83.

444.Tetsui, T., *Practical Use of Hot-Forged-Type Ti-42Al-5Mn and Various Recent Improvements.* Metals, 2021. **11**(9): p. 1361.

445.Azhagan, M.T., B. Mohan, and A. Rajadurai, *Comparative Study of Squeeze Casting of AA6061 With and Without Employing Ultrasonic Cavitations.* Applied Mechanics and Materials, 2014. **541-542**: p. 349-353.

446.Qi, Y., et al., *Preparation and Properties of Special Vehicle Cover via a Novel Squeeze Casting Quantitative Feeding System of Molten Metal.* Metals, 2020. **10**(2): p. 266.

447.Men, Z., et al., *Analysis on the Free Forging Process of the Heavy Rack for Ship Lift.* Applied Mechanics and Materials, 2017. **865**: p. 105-108.

448.Gryguc, A., et al., *A Method for Comparing the Fatigue Performance of Forged AZ80 Magnesium.* Metals, 2021. **11**(8): p. 1290.

449.Dziubińska, A., *The New Technology of Die Forging of Automotive Connecting Rods From EN AB-71100 Aluminium Alloy Cast Preforms.* Materials, 2023. **16**(7): p. 2856.

450.Böhmichen, U., et al., *From Casting to Forging—The Combined Simulation for a Steel Component.* Engineering Reports, 2021. **4**(7-8).

451.Zhang, Q., et al., *Research on Integrated Casting and Forging Process of Aluminum Automobile Wheel.* Advances in Mechanical Engineering, 2014. **6**: p. 870182.

452.Yu, Z., et al., *Numerical Model Simulation of the Double-Roll Rotary Forging of Large Diameter Thin-Walled Disk.* Metals, 2021. **11**(11): p. 1767.

453.Zhao, Z., et al., *Microstructure and Corrosion Behavior of Cold-Sprayed Zn-Al Composite Coating.* Coatings, 2020. **10**(10): p. 931.

454.Prasai, D., et al., *Graphene: Corrosion-Inhibiting Coating.* Acs Nano, 2012. **6**(2): p. 1102-1108.

455.Ke, Y., et al., *Urushiol Titanium <scp>polymer-based</Scp> Composites Coatings for Anti-corrosion and Antifouling in Marine Spray Splash Zones.* Journal of Applied Polymer Science, 2021. **138**(34).

456.Wang, Y., et al., *Modification and Corrosion Resistance of Halloysite Carrier With Metal Nanoinhibitor in Marine Corrosion Environment.* Anti-Corrosion Methods and Materials, 2022. **69**(4): p. 371-379.

457.Zhang, J., et al., *Non-Destructive Evaluation of Coating Thickness Using Water Immersion Ultrasonic Testing.* Coatings, 2021. **11**(11): p. 1421.

458.Han, L., Q. Pang, and X. Yu, *Application of Organic Coating in Marine Anticorrosion.* Highlights in Science Engineering and Technology, 2023. **58**: p. 131-141.

459.Uzun, D., et al., *A CFD Study: Influence of Biofouling on a Full-Scale Submarine.* Applied Ocean Research, 2021. **109**: p. 102561.

460.Cui, M., B. Wang, and Z. Wang, *Nature-Inspired Strategy for Anticorrosion.* Advanced Engineering Materials, 2019. **21**(7).

461.Liu, L., et al., *Double-Layer Superhydrophobic Coating: Enhanced Anti-Corrosion and Anti-Fouling Resistance Through Star Cross-Linking of FPI/POSS@POA/F-SiO 2.* 2024.

462.Saifullah, A., et al., *Mechanical Properties of Layered-Carbon Fiber Reinforced With Vacuum Infusion Process.* Journal of Energy Mechanical Material and Manufacturing Engineering, 2021. **6**(1): p. 33-40.

463.Buchacz, A., et al., *An Investigation of the Influence of a Fiber Arrangement of a Laminate on the Values of Stresses in the Composite Panel of a Modified Freight Wagon Using the FEM Method.* Matec Web of Conferences, 2017. **112**: p. 04015.

464.Salman, S.D., W.S.W. Hassim, and Z. Leman, *Experimental Comparison Between Two Types of Hybrid Composite Materials in Compression Test.* Manufacturing Science and Technology, 2015. **3**(4): p. 119-123.

465.Zhou, Y. and C. Zheng, *Impact Resistance Performance and Structural Optimization of Lightweight Composite Target Plate.* Polymer Composites, 2024. **45**(7): p. 6482-6497.

466.Xie, X., et al., *Experiments on Vibration Transmission Control in a Shaft-Hull System Excited by Propeller Forces via an Activemulti-Strut Assembly.* Proceedings of the Institution of Mechanical Engineers Part M Journal of Engineering for the Maritime Environment, 2022. **236**(3): p. 688-700.

467.Gatin, I., et al., *CFD Study on the Influence of Exostructure Elements on the Resistance of a Submarine.* Journal of Marine Science and Engineering, 2022. **10**(10): p. 1542.

468.Chrismianto, D., et al., *Development of Cubic Bezier Curve and Curve-Plane Intersection Method for Parametric Submarine Hull Form Design to Optimize Hull Resistance Using CFD.* Journal of Marine Science and Application, 2015. **14**(4): p. 399-405.

469.Patri, M., *Naval Materials and Energy Systems.* Technology Focus, 2020. **28**(6).

470.Miasnikov, E., *The Future of Russia's Strategic Nuclear Forces: Discussions and Arguments.* The Center For Arms Control, Energy, and Environmental Studies: at Moscow Institute of Physics and Technology, 1995.

471.Yuan, B., et al., *Underwater Acoustic Properties of Graphene Nanoplatelet-Modified Rubber.* Journal of Reinforced Plastics and Composites, 2018. **37**(9): p. 609-616.

472.Caresta, M. and N. Kessissoglou, *Acoustic Signature of a Submarine Hull Under Harmonic Excitation.* Applied Acoustics, 2010. **71**(1): p. 17-31.

473.Liu, Y., Y. Li, and D. Shang, *The Generation Mechanism of the Flow-Induced Noise From a Sail Hull on the Scaled Submarine Model.* Applied Sciences, 2018. **9**(1): p. 106.

474.Xu, W. and Z. Li, *Study on Two-Stage Mounting Systems Having Distributed Intermediate Mass.* Journal of Vibroengineering, 2020. **22**(2): p. 313-321.

475.Qiu, Y., W. Xu, and Z. Li, *Performance Evaluation of a Novel Intelligent Distributed Mounting System for Marine Mechanical Equipment.* International Journal of Advanced Robotic Systems, 2019. **16**(6).

476.Hooper, C., *Virginia Class: When does hull coating separation endanger the boat?* 2010, Next Navy.

477.Xia, Q.Q., et al., *Research on Metal-Rubber Composite Damping Steel Plate Application in Acoustic Bridge for Double Cylindrical Shell.* Advanced Materials Research, 2012. **490-495**: p. 3505-3510.

478.Takahashi, K., J. Arai, and T. Mori, *Numerical Study on Multiple-Blade-Rate Unsteady Propeller Forces for Underwater Vehicles.* Journal of Ship Research, 2021. **66**(04): p. 349-368.

479.Andersen, P., J.J. Kappel, and E. Spangenberg. *Aspects of propeller developments for a submarine.* in *Proceedings of the First International Symposium on Marine Propulsors–smp*. 2009.

480.nasiri, m. and m. nazemizadeh, *Hydromagnetic Propulsion System in Marine Applications: A Literature Review.* Mechanic of Advanced and Smart Materials, 2022. **2**(1): p. 13-34.

481.Yeo, S.-J., et al., *Integrated Analysis of Flow-Induced Noise From Submarine Under Snorkel Condition.* Proceedings of the Institution of Mechanical Engineers Part M Journal of Engineering for the Maritime Environment, 2020. **234**(4): p. 771-784.

482.Zhu, Z., et al., *A Review of Underwater Acoustic Metamaterials for Underwater Acoustic Equipment.* Frontiers in Physics, 2022. **10**.

483.Cao, H. and L. Wen, *High-Precision Numerical Research on Flow and Structure Noise of Underwater Vehicle.* Applied Sciences, 2022. **12**(24): p. 12723.

484.Jawahar, A., et al., *Advanced Submarine Integrated Weapon Control System.* Indian Journal of Science and Technology, 2015. **8**(35).

485.Su, P., et al., *Research Status and Development Trend of Acoustic Array Integrating Technology.* 2023.

486.Luo, X., et al., *A Survey of Underwater Acoustic Target Recognition Methods Based on Machine Learning.* Journal of Marine Science and Engineering, 2023. **11**(2): p. 384.

487.Kumar, S., et al., *Submarine Acoustic Target Strength Modeling at High-Frequency Asymptotic Scattering.* Ieee Access, 2024. **12**: p. 4859-4870.

488.Jiang, S., *On Securing Underwater Acoustic Networks: A Survey.* Ieee Communications Surveys & Tutorials, 2019. **21**(1): p. 729-752.

489.Likun, M.E.I. and C. Zhili, *Based on YOLOv5 Lightweight Submarine Target Detection Algorithm.* 2023.

490.Mei, L. and Z. Chen, *An Improved YOLOv5-Based Lightweight Submarine Target Detection Algorithm.* Sensors, 2023. **23**(24): p. 9699.

491.Barr, S., J. Wang, and B. Liu, *An Efficient Method for Constructing Underwater Sensor Barriers.* Journal of Communications, 2011. **6**(5).

492.Liu, Z.-K. and Z. Tang, *Numerical Analysis of Multi-Scale Pressure Pulsation on the Energy Accumulation for Submarine-Based Tracking and Pointing Systems.* Measurement and Control, 2021. **54**(3-4): p. 196-208.

493.Schaefer, D., et al., *Above Water Electric Potential Signatures of Submerged Naval Vessels.* Journal of Marine Science and Engineering, 2019. **7**(2): p. 53.

494.Chen, Y. and J. Yuan, *Methods of Differential Submarine Detection Based on Magnetic Anomaly and Technology of Probes Arrangement.* 2015.

495.Zhang, Z.Y. and J. Yi, *Measuring the Submarine's Induced Magnetic Field by Geomagnetic Simulation Method.* Advanced Materials Research, 2014. **981**: p. 579-584.

496.Kim, Y., Y.H. Kim, and K. Shin, *Efficiency of Exponential Deperm Protocol.* Journal of Magnetics, 2013. **18**(3): p. 326-330.

497.Nguyen, T.S., et al., *Ships Magnetic Anomaly Computation With Integral Equation and Fast Multipole Method.* Ieee Transactions on Magnetics, 2011. **47**(5): p. 1414-1417.

498.Badr, A., A.A. Elserougi, and A.A. Hossam-Eldin, *Analyzing the Effect of Flux Intensification Due to Propeller Shaft and Rudder Upon the 3D Induced Magnetic Signature of Marine Vessels.* Ieee Access, 2022. **10**: p. 131627-131639.

499.Madrid, A.V. and A.Á. Melcón, *Process for Compensating Local Magnetic Perturbations on Ferromagnetic Surfaces.* Journal of Electromagnetic Analysis and Application, 2012. **04**(10): p. 387-399.

500.SchoolTube Community, *Stealth Submarines: The Science Behind Silent Warfare.* 2024, SchoolTube Community.

501.Vice Admiral A.K. Chawla (Retd), *Unseen & Unheard: The Role of Stealth.* 2023, SP's Naval Forces.

502.Sutton, H.I., *Chinese Submarine Is First To Exploit New Stealth Technology.* 2023, Naval News.

503.Dehghani, R. and H.M. Khanlo, *A Regressor-Free Robust Adaptive Controller for Autonomous Underwater Vehicles.* Proceedings of the Institution of Mechanical Engineers Part M Journal of Engineering for the Maritime Environment, 2016. **231**(2): p. 569-582.

504.Wang, J., T. Xu, and Z. Wang, *Adaptive Robust Unscented Kalman Filter for AUV Acoustic Navigation.* Sensors, 2019. **20**(1): p. 60.

505.Ma, H. and X. Mu, *Adaptive Navigation Algorithm With Deep Learning for Autonomous Underwater Vehicle.* Sensors, 2021. **21**(19): p. 6406.

506.He, B., et al., *Autonomous Navigation for Autonomous Underwater Vehicles Based on Information Filters and Active Sensing.* Sensors, 2011. **11**(11): p. 10958-10980.

507.Ying, F., et al., *Maximum Correntropy Based Unscented Particle Filter for Cooperative Navigation With Heavy-Tailed Measurement Noises.* Sensors, 2018. **18**(10): p. 3183.

508.Rypkema, N.R., E.M. Fischell, and H. Schmidt, *One-Way Travel-Time Inverted Ultra-Short Baseline Localization for Low-Cost Autonomous Underwater Vehicles.* 2017.

509.Palomer, A., P. Ridao, and D. Ribas, *Inspection of an Underwater Structure Using Point-cloud SLAM With an AUV and a Laser Scanner.* Journal of Field Robotics, 2019. **36**(8): p. 1333-1344.

510.Zhang, T., et al., *AUV Positioning Method Based on Tightly Coupled SINS/LBL for Underwater Acoustic Multipath Propagation.* Sensors, 2016. **16**(3): p. 357.

511.Villar, S.A., et al., *Evaluation of an Efficient Approach for Target Tracking From Acoustic Imagery for the Perception System of an Autonomous Underwater Vehicle.* International Journal of Advanced Robotic Systems, 2014. **11**(2).

512.Lyu, W., X. Cheng, and J. Wang, *Adaptive Federated IMM Filter for AUV Integrated Navigation Systems.* Sensors, 2020. **20**(23): p. 6806.

513.Zhan, B., et al., *Three-Dimensional Path Planning for AUVs Based on Standard Particle Swarm Optimization Algorithm.* Journal of Marine Science and Engineering, 2022. **10**(9): p. 1253.

514.Soares, J.M., et al., *Joint ASV/AUV Range-Based Formation Control: Theory and Experimental Results.* 2013.

515.Yan, X., et al., *Formation Control and Obstacle Avoidance Algorithm of a Multi-Usv System Based on Virtual Structure and Artificial Potential Field.* Journal of Marine Science and Engineering, 2021. **9**(2): p. 161.

516.Chen, C.-W., Y. Chen, and Q.-W. Cai, *Hydrodynamic-Interaction Analysis of an Autonomous Underwater Hovering Vehicle and Ship With Wave Effects.* Symmetry, 2019. **11**(10): p. 1213.

517.Wang, Y., et al., *Investigation and Optimization of Appendage Influence on the Hydrodynamic Performance of AUVs.* Journal of Marine Science and Technology, 2018. **24**(1): p. 297-305.

518.Miller, L.I., et al., *Drag Reduction and Power Optimization Due to an Innovative, Toroidal Hull Form of an AUV.* Iop Conference Series Materials Science and Engineering, 2023. **1288**(1): p. 012039.

519.Omeke, K.G., et al., *Energy Optimisation Through Path Selection for Underwater Wireless Sensor Networks.* 2020.

520.Gao, T., et al., *Hull Shape Optimization for Autonomous Underwater Vehicles Using CFD.* Engineering Applications of Computational Fluid Mechanics, 2016. **10**(1): p. 599-607.

521.Guo, Y., et al., *Research on the Influence of Turbulent Flow Induced by Dunes on AUVs.* Applied Sciences, 2023. **13**(18): p. 10273.

522.Wang, Z., et al., *Development of an Autonomous Underwater Helicopter With High Maneuverability.* Applied Sciences, 2019. **9**(19): p. 4072.

523.Chen, C.-W. and Y. Lu, *Computational Fluid Dynamics Study of Water Entry Impact Forces of an Airborne-Launched, Axisymmetric, Disk-Type Autonomous Underwater Hovering Vehicle.* Symmetry, 2019. **11**(9): p. 1100.

524.Mitra, A., J.P. Panda, and H. Warrior, *The Effects of Free Stream Turbulence on the Hydrodynamic Characteristics of an AUV Hull Form.* Ocean Engineering, 2019. **174**: p. 148-158.

525.Joung, T.-H., et al., *Shape Optimization of an Autonomous Underwater Vehicle With a Ducted Propeller Using Computational Fluid Dynamics Analysis.* International Journal of Naval Architecture and Ocean Engineering, 2012. **4**(1): p. 44-56.

526.Sun, T., et al., *Design and Optimization of a Bio-Inspired Hull Shape for AUV by Surrogate Model Technology.* Engineering Applications of Computational Fluid Mechanics, 2021. **15**(1): p. 1057-1074.

527.Alkan, B., *Hydrodynamic Design Optimization of an Autonomous Underwater Vehicle Based on Response Surface Methodology.* 2020.

528.Lin, Y., et al., *Study on the Motion Stability of the Autonomous Underwater Helicopter.* Journal of Marine Science and Engineering, 2022. **10**(1): p. 60.

529.DeVries, L., et al., *Hull Shape Actuation for Speed Regulation in an Underwater Vehicle.* Journal of Mechanisms and Robotics, 2019. **12**(1).

530.Xu, X.-S. and L. H, *An Improved Collaborative Algorithm With Artificial Neural Network in Multidisciplinary Design Optimization of AUV.* 2017.

531.Tang, X., *Head Optimization Design for Small Commercial AUV Based on CFD.* 2022.

532.Yan, G., G. Pan, and Y. Shi, *Ricochet Characteristics of AUVs During Small-Angle Water Entry Process.* Mathematical Problems in Engineering, 2019. **2019**(1).

533.Helmaoui, M., et al., *Layout Effect of Two Autonomous Underwater Vehicles on the Hydrodynamics Performances.* Wseas Transactions on Systems, 2020. **19**: p. 47-54.

534.Lin, H., et al., *Numerical Investigation on Hydrodynamic Performance of a Portable AUV.* Journal of Marine Science and Engineering, 2021. **9**(8): p. 812.

535.Zhu, H., et al., *The Design and Application of a Vectored Thruster for a Negative Lift-Shaped AUV.* Actuators, 2024. **13**(6): p. 228.

536.Font, D., et al., *Design and Implementation of a Biomimetic Turtle Hydrofoil for an Autonomous Underwater Vehicle.* Sensors, 2011. **11**(12): p. 11168-11187.

537.Singh, S., et al., *Modeling and Control Design for an Autonomous Underwater Vehicle Based on Atlantic Salmon Fish.* Ieee Access, 2022. **10**: p. 97586-97599.

538.Gasparoto, H.F., et al., *Advances in Reconfigurable Vectorial Thrusters for Adaptive Underwater Robots.* Journal of Marine Science and Engineering, 2021. **9**(2): p. 170.

539.Roper, D., et al., *A Review of Developments Towards Biologically Inspired Propulsion Systems for Autonomous Underwater Vehicles.* Proceedings of the Institution of Mechanical Engineers Part M Journal of Engineering for the Maritime Environment, 2011. **225**(2): p. 77-96.

540.Phillips, A.B., et al., *Nature in Engineering for Monitoring the Oceans: Comparison of the Energetic Costs of Marine Animals and AUVs.* 2012: p. 373-405.

541.Russo, R., et al., *Biomechanical Model of Batoid (Skates and Rays) Pectoral Fins Predicts the Influence of Skeletal Structure on Fin Kinematics: Implications for Bio-Inspired Design.* Bioinspiration & Biomimetics, 2015. **10**(4): p. 046002.

542.Heinen, Y., I. Tanev, and T. Kimura, *The Effect of a Limited Underactuated Posterior Joint on the Speed and Energy Efficiency of a Fish Robot.* Applied Sciences, 2024. **14**(12): p. 5010.

543.Zamora-Méndez, A., T.J. Leo, and M.A. Herreros, *Current State of Technology of Fuel Cell Power Systems for Autonomous Underwater Vehicles.* Energies, 2014. **7**(7): p. 4676-4693.

544.You, F., et al., *State of Charge Estimation of Lithium-Ion Batteries Based on an Adaptive Iterative Extended Kalman Filter for AUVs.* Sensors, 2022. **22**(23): p. 9277.

545.Faria, C., S. Priya, and D.J. Inman, *Modeling and Control of a Jellyfish-Inspired AUV.* 2013.

546.Anevlavi, D., E. Filippas, and K. Belibassakis, *Hydrodynamic Optimization of Actively Deforming Flapping-Foil Thrusters for Auv Propulsion.* 2023: p. 206-222.

547.Leonard, N.E., et al., *Coordinated Control of an Underwater Glider Fleet in an Adaptive Ocean Sampling Field Experiment in Monterey Bay.* Journal of Field Robotics, 2010. **27**(6): p. 718-740.

548.Dinc, M. and C. Hajiyev, *Integration of Navigation Systems for Autonomous Underwater Vehicles.* Journal of Marine Engineering & Technology, 2015. **14**(1): p. 32-43.

549.Li, P., et al., *A Robust INS/USBL/DVL Integrated Navigation Algorithm Using Graph Optimization.* Sensors, 2023. **23**(2): p. 916.

550.Wan, J., et al., *Fractional-Order PID Motion Control for AUV Using Cloud-Model-Based Quantum Genetic Algorithm.* Ieee Access, 2019. **7**: p. 124828-124843.

551.Ajmal, M., M. Labeeb, and D.V. Dev, *Fractional Order PID Controller for Depth Control of Autonomous Underwater Vehicle Using Frequency Response Shaping Approach.* 2014.

552.Yu, M., Q. Zhi, and H. Chen, *Kalman-Pd Control for Depth System of Autonomous Underwater Vehicle.* Applied Mechanics and Materials, 2014. **530-531**: p. 990-998.

553.Hu, Z., et al., *Trajectory Tracking and Re-Planning With Model Predictive Control of Autonomous Underwater Vehicles.* Journal of Navigation, 2018. **72**(2): p. 321-341.

554.Fischell, E.M., N.R. Rypkema, and H. Schmidt, *Relative Autonomy and Navigation for Command and Control of Low-Cost Autonomous Underwater Vehicles.* Ieee Robotics and Automation Letters, 2019. **4**(2): p. 1800-1806.

555.Hollinger, G.A., et al., *Underwater Data Collection Using Robotic Sensor Networks.* Ieee Journal on Selected Areas in Communications, 2012. **30**(5): p. 899-911.

556.Islam, K.Y., et al., *Green Underwater Wireless Communications Using Hybrid Optical-Acoustic Technologies.* Ieee Access, 2021. **9**: p. 85109-85123.

557.Han, X., et al., *The Research Progress in Underwater Wireless Optical Communication Technology.* 2023.

558.Noh, H., K. Kang, and J.-Y. Park, *Risk Analysis of Autonomous Underwater Vehicle Operation in a Polar Environment Based on Fuzzy Fault Tree Analysis.* Journal of Marine Science and Engineering, 2023. **11**(10): p. 1976.

559.Ashry, I., et al., *CNN-based Detection of Red Palm Weevil Using Optical-Fiber-Distributed Acoustic Sensing.* 2022.

560.Shen, T., et al., *Research on a Blue–Green LED Communication System Based on an Underwater Mobile Robot.* Photonics, 2023. **10**(11): p. 1238.

561.Ataner, E., et al., *İnsansız Su Altı Araçlarında Haberleşme Ve Güç Sistemlerinin Tasarımı.* European Journal of Science and Technology, 2020.

562.Pan, X., et al., *RAP-MAC: A Robust and Adaptive Pipeline MAC Protocol for Underwater Acoustic String Networks.* Remote Sensing, 2024. **16**(12): p. 2195.

563.Campbell, K.J., S. Kinnear, and A. Thame, *AUV Technology for Seabed Characterization and Geohazards Assessment.* The Leading Edge, 2015. **34**(2): p. 170-178.

564.Hofmann, A.F., et al., *High-Resolution Topography-Following Chemical Mapping of Ocean Hypoxia by Use of an Autonomous Underwater Vehicle: The Santa Monica Basin Example.* Journal of Atmospheric and Oceanic Technology, 2013. **30**(11): p. 2630-2646.

565.Unterseh, S. and W. Letaief, *Time-Lapse Attempt Using AUV Survey Data in Geohazards Prone Area, Results and Lessons Learned.* 2017.

566.Jakuba, M.V., et al. *Toward automatic classification of chemical sensor data from autonomous underwater vehicles.* in *2011 IEEE/RSJ International Conference on Intelligent Robots and Systems*. 2011. IEEE.

567.Weng, Y., et al., *Autonomous underwater vehicle link alignment control in unknown environments using reinforcement learning.* Journal of Field Robotics, 2024.

568.Flores, C.H. and U.S. ten Brink, *Photogrammetry of the Deep Seafloor from Archived Unmanned Submersible Exploration Dives.* Journal of Marine Science and Engineering, 2024. **12**(8): p. 1250.

569.Rogowski, P., et al., *Ocean outfall plume characterization using an autonomous underwater vehicle.* Water Science and Technology, 2013. **67**(4): p. 925-933.

570.Xia, Z., et al., *Revolutionizing marine Protection: Innovative designs for In-Situ applicable underwater antifouling super-oleophobic coatings.* Chemical Engineering Journal, 2024. **496**: p. 154157.

571.Hwang, M.S., M.K. Park, and H.S. Lee, *A Study on the Visual Effects Production Process for Efficient Underwater Explosion CG Visualization.* Journal of Coastal Research, 2024. **116**(SI): p. 518-522.

572.Mahmoodi, K.A. and M. Uysal. *AUV trajectory optimization for an optical underwater sensor network in the presence of ocean currents.* in *2021 IEEE International Black Sea Conference on Communications and Networking (BlackSeaCom)*. 2021. IEEE.

573.Woithe, H.C. and U. Kremer, *Feature based adaptive energy management of sensors on autonomous underwater vehicles.* Ocean Engineering, 2015. **97**: p. 21-29.

574.González-García, J., et al., *Autonomous underwater vehicles: Localization, navigation, and communication for collaborative missions.* Applied sciences, 2020. **10**(4): p. 1256.

575.Xing, T.Y., et al., *Improved Artificial Potential Field Algorithm Assisted by Multisource Data for AUV Path Planning.* Sensors, 2023. **23**(15): p. 6680.

576.Wibisono, A., et al., *An Autonomous Underwater Vehicle Navigation Technique for Inspection and Data Acquisition in UWSNs.* Ieee Access, 2024. **12**: p. 8641-8654.

577.Danielis, P., et al., *Integrated Autonomous Underwater Vehicle Path Planning and Collision Avoidance Algorithms.* Transactions on Maritime Science, 2024. **13**(1).

578.Zhao, Y., et al., *Research on Modeling Method of Autonomous Underwater Vehicle Based on a Physics-Informed Neural Network.* Journal of Marine Science and Engineering, 2024. **12**(5): p. 801.

579.Okereke, C.E., et al., *An Overview of Machine Learning Techniques in Local Path Planning for Autonomous Underwater Vehicles.* Ieee Access, 2023. **11**: p. 24894-24907.

580.Zhu, Z., et al., *An Efficient Multi-Auv Cooperative Navigation Method Based on Hierarchical Reinforcement Learning.* Journal of Marine Science and Engineering, 2023. **11**(10): p. 1863.

581.Atyabi, A., S. MahmoudZadeh, and S. Nefti-Meziani, *Current Advancements on Autonomous Mission Planning and Management Systems: An AUV and UAV Perspective.* Annual Reviews in Control, 2018. **46**: p. 196-215.

582.Zhang, L., et al., *Cooperative Navigation Based on Cross Entropy: Dual Leaders.* Ieee Access, 2019. **7**: p. 151378-151388.

583.Hernández, J.D., et al., *Online Motion Planning for Unexplored Underwater Environments Using Autonomous Underwater Vehicles.* Journal of Field Robotics, 2018. **36**(2): p. 370-396.

584.Meurer, C., et al., *Differential Pressure Sensor Speedometer for Autonomous Underwater Vehicle Velocity Estimation.* Ieee Journal of Oceanic Engineering, 2020. **45**(3): p. 946-978.

585.Wang, C., et al., *AUV Planning and Calibration Method Considering Concealment in Uncertain Environments.* Frontiers in Marine Science, 2023. **10**.

586.Ridao, P., et al., *Intervention AUVs: The Next Challenge.* Annual Reviews in Control, 2015. **40**: p. 227-241.

587.Carreras, M., et al., *Sparus II AUV—A Hovering Vehicle for Seabed Inspection.* Ieee Journal of Oceanic Engineering, 2018. **43**(2): p. 344-355.

588.Guevara, A.M., T.J. Leo, and M.A. Herreros, *Fuel Cell Power Systems for Autonomous Underwater Vehicles: State of the Art.* 2014.

589.Roper, D., et al., *Evaluating the Use of Lithium Sulphur Batteries for a Deep Ocean Pressure Balanced AUV Energy Source.* 2016.

590.Li, B., et al., *Study on Battery Thermal Management of Autonomous Underwater Vehicle by Bionic Wave Channels With Liquid Cooling.* International Journal of Energy Research, 2021. **45**(9): p. 13269-13283.

591.Potrykus, S., et al., *Advanced Lithium-Ion Battery Model for Power System Performance Analysis.* Energies, 2020. **13**(10): p. 2411.

592.Ji, D.-H., et al., *Design and Analysis of the High-Speed Underwater Glider With a Bladder-Type Buoyancy Engine.* Applied Sciences, 2023. **13**(20): p. 11367.

593.Zhang, F., D. Xu, and C. Cheng, *An Underwater Distributed SLAM Approach Based on Improved GMRBnB Framework.* Journal of Marine Science and Engineering, 2023. **11**(12): p. 2271.

594.Xing, T., et al., *A Multi-Source-Data-Assisted AUV for Path Cruising: An Energy-Efficient DDPG Approach.* Remote Sensing, 2023. **15**(23): p. 5607.

595.Zhi, L. and Y. Zuo, *Collaborative Path Planning of Multiple AUVs Based on Adaptive Multi-Population PSO.* Journal of Marine Science and Engineering, 2024. **12**(2): p. 223.

596.Gao, H., et al., *Distributed Path Tracking for Autonomous Underwater Vehicles Based on Pseudo Position Feedback.* Journal of Marine Science and Engineering, 2022. **10**(10): p. 1477.

597.Albarakati, S., et al., *Multiobjective Risk-Aware Path Planning in Uncertain Transient Currents: An Ensemble-Based Stochastic Optimization Approach.* Ieee Journal of Oceanic Engineering, 2021. **46**(4): p. 1082-1098.

598.Anderson, J. and G.A. Hollinger, *Communication Planning for Cooperative Terrain-Based Underwater Localization.* Sensors, 2021. **21**(5): p. 1675.

599.Fish, F.E., *Advantages of Aquatic Animals as Models for Bio-Inspired Drones Over Present AUV Technology.* Bioinspiration & Biomimetics, 2020. **15**(2): p. 025001.

600.Mondal, K., *Autonomous Underwater Vehicles: Recent Developments and Future Prospects.* International Journal for Research in Applied Science and Engineering Technology, 2019. **7**(11): p. 215-222.

601.Li, Q., et al., *Interface Engineering Enhances the Photovoltaic Performance of Wide Bandgap FAPbBr₃ Perovskite for Application in Low-Light Environments.* Advanced Functional Materials, 2023. **33**(40).

602.Hasan, M.B., *Yaw Angle Tracking Control Design of Underactuated AUV by Using State Dependent Riccati Equations (SDRE)-LQT.* Jeee-U (Journal of Electrical and Electronic Engineering-Umsida), 2019. **3**(2): p. 325-337.

603.Wang, T., et al., *DOB-Net: Actively Rejecting Unknown Excessive Time-Varying Disturbances.* 2020.

604.Li, B., P. Wang, and L. Du, *Path Planning Technologies for Autonomous Underwater Vehicles-a Review.* Ieee Access, 2019. **7**: p. 9745-9768.

605.Fernández, R.P., et al., *Nonlinear Attitude Control of a Spherical Underwater Vehicle.* Sensors, 2019. **19**(6): p. 1445.

606.Fu, Z. and Z. Jia, *Integrating Seabed Topography and Ocean Current Dynamics for Submarine Trajectory Analysis.* Highlights in Science Engineering and Technology, 2024. **104**: p. 180-187.

607.Zhang, T., Q. Li, and X. Liu, *An Object Detection and Classification Method for Underwater Visual Images Based on the Bag-of-Words Model.* Proceedings of the Institution of Mechanical Engineers Part M Journal of Engineering for the Maritime Environment, 2022. **237**(2): p. 487-497.

608.Ghafoor, H. and Y. Noh, *An Overview of Next-Generation Underwater Target Detection and Tracking: An Integrated Underwater Architecture.* Ieee Access, 2019. **7**: p. 98841-98853.

609.Chen, L., et al., *Underwater Target Recognition Based on Improved YOLOv4 Neural Network.* Electronics, 2021. **10**(14): p. 1634.

610.Hożyń, S., *A Review of Underwater Mine Detection and Classification in Sonar Imagery.* Electronics, 2021. **10**(23): p. 2943.

611.Polymenis, I., et al., *Virtual Underwater Datasets for Autonomous Inspections.* Journal of Marine Science and Engineering, 2022. **10**(9): p. 1289.

612.Stevens, R.H. and T. Galloway, *Are Neurodynamic Organizations a Fundamental Property of Teamwork?* Frontiers in Psychology, 2017. **8**.

613.Cao, X., *A Leader–follower Formation Control Approach for Target Hunting by Multiple Autonomous Underwater Vehicle in Three-Dimensional Underwater Environments.* International Journal of Advanced Robotic Systems, 2019. **16**(4): p. 172988141987066.

614.Stevens, R.H. and T. Galloway, *Teaching Machines to Recognize Neurodynamic Correlates of Team and Team Member Uncertainty.* Journal of Cognitive Engineering and Decision Making, 2019. **13**(4): p. 310-327.

615.Ming, L., et al., *Cognition-Based Hybrid Path Planning for Autonomous Underwater Vehicle Target Following.* International Journal of Advanced Robotic Systems, 2019. **16**(4): p. 172988141985755.

616.Zhang, M., et al., *Localization Uncertainty Estimation for Autonomous Underwater Vehicle Navigation.* Journal of Marine Science and Engineering, 2023. **11**(8): p. 1540.

617.Christensen, L., et al., *Recent Advances in AI for Navigation and Control of Underwater Robots.* Current Robotics Reports, 2022. **3**(4): p. 165-175.

618.Yang, H., et al., *Underwater Acoustic Research Trends With Machine Learning: Passive SONAR Applications.* Journal of Ocean Engineering and Technology, 2020. **34**(3): p. 227-236.

619.Merveille, F.F.R., B. Jia, and Z. Xu, *Advancements in Underwater Navigation: Integrating Deep Learning and Sensor Technologies for Unmanned Underwater Vehicles.* 2024.

620.Fayyad, J., et al., *Deep Learning Sensor Fusion for Autonomous Vehicle Perception and Localization: A Review.* Sensors, 2020. **20**(15): p. 4220.

621.Oludayo Olatoye Sofoluwe, N., et al., *AI-enhanced Subsea Maintenance for Improved Safety and Efficiency: Exploring Strategic Approaches.* International Journal of Science and Research Archive, 2024. **12**(1): p. 114-124.

622.Osayi Philip Igbinenikaro, N., N. Oladipo Olugbenga Adekoya, and N. Emmanuel Augustine Etukudoh, *A Comparative Review of Subsea Navigation Technologies in Offshore Engineering Projects.* International Journal of Frontiers in Engineering and Technology Research, 2024. **6**(2): p. 019-034.

623.Zhang, T., et al., *Current Trends in the Development of Intelligent Unmanned Autonomous Systems.* Frontiers of Information Technology & Electronic Engineering, 2017. **18**(1): p. 68-85.

624.None, N., et al., *Artificial Intelligence and Machine Learning Enhance Robot Decision-Making Adaptability and Learning Capabilities Across Various Domains.* 2024. **1**(3): p. 14-27.

625.Kozhubaev, Y., et al., *Controlling of Unmanned Underwater Vehicles Using the Dynamic Planning of Symmetric Trajectory Based on Machine Learning for Marine Resources Exploration.* Symmetry, 2023. **15**(9): p. 1783.

626.Wang, Z., et al., *Bio-Inspired Cooperative Control Scheme of Obstacle Avoidance for UUV Swarm.* Journal of Marine Science and Engineering, 2024. **12**(3): p. 489.

627.Liao, Y., et al., *Research on Disturbance Rejection Motion Control Method of USV for UUV Recovery.* Journal of Field Robotics, 2022. **40**(3): p. 574-594.

628.Gao, H., et al., *Energy Management of Hybrid Power System in UUV Based on Pontryagin's Minimum Principle.* E3s Web of Conferences, 2020. **182**: p. 03001.

629.Martínez de Alegría, I., et al., *Wireless Power Transfer for Unmanned Underwater Vehicles: Technologies, Challenges and Applications.* Energies, 2024. **17**(10): p. 2305.

630.Zhang, Z., et al., *Design and Implementation of a Modular UUV Simulation Platform.* Sensors, 2022. **22**(20): p. 8043.

631.Zhu, D., T. Yan, and S.X. Yang, *Motion Planning and Tracking Control of Unmanned Underwater Vehicles: Technologies, Challenges and Prospects.* Intelligence & Robotics, 2022. **2**(3): p. 200-222.

632.Farré, A.B., et al., *Commercial Arctic Shipping Through the Northeast Passage: Routes, Resources, Governance, Technology, and Infrastructure.* Polar Geography, 2014. **37**(4): p. 298-324.

633.Barker, L.D.L., et al., *Scientific Challenges and Present Capabilities in Underwater Robotic Vehicle Design and Navigation for Oceanographic Exploration Under-Ice.* Remote Sensing, 2020. **12**(16): p. 2588.

634.Norgren, P. and R. Skjetne, *Using Autonomous Underwater Vehicles as Sensor Platforms for Ice-Monitoring.* Modeling Identification and Control a Norwegian Research Bulletin, 2014. **35**(4): p. 263-277.

635.Randeni, S., et al., *A High-resolution AUV Navigation Framework With Integrated Communication and Tracking for Under-ice Deployments.* Journal of Field Robotics, 2022. **40**(2): p. 346-367.

636.Choi, T. and A. Lajeunesse, *Some Design Considerations for Arctic-Capable Submarines.* 2020.

637.Brandt, H., et al., *A Multi-Purpose Submarine Concept for Arctic Offshore Operations.* 2015.

638.Janout, M., et al., *Episodic Warming of Near-bottom Waters Under the Arctic Sea Ice on the Central Laptev Sea Shelf.* Geophysical Research Letters, 2016. **43**(1): p. 264-272.

639.Barro, R.D. and D.C. Lee, *Excitation Response Estimation of Polar Class Vessel Propulsion Shafting System.* Transactions of the Korean Society for Noise and Vibration Engineering, 2011. **21**(12): p. 1166-1176.

640.Barro, R.D. and D.C. Lee, *Transient Torsional Vibration Analysis of Ice-Class Propulsion Shafting System Driven by Electric Motor.* Transactions of the Korean Society for Noise and Vibration Engineering, 2014. **24**(9): p. 667-674.

641.Stroeve, J., A.P. Barrett, and A. Schweiger, *Using Records From Submarine, Aircraft and Satellite to Evaluate Climate Model Simulations of Arctic Sea Ice Thickness.* 2014.

642.Bess, Z., et al., *Fish Diversity and Use of Nearshore and <scp>Open-Water</Scp> Habitats in Terminal Lakes.* Fisheries, 2024. **49**(4): p. 159-168.

643.Evtimova, V. and I. Donohue, *Quantifying Ecological Responses to Amplified Water Level Fluctuations in Standing Waters: An Experimental Approach.* Journal of Applied Ecology, 2014. **51**(5): p. 1282-1291.

644.Polte, P., et al., *Ontogenetic Loops in Habitat Use Highlight the Importance of Littoral Habitats for Early Life-Stages of Oceanic Fishes in Temperate Waters.* Scientific Reports, 2017. **7**(1).

645.Zohary, T. and A. Gasith, *The Littoral Zone.* 2014: p. 517-532.

646.Dupuch, A., et al., *Does Predation Risk Influence Habitat Use by Northern Redbelly Dace<i>Phoxinus Eos</I>at Different Spatial Scales?* Journal of Fish Biology, 2009. **74**(7): p. 1371-1382.

647.Evtimova, V. and I. Donohue, *Water-level Fluctuations Regulate the Structure and Functioning of Natural Lakes.* Freshwater Biology, 2015. **61**(2): p. 251-264.

648.Fanini, L., et al., *The Extended Concept of Littoral Active Zone Considering Soft Sediment Shores as Social-Ecological Systems, and an Application to Brittany (North-Western France).* Estuarine Coastal and Shelf Science, 2021. **250**: p. 107148.

649.Leo, F.C.D., et al., *Submarine Canyons: Hotspots of Benthic Biomass and Productivity in the Deep Sea.* Proceedings of the Royal Society B Biological Sciences, 2010. **277**(1695): p. 2783-2792.

650.Ramirez-Llodra, E., et al., *Man and the Last Great Wilderness: Human Impact on the Deep Sea.* Plos One, 2011. **6**(8): p. e22588.

651.McClain, C.R. and J. Barry, *Habitat Heterogeneity, Disturbance, and Productivity Work in Concert to Regulate Biodiversity in Deep Submarine Canyons.* Ecology, 2010. **91**(4): p. 964-976.

652.Pham, C.K., et al., *Marine Litter Distribution and Density in European Seas, From the Shelves to Deep Basins.* Plos One, 2014. **9**(4): p. e95839.

653.Fakiris, E., et al., *Multi-Frequency, Multi-Sonar Mapping of Shallow Habitats—Efficacy and Management Implications in the National Marine Park of Zakynthos, Greece.* Remote Sensing, 2019. **11**(4): p. 461.

654.Jobstvogt, N., et al., *How Can We Identify and Communicate the Ecological Value of Deep-Sea Ecosystem Services?* Plos One, 2014. **9**(7): p. e100646.

655.Fagervold, S.K., et al., *Microbial Communities in Sunken Wood Are Structured by Wood-Boring Bivalves and Location in a Submarine Canyon.* Plos One, 2014. **9**(5): p. e96248.

656.Úrgeles, R. and A. Camerlenghi, *Submarine Landslides of the Mediterranean Sea: Trigger Mechanisms, Dynamics, and Frequency-Magnitude Distribution.* Journal of Geophysical Research Earth Surface, 2013. **118**(4): p. 2600-2618.

657.García-Davis, S., et al., *Bioprospecting Antiproliferative Marine Microbiota From Submarine Volcano Tagoro.* Frontiers in Marine Science, 2021. **8**.

658.Townsend, M., S.F. Thrush, and M.J. Carbines, *Simplifying the Complex: An 'Ecosystem Principles Approach' to Goods and Services Management in Marine Coastal Ecosystems.* Marine Ecology Progress Series, 2011. **434**: p. 291-301.

659.Nestorowicz, I.-M., et al., *Identifying Habitats of Conservation Priority in the São Vicente Submarine Canyon in Southwestern Portugal.* Frontiers in Marine Science, 2021. **8**.

660.Kenchington, R. and J.C. Day, *Zoning, a Fundamental Cornerstone of Effective Marine Spatial Planning: Lessons Learnt From the Great Barrier Reef, Australia.* Journal of Coastal Conservation, 2011. **15**(2): p. 271-278.

661.Raffaelli, D., *Biodiversity and Ecosystem Functioning: Issues of Scale and Trophic Complexity.* Marine Ecology Progress Series, 2006. **311**: p. 285-294.

662.Taormina, B., et al., *A Review of Potential Impacts of Submarine Power Cables on the Marine Environment: Knowledge Gaps, Recommendations and Future Directions.* Renewable and Sustainable Energy Reviews, 2018. **96**: p. 380-391.

663.Santora, J.A., et al., *Submarine Canyons Represent an Essential Habitat Network for Krill Hotspots in a Large Marine Ecosystem.* Scientific Reports, 2018. **8**(1).

664.Oehler, T., et al., *DSi as a Tracer for Submarine Groundwater Discharge.* Frontiers in Marine Science, 2019. **6**.

665.Nordin, M., *In Search of the Best Design – A Systems Analysis Methodology for Submarine Design.* 2014: p. 1-12.

666.Gaggero, S., et al., *Numerical Approaches for Submarine Hydrodynamic Design and Performance Analysis.* 2022.

667.Helal, M. and E. Fathallah, *Finite Element Analysis and Design Optimization of a Non-Circular Sandwich Composite Deep Submarine Pressure Hull.* Materials Testing, 2020. **62**(10): p. 1025-1032.

668.Shelton, J.H., V. Heisler, and K. Sebacher, *Formulas and Guidelines for Deriving Functional System Requirements From a Systems Engineering Model.* Incose International Symposium, 2021. **31**(1): p. 111-143.

669.Piaggio, B., et al., *Submarine Manoeuvrability Design: Traditional Cross-Plane vs. X-Plane Configurations in Intact and Degraded Conditions.* Journal of Marine Science and Engineering, 2022. **10**(12): p. 2014.

670.Cole, H., *The Influence of the Facility Nuclear Safety Case on the Design of Naval Refit Support Equipment.* 2018.

671.Goff, C.I., et al., *Maximizing Platform Value: Increasing VIRGINIA Class Deployments.* Naval Engineers Journal, 2011. **123**(3): p. 119-139.

672.Pak, J.-M., et al., *Effectiveness Analysis for a Lightweight Torpedo Considering Evasive Maneuvering and TACM of a Target.* Journal of the Korea Society for Simulation, 2011. **20**(4): p. 1-11.

673.Westwood, J.D., *Global Prospects for AUVs.* 2001.

674.Dc, N.P.W.N.Y., *SSN 774 Virginia Class Submarine (SSN 774).* 2013.

675.Zimmerman, P., et al., *Considerations and Examples of a Modular Open Systems Approach in Defense Systems.* The Journal of Defense Modeling and Simulation Applications Methodology Technology, 2018. **16**(4): p. 373-388.

676.Knox, C., D.D. Reid, and T.M. Winters, *A Qualitative Study of Affordability: Virginia and San Antonio Class Programs.* 2014.

677.Johnson, D., et al., *Managing Change on Complex Programs: VIRGINIA Class Cost Reduction.* Naval Engineers Journal, 2009. **121**(4): p. 79-94.

678.Aapaoja, A., H. Haapasalo, and P. Söderström, *Early Stakeholder Involvement in the Project Definition Phase: Case Renovation.* Isrn Industrial Engineering, 2013. **2013**: p. 1-14.

679.Lawanga, P.W.A.H. and Y.G. Sandanayake, *Lean Design Management Practices Associated With Stakeholder Management During Pre-Construction Stage in Sri Lanka.* 2021: p. 392-402.

680.Hannan, A., et al., *Estimation of Highway Project Cost Using Probabilistic Technique.* Destech Transactions on Engineering and Technology Research, 2017(ictim).

681.McCuen, T., *BIM and Cost Estimating: A Change in the Process for Determining Project Costs.* 2015: p. 63-81.

682.Wu, W., *Application of Project Whole Process Cost Control in Construction Project Cost Audit.* Forest Chemicals Review, 2021: p. 240-246.

683.Daluwatte, L. and M. Ranasinghe, *Effects of Inputs From the Preconstruction Activities on the Design Phase of Construction Projects.* Engineer Journal of the Institution of Engineers Sri Lanka, 2018. **51**(2): p. 31.

684.Psallidas, K., C.A. Whitcomb, and J.C. Hootman, *Forecasting System-Level Impacts of Technology Infusion in Ship Design.* Naval Engineers Journal, 2010. **122**(1): p. 21-31.

685.Wang, Y.L., G.X. Wang, and Y. Yan, *Scientific Research Project Cost Estimating Method and System Based on Improved BP Neural Network.* Advanced Materials Research, 2013. **756-759**: p. 1696-1700.

686.O'Rourke, R., *Navy Next-Generation Attack Submarine (SSN[X]) Program: Background and Issues for Congress.* 2024.

687.Sumadinata, R.W.S., *Analysis of the AUKUS Agreement on Security in the South China Sea Region.* Neo Journal of Economy and Social Humanities, 2023. **1**(4): p. 265-271.

688.Terebov, O., *The Present-Day Condition and Prospects for Development of the AUKUS Program.* Russia and America in the 21st Century, 2023(6): p. 0.

689.McKenzie, S. and E. Massingham, *AUKUS: The Regulation of the Ocean and the Legal Dangers of Working Together.* Ocean Yearbook Online, 2023. **37**(1): p. 136-170.

690.Arthamevya Zherlindya Putri Darmawan, N. and N. Ratu Salmazahra Karmilawaty, *Proposal Nuclear Naval Propulsion Dari Indonesia Untuk PBB Sebagai Respons Atas Dinamika Keamanan Di Indo-Pasifik.* 2023. **1**(1): p. 39-59.

691.Novita, A.A.D., *AUKUS Alliance: United States Strategic Interest in Indo-Pacific.* Jurnal Diplomasi Pertahanan, 2022. **8**(1).

692.Poshedin, O., *A New Enhanced Trilateral Security Partnership Between Australia, the United Kingdom, and the USA (AUKUS): Reasons for Creation, Consequences for International Security.* Problems of World History, 2022(19): p. 123-142.

693.Korwa, J.R.V. and M.S.F. Wambrauw, *A Constructivist Analysis of the Establishment of the AUKUS Security Pact and Its Implications for Regional Stability in the Indo-Pacific.* Jurnal Hubungan Internasional, 2023. **16**(1): p. 19-35.

694.Hanggarini, P., et al., *AUKUS: Unresolved Threats to the International System?* Multidisciplinary Reviews, 2023. **6**(3): p. 2023027.

695.Cheng, M., *AUKUS: The Changing Dynamic and Its Regional Implications.* European Journal of Development Studies, 2022. **2**(1): p. 1-7.

696.Sobarini, E., *AUKUS Pact in the Perspective of Security Dilemma.* International Journal of Social Science and Human Research, 2021. **04**(12).

697.McKenzie, S. and E. Massingham, *AUKUS, the Regulation of the Ocean and the Legal Perils of Working Together.* 2022.

698.Rusli, M.H.M., *The Legality of Passage of the Nuclear-Powered Submarines: Are Malaysia and Indonesia in Catch-22?* Moscow Journal of International Law, 2023(4): p. 34-43.

699.Gálvez, E.D. and S.F. Capuz-Rizo, *Assessment of Global Sensitivity Analysis Methods for Project Scheduling.* Computers & Industrial Engineering, 2016. **93**: p. 110-120.

700.Sayre, T. and C.M. Graham, *The Effects of Task Interdependency on Cooperative and Competitive Group Processes: An Example of Scheduling Audit Engagements.* Human Resource Management Research, 2013. **2**(5): p. 74-82.

701.Ramkumar, K.G. and P. Panneerselvam, *Indian Navy's Submarine Development Programme: A Critical Assessment.* Journal of Asian Security and International Affairs, 2023. **10**(3): p. 395-416.

702.Bagshaw, K.B., *NEW PERT and CPM in Project Management With Practical Examples.* American Journal of Operations Research, 2021. **11**(04): p. 215-226.

703.Ahmed, A., B. Kayis, and S. Amornsawadwatana, *A Review of Techniques for Risk Management in Projects.* Benchmarking an International Journal, 2007. **14**(1): p. 22-36.

704.Pimchangthong, D. and V. Boonjing, *Effects of Risk Management Practices on IT Project Success.* Management and Production Engineering Review, 2017. **8**(1): p. 30-37.

705.Raz, T., A.J. Shenhar, and D. Dvir, *Risk Management, Project Success, and Technological Uncertainty.* R and D Management, 2002. **32**(2): p. 101-109.

706.Kristensen, H.M. and M. Korda, *Indian Nuclear Forces, 2020.* Bulletin of the Atomic Scientists, 2020. **76**(4): p. 217-225.

707.Joshi, Y., *From Ambivalence to Resurgence: India's Journey as a Nuclear Power.* India Quarterly a Journal of International Affairs, 2022. **78**(2): p. 350-370.

708.None, N., et al., *A Step-by-Step Hybrid Approach Based on Multi-Criteria Decision-Making Methods and a Bi-Objective Optimization Model to Project Risk Management.* Decision Making Applications in Management and Engineering, 2024. **7**(1): p. 442-472.

709.Ali Alshehhi, H.S.M., R.S. Mat Sidek, and E.A. Rozali, *The Function of Risk Management in Improving Construction Project Implementation Efficiency.* International Journal of Academic Research in Accounting Finance and Management Sciences, 2024. **14**(1).

710.Anastasiu, L., C. Campian, and N. Roman, *Boosting Construction Project Timeline: The Case of Critical Chain Project Management (CCPM).* Buildings, 2023. **13**(5): p. 1249.

711.Scally, W.A., *Advanced Multi-Project Management: Achieving Outstanding Speed and Results With Predictability.* Project Management Journal, 2013. **44**(6): p. e1-e1.

712.Bueno, J. and R.G. Parra, *Enhancing Construction and Urban Planning Outcomes: An Examination of Project Management Institute Methodology.* Journal of Urban Development and Management, 2023. **2**(2): p. 95-103.

713.Zheng, X., et al., *Research on 3D Shape Monitoring System of Submarine Cable.* Journal of Physics Conference Series, 2023. **2474**(1): p. 012062.

714.Xie, Y. and C. Wang, *Vulnerability of Submarine Cable Network of Mainland China: Comparison of Vulnerability Between Before and After Construction of Trans-Arctic Cable System.* Complexity, 2021. **2021**: p. 1-14.

715.Lu, Z., et al., *Research on Improving the Working Efficiency of Hydraulic Jet Submarine Cable Laying Machine.* Journal of Marine Science and Engineering, 2021. **9**(7): p. 745.

716.Meng, K., M. Li, and J. Zhou, *Study on Control of Cumulative Risk for Submarine Tunnel Shield Construction.* Advances in Civil Engineering, 2023. **2023**: p. 1-18.

717.Wang, Z., et al., *Risk Zonation of Submarine Geological Hazards in the Chengdao Area of the Yellow River Subaqueous Delta.* Frontiers in Marine Science, 2023. **10**.

718.Xu, X., et al., *Subject Behavior of Collaborative Innovation in Civil-Military Integration: An Evolutionary Game Analysis.* Mathematical Problems in Engineering, 2021. **2021**: p. 1-7.

719.Wadjdi, A.F., J. Tambayong, and E.M.T. Sianturi, *Enhancing National Defense Capabilities Through Collaborative Programs: Insights and Policy Recommendations for Indonesia.* Insights Into Regional Development, 2023. **5**(3): p. 10-23.

720.Mendonça, A., et al., *Incorporating a Risk Assessment Procedure Into Submarine Outfall Projects and Application to Portuguese Case Studies.* Coastal Engineering Proceedings, 2012(33): p. 18.

721.Mendonça, A., et al., *Risk Assessment in Submarine Outfall Projects: The Case of Portugal.* Journal of Environmental Management, 2013. **116**: p. 186-195.

722.Lavallee, T.M., *Civil-Military Integration: The Politics of Outsourcing National Security.* Bulletin of Science Technology & Society, 2010. **30**(3): p. 185-194.

723.Belin, J., et al., *Defense Firms Adapting to Major Changes in the French R&D Funding System.* Defence and Peace Economics, 2018. **30**(2): p. 142-158.

724.Morris, B.A., *A Framework for Identifying and Managing New Operational Requirements During Naval Vessel Batch-Building Programs.* Incose International Symposium, 2021. **31**(1): p. 713-727.

725.Doerry, N., *Designing Electrical Power Systems for Survivability and Quality of Service.* Naval Engineers Journal, 2007. **119**(2): p. 25-34.

726.Anghinolfi, D., et al., *A System Supporting the Evaluation of the Operational Effectiveness of Naval Tasks Based on Agent Simulation.* Scalable Computing Practice and Experience, 2014. **15**(3).

727.Bemelmans, J., H. Voordijk, and B. Vos, *Supplier-contractor Collaboration in the Construction Industry.* Engineering Construction & Architectural Management, 2012. **19**(4): p. 342-368.

728.Kim, M.G., et al., *Environmental Capabilities of Suppliers for Green Supply Chain Management in Construction Projects: A Case Study in Korea.* Sustainability, 2016. **8**(1): p. 82.

729.Mirawati, N.A., S.a. Othman, and M.I. Risyawati, *Supplier-Contractor Partnering Impact on Construction Performance: A Study on Malaysian Construction Industry.* Journal of Economics Business and Management, 2015. **3**(1): p. 29-33.

730.Sariola, R., *Utilizing the Innovation Potential of Suppliers in Construction Projects.* Construction Innovation, 2018. **18**(2).

731.Woo, C., et al., *Suppliers' Communication Capability and External Green Integration for Green and Financial Performance in Korean Construction Industry.* Journal of Cleaner Production, 2016. **112**: p. 483-493.

732.Bemelmans, J., et al., *Assessing Buyer-Supplier Relationship Management: Multiple Case-Study in the Dutch Construction Industry.* Journal of Construction Engineering and Management, 2012. **138**(1): p. 163-176.

733.Sundquist, V., K. Hulthén, and L.E. Gadde, *From Project Partnering Towards Strategic Supplier Partnering.* Engineering Construction & Architectural Management, 2018. **25**(3): p. 358-373.

734.Segerstedt, A. and T. Olofsson, *Supply Chains in the Construction Industry.* Supply Chain Management an International Journal, 2010. **15**(5): p. 347-353.

735.Eckerd, A. and K.F. Snider, *Does the Program Manager Matter? New Public Management and Defense Acquisition.* The American Review of Public Administration, 2016. **47**(1): p. 36-57.

736.Karnes, F.L.J.L. and C.R.F. Mortlock, *Aligning Program Management Competencies to Industry Standards.* Defense Acquisition Research Journal, 2021. **28**(98): p. 366-419.

737.Bradley, J., et al., *Evaluating Australia's Most Complex System-of-systems, the Future Submarine: A Case for Using New Complex Systems Governance.* Incose International Symposium, 2017. **27**(1): p. 187-199.

738.Monfre, F.O. and C.d. Costa, *Hardware-in-the-Loop Emulation of a Control and Longitudinal Compensation System of a Submarine.* International Journal of Advanced Engineering Research and Science, 2019. **6**(8): p. 171-178.

739.McLaurin, D., A. Aston, and J.C.D. Brand, *Prevention of Offshore Wind Power Cable Incidents by Employing Offshore Oil/Gas Common Practices.* 2021.

Index

V

W

Y

Made in the USA
Coppell, TX
06 September 2026

89200763R10282